Polyamino Acids, Polypeptides, and Proteins

Polyamino Acids, Polypeptides, and Proteins

Proceedings of an International Symposium held at the University of Wisconsin, 1961

Edited by

MARK A. STAHMANN

The University of Wisconsin Press

MADISON 1962

Published by The University of Wisconsin Press
430 sterling court, madison 6, wisconsin

Printed in the United States of America by
George Banta Company, Inc., Menasha, Wisconsin

Library of Congress Catalog Card Number 62-12893

Preface

THIS symposium, the first international discussion devoted to polyamino acids, polypeptides, and proteins, was held at the University of Wisconsin on June 19 to 24, 1961. The program was planned by Drs. Elkan R. Blout, Paul Doty, Sidney W. Fox, Ephraim Katchalski, and Mark A. Stahmann. This committee felt that a meeting in which participants only report on recent work would add little, for this information would normally be forthcoming in scientific publications. However, they considered that much good would come from an International Discussion-Symposium in which a small group of investigators from many disciplines could collectively consider important problems and methods for their solution. To facilitate considered discussion, speakers prepared duplicated copies of their papers which were mailed to most participants before the symposium.

Each session began with a lecture by a selected participant who was asked to review the present state of our knowledge in a given area, to emphasize the important problems, and to critically evaluate methods for their solution. Short summarizing reports of the related papers were then given; each was followed by a discussion period. It was thought that to record all that was said in the discussion would discourage spontaneous comment and so we have published only those comments which were later submitted in writing by the participants. Consequently, the discussion sections are frequently incomplete and represent only a small part of what was actually said.

The papers may be divided into three general groups. The first group (Parts I and II of this book) consisted of nine papers that concerned the synthesis of polyamino acids or polypeptides and the mechanism of the polymerization reaction. In the second group (Parts III and IV) were sixteen papers that dealt with the physical or chemical properties of polyamino acids and the structure of proteins and polypeptides. The third group of eleven papers (Part V) was devoted to the biological properties of polyamino acids, particularly their role in enzyme research, their action on disease-producing agents, and their immunochemical properties. In all, thirty-six papers were presented, twelve by scientists from overseas. We extend our thanks to them, and indeed to all the speakers and participants, for their efforts in preparing the preprints and their contributions to the discussions.

The symposium began with an informal evening reception and dinner at which Conrad A. Elvehjem, President of the University of Wisconsin, and Farrington Daniels, Vice-President of the National Academy of Sciences, gave messages of welcome to the participants. Professor Ephraim Katchalski of the Weismann Institute delivered the keynote address outlining the purposes and objectives of the symposium. Sessions were held in the Wisconsin Center Building each morning and afternoon from Tuesday through Friday. A large part of each session was devoted to spontaneous discussions.

The impetus for the formation of this symposium came as a result of suggestions by Dr. Clem O. Miller, Executive Secretary of the Division of Chemistry and Chemical Technology of the National Research Council, and Dr. Sidney W. Fox, Chairman of the Division of Biological Chemistry of the American Chemical Society. It was realized that much significant research on polyamino acids and related substances is currently being done in many fields and that a forum devoted to discussions of this research and problems for the future would be highly desirable.

The symposium was made possible by grants from the National Institutes of Health to the National Academy of Sciences for transportation and publication expenses, from the Research Committee of the University of Wisconsin with funds supplied by the Wisconsin Alumni Research Foundation that allowed the University of Wisconsin to be the host and to provide meals and lodging for the speakers, and by a MATS transportation grant for Japanese participants from the Office of Naval Research. We wish to acknowledge the generosity of these sponsors and to express our appreciation to them, to Mr. R. P. Lee, Director of Wisconsin Center, and to Mr. G. F. Gurda, Assistant Director of Residence Halls, and their associates for their aid in making available the needed facilities for holding the symposium.

We felt that the symposium was a distinct success, particularly as evidenced by the high quality of the papers and the thoughtful discussion that followed their presentation. Credit for this success should go to the speakers who prepared preprints, to the participants who studied them and came prepared to discuss the papers, and to my colleagues on the program-planning committee who spent considerable time in planning the program and selecting the participants. I want to thank Dr. Blout, Professor Doty, Professor Fox, Professor Katchalski, and Dr. Miller for their dedication to the success of this symposium.

M. A. Stahmann

January, 1962
Madison, Wisconsin

Contents

Contributors and Participants

Christian B. Anfinsen

Laboratory of Cellular Physiology and Metabolism, National Heart Institute, National Institutes of Health, Bethesda, Md.

Jon Applequist

Dept. of Chemistry, Columbia University, New York 27, N. Y.

C. H. Bamford

Courtaulds, Ltd., Research Laboratory, Lower Cookham Road, Maidenhead, Berkshire, England

R. R. Becker

Biology Division, Oak Ridge National Laboratory, Oak Ridge, Tenn.

W. W. Beeman

Dept. of Physics, University of Wisconsin, Madison 6, Wis.

Arieh Berger

Dept. of Biophysics, Weizmann Institute of Science, Rehovoth, Israel

Sidney Bernhard

Section on Physical Chemistry, National Institute of Mental Health, Bethesda 14, Md.

H. Block

Courtaulds, Ltd., Research Laboratory, Lower Cookham Road, Maidenhead, Berkshire, England

Elkan R. Blout

Children's Cancer Research Foundation, Children's Hospital Medical Center, and Harvard Medical School, Boston 15, Mass., and Polaroid Corporation, Cambridge 39, Mass.

R. C. Blume

Pioneering Research Laboratory, Experimental Station, E. I. DuPont de Nemours Co., Wilmington 98, Del.

F. Boardman
Dept. of Chemistry, Polytechnic Institute of Brooklyn, Brooklyn, N. Y.

E. M. Bradbury
Courtaulds, Ltd., Research Laboratory, Lower Cookham Road, Maidenhead, Berkshire, England

Ludwig Brand
Graduate Dept. of Biochemistry, Brandeis University, Waltham 54, Mass.

Dorothy J. Buchanan-Davidson
2407 Arboretum Drive, Madison, Wis.

Peter Cammarata
G. D. Searle and Co., P.O. Box 5110, Chicago 80, Ill.

M. Cesari
Centre de Recherches sur les Macromolécules, 6, rue Boussingault, Strasbourg, France

Kenneth W. Cochran
Dept. of Epidemiology and Virus Laboratory, University of Michigan, Ann Arbor, Mich.

Ezra Daniel
Dept. of Biophysics, Weizmann Institute of Science, Rehovoth, Israel

Paul Doty
Dept. of Chemistry, Harvard University, Cambridge 38, Mass.

A. R. Downie (deceased)
Courtaulds, Ltd., Research Laboratory, Lower Cookham Road, Maidenhead, Berkshire, England

E. L. Durrum
Beckman Instruments, Inc., Beckman Spinco Division, 1117 California Ave., Palo Alto, Calif.

A. Elliott
Courtaulds, Ltd., Research Laboratory, Lower Cookham Rd., Maidenhead, Berkshire, England

S. Farber
Children's Cancer Research Foundation, Children's Hospital Medical Center and Harvard Medical School, Boston 15, Mass.

GERALD D. FASMAN
Children's Cancer Research Foundation, Children's Hospital Medical Center and Harvard Medical School, Boston 15, Mass. *Present address:* Graduate Department of Biochemistry, Brandeis University, Waltham 54, Mass.

J. D. FERRY
Dept. of Chemistry, University of Wisconsin, Madison 6, Wis.

SIDNEY W. FOX
Dept. of Chemistry and Oceanographic Institute, Florida State University, Tallahassee, Fla.

THOMAS J. GILL, III
Dept. of Chemistry, Harvard University, Cambridge 38, Mass.

M. GOODMAN
Dept. of Chemistry, Polytechnic Institute of Brooklyn, Brooklyn, N. Y.

WALTER B. GRATZER
Dept. of Chemistry, Harvard University, Cambridge 38, Mass.

ANITA HALPERN
Dept. of Chemistry, University of California, Berkeley, Calif.

W. E. HANBY
Courtaulds, Ltd., Research Laboratory, Lower Cookham Road, Maidenhead, Berkshire, England

KAORU HARADA
Dept. of Chemistry and Oceanographic Institute, Florida State University, Tallahassee, Fla.

KLAUS HOFMANN
Biochemistry Dept., University of Pittsburgh School of Medicine, Pittsburgh, Pa.

M. IDELSON
Children's Cancer Research Foundation, Children's Hospital Medical Center and Harvard University, Harvard Medical School, Boston 15, Mass.

KAZUTOMO IMAHORI
Dept. of Chemistry, College of General Education, University of Tokyo, Komaba, Meguro, Tokyo, Japan

EPHRAIM KATCHALSKI
Dept. of Biophysics, Weizmann Institute of Science, Rehovoth, Israel

J. C. Kendrew
Medical Research Council Unit, Cavendish Laboratory, University of Cambridge, Cambridge, England

E. Klein
Children's Cancer Research Foundation, Children's Hospital Medical Center, Boston 15, Mass.

Kenneth D. Kopple
Dept. of Chemistry, University of Chicago, Chicago 37, Ill.

Joseph Kovacs
Dept. of Chemistry, St. John's University, New York, N. Y.

Gottfried Krampitz
Institut für Anatomie und Physiologie der Haustiere, The University of Bonn, Bonn, Germany

I. Listowsky
Dept. of Chemistry, Polytechnic Institute of Brooklyn, Brooklyn, N. Y.

Christiane de Lozé
Children's Cancer Research Foundation, Children's Hospital Medical Center and Harvard University, Boston 15, Mass.

Vittorio Luzzati
Centre de Recherches sur les Macromolécules, 6, rue Boussingault, Strasbourg, France

McGregor, W. H.
Union Carbide Research Institute, 32 Depot Plaza, White Plains, N. Y.

F. Masson
Centre de Recherches sur les Macromolécules, 6, rue Boussingault, Strasbourg, France

Paul H. Maurer
Dept. of Microbiology, Seton Hall College of Medicine and Dentistry, Jersey City 4, N. J.

Roy R. Miller
Dept. of Chemistry, University of Chicago, Chicago 37, Ill.

Wilmer G. Miller
Dept. of Chemistry, University of Iowa, Iowa City, Iowa

THOMAS K. MIWA
Northern Regional Laboratory, U. S. Dept. of Agriculture, Peoria, Ill.

TATSUO MIYAZAWA
Institute for Protein Research, Osaka University, Kitaku, Osaka, Japan

THOMAS C. MULLER
Dept. of Chemistry, University of Chicago, Chicago 37, Ill.

M. NARROD
Children's Cancer Research Foundation, Children's Hospital Medical Center, Boston 15, Mass.

JUNZO NOGUCHI
Dept. of Chemistry, Faculty of Science, Kanazawa University, Kanazawa, Japan. *Present address:* Department of Polymer Science, Faculty of Science, Hokkiado University, Sapporo, Japan

ABRAHAM PATCHORNIK
Dept. of Biophysics, Weizmann Institute of Science, Rehovoth, Israel

LEONARD PELLER
Dept. of Chemistry, University of Wisconsin, Madison 6, Wis.

G. E. PERLMANN
The Rockefeller Institute, New York 21, N. Y.

DAVID PLATT
Biochemistry Dept., University of Pittsburgh School of Medicine, Pittsburgh 13, Pa.

DUANE L. ROHLFING
Dept. of Chemistry and Oceanographic Institute, Florida State University, Tallahassee, Fla.

I. G. ROSEN
Dept. of Chemistry, Polytechnic Institute of Brooklyn, Brooklyn, N. Y.

JAMES RYAN
Pilot Chemicals, Watertown, Mass.

TOMOO SAITO
Dept. of Chemistry, Faculty of Science, Kanazawa University, Kanazawa, Japan

S. SAKAKIBARA
Dept. of Biochemistry, Cornell University, Ithaca, N. Y.

SETSUJI SAKURAI
Ajinomoto Company, Inc., Central Research Laboratory, 2964, Suzuki-cho, Kawasaki, Japan

HAROLD A. SCHERAGA
Dept. of Chemistry, Cornell University, Ithaca, N. Y.

E. E. SCHMITT
Dept. of Chemistry, Polytechnic Institute of Brooklyn, Brooklyn, N. Y.

MICHAEL SELA
Dept. of Biophysics, Weizmann Institute of Science, Rehovoth, Israel

WILLIAM T. SIMPSON
Dept. of Chemistry, University of Washington, Seattle, Washington

DONALD E. SLAGEL
Dept. of Pathology, University of Wisconsin, Madison 6, Wis.

MARVIN A. SMITH
Dept. of Biochemistry, University of Wisconsin, Madison 6, Wis.

HERBERT A. SOBER
National Cancer Institute, National Institutes of Health, Bethesda 14, Md.

G. SPACH
Centre de Recherches sur les Macromolécules, 6, rue Boussingault, Strasbourg, France

MARK A. STAHMANN
Dept. of Biochemistry, University of Wisconsin, Madison 6, Wis.

M. A. STAKE
Dept. of Chemistry, Polytechnic Institute of Brooklyn, Brooklyn, N. Y.

I. Z. STEINBERG
Dept. of Chemistry, Cornell University, Ithaca, N. Y.

JOHN W. STEWART
Dept. of Biochemistry, University of Wisconsin, Madison 6, Wis.

LUBERT STRYER
Children's Cancer Research Foundation and Harvard Medical School, Boston 15, Mass. *Present address:* Department of Physics, Harvard University, Cambridge 38, Mass.

J. TANAKA

Dept. of Chemistry, College of General Education, University of Tokyo, Komaba, Meguro, Tokyo, Japan

DUANE A. TEWKSBURY

Dept. of Biochemistry, University of Wisconsin, Madison 6, Wis.

IGNACIO TINOCO, JR.

Dept. of Chemistry, University of California, Berkeley, Calif.

N. W. TSCHOEGL

Bread Research Institute of Australia, Epping Highway, North Ryde, N.S.W., Australia

J. M. VINCENT

Centre de Recherches sur le Macromolécules, 6, rue Boussingault, Strasbourg, France

AKIYOSHI WADA

Dept. of Chemistry, Faculty of Science, Ochanomizu University, 66 Akasaka Shinsaka, Bunkyo-ku, Tokyo, Japan. *Present address:* Laboratory of Molecular Biophysics, Department of Physics, Tokyo University, Bunkyo-ku, Tokyo, Japan

SUDHAKAR S. WAGLE

Dept. of Biochemistry, University of Wisconsin, Madison 6, Wis.

DON WETLAUFER

Dept. of Biochemistry, School of Medicine, Indiana University, 1100 West Michigan St., Indianapolis 7, Ind.

R. E. WILFONG

Carothers Research Laboratory, Experimental Station. E. I. DuPont de Nemours Co., Wilmington, Del.

J. W. WILLIAMS

Dept. of Chemistry, University of Wisconsin, Madison 6, Wis.

HARUAKI YAJIMA

Biochemistry Dept., University of Pittsburgh School of Medicine, Pittsburgh, Pa.

JEN TSI YANG

University of California Medical Center, Cardiovascular Research Institute, San Francisco, Calif.

Arieh Yaron
Dept. of Biophysics, Weizmann Institute of Science, Rehovoth, Israel

Bruno H. Zimm
Dept. of Chemistry, School of Science and Engineering,
Universitv of California, San Diego, La Jolla, Calif.

PART I

Synthesis and Chemical Properties of Poly-α-Amino Acids and Polypeptides

PAPER 1

Synthesis and Chemical Properties of Polypeptides

ELKAN R. BLOUT

The important developments in the synthesis of polypeptides of high molecular weight, and the study of their physical-chemical properties as well as their use as protein models, stem directly from the synthesis some fifty-five years ago of the first α-amino acid N-carboxyanhydride by Leuchs (*1*). The early investigations of the properties of oxazolidine-2,5-diones revealed that carbon dioxide was evolved by the treatment of these compounds with water and, upon heating, such anhydrides polymerized with the formation of polypeptides of low molecular weight. However, only in the last fifteen years have comprehensive investigations of the polymerization of α-amino acid N-carboxyanhydrides (NCA's) been pursued (*2*). This work has indicated that poly-α-amino acids of intermediate molecular weight (10,000 to 50,000) may be prepared by primary amine initiated polymerizations, but very high molecular weight polymers (100,000 to 1,500,000) may be prepared through initiation of the polymerization by secondary and tertiary amines and by the use of strong bases. The two-step synthesis of polypeptides from α-amino acids via NCA's followed by sodium-methoxide or sodium-hydroxide initiated polymerization yields polymers with 200–6,000 amino-acid residues in the chain (*3*) according to the general scheme shown below.

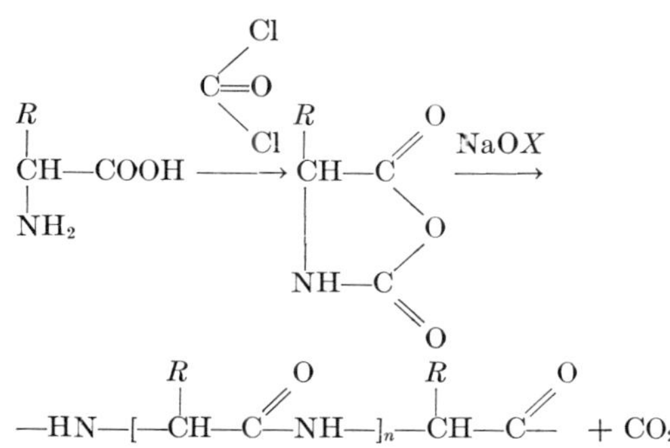

It is not the purpose of this communication to discuss the mechanism of amino-acid polymerization reactions since that topic is the subject of subsequent papers in this book. It is now clear, however, that from the point of view of obtaining high molecular weight materials, in good yield, the strong-base initiated polymerization of α-amino acid N-carboxyanhydrides is the preferred technique. It is important to note that this method not only yields polymers with a relatively narrow molecular weight distribution but also yields polypeptides (as far as now known) without racemization having occurred.

The publication in 1958 of a comprehensive review on the synthesis of polyamino acids by Katchalski and Sela (*2b*) makes unnecessary the relisting of polypeptides prepared up to that date. Therefore, in this paper we will confine our attention to a listing of homopolypeptides synthesized since 1958 and a discussion of some synthetic methods which might be applicable to the preparation of more complicated polypeptides. In addition, an attempt will be made to relate some chemical and physical properties of polyamino acids to their amino-acid composition; and finally, some unsolved problems in polypeptide synthesis will be discussed.

Synthesis of polypeptides

New synthetic polypeptides.—By the use of α-amino acid N-carboxyanhydrides as starting materials polypeptide polymers have been prepared from almost all naturally-occurring α-amino acids and many modifications thereof. In Table 1.1 are listed some new polypeptide homopolymers synthesized since the publication of the last review of this subject.

Although the syntheses of many polyamino-acid polymers have now been described, there are still significant gaps in our knowledge of the physical and chemical properties of these materials. For example, in many cases viscosities have not been reported so that even approximate estimates of molecular weight cannot be made. Furthermore, since the physical-chemical properties of polypeptides depend to some extent upon their molecular weights, especially in the lower molecular weight regions, physical-chemical data on these materials should be related to molecular weights.

Since the strong-base catalyzed polymerization of α-amino acid NCA's appears to be the most generally applicable technique for the preparation of high molecular weight polypeptides, a few words about the characteristics of this reaction are perhaps pertinent at this point. Recently reported work on the mechanism of this polymerization (*4*) indicates that protons are not required and that there appears to be an initial autocatalytic period followed by a first-order polymerization reaction. Also, the rate of the first-order reaction is at least 100 times faster than that of amine-initiated polymerizations. The kinetic results suggest that the effective initiator for the polymerization is not the strong base but a low molecular weight reaction product resulting from the strong base and the amino-acid *N*-carboxyanhydride. Although the reaction is not understood in all its aspects, a possible polymerization scheme involving electron transfer has been suggested (*4*).

Random copolypeptides.—Truly random copolymers can only be obtained from NCA's under very specific conditions—that is, those in which the rate of addition of two (or more) NCA's to a growing peptide chain is the same irrespective of which amino-acid residue is on the growing end of the peptide chain. Furthermore, since the conformation of the peptide chain may affect the rate of polymerization, it is possible that amino-acid residues other than the terminal one will affect the preference for the next residue to be added. The possibilities

Table 1.1

Polypeptide from	[η] and solvent		Estimated MW	Estimated DP	Reference
L-asparagine					*a*
S-alkyl-L-cysteines					*b*
S-alkyl-D,L-homocysteines					*c*
L-methionine	0.22	TFA	18,000	135	*d*
Lanthionine					*e*
L-serine	0.10	H_2O	11,300	130	*f*
O-acetyl-L-serine	0.19	DCA	16,800	130	*f*
O-phenyl-D,L-homoserine			21,000–26,000	120–150	*g*
L-threonine					*h*
O-acetyl-L-threonine	0.12*	DCA	13,000	91	*h*
O-methyl-L-threonine	0.12*	DCA	10,000	91	*h*
L-cyclohexylalanine			4,900	32	*i*
L-nitroarginine				5–6	*j*
anhydro-L-aspartic acid (L-succinimide)					*k*
D,L-anhydro-glutamic acid					*l*

* Specific viscosity at a concentration of 0.2 per cent.
[a] J. Noguchi, T. Saito, and M. Asai, *J. Chem. Soc. Japan* **81**, 620 (1960).
[b] M. Frankel, D. Gertner, H. Jacobson, and A. Zilkha, *J. Chem. Soc.* **1960**, 1390.
[c] M. Frankel and D. Gertner, *ibid.*, **1961**, 463.
[d] E. R. Blout, C. de Lozé, S. M. Bloom, and G. D. Fasman, *J. Am. Chem. Soc.* **82**, 3787 (1960).
[e] M. Frankel and D. Gertner, *J. Chem. Soc.* **1961**, 459.
[f] G. D. Fasman and E. R. Blout, *J. Am. Chem. Soc.* **82**, 2262 (1960).
[g] M. Frankel and Y. Knobler, *J. Chem. Soc.* **1958**, 3733.
[h] M. Volz, G. D. Fasman and E. R. Blout, unpublished results.
[i] M. Sela and R. Arnon, *J. Am. Chem. Soc.* **82**, 2625 (1960).
[j] T. Hayakawa, T. Mizuno, and J. Noguchi, *J. Chem. Soc. Japan* **81**, 618 (1960).
[k] J. Noguchi, T. Saito, and M. Asai, *ibid.* **81**, 620 (1960).
[l] M. Asai, K. Imai, and J. Noguchi, *ibid.* **81**, 624 (1960).

of readily obtaining truly random copolypeptides, especially those containing more than two amino acids, are not very encouraging, since little data of this kind has been reported, and that which is available indicates that the rate of polymerization of two closely related anhydrides, γ-benzyl-L-glutamate NCA and ε-carbobenzyloxy-L-lysine NCA, are different (*5*).

Block copolypeptides.—Using preformed polypeptides containing free amino groups as initiators it is possible to continue the polymerization of the peptide chain with other amino acid *N*-carboxyanhydrides and to obtain copolypeptides containing blocks of amino acids. Theoretically this technique could be used to prepare block copolypeptides of practically any composition and with practically any number of residues in the block. In practice, only a few such experiments have been reported since such materials would not be very interesting protein models. In addition, since such polymerizations are perforce amine initiated, side reactions which result in termination (*6*) well may limit the practical utility of this method in the preparation of high molecular weight or multicomponent block copolymers.

Another technique for making block copolymers has been described (*7*). This method involves the combining of molecules of preformed homopolypeptides having both terminal free amino and free carboxyl groups, and joining these preformed polymers by means of carbodiimide type reagents. This type of reaction has been carried out with two different preformed polypeptides in chloroform solution and has also been used to increase the molecular weight of poly-L-proline in water solution.

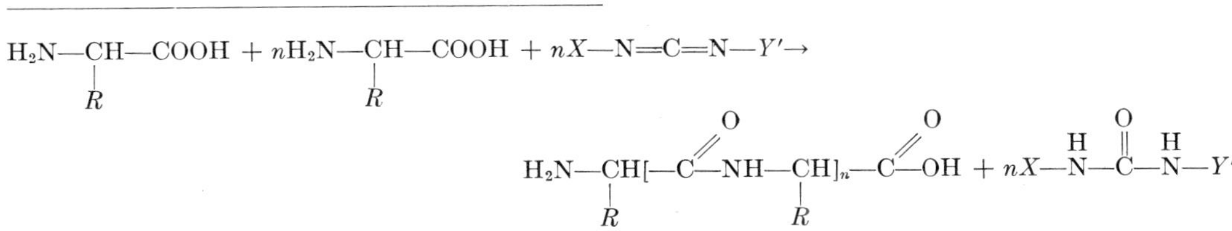

Multichain polyamino acids.—Multichain polyamino acids, that is, those whose molecules are composed of several linear polypeptide chains attached to a peptide core, have been obtained by initiation of *N*-carboxyamino acid anhydrides with multifunctional initiators. The use of proteins as multifunctional initiators was first described by Stahmann and Becker (*8*). Other investigators have used synthetic polypeptides containing multiple free amino groups as initiators for the preparation of multichain polyamino acids (*9*) and by this method a variety of water-soluble polypeptides of high molecular weight have been made. Because of the intrinsic "globular" shape of these molecules, it is thought that they might be important for some biological applications. However, it should be emphasized that natural globular proteins show their specific properties because of the specific type of peptide chain folding (that is, conformation) which occurs in these proteins and these properties are not results of branching of peptide chains such as exists in synthetic multichain polypeptides.

Application of new synthetic methods to heteropolypeptide preparation.—As indicated above, although amino acid NCA's have proven to be important and useful precursors in the synthesis of homopolypeptides (those composed of only one amino acid) there are several limitations on the types of heteropolypeptides (those composed of two or more amino acids) that can be prepared using this method. As a consequence it now appears that the application of synthetic techniques other than NCA polymerization should be investigated for the preparation of heteropolypeptides. For example, the carbodiimide technique originated by Sheehan and Hess (*10*) for the formation of peptide bonds has been investigated for the synthesis of homopolypeptides derived from glycine and from L-glutamate esters (*11, 12*). As of this time, however, only low molecular weight materials (DP 5–15) have been obtained probably because of precipitation of the polymer during the polymerization reaction as well as the occurrence of side reactions such as acyl-urea formation during the polymerization.

However, the preparation of new reagents and modifications of older techniques for synthesis of simple peptides is not without promise for the preparation of high molecular weight polypeptides. For example, the *p*-nitrophenyl ester method of Bodanszky (*13*) could conceivably be applied to the preparation of high molecular weight polypeptides by a modification of the present preparative procedure (shown here) for simple peptides.

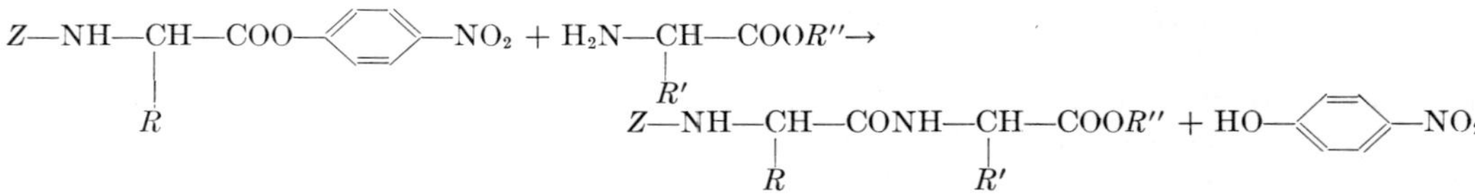

Modifications of this method using the *p*-nitrothiophenyl esters have been reported by English (*14*) and by Japanese workers (*15*). The latter have prepared polymers of low molecular weight (DP 10–12) by the self-condensation of pro-leu-gly-*p*-nitrothiophenyl ester.

Perhaps even more promising is the recent publication by Woodward, Olofson, and Mayer (*16*) of a new peptide synthesis involving the reaction of 3-unsubstituted isoxazolium salts with carboxylates to give "active" enol esters followed by reaction with an amino-acid ester to yield peptides. Although no reports of a polymerization involving this reagent have yet been forthcoming, it is conceivable that it can be used in such reactions.

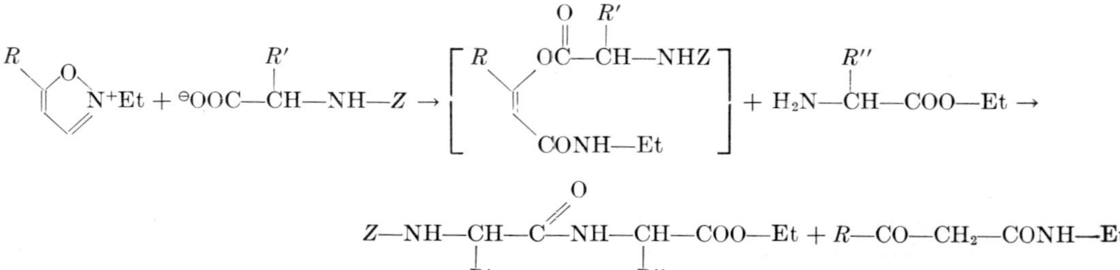

It is reported that the yields of simple peptides produced in this manner are high and the degree of racemization is low. The method appears to be general in its application since hydroxyproline and tyrosine can be used without protection of hydroxyl groups and good results have been obtained with both asparagine and glutamine derivatives.

A fourth peptide synthesis has just been described (*17*) which appears to be very simple and to have many possible applications. This method involves the reaction of *N*-substituted amino acid hydrazides with an amino acid ester in the presence of an oxidizing agent such as *N*-bromosuccinimide (NBS).

$$Z\text{—NH—}\underset{R}{\text{CH}}\text{—}\overset{O}{\text{C}}\text{—NH—NH}_2 + \text{H}_2\text{N—}\underset{R'}{\text{CH}}\text{—}\overset{O}{\text{C}}\text{—}OR'' \xrightarrow{\text{NBS}} Z\text{—NH—}\underset{R}{\text{CH}}\text{—}\overset{O}{\text{C}}\text{—NH—}\underset{R'}{\text{CH}}\text{—}\overset{O}{\text{C}}\text{—}OR + \text{N}_2$$

The reaction is complete within two minutes and very high yields of optically pure peptides were obtained. It would appear that this type of reaction could easily be applied to the polymerization of peptide hydrazides to yield peptides with known repeating sequences. In fact, a polymer derived from pro-gly-gly hydrazide has been prepared having a molecular weight of about 1,400 (*17*).

Properties of synthetic polypeptides

Molecular weights.—It is well known from work on di-, tri-, tetra-, etc., peptides that the physical properties of these materials change markedly as the number of residues is increased. For example, the solubility in water of tetraglycine is very much lower than that of glycine. Because of this it is most important that molecular weights be specified when an investigator is describing the properties of a synthetic polypeptide. Unfortunately, very few direct determinations of the molecular weights of polypeptides have been made. Perhaps the best determination of molecular weights are those of Doty *et al.* on poly-γ-benzyl-L-glutamate polymers (*18*). In this work the molecular weights were determined by light-scattering methods in two solvents and the data correlated with viscosity measurements in these solvents. Since one of the solvents, dichloroacetic acid, is a solvent which tends to disrupt both helical and β structures to form random coils, the determination of the viscosities of other polypeptides having random conformations in this solvent allows an *estimate* of their molecular weights using the molecular weight–viscosity relationship from the poly-γ-benzyl-L-glutamate data. Of course, for the proper use of this method it is essential that the polypeptide be in the random form in dichloroacetic acid (DCA) solution. Fortunately, DCA is a sufficiently strong H-bonding solvent so that most polypeptides are in the random form in it. Estimates of weight-average molecular weights of polypeptides using this simple technique are beginning to appear in the literature.

Osmotic pressure measurements of molecular weight, while often leading to absolute values, are tedious to obtain in many of the solvents in which synthetic polypeptides are soluble. On the other hand, end-group titration methods have been employed by some investigators. Such techniques are difficult to apply with precision at degrees of polymerization (DP) greater than about 200 but are undoubtedly useful in obtaining number average molecular weights at low DP's. However, because of the possibility of the occurrence of termination reactions which may change the nature of the end groups, such end-group determinations must certainly be applied with caution. Sedimentation techniques certainly should be applicable to the determination of molecular weights of synthetic polypeptides and indeed some work is being done using this method (*19*).

Available data on molecular weight for various synthetic polyamino acids indicates that amino-acid polymers tend to fall into two classes. First, those which yield high molecular weight (DP > 50,000) polymers; secondly, those which, so far, have yielded only lower molecular weight polymers. It is both interesting and important to note that those amino acids which yield high molecular weight polymers are those

Table 1.2

Polypeptide from	Specific visc.*	Estimated† DP_w	Estimated† MW_w	Conformation‡	Reference
ε-carbobenzyloxy-L-lysine	6.78	6390	1,680,000	α	*a*
γ-benzyl-L-glutamate	4.10	2700	600,000	α	*b*
L-methionine	1.18 (T)	910	120,000	α	*c*
L-leucine	1.13 (T)	850	96,000	α	*d*
S-benzyl-L-cysteine	.49	330	63,500	β	*d*
O-acetyl-L-serine	.14	82	10,600	β	*e*
L-valine	.29 (T)	185	18,500	β	*d*

* The specific viscosities were measured in dichloroacetic acid at a concentration of 0.2 per cent. Those viscosities marked with (T) were measured in trifluoroacetic acid.

† The weight average molecular weights, MW_w, and the weight average degrees of polymerization, DP_w, were estimated from the viscosity data using the molecular weight calibration for poly-γ-benzyl-L-glutamate in Ref. *18*.

‡ The conformations were determined in the solid state from infrared dichroism measurements. See Ref. *20*.

[a] G. D. Fasman, M. Idelson, and E. R. Blout, *J. Am. Chem. Soc.* **83**, 709 (1961).

[b] E. R. Blout and R. H. Karlson, *ibid.* **78**, 941 (1956).

[c] S. M. Bloom, G. D. Fasman, C. de Lozé, and E. R. Blout, *J. Am. Chem. Soc.* **84**, 458 (1962).

[d] Unpublished results.

[e] G. D. Fasman and E. R. Blout, *J. Am. Chem. Soc.* **82**, 2262 (1960).

which naturally tend to form helical strucsture in the polymerization solvents, whereas those which form lower molecular weight polymers show only β (extended) or random structures. Thus one is led, at least tentatively, to the conclusion that the formation of high molecular weights in synthetic polypeptides is a consequence of the helix-forming tendency of the particular amino acids of which the polypeptide is composed (*20*). Some representative examples of polypeptide polymers and the reported estimated maximum molecular weights are given in Table 1.2.

Solubilities.—The solubility of synthetic polypeptides depends not only on the molecular weight but also on the amino acids of which the polymer is composed. Many of the seemingly conflicting reports of solubilities of polypeptides may well be due to differences in molecular weights of the polymers investigated. Recent measurements of solubility of polypeptides of approximately the same molecular weight (DP$\sim$100) have indicated very great differences in solubility behavior (*21*). These differences are undoubtedly related to the nature of the side-chain groups, the packing of the side-chain groups, and the conformation of the polymer molecules. Some representative data are given in Table 1.3.

An interesting series of experiments has been reported in which it has been shown that copolymers appear to have greater solubility than homopolymers (*22*). This is not surprising in that copolymers having various side chains in the molecule would certainly allow the molecule, and especially the backbone of the molecule, to be more accessible to solvent. Finally, it should be noted that homopolymers made from D,L-amino acids have much greater solubility than homopolymers composed of amino acids having a single optical configuration (*23*).

Conformational changes with water.—It is now well established that ionic polypeptides undergo conformational changes with changes of *p*H and consequent ionization and deionization of side-chain groups. As has been noted several times the conformation of the synthetic polypeptide exerts a strong influence on its physical-chemical properties. In the last few years experiments with polypeptides having charged side-chain groups, such as poly-L-lysine hydrochloride (*24, 25*) and polysodium-L-glutamate (*26*), have indicated that the environment around the molecule also can affect the conformation of the molecule. In particular, by means of infrared dichroism measurements, it has been shown that these two ionic polypeptides undergo a reversible transformation from extended to α-helical conformations, as indicated in Figure 1.1, with varying water content. These observations suggest that such processes may operate in biological systems where varying water content may serve to invoke conformational changes in a portion of or in all of a protein molecule. In this manner it is possible that biologically "active sites" may thus become exposed or covered. Whether such changes in conformation can be caused by other simple environmental changes such as varying salt or ion concentration remains to be investigated.

Hydrogen-deuterium exchange.—If such compounds as alcohols, acids, and amides containing active hydrogens are dissolved in solvents having "active" deuterium atoms, such as deutero-dichloroacetic acid or D_2O, then an equilibrium is set up and the active hydrogens

Table 1.3

Polypeptide from	Solvent*				Conformation†
	Chloroform	Dioxane	DCA	TFA	
γ-benzyl-L-glutamate	sol.	sol.	sol.	sol.	α
L-methionine	sol.	insol.	sol.	sol.	α
L-alanine	insol.	insol.	sol.	sol.	α
L-leucine	insol.	insol.	insol.	sol.	α
O-acetyl-L-serine	insol.	insol.	sol.	sol.	β
L-valine	insol.	insol.	insol.	sol.	β

* The solubilities were measured at a concentration of 2 per cent.

† The conformations were determined in the solid state from infrared dichroism measurements. See Ref. *20*.

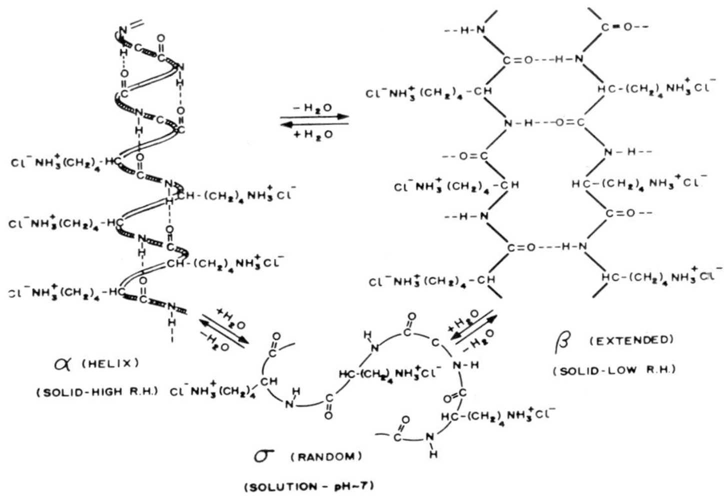

Fig. 1.1. Diagrammatic representation of conformations of poly-α, L-lysine · HCL.

of the solute will be exchanged for the active deuteriums of the solvent molecules. This type of exchange reaction has been investigated with polypeptides in organic solvents (*27*) and in aqueous systems (*28*). As a result of these investigations it is apparent that there is a large difference in the rate of amide hydrogen-deuterium exchange in the random conformation as compared with the α-helical conformation. This fact is easily explicable on the basis of the availability of the peptide backbone to the solvent molecules in the random conformation as compared with the α-helix. Data on polyglutamic acid in the helical and random conformations in aqueous systems indicate that the rate of exchange of the random conformation is at least 100 times greater than that of the helix (*28*). Thus at first consideration this method appears to offer a simple way to differentiate between helical, random, and possibly other conformations. However, it is important to be aware that factors other than conformation may affect the rate of hydrogen-deuterium exchange. Such factors include the presence or absence of ionizable groups and the presence or absence of hydrophobic side chains which would tend to shield the peptide backbone from the solvent. It is thus apparent that further investigation of this method is needed before definitive conformational evidence can be deduced from it.

Chemical reactions of polypeptides.—Very few chemical reactions of polypeptides have been studied. This is an area which holds great potential promise as it is theoretically possible to alter the physical, chemical, and perhaps biological properties of such materials by appropriate chemical modifications. Among the interesting work reported to date is the formation of dyes by coupling reactions into tyrosine-containing polypeptides (*29*) and the formation of covalently bonded polypeptide enzyme moieties by Katchalski and others (*30*). It appears that an interesting new chemistry will develop with the increasing realization of the importance of conformation upon chemical reactivity of synthetic polypeptides.

Some unsolved problems

Water-soluble nonionic helical polypeptides.—Although polypeptides from almost all naturally occurring amino acids have now been synthesized, it should be noted that there is yet no reported example of a high molecular weight, water-soluble, nonionic helical polypeptide composed only of one optical isomer. Such a material would be a highly desirable model compound for use in physical-chemical studies as well as having possible utility in various biological and biochemical work. It was expected that poly-L-serine would be such a material. This does not appear to be the case, since poly-L-serine of DP ~100 does not show the optical properties of the α-helix (*31*). Whether this is a consequence of the marginal stability of the α-helix in aqueous solution or is due to either side chain–main chain or side chain–side chain hydrogen bonding in poly-L-serine is now being investigated. The data so far obtained, however, suggest that the α-helical portions observed in water-soluble proteins probably derive at least part of their stability from bonds and forces other than the intramolecular hydrogen bonds of the α-helix.

Controlled sequences.—Now that it is possible to make synthetic polypeptides from any known amino acid in a range of molecular weights, and to make nearly random as well as block copolypeptides, the important remaining synthetic problem is that of preparing high molecular weight polypeptides with predetermined known sequences of amino acids. It does not appear that it will be possible to obtain such sequentially determined polypeptides through the use of the *N*-carboxyanhydride polymerization reaction. On the other hand, however, if the desired polypeptides are to be of molecular weights 10,000 to 50,000 (that is having ~100 to 500 amino-acid residues per molecule) and a polymerization reaction is not used, then it will be necessary to use a stepwise peptide-bond formation reaction that gives very high yields. If one adds to this problem that of maintaining the optical purity necessary for the ultimate use of such molecules in biological systems, then the problem becomes a truly formidable one. However, if one of the types of peptide-bond formation reactions described above is developed to the point where the yields of optically homogeneous material are high enough, then it is not impossible to visualize the application of a mechanized procedure for the synthesis of really high molecular weight polypeptides of predetermined and known sequence.

Finally, it is becoming increasingly apparent that, with the development of new and improved synthetic methods for the preparation of better defined and more complicated high molecular weight synthetic polypeptides, it is within our ken to begin the exciting job of relating chemical composition, macromolecular structure, and biological function in proteins.

References

1 H. Leuchs, *Ber. deut. chem. Ges.* **39,** 857 (1906).

2 For a comprehensive review of this subject see: (*a*) C. H. Bamford, A. Elliott, and W. E. Hanby, "Synthetic Polypeptides," Academic Press, New York, 1956. (*b*) E. Katchalski and M. Sela, *Advances in Protein Chem.* **13,** 243–492 (1958).

3 E. R. Blout, R. H. Karlson, P. Doty, and B. Hargitay, *J. Am. Chem. Soc.* **76,** 4492 (1954); E. R. Blout and R. H. Karlson, *ibid.* **78,** 941 (1956); G. D. Fasman, M. Idelson, and E. R. Blout, *ibid.* **83,** 709 (1961).

4 M. Idelson and E. R. Blout, *ibid.* **80,** 2387 (1958).

5 Y. Shalitin and E. Katchalski, *ibid.* **82,** 1630 (1960).

6 M. Sela and A. Berger, *ibid.* **77,** 1893 (1955).

7 E. R. Blout and M. E. DesRoches, *ibid.* **81,** 370 (1959).

8 M. A. Stahmann and R. R. Becker, *ibid.* **74,** 2695 (1952); R. R. Becker and M. A. Stahmann, *J. Biol. Chem.* **204,** 745 (1953).

9 M. Sela and E. Katchalski, *Experientia* **11,** 62 (1955); M. Sela, E. Katchalski, and M. Gehatia, *J. Am. Chem. Soc.* **78,** 746 (1956).

10 J. C. Sheehan and H. D. Hess, *ibid.* **77,** 1067 (1955).

11 J. H. Bradbury and D. C. Shaw, *Australian J. Chem.* **12,** 300 (1959).

12 V. Bruckner, M. Szekerke, and J. Kovacs, *Hoppe-Seyler's Z. physiol. Chem.* **309,** 25 (1957).

13 For references see M. Bodanszky, *Ann. N. Y. Acad. Sci.* **88,** 655 (1960).

14 J. A. Farrington, G. W. Kenner, and J. M. Turner, *Chem. & Ind.* **601** (1955); G. W. Kenner and J. M. Turner, *ibid.* **602** (1955).

15 H. Kitaoka, S. Sakakibara, and H. Tani, *Bull. Chem. Soc. Japan* **31,** 802 (1958).

16 R. B. Woodward, R. A. Olofson, and H. Mayer, *J. Am. Chem. Soc.* **83,** 1010 (1961).

17 Y. Wolman, P. N. Gallop, and A. Patchornik, *ibid.* **83,** 1263 (1961).

18 P. Doty, A. M. Holtzer, J. H. Bradbury, and E. R. Blout, *ibid.* **76,** 4493 (1954); P. Doty, J. H. Bradbury and A. M. Holtzer, *ibid.* **78,** 947 (1956).

19 J. C. Mitchell, A. E. Woodward, and P. Doty, *ibid.* **79,** 3955 (1957); E. Katchalski, personal communication.

20 E. R. Blout, C. de Lozé, S. M. Bloom, and G. D. Fasman, *J. Am. Chem. Soc.* **82,** 3787 (1960).

21 Unpublished results.

22 R. R. Becker and M. A. Stahmann, *J. Am. Chem. Soc.* **76,** 3707 (1954).

23 C. de Lozé, G. D. Fasman, and E. R. Blout, to be published.

24 E. R. Blout and H. Lenormant, *Nature* **179,** 960 (1957).

25 A. Elliott, B. R. Malcolm, and W. E. Hanby, *ibid.* **179,** 963 (1957).
26 H. Lenormant, A. Baudras, and E. R. Blout, *J. Am. Chem. Soc.* **80,** 6191 (1958).
27 A. Elliott and W. E. Hanby, *Nature* **182,** 654 (1958); D. G. H. Ballard, C. H. Bamford, and A. Elliott, *Macromolekulare Chem.* **35,** 222 (1960).
28 E. R. Blout, C. de Lozé, and A. Asadourian, *J. Am. Chem. Soc.* **83,** 1895 (1961).
29 M. Sela and E. Katchalski, *ibid.* **77,** 3663 (1955).
30 A. Bar-Eli and E. Katchalski, *Nature* **188,** 856 (1960).
31 R. D. Lundberg and P. Doty, *J. Am. Chem. Soc.* **79,** 3961 (1957).

Discussion of Paper 1: Synthesis and Chemical Properties

J. KOVACS

It may be worth while to mention our experience concerning the usefulness of the different methods in the preparation of polypeptides with predetermined known sequences of amino acids. For the polycondensation of γ-glutamyl-dipeptides, a modification of Noguchi's carbothiophenyl method proved to be one of the best regarding the yield and molecular weight of the final products. This modification involved the use of peptides with free amino and carboxyl groups which, after reaction with one mole of carbothiophenyl chloride, gave amino- and carboxyl-activated derivatives. These derivatives finally gave the desired polypeptides. (Details of these investigations are reported in *Acta Chim. Acad. Sci. Hung.* **12,** 363 [1957].)

K. D. KOPPLE

Apart from the obvious dangers in use of strong acid solvents such as DCA or TFA for benzyl esters, there is the possibility, in TFA at least, of cleavage of peptide bonds themselves. Polyleucine is about 20 per cent cleaved in refluxing TFA (71°) after 24 hours (K. D. Kopple and E. Bachli, *J. Org. Chem.* **24,** 2053 [1959]). A long extrapolation suggests that one bond in 100–200 might be broken on 24 hours' storage at 20° in this solvent. Therefore, for high molecular weight products, even in the absence of more sensitive groups, significant polymer degradation is possible when using TFA as solvent.

PAPER 2

Multichain Poly-α-Amino Acids

A. BERGER and A. YARON

Single-chain polyamino acids have received a great deal of attention in recent years, mainly in their capacity as protein models (*1*). They have yielded important information on problems of the structure of the polypeptide chain, its electrochemical behavior, its susceptibility to the attack of proteolytic enzymes, and its interaction with living systems. Few proteins, however, consist of molecules made of a single polypeptide chain. It was felt, therefore, that much additional information could be obtained from a more complex type of model compounds in which a number of polypeptide chains are combined in one molecule.

A number of structural possibilities for multichain polyamino acids can be envisaged. The obvious way to achieve the closest resemblance to a multichain protein would consist in crosslinking a predetermined number of peptide chains via suitable amino-acid residue side chains, e.g., by the oxidative crosslinking of cysteine-containing polyamino acids. The difficulties in obtaining in such a way materials of more or less uniform and predictable compositions are formidable. On the other hand, a model compound in which the component peptide chains are arranged in a way not found in native proteins might justify itself, if it consists of relatively uniform molecules and if the molecular parameters of size and shape can be controlled during synthesis.

The multichain polyamino acids described in this paper consist of polypeptide chains attached to a multifunctional core which in most cases is itself a polyamino acid. They are in fact branched molecules in which the individual branches are single, unbranched chains.

One way of obtaining branched polypeptides of this kind is to attach preformed polypeptide molecules via one of their terminal groups to reactive side chains of a polypeptide backbone. This was done by Bovarnick (*2*), who coupled the amino groups of a relatively short polyglutamic acid with the side chains of a higher molecular weight polyglutamic acid, after conversion of its carboxyl groups to azide groups via the polyester and the polyhydrazide. This approach, which has not been used since, may be very useful, mainly for the preparation of branched polypeptides from natural oligopeptides or from repeating sequence α-amino acid copolymers (*3*).

The alternative method consists in "growing" polypeptide side chains in situ, and this is the one that has been used for all the other multichain polyamino acids described in the literature. It takes advantage of the fact that primary (and secondary) amines, when initiating the polymerization of *N*-carboxyamino acid anhydrides, remain attached to the carboxyl end of the growing chain. Thus any molecule bearing several amino groups becomes, when used to initiate the polymerization, the "core" of a multichain polymer. A number of proteins (*4*) as well as gelatin (*5*) have thus been used to produce a variety of polypeptidyl proteins. Other papers in this book deal with this subject. In all these cases, however, only a small proportion of the amino-acid residues of the core, namely lysine and those bearing a free α-amino group, served as initiating sites. Consequently the properties of the products are determined by the nature of the core rather than by the side chains attached to it, and the molecular weights obtained are not much different from those of the parent proteins.

Purely synthetic multichain polyamino acids have been prepared by the use of polylysine, polyornithine, and copolymers containing those basic amino acids, as multivalent initiators for the polymerization of various *N*-carboxy amino-acid anhydrides (NCA's) (*6, 7, 8*).

When the proportion of amino groups in the initiator molecule is high, there occurs, obviously, a drastic increase of molecular weight on polymerization: an initiator consisting of a hundred lysyl residues can give rise to a multichain molecule with a molecular weight of a million. Moreover, proper choice of the initiator and of the polymerization conditions should enable one not only to control the size of the molecule but also to vary its shape. Thus a long core bearing short side-chains might be expected to resemble rodlike, prolate molecules while a starlike polymer made from a short initiator bearing long side-chains might behave like a spherical or an oblate particle.

Turning to the practical side of the problem, the question of the proper conditions for the branching reaction arises. Both polymerization reactions and organic reactions involving multifunctional polymers are particularly sensitive to reaction conditions. In our case we are dealing with a combination of both, namely, polymerization initiated by a polyfunctional macromolecule.

The amine-initiated polymerization of *N*-carboxy-α-amino acid anhydrides can be formulated as follows (*9*):

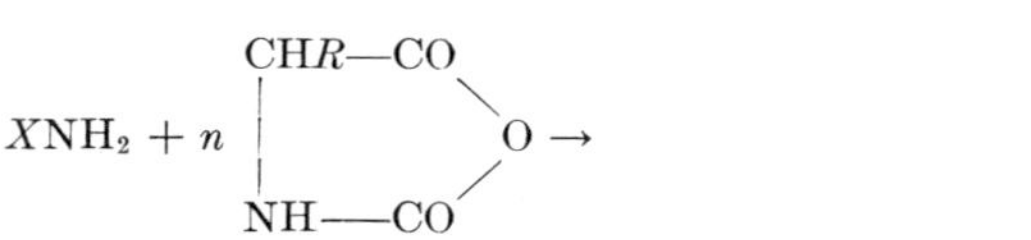

where XNH_2 represents the initiator molecule, or one of its reactive groupings in the case of a multivalent initiator. It may also represent the active end of a growing chain, being continuously regenerated during the polymerization process. Protonation of the amino group, $XNH_2 + AH \rightleftharpoons XNH_3^+ + A^-$, renders it inactive, and in the case of an equilibrium only the nonionized fraction is kinetically effective. In addition, any molecular species, YH, containing an active hydrogen can open the anhydride ring according to its characteristic velocity constant and thus initiate polymerization.

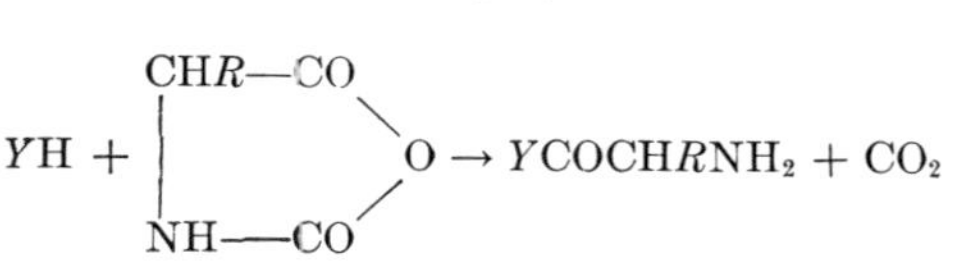

In order to obtain optimal conditions for this type of polymerization, of which the branching reaction is a particular case, one should obviously exclude solvents such as water, alcohols, or organic acids. It must also be borne in mind that other types of initiation of NCA polymerization do exist, namely, initiation by tertiary bases (*10*), by strong bases (*11*), and by salts (*12*). These reactions, the mechanism of which is as yet poorly understood, are operative when little or no water is present, and may compete with the desired branching reaction whenever the salt of a polyamine initiator is being deionized in such an environment.

In practice, however, the peculiar solubility properties of polyamines such as polylysine and polyornithine make it impossible to take advantage of the above conclusions, and some compromise must be reached. Polylysine, when isolated as the free base (*6*), is soluble only in aqueous acid and in formic acid. When, however, an equivalent of base is added to sufficiently dilute aqueous solution of a polylysine salt, the free base is not precipitated. Such a solution is too alkaline to give reasonably good yields of polymer, since NCA's are rapidly hydrolyzed under these conditions.

Polypeptidyl proteins and some of the branched polypeptides described in the literature have been prepared in aqueous solution (*4, 5, 6, 7, 8*), buffered at a *p*H value of 7 or slightly higher. Under these conditions about 20 per cent to 50 per cent, sometimes less and seldom more, of the NCA employed is incorporated into the branched product. Carbonate buffer was reported to be superior to phosphate buffer (*13*). This method has several drawbacks: yields, in some instances, may be so low as to make the process impractical (*6*); it is seldom possible to control the degree of incorporation. As a number of the NCA's are insoluble in aqueous medium, the reaction is often a heterogeneous one and certain features of the branching reaction, such as the relative rates of incorporation in the case of copolymers, will depend on uncertain factors, such as the kind of agitation employed. On the other hand, the products of side reactions are usually of low molecular weight and may be removed by dialysis, if a suitable solvent can be found.

Alternatively, the branching reaction with polylysine or polyornithine as the initiator can be performed in dioxane solution containing as little as 3 per cent water (*6*). In one case reported, polylysine was dissolved in a small amount of water containing an excess of triethylamine and the resulting solution was added to thirty volumes of dioxane containing the NCA of γ-benzyl glutamate. Incorporation in this case seemed to be much better (67 per cent after dialysis) than in aqueous phosphate buffer (16 per cent). As will be shown below, however, a large quantity (about 50 per cent) of nondialyzable linear polymer may be formed under these conditions.

Closer investigation of this nearly anhydrous method of growing γ-benzyl-glutamate side-chains on a polylysine backbone revealed (*14*) that the multichain fraction ($S \sim 8$ Svedberg units) was always accompanied by a nondialyzable linear fraction ($S \sim 2$ Svedberg units) even when less than one equivalent of tertiary base was used to deionize the primary amino groups of the initiator. The reaction product of a typical run, employing 40 moles of γ-benzylglutamate NCA per base mole of polylysine, could be separated by fractional precipitation (dimethylformamide-ether) into a branched portion (50 per cent by weight; Glu/Lys = 22) and a linear one which contained no lysine residues. The situation is further complicated by the fact that not all the polylysine employed could be accounted for by analysis of the reaction product. Some of it escaped reaction in the form of a sticky precipiate adhering to the walls of the reaction vessel.

Much better results were obtained when β-benzyl aspartate NCA and polylysine were employed in the 97 per cent dioxane system described above. Here the reaction mixture cleared up entirely after a short time, indicating the disappearance of free polylysine. Even at high anhydride to initiator ratios very little linear polymer was formed, and all of the lysine contained in the initiator was recovered in the branched polymer. At low ratios (3 to 10) the product was entirely homogeneous, incorporation being practically quantitative.

A branched polylysine, bearing an average of about three β-benzyl aspartate residues on its side chains, was prepared (*15*). After being isolated by precipitation and dialysis, this multichain polymer is soluble in a variety of organic solvents (e.g., dioxane, chloroform, dimethylformamide) and contains free amino groups approximately equal in number to its lysine residues. These amino groups, the α-amino groups of the terminal benzylaspartate residues of its side chains, can be utilized without additional treatment to initiate further polymerization of NCA's in anhydrous solvents. Some of the branched polymers obtained with the help of this "soluble initiator" (i.e., soluble in organic solvents) will be described below. In all cases incorporation of various NCA's was practically quantitative. It seems, therefore, that on this "soluble initiator" one can grow side chains of well-defined and predetermined average length, which (because of the absence of a linear fraction) can be calculated from the relative amount of the various amino-acid residues which make up the multichain molecule.

In order to define fully the branched polymer, reliable information as to the length of the polylysine backbone is necessary. Polylysine is prepared from poly-ε-carbobenzoxylysine by treatment with hydrogen bromide in acetic acid or by means of similar reagents. The relationship between the hydrodynamic properties (e.g., viscosity) of polycarbobenzoxylysine and its average degree of polymerization has been extensively investigated (*16*) and the conclusions reached seem to have a sound basis. A direct determination of the molecular weight of polylysine seems to be more difficult on theoretical grounds. This polyelectrolyte, as mentioned above, is soluble only in its ionized form, and configurational changes as well as the strong interactions of a charged molecule with its environment complicate the interpretation of physical measurements. Investigations on this subject have been published (*17*) but it seemed to us that a different approach might be helpful. The idea was to convert polylysine, under conditions not causing cleavage of the polypeptide chain, into derivatives, the molecular weight of which could be measured with certainty. This was done in two independent ways, relying on viscosity measurements in the one and on osmotic pressure measurements in the other.

It was found that polylysine could be fully carbobenzoxylated under mildly alkaline conditions. The viscosities in dimethylformamide of a number of recarbobenzoxylated samples were measured, and their degrees of polymerization thus determined were assumed to be equal to those of the respective polylysines.

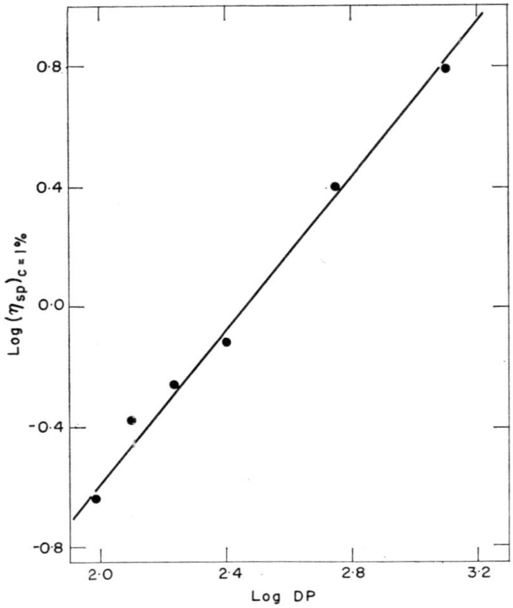

Fig. 2.1. The dependence of the specific viscosity on the degree of polymerization of poly-L-lysine hydrobromide in salt-free aqueous solution at 26°, at a concentration of 1 per cent. The degrees of polymerization were determined by measuring the viscosity in DMF of the recarbobenzoxylated polylysine samples (— ● —).

Agreement between these values and those obtained from the osmotic measurements described in the next paragraph was excellent. In order to facilitate routine determinations of the degree of polymerization of polylysine hydrobromide samples, an empirical curve was also constructed, correlating the viscosity of the polymer in 1 per cent aqueous salt-free solution with the degree of polymerization as determined by the recarbobenzoxylation procedure (see Fig. 2.1).

To some polylysine samples short (2–3 residue) β-benzylaspartate side-chains were attached as described above. These "soluble initiator" preparations could be debenzylated under relatively mild conditions, a carboxyl group appearing on each aspartyl residue. The molecular weights of these polyampholytic derivatives can be determined under near-isoelectric conditions (pH 3) and at high salt concentration (1 M $NaNO_3$), or by the extrapolation to zero polymer concentration of π/c vs. c curves obtained in different salt concentrations (0.1 to 2 M Na acetate, pH 7) (Fig. 2.2). Potentiometric titration of the ampholyte (Fig. 2.3) shows the relative amount of amino and carboxyl groups in a polyampholyte sample (average of 2.5 aspartyl residues per side chain). It may be seen that the basic groups have the pK of an α-amino group, shifted to higher values because of the proximity of ionized carboxylate groups. The ε-amino groups of a linear copolymer of lysine and aspartic acid, shown for comparison, have a pK about 1.5 units higher.

In order to illustrate the use of poly-β-benzylaspartyl-polylysine with short side-chains (the "soluble initiator"), a few examples of branched polyamino acids with well-defined molecular parameters are given below. In all cases n is the degree of polymerization of the polylysine backbone, m' the number of aspartyl residues per side chain, and m the number of other amino-acid residues attached to each side chain.

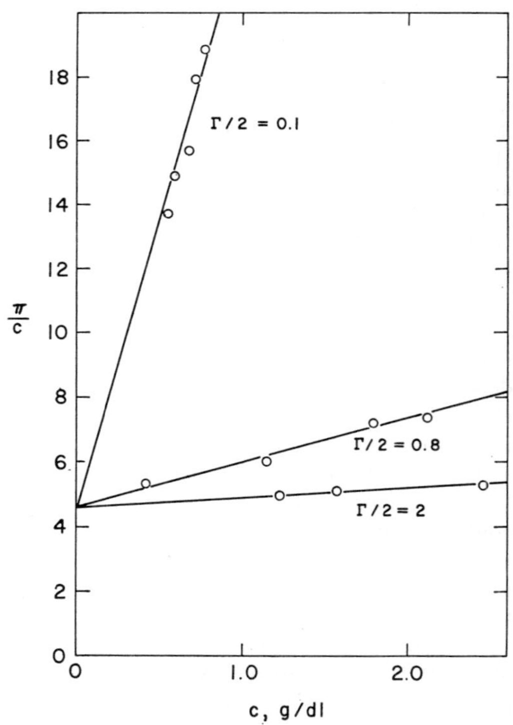

Fig. 2.2. The osmotic pressure of multichain polyaspartyl-polylysine ($n = 170$, $m' = 2.1$). Osmotic pressure, π, is given in cm decalin at 30°; polymer concentration, c, in gm/100 ml. (Measurements by Dr. E. Riesel.)

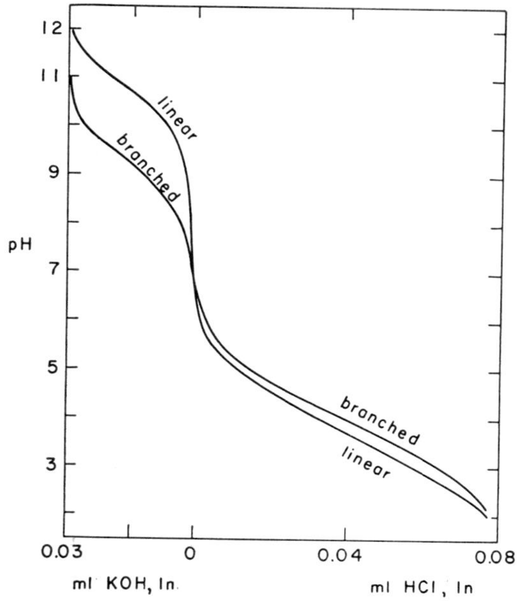

Fig. 2.3. Potentiometric titration of multichain polyaspartyl-polylysine (16 mg; $n=70$, $m'=2.5$), and of a linear copolymer of aspartic acid and lysine in 2 M KCl.

Table 2.1—Hydrodynamic properties of some multichain polyamino acids. n = number of L-lysyl residues in backbone, m' = number of β-benzyl-L-aspartate residues per side chain, m = number of γ-benzyl-L-glutamate residues per side chain.

n	m'	m	M.W.	S_0 Sv. U.	$[\eta]$* dl/gm	$[\eta]$† dl/gm
60	3.0	0	44,000	4.3	0.06	—
140	3.3	0	110,000	6.2	0.16	—
60	3.0	11	190,000	6.4	0.25	0.40
140	3.3	10	420,000	11.4	0.29	0.78

* Measured in dimethylformamide.
† Measured in dichloroacetic acid.

Table 2.1 shows some hydrodynamic characteristics of two "soluble initiator" samples and of two multichain poly-γ-benzylglutamate polymers derived from them (*18*). It can be seen that, as expected, viscosities are low and sedimentation coefficients rather high. The γ-benzylglutamate side chains seem to exist in an α-helical form in dimethylformamide as can be concluded from optical rotatory dispersion data, which give a b_0 (*19*) value of -506 ($[\alpha]_D = 0°$). In dichloroacetic acid the intrinsic viscosity rises from 0.25 and 0.29 to 0.40 and 0.78 respectively and rotatory dispersion becomes normal with a λ_c of 250 ($[\alpha]_D = -15°$), indicating a breakdown of the helices. A thorough theoretical analysis of the hydrodynamic properties of molecules like those of branched polyamino acids is not available as yet. The closest treatment is that of Zimm and Kilb (*20*), dealing with "starlike" multichain polymers of the random-coil type. Preliminary work has indicated that for the cases recorded in Table 2.1 an application of Kuhn's string-of-beads model for flexible rods (*21*) leads to quite reasonable molecular dimensions. Also, when the experimental values for $[\eta]$, S_0 and the known molecular weight are introduced into the Scheraga-Mandelkern equation (*22*), β values of 2.4 to 2.9 are obtained which are consistent with the elongated character of these molecules. Of particular interest here is the evaluation, both experimental and theoretical, of the contribution of the hydrodynamic drag of the side chains to the over-all viscosity, as well as the question whether the backbone itself can assume helical and random configuration. Work on these problems is in progress.

Multichain polyglutamic acid was prepared by debenzylation of the polybenzyl ester by means of hydrogen bromide. The observation that this treatment did not cause a change in the ratio of glutamic acid to lysine proves that no degradation of the side chains occurs. The reduced viscosity as a function of concentration of the polysodium salt of branched polyglutamic acid ($n=70$, $m'=3$, $m=19$, M.W. 180,000) in water and in phosphate buffer pH 7 of 0.6 ionic strength, is shown in Fig. 2.4. For comparison the corresponding data for linear sodium polyglutamate (M.W. 110,000) are also given.

The maximum found for the branched polymer has been observed also with linear polyelectrolytes such as carboxymethyl cellulose (*23*), although at much lower concentration (0.005 gm/100 ml). The fact that the viscosity increases with concentration was explained by the increasing electrostatic interactions between the domains of the macromolecules. With linear molecules these domains are large, hence the rapid increase and the high values of the viscosity. Decrease of the viscosity occurs when interpenetration of these domains leads to increased ionic strength within them, accompanied by screening phenomena and increased coiling of the molecular threads. The

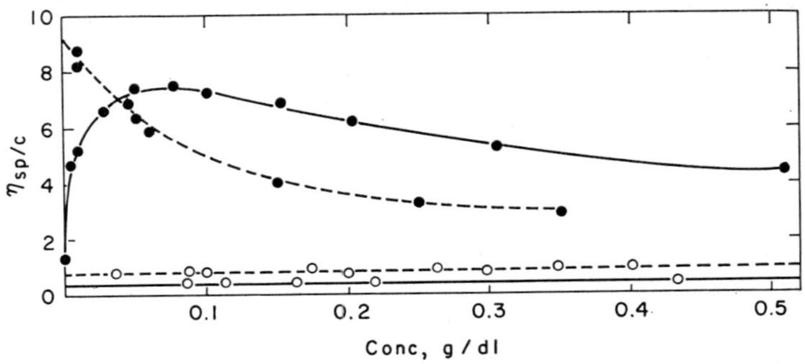

Fig. 2.4. The viscosity of branched poly-L-glutamic acid (sodium salt) (n 70, $m' = 3$, $m = 19$, M.W. 180,000) in water (—●—●—●—) and in phosphate buffer pH 7, ionic strength 0.6 (—○—○—○—) and of linear poly-L-glutamic acid (sodium salt) in water (---●---●---) and in the above buffer (---○---○---). Polymer concentration c, in gm/dl.

fact that the molecule of a branched polymer is confined to a much smaller domain than that of a linear one of equal molecular weight explains both the more gradual increase of the viscosity with concentration and the lower value at its maximum, as well as the fact that decrease of viscosity sets in only at much higher concentration. This decrease is very slight because coiling of the short side-chains has only a small effect on the total hydrodynamic resistance of the molecules.

A water-soluble nonionic derivative of branched polyglutamic acid, branched poly(*N*-3-hydroxypropyl)glutamine, could be obtained by reacting the polybenzyl ester with 3-aminopropanol-l (*24*). This polymer is also soluble in a variety of organic solvents. Rotatory dispersion measurements indicate that this chain is helical in dimethylformamide and random in water.

Using a "soluble initiator" made of DL-lysyl and DL-β-benzylaspartate residues, a series of branched polymers bearing poly-L-proline side chains ($m = 0.5$, 1, 2, 3, 4, 5, 6, and 12) was prepared (*25*). This was done in order to determine the minimum chain length that could make a structural contribution (due to polyproline I or II helices) to the optical rotation of the polymer (*26*). Whereas direct synthesis of these proline oligopeptides proved impractical, the branched molecules of the desired structure were obtained with ease. Fig. 2.5, which shows the change in optical rotation with chain length, indicates that a helical structure exists in chains with more than five residues. From the rather abrupt transition occurring between the values $m = 4$ and $m = 6$, one may conclude that the statistical distribution of chain length in the branched molecules is a rather sharp one, in accord with the results of the theoretical analysis (*27*) of the branching reaction involving NCA polymerization.

Finally it is of interest that the "soluble initiator" can be used also to build up branched polymeric molecules by means of the classical techniques of stepwise peptide synthesis. Thus, for instance an azide or mixed anhydride of an *N*-carbobenzoxyamino acid can be coupled with the amino groups of the initiator under anhydrous conditions, and the product, after removal of the blocking groups, can be used for a second step and so on. Purification of the in-

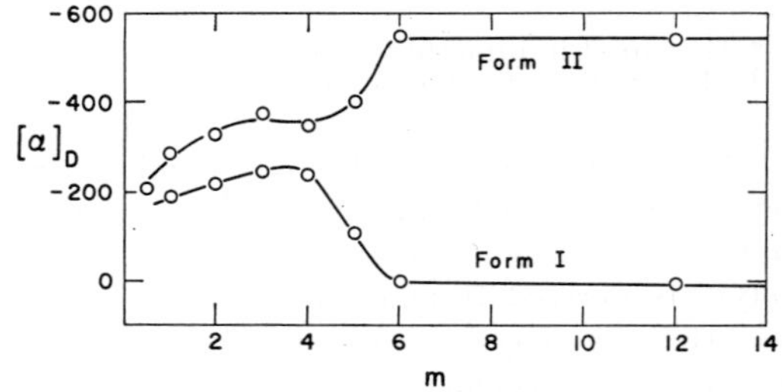

Fig. 2.5. The specific optical rotation per proline residue of branched polymers with poly-L-proline side chains. The upper curve (Form II) was obtained after complete mutarotation in the presence of formic acid. The lower curve (Form I) was obtained by extrapolation of the mutarotation curves to zero time. The abscissa, m, gives the average number of prolyl residues per side chain.

termediates is facilitated by the ease with which the polymers can be precipitated, and by the possibility of using dialysis. An example of this kind of reaction is the preparation of a branched polymer in which α-*N*-carbobenzoxy-L-histidine azide was coupled with the "soluble initiator." Analysis of the product showed that all the amino groups of the initiator had reacted. After removal of both carbobenzoxy groups and β-benzyl ester groups, an ampholytic polymer was obtained, and its titration curve indicated the presence of carboxyl, imidazole, and α-amino groups, without resolving these groups clearly.

In conclusion it might be of general interest to indicate the fields of research in which branched polyamino acids could be particularly useful, because of the ease with which compounds of high molecular weight and well-defined molecular shape can be made. Advantage may be taken of their nondialyzability in the investigation of binding phenomena at polymeric chains, at single amino-acid residues, and at oligopeptides of defined sequence, by means of the equilibrium dialysis technique. They may be of value in electrochemical research concerned with systems of high charge density. They may shed light on problems of interaction of polypeptide chains, because of their spatial proximity in these molecules. And they may prove to possess unusual biological properties, as has been shown in the case of the first fully synthetic antigen (*8*).

Acknowledgment

This investigation was supported by a research grant (FG-Is-104-60) from the United States Department of Agriculture.

References

1 E. Katchalski and M. Sela, *Advances in Protein Chem.* **13,** 243 (1958); M. Sela and E. Katchalski, *ibid.* **14,** 391 (1959).

2 M. Bovarnick, F. Eisenberg, Jr., D. J. O'Connell, J. Victor, and P. Ovades, *J. Biol. Chem.* **207,** 593 (1954).

3 Y. Wolman, Ph.D. thesis, the Hebrew University, Jerusalem, Israel (1961).

4 R. R. Becker and M. A. Stahmann, *J. Biol. Chem.* **204,** 745 (1953); A. Stracher and R. R. Becker, *J. Am. Chem. Soc.* **81,** 1432 (1959).

5 M. Sela and R. Arnon, *Biochem. J.* **75,** 91 (1960); **75,** 103 (1960); **77,** 394 (1960).

6 M. Sela, E. Katchalski, and M. Gehatia, *J. Am. Chem. Soc.* **78,** 746 (1956).

7 K. F. Weinke, Ph.D. thesis, the University of Wisconsin, Madison, Wis. (1956).

8 M. Sela and R. Arnon, *Biochim. et Biophys. Acta* **40,** 382 (1960).

9 S. G. Waley and J. Watson, *Proc. Roy. Soc.* **A199,** 499.

10 E. R. Blout and R. H. Karlson, *J. Am. Chem. Soc.* **78,** 941 (1956).

11 E. R. Blout, R. H. Karlson, P. Doty, and B. Hargitay, *ibid.* **76,** 4492 (1954).

12 D. G. H. Ballard, C. H. Bamford, and F. J. Weymouth, *Nature* **174,** 173 (1954).

13 H. Tsuyuki, H. J. Van Kley, and M. A. Stahmann, *J. Am. Chem. Soc.* **78,** 764 (1956).

14 A. Yaron, Ph.D. thesis, the Hebrew University, Jerusalem, Israel (1961).

15 A. Yaron and A. Berger, *Bull. Research Council Israel* **7 A,** 96 (1958).

16 E. Daniel, Ph.D. thesis, the Hebrew University, Jerusalem, Israel (1960); E. Daniel and E. Katchalski, *Bull. Research Council Israel* **9 A,** 116 (1960); E. Katchalski and E. Daniel, this book, Paper 16.

17 J. B. Appelquist, Ph.D. thesis, Harvard University, Cambridge, Mass. (1959); G. D. Fasman, M. Idelson, and E. R. Blout, *J. Am. Chem. Soc.* **83,** 709 (1961).

18 A. Yaron and A. Berger, *Bull. Research Council Israel* **9 A,** 117 (1960).

19 K. S. Norland, G. D. Fasman, and E. R. Blout, *J. Am. Chem. Soc.* **82,** 2268 (1960); W. Moffitt, *J. Chem. Phys.* **25,** 467 (1956).

20 B. H. Zimm and R. W. Kilb, *J. Polymer Sci.* **37,** 19 (1959).

21 W. Kuhn, *Helv. Chim. Acta* **28,** 97 (1945).

22 H. A. Scheraga and L. Mandelkern, *J. Am. Chem. Soc.* **75,** 179 (1953).

23 M. Fujita and F. Homma, *J. Colloid Sci.* **9,** 591 (1954).

24 N. Lupu, A. Yaron, M. Sela, and A. Berger, *Bull. Research Council Israel* **10A,** 47 (1961).

25. A. Yaron and A. Berger, *ibid.* **10A,** 46 (1961).

26 I. Z. Steinberg, W. F. Harrington, A. Berger, M. Sela, and E. Katchalski, *J. Am. Chem. Soc.* **82,** 5263 (1960).

27 E. Katchalski, M. Gehatia, and M. Sela, *ibid.* **77,** 6175 (1955).

PAPER 3

Synthesis of Complex Polypeptides Possessing a Defined Arrangement of Their Amino-Acid Residues

KLAUS HOFMANN and HARUAKI YAJIMA

Advances in understanding of the chemistry of the hormones which are elaborated by the pituitary gland, spearheaded by du Vigneaud's synthesis of oxytocin (*1, 2*) and the vasopressins (*3, 4, 5*), have exerted pronounced influence on recent developments in peptide chemistry. Elucidation of the complete amino-acid sequence of porcine corticotropin (ACTH) by Bell (*6–11*) and White (*12*) and their collaborators and the establishment of the structure of α-MSH (melanocyte stimulating hormone) by Harris and Lerner (*13, 14*) are other landmarks in this field.*

The physiological function of polypeptide hormones seems to depend critically on a specific arrangement of the constituent amino-acid residues along their peptide chains. Understanding of the essential structural prerequisites for biological function may be gained from studies relating primary structure to physiological activity.

Recognition that a large spectrum of physiological activities may be produced, based on the peptide architectural principle, has led to an unprecedented renaissance in peptide chemistry.

Refinement in synthetic methodology, the development of techniques allowing the sharp separation of peptides possessing closely similar properties, and the availability of new highly sensitive analytical tools for assessing homogeneity have set the stage for production in the test tube of highly complex stereochemically homogeneous peptides possessing a defined arrangement of their amino-acid residues (*15, 16*).

In our own laboratory we have synthesized a tricosapeptide amide (1–23) (Structure I, Figure 3.1) corresponding to the arrangement of the first 23 amino-acid residues in corticotropin and have observed that this molecule possesses essentially the same biological activity as the natural hormone (*17*). The synthesis of this peptide which is composed of 23 amino-acid residues derived from 12 different amino acids will be reviewed. Discussion of various aspects of the problem will serve as a basis for critical evaluation of certain methods of peptide synthesis and purification.

The over-all plan

The remarkable stability toward acid of adrenocorticotropic activity in crude extracts and of pure corticotropin was recorded by its earliest students (*18*). Thus prolonged heating

* Except where noted otherwise, the amino acids and peptides mentioned in this discussion (except glycine) are of the L-variety. In the interest of space conservation the customary L-designations for individual amino-acid residues are omitted. The numbers in parentheses following peptide names designate their location in the corticotropin sequence.

Ascorbic-acid-depleting activity was determined in 24-hour hypophysectomized rats according to the method of U.S. Pharmacopoeia XV against the USP reference standard. The plasma corticosterone levels were determined 15 minutes following administration, R. Guillemin, G. W. Clayton, J. D. Smith, and H. S. Lipscomb, *Endocrinol.* **63**, 349 (1958). The free steroid was separated chromatographically and assayed by a modification of the method of H. Kalant, *Biochem. J.* **69**, 93 (1958). We are much indebted to Dr. Joseph D. Fisher of Armour Pharmaceutical Company, Kankakee, Illinois for these biological assays.

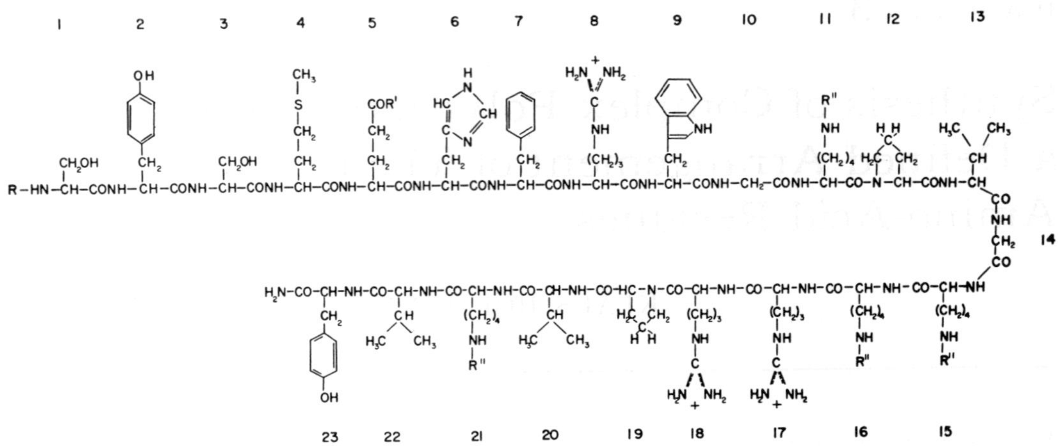

Fig. 3.1. Structure of the tricosapeptide amide (1–23) (I) and of its partially protected precursor (II). I: $R=H$; $R'=OH$; $R''=H$. II: $R=$Acetyl; $R'=NH_2$; $R''=$Formyl.

at 100° in 0.1 N hydrochloric acid exerts little, if any, effect on the biological activity of these preparations but short exposure to dilute alkali results in rapid inactivation. Indeed, it was inferred from partial hydrolysis studies that the bond between the proline in position 24 and the aspartic acid residue in position 25 in the 39 amino-acid porcine corticotropin molecule was cleaved and that the remaining tetracosapeptide fragment (1–24) represents the biologically active moiety of the hormone (*9*).*

Guided by this information and based on the assumption that the C-terminal proline in the tetracosapeptide was not essential for biological activity, we synthesized the protected tricosapeptide amide (1–23) (Structure II, Figure 3.1) (*17*). The various protecting groups (*N*-acetyl,N^{ϵ}-formyl and glutamine carboxamide), it was reasoned, should be removable from the molecule by exposure to dilute acid under conditions which were not expected to cause significant fission of peptide bonds. We believe this to be a novel approach to peptide synthesis.

Conditions for removal of the various protectors were worked out with model peptides (*19*).

Selection of suitable subunits from which to construct highly complex peptides without risk of racemization is of critical importance. A minor degree of racemization of every amino-acid residue making up such a molecule may afford a biologically inactive final product despite the fact that it possesses the correct amino-acid composition and sequence.

Inspection of the structure of the tricosapeptide amide (Structure I, Figure 3.1) shows the presence of glycine residues at positions 10 and 14. Suitably protected peptides terminating in a C-terminal glycine residue are desirable subunits from which to assemble large peptides, since absence of an asymmetric center in glycine precludes racemization during activation. The validity of this principle was demonstrated for the first time in our synthesis of a derivative of α-MSH (Structure II, Figure 3.2) which possesses the full biological activity of the natural hormone (Structure I, Figure 3.2) (*20*).

Proline offers promise of becoming another suitable C-terminus for elongation of peptide chains since the presence of the imino group precludes azlactone formation and hence racemization. However, examples of the application of this principle are not available.

A number of peptide-bond–forming procedures were used in the synthesis of the tricosapeptide amide. Of these, the azide (*21, 22*) and carbodiimide (DCC) (*23*) procedures served in the assembly of larger building blocks, whereas the mixed anhydride procedure (*24–26*) was

* The tetracosapeptide differs from the tricosapeptide (Structure I, Figure 3. 1) by a C-terminal proline residue.

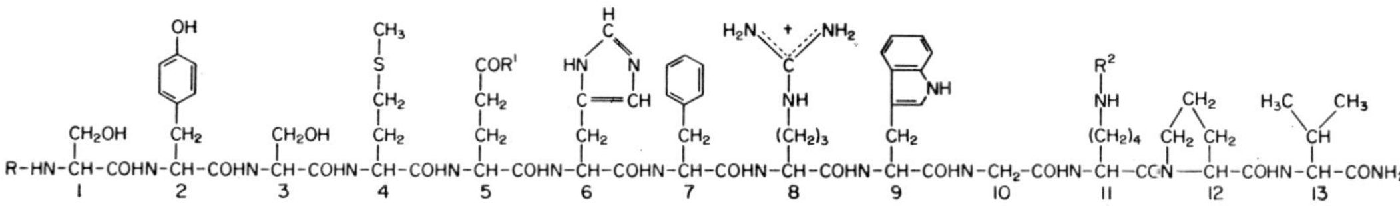

Fig. 3.2. Structure of α-MSH (I) and of some of its analogs. I: R = Acetyl; R' = OH; R^2 = H. II: R = Acetyl; R' = NH_2; R^2 = Formyl. III: R = H; R' = OH; R^2 = H. IV: R = Carbobenzoxy; R' = NH_2; R^2 = Tosyl.

used *in extenso* to construct various subunits. Only C-terminal glycine carboxyls were activated by DCC since racemization has been shown to occur when this reagent activates carboxyl groups in certain acylated peptides (*27–29*).

Low yields of the desired peptides are experienced occasionally when attempting to bring about peptide-bond formation between a suitably protected acylpeptide terminating with glycine and a peptide ester or amide. Favored formation of the acylurea derivative rather than the peptide seems to provide an explanation for this behavior. The same coupling reactions may be achieved in excellent yields when N,N'-carbonyldiimidazole (*30–32*) serves as the carboxyl activator. This situation prevailed in the synthesis of N^{α}-carbobenzoxy-N^{ϵ}-formyllysylprolylvalylglycyl-N^{ϵ}-formyllysyl-N^{ϵ}-formyllysine methyl ester from N^{α}-carbobenzoxy-N^{ϵ}-formyllysylprolylvalylglycine and methyl N^{ϵ}-formyllysyl-N^{ϵ}-formyllysinate (*33*) (see Fig. 3.10).

Stepwise elongation from the carboxyl toward the amino end provides a sound approach for construction of stereochemically homogeneous peptides. But the many steps involved limit this approach to the synthesis of subunits in the octa- to deca-peptide range. Activation of individual acylamino acids may be achieved in a practical manner by the DCC, the mixed anhydride, and the p-nitrophenyl ester procedures. The latter method proved of outstanding value in recent syntheses of the protected nonapeptide amide precursors of oxytocin (*34*) and lysine vasopressin (*35*).

Individual steps proceed in high yield and the final product and intermediates appear to possess a high degree of homogeneity. However, these advantages are somewhat counterbalanced by the additional steps required for production of the pure activated ester intermediates, and by the difficulties which are frequently encountered in freeing the reaction product from traces of unreacted p-nitrophenyl ester.

We have employed the mixed anhydride procedure for construction of the partially protected heptapeptide amide (17–23) (Structure VI, Figure 3.12) with excellent results (*33*). The hydrogen bromide technique (*36*) for removing of carbobenzoxy groups is not applicable to this synthesis since the N^{ϵ}-formyl moiety undergoes hydrolysis under these conditions. Carbobenzoxy groups were removed by hydrogenolysis over palladium in presence of acetic acid.

Two new reagents for activation of carboxyl groups have been described recently. Woodward *et al.* (*37*) prepared a series of simple peptides by the use of N-ethyl-5-phenylisoxazolium-3′-sulfonate and Nefkens and Tesser (*38*) employ esters of N-hydroxyphthalimide in an extension of the activated ester procedure for formation of peptide bonds.

Synthetic routes to peptides containing basic amino-acid residues

Incorporation of arginine residues into peptides is beset with considerable experimental difficulties and satisfactory procedures for synthesis of peptides containing this basic amino acid are of rather recent origin. The recognition in 1952 that ACTH is a basic peptide which contains arginine stimulated our own interest in arginine-containing peptides. A few poorly characterized representatives of this class of compounds, with the amino group of the arginine involved in peptide linkage, were known at that time but arginyl peptides, i.e., peptides in which the carboxyl group of arginine provides the carbonyl portion of the peptide linkage, were unavailable. In 1953 three groups of investigators (*39–41*), practically simultaneously but using different routes, provided solutions to this problem. Our procedure (*42*) is based on N^{ω}-nitroarginine which is readily prepared from arginine by treatment with a mixture of concentrated nitric and fuming sulfuric acids. Introduction of the strongly electronegative nitro group depresses the basic character of the guanido group, presumably because of interference with the guanidinium ion resonance.

In 1934, Bergmann *et al.* (*43*) prepared N^{ω}-nitroarginine and explored its applicability to the synthesis of arginyl peptides. Although these investigators were unable to activate the carboxyl group of N^{α}-carbobenzoxy-N^{ω}-nitroarginine by the procedures then available, they demonstrated that N^{α}-carbobenzoxy-N^{ω}-nitroarginine gave unracemized arginine on catalytic hydrogenation. The nitro group is eliminated in the form of an ammonium ion during hydrogenolysis. This key experiment established the potential of the nitro group as a protector of the guanido function of arginine. Our observation (*39, 42*) that N^{α}-carbobenzoxy-N^{ω}-nitroarginine forms mixed anhydrides which react smoothly with amino-acid or peptide esters provided a generally applicable route to arginyl peptides which has since found extensive use not only in our own work but for the preparation of such other complex biologically active peptides as hypertensin (*29, 44, 45*), bradykinin (*46*) and a nonadecapeptide (1–19) related to corticotropin (*47*). The essential steps of the procedure are shown in Figure 3.3. Arginine (I) is converted into N^{ω}-nitroarginine (II) which is carbobenzoxylated to give (III). Reaction of a mixed anhydride (IV) of (III) with an amino-acid ester affords the N^{α}-carbobenzoxy-N^{ω}-nitroarginylpeptide ester (V) which is then saponified (VI). Hydrogenation of (VI) over a palladium catalyst, preferably in presence of acetic acid, affords the acetate salt of the arginyldipeptide (VII). Activation of the carboxyl group of (III) may also be effected by the use of DCC (*29*). Peptides containing a C-terminal arginine residue are readily prepared by interaction of N^{ω}-nitroarginine methyl ester and activated acylamino acids, followed by saponification and hydrogenation (*42*). The absorption maximum at 270 mμ of nitroarginine and of peptides containing this residue facilitates location of these compounds during purification.

Protonization of the guanido group of N^{α}-carbobenzoxyarginine converts this zwitter ion

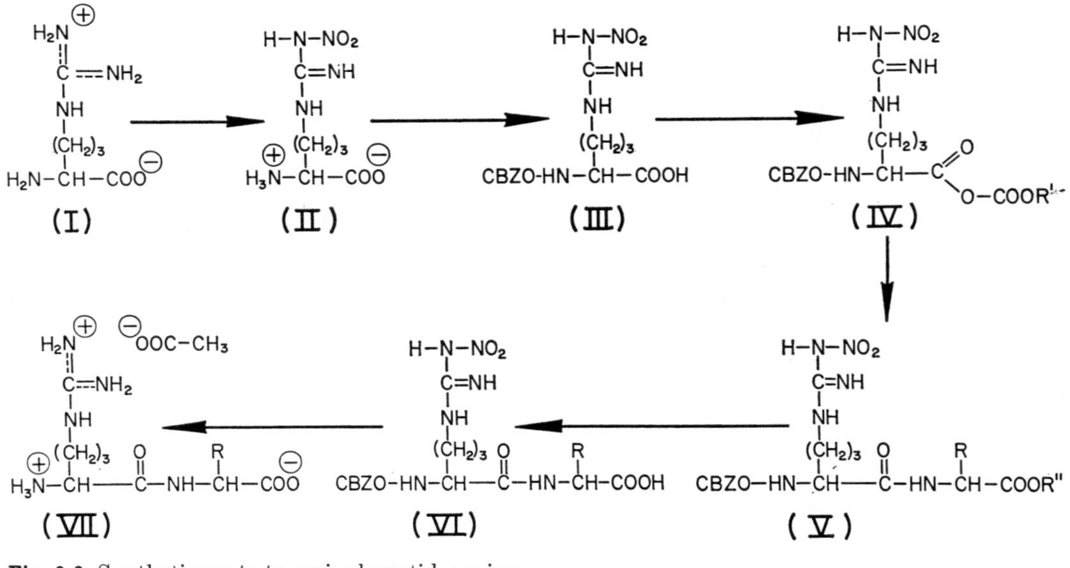

Fig. 3.3. Synthetic route to arginyl peptides using the nitroarginine procedure.

into a salt in which the carboxyl group exhibits normal functional behavior. Anderson (*40*) demonstrated for the first time that the carboxyl group of protonized N^{α}-carbobenzoxyarginine can be activated to undergo peptide-bond formation. He prepared N^{α}-carbobenzoxyarginylleucine methyl ester hydrobromide from N^{α}-carbobenzoxyarginine hydrobromide and leucine methyl ester, using tetraethylpyrophosphite in diethylphosphite for carboxyl activation. Hydrolysis of the ester group with hydrogen bromide followed by catalytic hydrogenation to remove the carbobenzoxy group gave arginylleucine hydrobromide hydrate. The synthesis of arginine vasopressin by du Vigneaud and collaborators (*5*) exemplifies an application of this procedure. It remained for Boissonnas and collaborators (*48–50*) to show that the DCC reagent activates the carboxyl group of N^{α}-protected N^{ω}-protonized arginine derivatives. The preparation of N^{α}-carbobenzoxyarginyltryptophan methyl ester hydrochloride may serve as an illustration of their procedure. A mixture of N^{α}-carbobenzoxyarginine and of tryptophan methyl ester hydrochloride is dissolved in dimethylformamide containing diethylphosphite and DCC is added to bring about formation of the peptide bond. Under these conditions protonization of the guanido group involves a proton transfer from the ammonium function of the amino-acid ester salt to the guanido function with simultaneous formation of an amino group capable of reacting with the activated carboxyl.

In addition to the nitro group, the *p*-nitrophenyloxycarbonyl (*41*), phenyloxycarbonyl (*51*), and tosyl functions (*52*) have found application as protectors of the guanido group of arginine. The first two groups are removed by catalytic hydrogenation in an acidic milieu, whereas reduction with sodium in liquid ammonia is required for removal of the tosyl moiety. The tosyl procedure seems to be the least desirable one since the final products are contaminated with inorganic salts, and special desalting procedures are required to free them from this contamination. Also, some peptides undergo fragmentation when exposed to sodium in liquid ammonia.

At first glance it would appear that working with protonized arginine would provide the most direct route to arginine peptides since it eliminates the need for additional blocking and deblocking steps. However, this advantage has to be weighed against certain difficulties which are encountered with peptide salts containing free arginine. Many of these highly water-soluble substances fail to crystallize and purification to a state of analytical homogeneity may require involved procedures. Conversely, peptides containing N^{ω}-protected arginine, particularly those derived from N^{ω}-nitroarginine, are nonhygroscopic solids which are soluble in organic solvents. Many have been obtained in analytical purity by crystallization.

Thus, it is frequently desirable to retain the arginine in protected form throughout a synthesis and to remove the blocking group as the final step. As will be shown in a later section, we have employed both the nitroarginine procedure and protonization in the synthesis of the tricosapeptide amide (1–23).

Guanylation of ornithine peptides, successfully used in preparation of polyarginine (*53*), has thus far not found application in the synthesis of peptides containing arginine and exhibiting a defined sequential arrangement of amino-acid residues.

The N^{ϵ}-amino group of lysine offers problems akin to those encountered with the guanido function of arginine. The carbobenzoxy and tosyl groups have been used most frequently for protection of its ϵ-amino group (*54, 55*). The resistance of the N^{ϵ}-tosyl group to conditions which result in removal of the carbobenzoxy group allows construction of peptide chains containing N^{ϵ}-tosyllysine by the carbobenzoxy method. The corresponding lysylpeptides are obtained following removal of the tosyl group by sodium reduction in liquid ammonia. This scheme, which was so successfully used in various syntheses of lysine varopressin (*3, 4, 35*), failed when applied to the synthesis of certain derivatives of α-MSH.

Attempts to remove the N^{ϵ}-tosyl-, and N^{α}-carbobenzoxy groups from the protected tridecapeptide amide (Structure IV, Figure 3.2) by by the sodium in liquid ammonia procedure resulted in marked disruption of peptide bonds with formation of a complex mixture which proved to be extremely difficult to separate (*19*). Similar observations were made by Schwyzer and collaborators in unsuccessful attempts to synthesize bovine β-MSH (*47*).

These failures prompted a search for superior protecting groups for the N^{ϵ}-amino group of

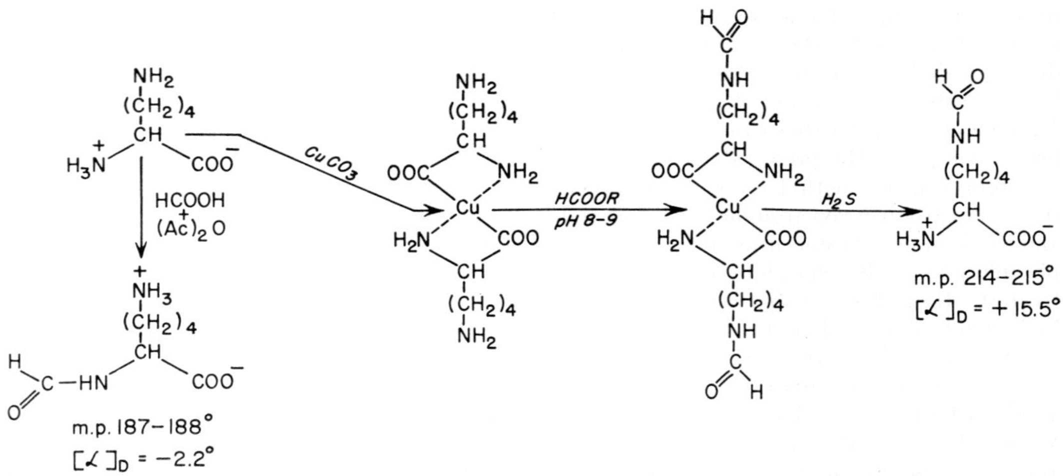

Fig. 3.4. Synthetic route to N^{ϵ}-formyllysine.

lysine.* Of particular interest are groups which resist hydrogenolysis and exhibit satisfactory stability to withstand the conditions encountered in the usual peptide-bond forming steps but are labile enough to be removable under the mildest possible conditions from the completed peptide chain. The N^{ϵ}-formyl group seemed a likely candidate to satisfy these criteria and the method illustrated in Figure 3.4 was developed for preparation of N^{ϵ}-formyllysine (*56*).

The copper chelate of lysine, on exposure to ethyl formate at a *p*H of 8 to 9, affords the less soluble chelate of N^{ϵ}-formyllysine which on decomposition with hydrogen sulfide gives N^{ϵ}-formyllysine. Traces of free lysine are removed from the final product by treatment with Dowex-50. N^{ϵ}-Formyllysine may be carbobenzoxylated and N^{α}-carbobenzoxy-N^{ϵ}-formyllysine is readily incorporated into peptides by the mixed anhydride procedure. The methyl ester hydrochloride of N^{ϵ}-formyllysine is prepared via N^{α}-carbobenzoxy-N^{ϵ}-formyllysine methyl ester followed by catalytic hydrogenation in presence of acetic acid. The acetate salt of the methyl ester is then transformed to the hydrochloride.

Studies with model peptides corresponding to sequences encountered in the ACTH molecule verified expectations pertaining to the lability of the N^{ϵ}-formyl group. Exposure to hot dilute hydrochloric acid effects cleavage of the formyl moiety at the same time causing only minor hydrolysis of peptide bonds. These findings were of prime importance for the ultimate success of our ACTH-peptide synthesis.

The *t*-butyloxycarbonyl group (*57–60*) has recently been employed as a protector of the ϵ-amino group of lysine (*61*). This labile group is readily removed by exposure to strong acids.

The general significance of N^{ϵ}-tritryllysine (*62*) for the synthesis of complex polypeptides containing lysine remains to be established.

Protection of carboxyl groups

Esterification is employed routinely to suppress ionization of carboxyl groups in the conventional routes to peptides. Since exposure to alkali, which is necessary for regeneration of the free carboxyl group, may bring about racemization, protection by esterification is not a safe procedure. The use of the benzyl esters which can be cleaved by hydrogenolysis avoids this difficulty.

In our own work we have resorted to the amide group as a protector of carboxyls, both

* The synthesis of a nonadecapeptide corresponding to positions 1–19 of the corticotropin sequence has been reported by Li *et al.*, *J. Am. Chem. Soc.* **82**, 5760 (1960). It is surprising that these investigators succeeded in deblocking by reduction with sodium in liquid ammonia a protected nonadecapeptide precursor which contains three N^{ω}-tosylarginine and three N^{ϵ}-tosyllysine residues.

as concerns the C-terminus and the γ-carboxyl of glutamic acid. This selection was made with the expectation that biological activity of the amide derivatives would be essentially identical to that of the respective carboxylic acids.

The presence of a C-terminal glutamine in the N-terminal pentapeptide precludes the use of ester protection during synthesis, since esters of acylated glutamine undergo lactam formation on exposure to alkali (*63*). Thus, the preparation of this portion of the molecule was accomplished by a route involving the triethylammonium salts of the C-terminal carboxyl group (*64*).

The *t*-butyl esters recently introduced into peptide chemistry (*66*) offer advantages over other carboxyl-protecting procedures, since they are readily hydrolyzed by exposure to strong acids.

Purification and assessment of homogeneity

Synthetic procedures for production of polypeptide hormones must allow the preparation of these biologically important products in such quantities as permit extensive evaluation of their physiological actions. Purification of intermediates and final products presents the main difficulty in producing these substances in quantity. In order to facilitate this task it becomes of considerable importance to control the various synthetic steps so as to insure formation of initial reaction products possessing the highest possible purity. The cleaner the crude products, the easier becomes their preparation in analytical purity. Multistep syntheses requiring purification of every intermediate by the countercurrent procedure are highly impractical. The countercurrent procedure, an excellent tool indeed for purification and for assessment of homogeneity of peptides, becomes impractical when employed for routine purification of these compounds in gram quantities. Also, separation of mixtures composed of basic peptides is frequently difficult to achieve by the countercurrent technique. In our work purification of many intermediates is achieved by carefully selected washing procedures and by chromatography on carboxymethylcellulose (CMC) (*67*). We find chromatography on CMC of particular value for purification of peptides containing arginine. Ammonium acetate solutions of varying ionic strengths and *p*H's were employed for development of the columns. Ammonium acetate is removed from the final products by repeated lyophilization from dilute acetic acid or water solutions. Ion exchange chromatography is readily adaptable to purification of gram lots of these peptides.

The criteria for homogeneity of the various peptides were essentially those described in connection with our syntheses of α-MSH analogs (*16*). They are based on formation of a single spot on paper chromatograms, preferably in more than one solvent system, and correct amino-acid ratios in acid hydrolysates. Complete digestibility by leucine aminopeptidase (LAP) with formation of the constituent amino acids in the ratios predicted by theory was accepted as evidence for stereochemical homogeneity.

In earlier studies we have estimated the amino-acid composition of acid and LAP digests by quantitative paper chromatography. These determinations are now performed routinely by the procedure of Spackmann, Stein, and Moore (*68*). It is informative to compute the average recovery of amino acids in the acid or LAP hydrolysates. In conjunction with total nitrogen and ammonia nitrogen figures, these values provide a sensitive measure for contamination by ammonium salts and non-nitrogenous materials and in addition give information regarding the degree of hydration. Average amino-acid recoveries in excess of 85 per cent are observed with the peptides and peptide derivatives which were prepared in connection with the synthesis of the tricosapeptide amide.

The importance for critical evaluation of the homogeneity of intermediates and final products in synthetic work involving the production of highly complex biologically active polypeptides cannot be overemphasized.

Deblocking with dilute hydrochloric acid

The synthesis in 1955 of the pentapeptide seryltyrosylserylmethionylglutamic acid by Hofmann and Jöhl (*64*, *69*) (Figure 3.5), and the demonstration of its identity with the N-terminus of porcine corticotropin initiated a series of systematic studies on peptides corresponding to sequences present in that hormone. The successful synthesis in 1960 (*17*) of

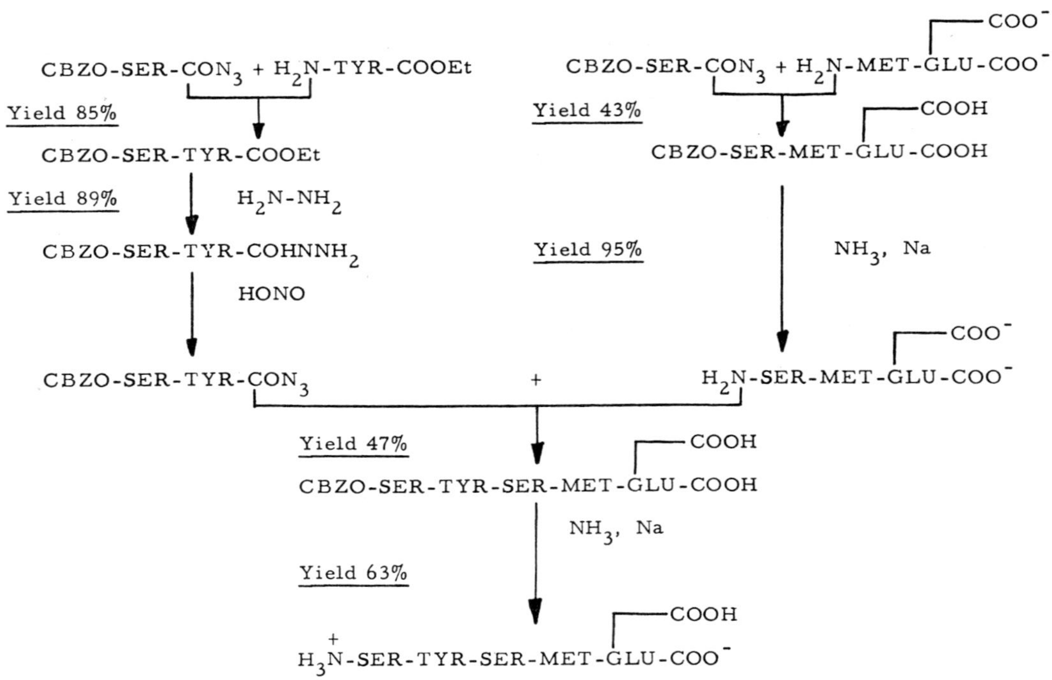

Fig. 3.5. Synthetic route to N-terminal pentapeptide.

the tricosapeptide amide (1–23) (Structure I, Figure 3.1) is an outgrowth of these fundamental investigations.

The original Hofmann-Jöhl procedure was adapted to the preparation of the glutamine analog of the pentapeptide and the latter affords *N*-acetylseryltyrosylserylmethionylglutamine on acetylation (*70*). The azide of this acetyl derivative (subunit A) has found application for introducing the N-terminus into complex peptides which are related to the corticotropins.

Exposure to hot 0.5 N hydrochloric acid converts *N*-acetylseryltyrosylserylmethionylglutamine into seryltyrosylserylmethionylglutamic acid with elimination of the *N*-acetyl-, and glutamine amide groups (*19*). Invariably some fragmentation of the peptide chain accompanies the deblocking reaction but the conversion, illustrated in Figure 3.6, proceeds in a yield of 60 per cent. The structure of the deblocked pentapeptide follows from its identity with an authentic preparation and its complete digestibility by LAP. The presence in the enzymic digest of glutamic acid and not glutamine demonstrated conclusively that the amide group was split by the acid treatment. The somewhat unexpected lability of the *N*-acetyl group may be attributable to an N→O shift.

The observation that three groups (*N*-acetyl, glutamine carboxamide, and N^{ϵ}-formyl of lysine) are removable from model peptides under conditions which do not bring about major fission of peptide bonds (*19*) prompted studies on the stability to dilute hydrochloric acid of more complex protected peptides corresponding to parts of the corticotropin sequence.

The tridecapeptide amide (1–13)

The preparation of the protected tridecapeptide amide (Structure II, Figure 3.2) (*20*), a derivative of the pituitary hormone α-MSH, illustrates applications of some of the aforementioned principles to the synthesis of a more complex peptide derivative. As is shown in Figure 3.7 the protected tridecapeptide amide forms when the previously mentioned subunit A is allowed to react with the partially pro-

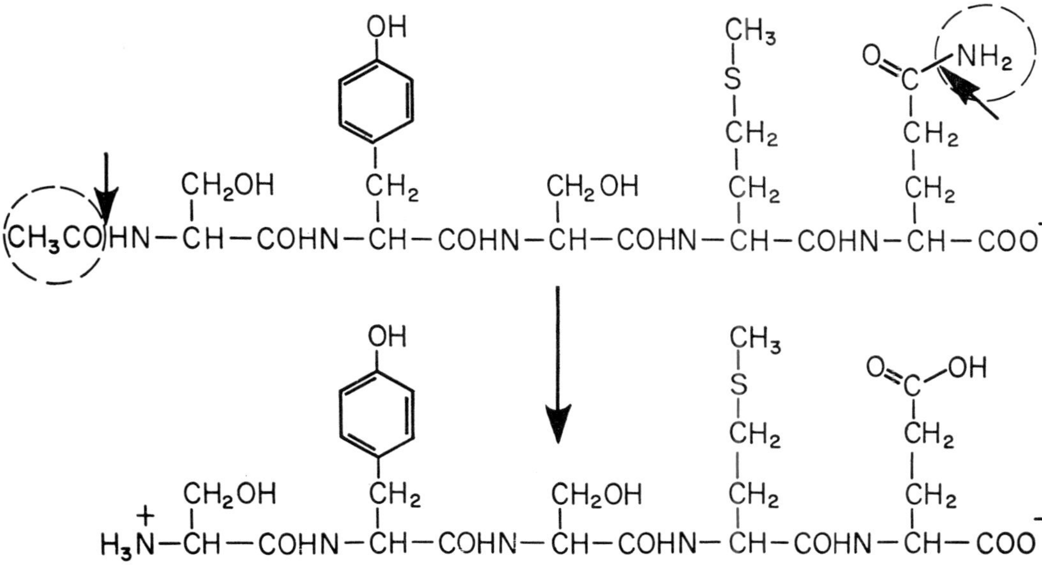

$[\alpha]_D = -22.7°$; $R_f = 0.51$

REF. SAMPLE $[\alpha]_D = -20.6°$; $R_f = 0.51$

Fig. 3.6. Deblocking of protected pentapeptide with 0.5 N hydrochloric acid.

tected octapeptide amide (subunit B). The synthesis of the latter substance as illustrated in Figure 3.8 provides examples of the use of N^ω-nitroarginine and of N^ϵ-formyllysine in peptide synthesis. In addition, it exemplifies the utility of C-terminal glycine for elongation of a peptide chain from the carboxyl end without racemization. The highly purified protected tridecapeptide amide possesses the full in vitro melanocyte-expanding activity of natural α-MSH and brings about marked darkening of the skin when administered to man (*71*).

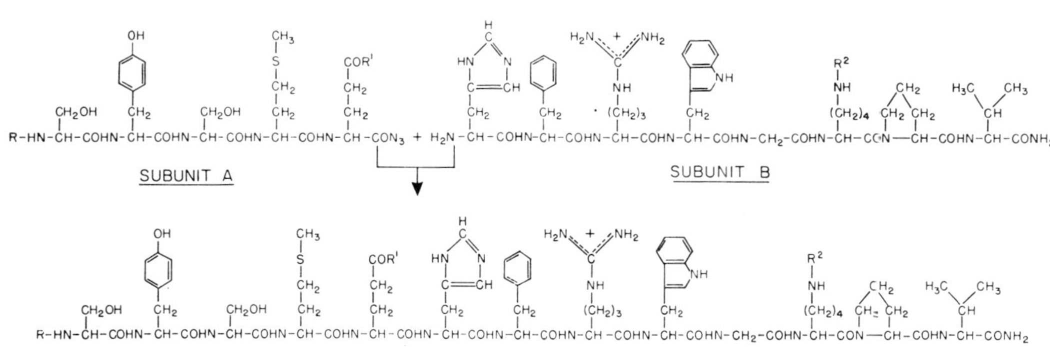

Fig. 3.7. Synthetic route to biologically active analog of α-MSH. R = Acetyl; R' = NH_2; R^2 = Formyl.

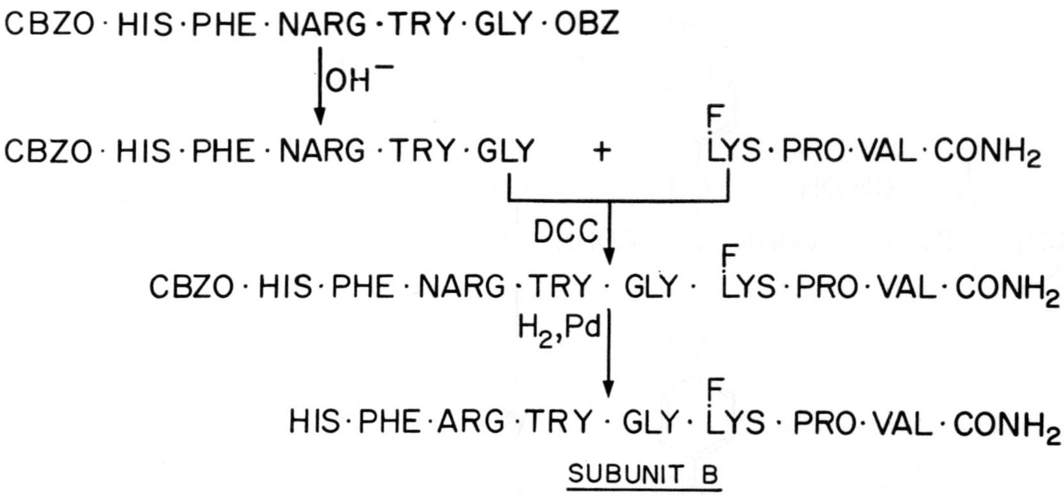

Fig. 3.8. Synthetic route to Subunit B. CBZO = Carbobenzoxy; NARG = N^{ω}-Nitroarginine; OBZ = Benzyl ester; F = N^{ϵ}-Formyl.

The unprotected tridecapeptide amide (Structure III, Figure 3.2) is obtained when the protected material is exposed to 0.5 N hydrochloric acid under conditions which were shown to remove the same protecting groups from model peptides (*19*). The purified deblocked material satisfied our criteria for homogeneity and was completely digestible by LAP, but resisted the action of carboxypeptidase. The latter finding, in conjunction with the presence in the acid hydrolysate of one equivalent of ammonium ion, corroborates the stability of the C-terminal amide group to the treatment with acid. Lability to LAP and the presence in the enzymic digest of glutamic acid and free lysine provide conclusive evidence for the structure of this tridecapeptide amide.

The hexadecapeptide amide (1–16)

The observation that the aforementioned tridecapeptide amide (1–13) exhibits low, but reproducibly significant, in vivo adrenocorticotropic activity (<0.1 U/mg) prompted further extension of the peptide chain in order to explore the importance for biological activity of the lysine residues in positions 15 and 16. To this end the hexadecapeptide amide (1–16) was synthesized by a scheme which is illustrated in Figure 3.9 with subunit A again serv-

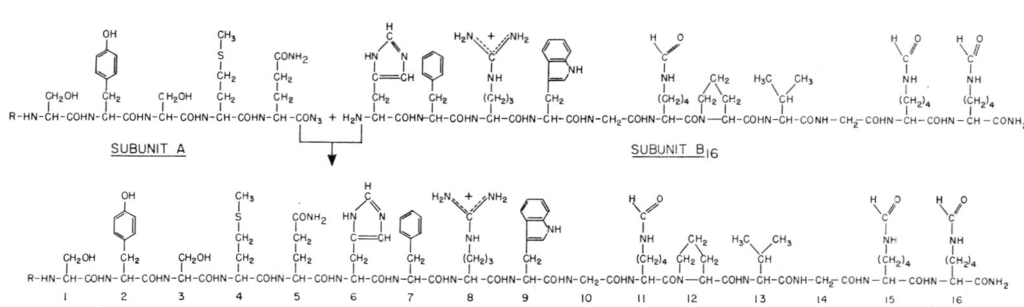

Fig. 3.9. Synthetic route to the protected hexadecapeptide amide (1–16). *R* = Acetyl.

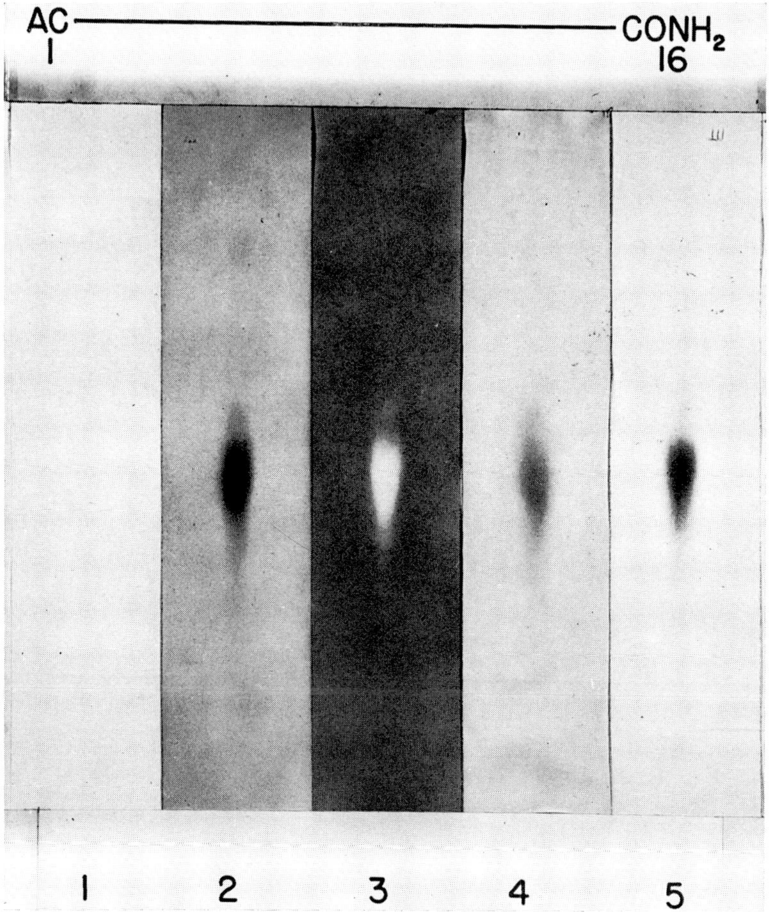

Plate 3.1. Unidimensional paper chromatogram of partially protected hexadecapeptide amide (1–16) (Figure 3.10). Bands 1 to 5 were developed respectively with the ninhydrin, Pauly, methionine, Sakaguchi, and Ehrlich reagents (Partridge).

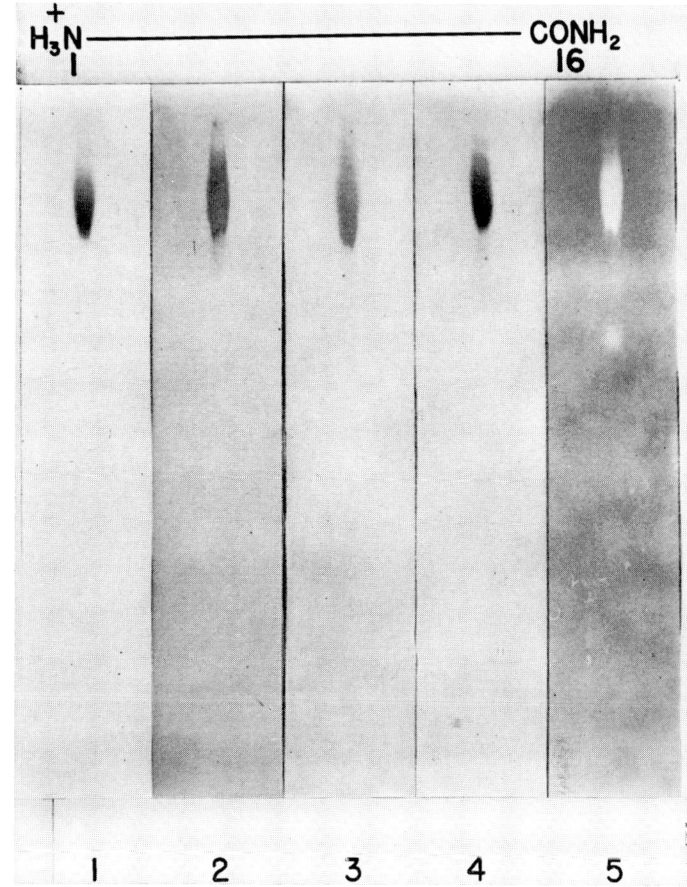

Plate 3.2. Unidimensional paper chromatogram of free hexadecapeptide amide (1–16). Bands 1 to 5 were developed respectively with the ninhydrin, Pauly, Sakaguchi, Ehrlich, and methionine reagents. (Solvent system Pyridine, Acetic acid, *n*-Butanol, water, 20:6:30:24.)

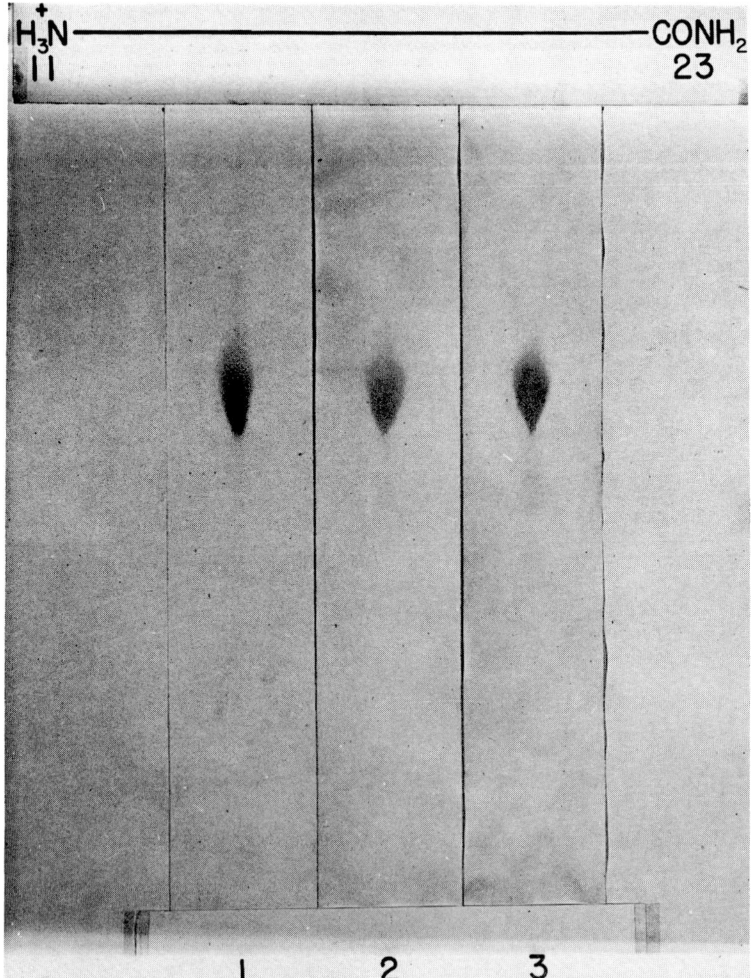

Plate 3.3. Unidimensional paper chromatogram of partially protected subunit (11–23) (Figure 3.14). Bands 1 to 3 were developed respectively with the ninhydrin, Pauly, and Sakaguchi reagents (Partridge).

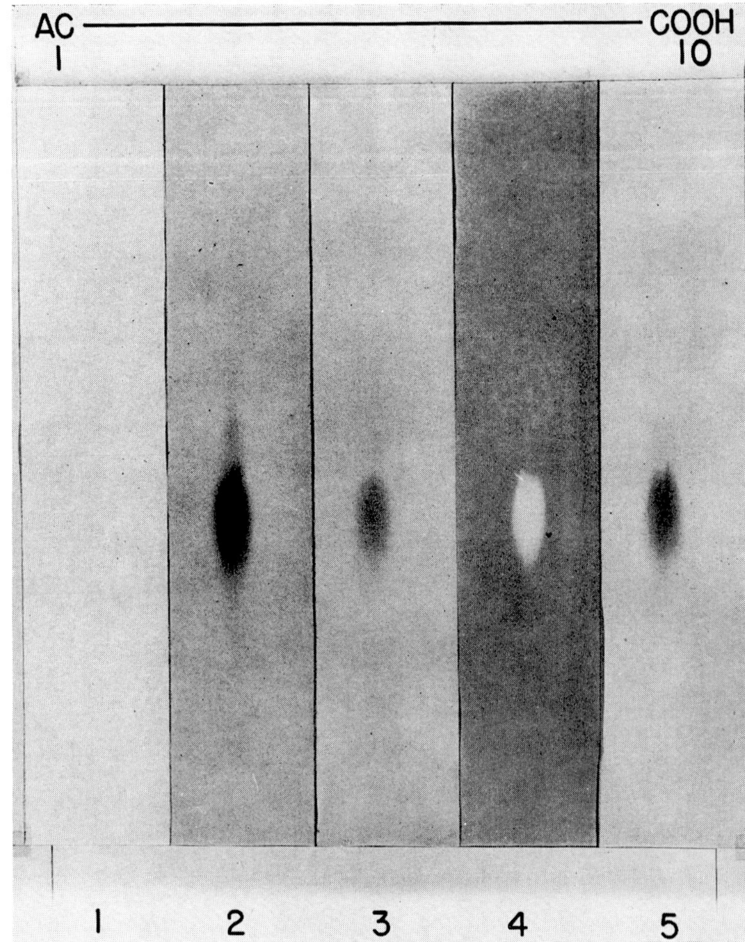

Plate 3.4. Unidimensional paper chromatogram of acetyldecapeptide (1–10). Bands 1 to 5 were developed respectively with the ninhydrin, Pauly, Sakaguchi, methionine, and Ehrlich reagents.

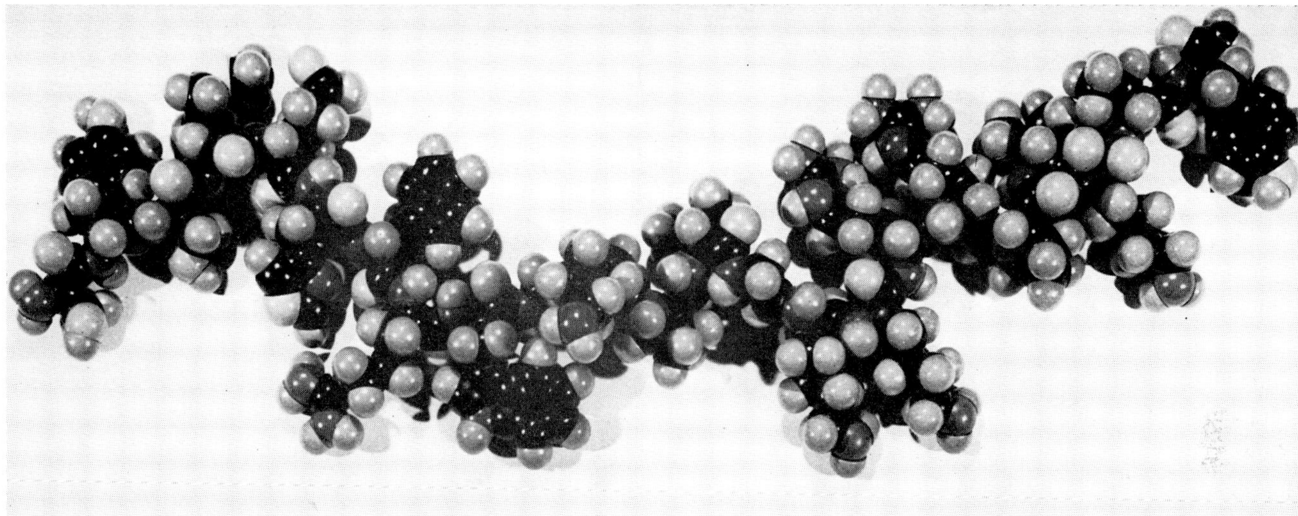

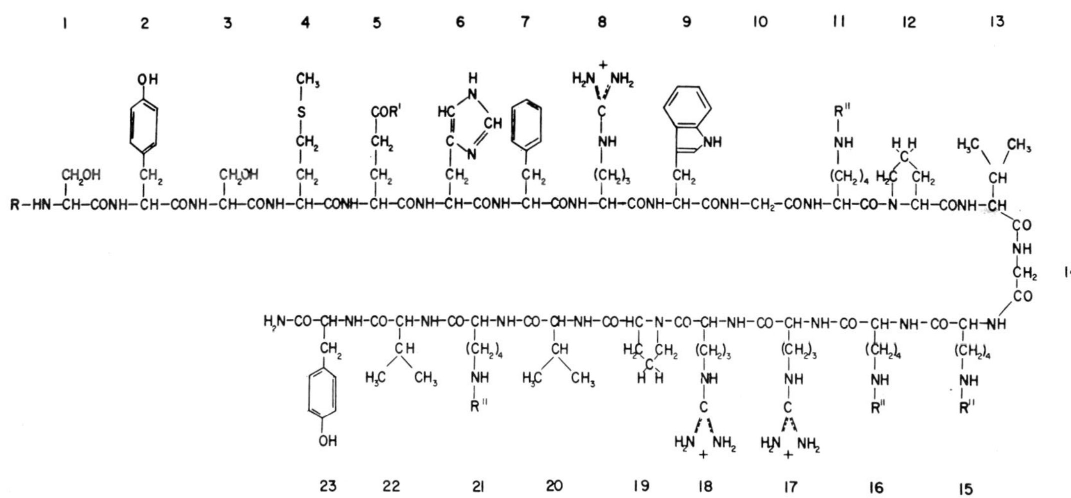

Plate 3.5. Scale model of the tricosapeptide amide (1–23).

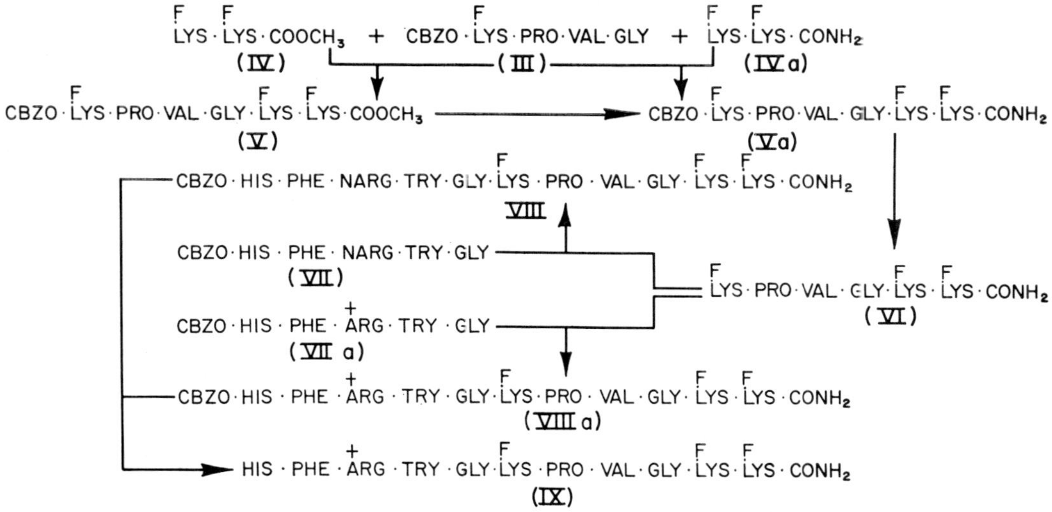

Fig. 3.10. Synthetic route to Subunit B_{16}. CBZO = Carbobenzoxy; F = N^{ϵ}-Formyl; NARG = N^{ω}-Nitroarginine.

ing to incorporate the N-terminus. The route to the partially protected undecapeptide amide (subunit B_{16}) provides additional examples of the use of N^{ϵ}-formyllysine in peptide synthesis. The various routes illustrated in Figure 3.10 were explored during the development of a productive procedure for preparing this peptide. Various properties (R_f values, optical rotations, and amino-acid composition of LAP digests) of material prepared by these alternate routes are in good agreement.

N^{α}-Carbobenzoxy-N^{ϵ}-formyllysylprolylvalylglycine (III) is reacted by the use of N,N'-carbonyldiimidazole with the methyl ester of N^{ϵ}-formyllysyl-N^{ϵ}-formyllysine (IV) to give N^{α}-carbobenzoxy-N^{ϵ}-formyllysylprolylvalylglycyl-N^{ϵ}-formyllysyl-N^{ϵ}-formyllysine methyl ester (V) (*33*). Exposure to methanolic ammonia converts this methyl ester into the amide (V a). The same amide is obtained when coupling (III) with the amide of N^{ϵ}-formyllysyl-N^{ϵ}-formyllysine (IV a); this latter procedure affords better over-all yields than does the former approach. Hydrogenolysis converts (V a) into the partially protected hexapeptide (VI). Two routes were explored for preparation of the partially protected undecapeptide amide (IX). The scheme which utilizes compound (VII a) as the acylating reagent for (VI) was adapted for production of the compound on a larger scale. Activation of the C-terminal glycine in (VII a) can be effected by either the DCC or the N,N'-carbonyldiimidazole procedure with practically equally satisfactory re-

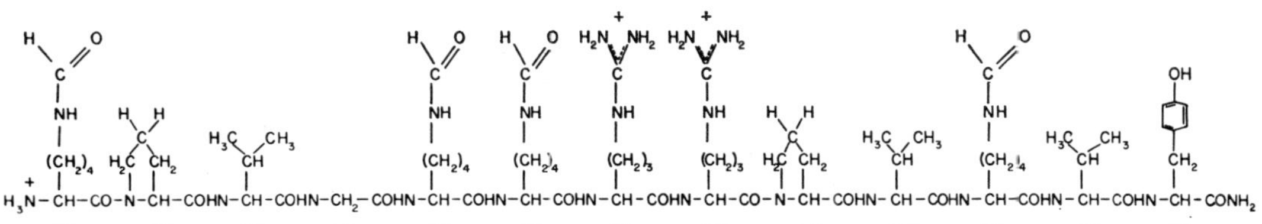

Fig. 3.11. Chemical structure of the partially protected subunit (11–23).

sults. The partially protected undecapeptide amide (IX) which was homogeneous, as judged by our criteria, reacted with subunit A to give the protected hexadecapeptide amide (1–16) (Figure 3.9). A chromatogram of analytically pure material (R_f 0.55 in Partridge system) is illustrated in Plate 3.1. Exposure to hot 0.5 N hydrochloric acid converts the protected hexadecapeptide amide into a mixture of compounds from which the unprotected hexadecapeptide amide is isolated by a combination of CMC chromatography and cellulose-block electrophoresis. A paper chromatogram of the highly purified hexadecapeptide amide (1–16) which migrates at the rate of histidine in the system pyridine, acetic acid, *n*-butanol, water, 20:6:30:24, is illustrated in Plate 3.2. There can be little doubt regarding the stereochemical homogeneity of this peptide amide since it is cleanly digestible with LAP with formation of the constituent amino acids. The product exhibits adrenocorticotropic activity in vivo at the level of <0.1 U/mg.

The tricosapeptide amide (1–23)

The low adrenocorticotropic activity of the hexadecapeptide amide (1–16) which is not significantly higher than that of the tridecapeptide (1–13) and the inferred full activity of the tetracosapeptide (1–24) points to the critical importance for full activity of a structural element located between positions 16 and 24. Particular attention focuses on the lysyllysylarginylarginine fragment in positions 15–18. The tricosapeptide amide (1–23) (Structure I, Figure 3.1) was synthesized with the expectation that it would possess full biological activity.

Two partially protected subunits, a decapeptide terminating with glycine covering positions 1–10 and a tridecapeptide amide covering positions 11–23, were used for construction of the protected tricosapeptide amide (Structure II, Figure 3.1). It will be noted from inspection of Figure 3.11 that subunit (11–23) contains four lysine and two arginine units and thus presents a problem of challenging complexity from the point of view of both synthesis and purification. Our route to this fragment is illustrated in Figure 3.12.

Prolylvalyl-N^{ϵ}-formyllysylvalyltyrosine amide (II) is prepared by stepwise elongation of the peptide chain of valyltyrosine amide with mixed anhydrides of the desired carbobenzoxyamino acids serving as individual building blocks. The intermediate carbobenzoxypeptide amides are deblocked by hydrogenolysis in presence of acetic acid. The first arginine residue is incorporated by the nitroarginine procedure (*42*). The ensuing hexapeptide amide

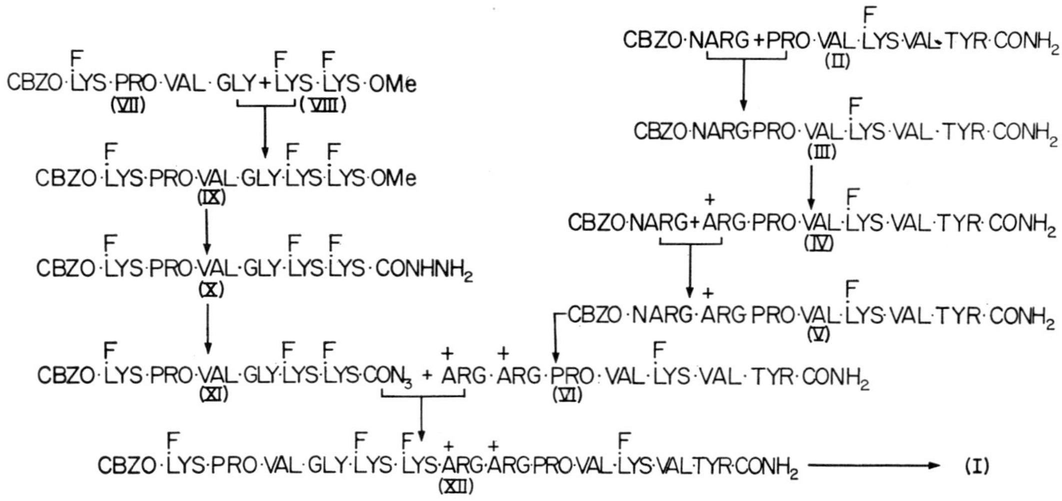

Fig. 3.12. Synthetic route to partially protected subunit (11–23). CBZO = Carbobenzoxy; **NARG** = N^{ω}-Nitroarginine; F = N^{ϵ}-Formyl.

(IV) is again acylated with a mixed anhydride of N^α-carbobenzoxy-N^ω-nitroarginine in aqueous tetrahydrofuran to give (V) in high yields. An excess of the mixed anhydride is employed in this step to ensure complete acylation of the hexapeptide amide since N^α-carbobenzoxy-N^ω-nitroarginine is removed quantitatively from the reaction product by simple extraction with ethyl acetate. Hydrogenation of (V) affords a crude heptapeptide amide (VI). The azide of N^α-carbobenzoxy-N^ϵ-formyllysylprolylvalylglycyl-N^ϵ-formyllysyl-N^ϵ-formyllysine (XI) (*33*) reacts with the heptapeptide amide (VI) to give the protected tridecapeptide amide (XII) which is decarbobenzoxylated by hydrogenolysis to afford (I).

A paper chromatogram of the analytically pure partially protected tridecapeptide amide (11–23), which is completely digestible by LAP, is shown in Plate 3.3.

The characteristic absorption maximum at 275 mμ of the tyrosine residue provides a convenient means of following the various synthetic C-terminal peptides during purification. It is to be noted that both protonization and nitro group blocking served to protect the arginine residues during the various manipulations involved in construction of the partially protected tridecapeptide amide (11–23) (Figure 3.11).

Preparation of the protected decapeptide (1–10) is patterned along the lines of the previously discussed syntheses of the protected tri- and hexadecapeptide sequences. As shown in Figure 3.13, it involves coupling of subunit A (I) with histidylphenylalanylarginyltryptophylglycine (II) (*19, 72*). Purification of the protected decapeptide to analytical homogeneity requires countercurrent distribution. This is the sole step in the synthesis of the tricosapeptide amide which demands the use of this method of purification. A paper chromatogram of the analytically pure product is shown in Plate 3.4. For assessing stereochemical homogeneity a sample of the protected decapeptide was deblocked by treatment with 0.5 N hydrochloric acid affording a free decapeptide which satisfied our criteria for homogeneity and was completely digestible with LAP.

Treatment with DCC of an equimolar mixture of the hydrochloride of the protected decapeptide and of the dihydrochloride of the partially protected tridecapeptide amide (11–23) in dimethylformamide followed by chromatography on CMC gave the protected tricosapeptide amide (1–23). This step is illustrated in Figure 3.14. Analyses of acid hydrolysates of the analytically pure protected tricosapeptide amide gave the correct amino acid composition with average amino acid recoveries ranging from 86 per cent to 100 per cent. Since this peptide is obtained from two subunits of established purity by a coupling step which is unlikely to bring about racemization, there can be little doubt pertaining to its stereochemical homogeneity.

Although highly effective in expanding melanocytes (2.0×10^8 MSH units/gm) the protected tricosapeptide amide was totally inactive as far as in vivo adrenocorticotropic activity is concerned. This situation prevailed, regardless of whether administration was by

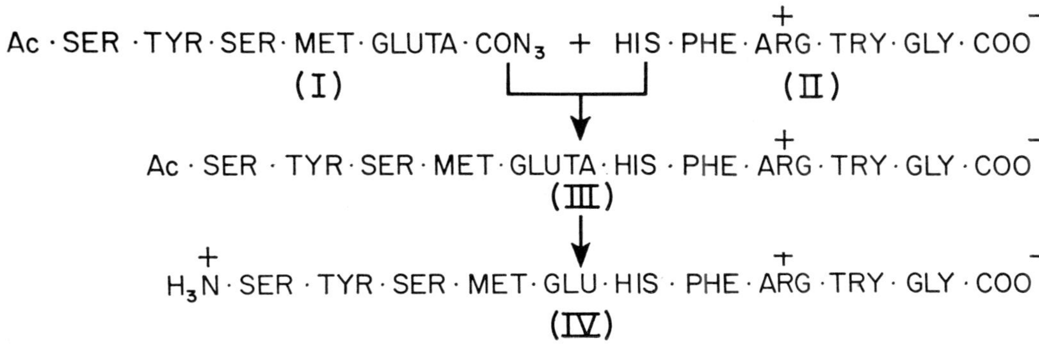

Fig. 3.13. Synthetic route to N-terminal decapeptide.

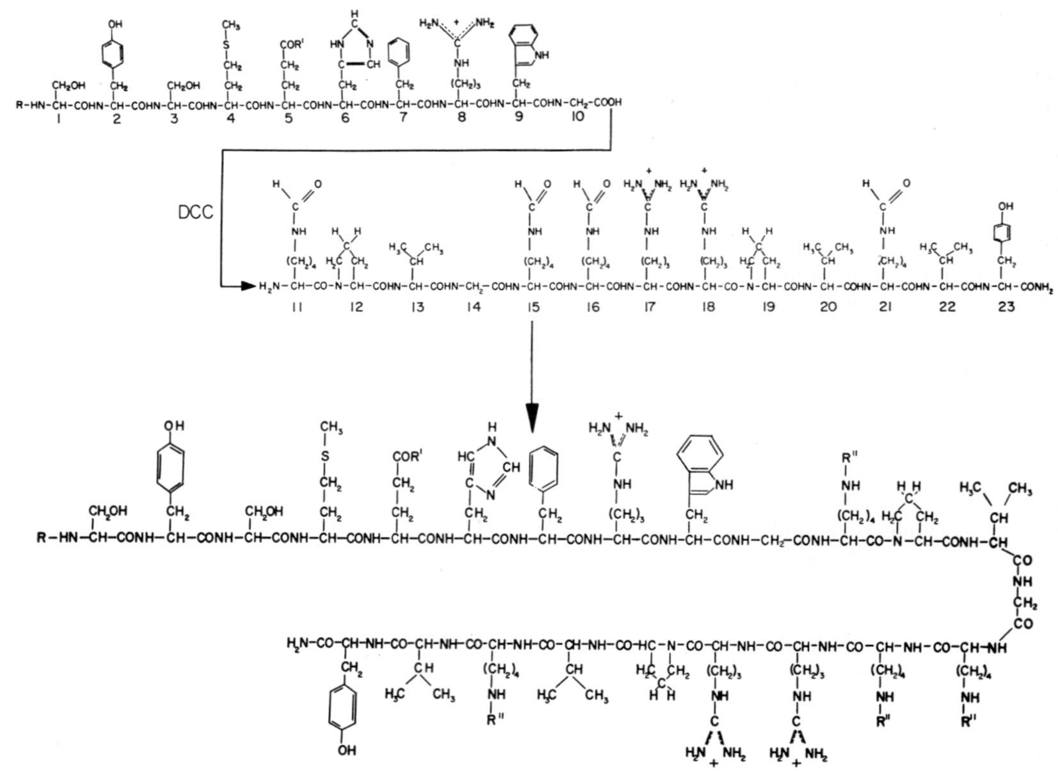

Fig. 3.14. Final step in synthesis of tricosapeptide amide.

the intravenous or by the subcutaneous routes. Solutions of this peptide were then heated in 0.5 N hydrochloric acid for 60 to 80 minutes, the chloride ions were exchanged for acetate ions, and the solutions were lyophilized. Biological assays of the residue showed in vivo adrenocorticotropic activity of 30–40 units per mg. Highly purified material which assayed 103 ± 10.3 units per mg is obtained from the crude product by CMC chromatography. This material exhibits essentially the same biological activity as natural ACTH. The peptide migrates at a rate somewhat slower than histidine in 24-hour Partridge paper chromatograms and Stein-Moore analysis of an acid hydrolysate gave the correct amino-acid composition. The average recovery of individual amino acids was 95 per cent.

Conclusions and outlook

The methodology which led to production of gram quantities of a fully active synthetic corticotropin can now be employed to solve other questions regarding the biological function of this molecule.

Particularly intriguing along these lines is the assessment of the essential nature of every amino-acid residue. Results of such investigations should provide a more complete picture regarding the exact relation between primary structure and biological function. We expect to provide some of these answers which are at the very basis of an understanding of the biological function of the polypeptide hormones.

Considerably more difficult to answer but of equal significance are questions pertaining to the role of secondary structure for hormonal activity. A brief look at a scale model of the fully active tricosapeptide amide molecule as shown in Plate 3.5 may be instructive at this point. Does the peptide chain assume a preferred conformation in solution or is it randomly coiled? If randomly coiled, how does the molecule assume the geometrical arrangement necessary to bring it into juxtaposition with the receptor site? It seems rather certain, but difficult to prove, that in its active state in the cell the corticotropin molecule will assume a geometrically defined arrangement.

Investigation of this arrangement and understanding of the orienting forces offer challenging problems for future inquiry.

Acknowledgment

Supported by grants from the U.S. Public Health Service, the National Science Foundation, the American Cancer Society, and Armour Pharmaceutical Company, Kankakee, Illinois. The authors wish to express their most sincere thanks to Drs. T. Liu, S. Lande, and N. Yanaihara, and to Mrs. C. Yanaihara and Mr. J. Humes for their contributions to the success of this investigation. They also wish to acknowledge their gratitude to Drs. M. R. Wright and A. B. Lerner of the Department of Medicine, Yale University School of Medicine, for the MSH assays which were carried out by the in vitro method of R. Shizume, A. B. Lerner, and T. B. Fitzpatrick, *Endocrinol.* 54, 553 (1954).

References

1 V. du Vigneaud, C. Ressler, J. M. Swan, C. W. Roberts, P. G. Katsoyannis, and S. Gordon, *J. Am. Chem. Soc.* **75,** 4879 (1953).
2 V. du Vigneaud, C. Ressler, J. M. Swan, C. W. Roberts, and P. G. Katsoyannis, *ibid.* **76,** 3115 (1954).
3 M. F. Bartlett, A. Jöhl, R. Roeske, R. J. Stedman, F. H. C. Stewart, D. N. Ward, and V. du Vigneaud, *ibid.* **78,** 2905 (1956).
4 V. du Vigneaud, M. F. Bartlett, and A. Jöhl, *ibid.* **79,** 5572 (1957).
5 V. du Vigneaud, D. T. Gish, P. G. Katsoyannis, and G. P. Hess, *ibid.* **80,** 3355 (1958).
6 P. H. Bell, *ibid.* **76,** 5565 (1954).
7 K. S. Howard, R. G. Shepherd, E. A. Eigner, D. S. Davies, and P. H. Bell, *ibid.* **77,** 3419 (1955).
8 R. G. Shepherd, K. S. Howard, P. H. Bell, A. R. Cacciola, R. G. Child, M. C. Davies, J. P. English, B. M. Finn, J. H. Meisenhelder, A. W. Moyer, and J. van der Scheer, *ibid.* **78,** 5051 (1956).
9 P. H. Bell, K. S. Howard, R. G. Shepherd, B. M. Finn, and J. H. Meisenhelder, *ibid.* **78,** 5059 (1956)
10 R. G. Shepherd, S. D. Willson, K. S. Howard, P. H. Bell, D. S. Davies, S. B. Davis, E. A. Eigner, and N. E. Shakespeare, *ibid.* **78,** 5067 (1956).
11 R. A. Brown, M. Davies, M. Englert, and H. R. Cox, *ibid.* **78,** 5077 (1956)
12 W. F. White and W. A. Landmann, *ibid.* **77,** 1711 (1955).
13 J. I. Harris and A. B. Lerner, *Nature* **179,** 1346 (1957).
14 J. I. Harris, *Biochem. J.* **71,** 451 (1960).
15 For reviews see: J. S. Fruton, *Advances in Protein Chem.* **5,** 1 (1949); M. Goodman and G. W. Kenner, *ibid.* **12,** 466 (1957); A. H. Cook and G. Harris in "Progress in Organic Chemistry," J. W. Cook (ed.), Butterworths Scientific Publications, London, 1958, p. 140; R. Schwyzer, *Ann. Rev. Biochem.* **29,** 183 (1960); P. G. Katsoyannis, *J. Polymer Sci.* **49,** 51 (1961).
16 K. Hofmann, *Ann. N. Y. Acad. Sci.* **88,** 689 (1960).
17 K. Hofmann, H. Yajima, N. Yanaihara, T. Liu, and S. Lande, *J. Am. Chem. Soc.* **83,** 487 (1961).
18 For a review see: E. B. Astwood, M. S. Raben, and R. W. Payne, *Recent Progr. in Hormone Research* **7,** 1 (1952).
19 K. Hofmann and H. Yajima, *J. Am. Chem. Soc.* **83,** 2289 (1961).
20 K. Hofmann, H. Yajima, and E. T. Schwartz, *ibid.* **82,** 3732 (1960).
21 T. Curtius, *J. prakt. Chem.* **70,** 57 (1902).
22 T. Curtius, *Ber. deut. chem. Ges.* **35,** 3226 (1902).
23 J. C. Sheehan and G. Hess, *J. Am. Chem. Soc.* **77,** 1067 (1955)
24 T. Wieland and R. Sehring, *Ann. Chem. Liebigs* **569,** 122 (1950)
25 R. Boissonnas, *Helv. Chim. Acta* **34,** 874 (1951).
26 J. R. Vaughan, Jr., *J. Am. Chem. Soc.* **73,** 3547 (1951).
27 K. Hofmann, M. E. Woolner, G. Spühler, and E. T. Schwartz, *ibid.* **80,** 1486 (1958).
28 G. W. Anderson and F. M. Callahan, *ibid.*, **80,** 2902 (1958).
29 H. Schwarz and F. M. Bumpus, *ibid.*, **81,** 890 (1959).
30 H. A. Staab, *Ann. Chem. Liebigs.* **609,** 75 (1957).
31 G. W. Anderson and R. Paul, *J. Am. Chem. Soc.* **80,** 4423 (1958).
32 R. Paul and G. W. Anderson, *ibid.* **82,** 4596 (1960).
33 K. Hofmann, T. Liu, H. Yajima, N. Yanaihara, and S. Lande, *ibid.* **83,** 2294 (1961).
34 M. Bodanszky and V. du Vigneaud, *ibid.* **81,** 5688 (1959).
35 M. Bodanszky, J. Meienhofer, and V. du Vigneaud, *ibid.* **82,** 3195 (1960).
36 D. Ben-Ishai and A. Berger, *J. Org. Chem.* **17,** 1564 (1952).
37 R. B. Woodward, R. A. Olofson, and H. Mayer, *J. Am. Chem. Soc.* **83,** 1010 (1961).
38 G. H. L. Nefkens and G. I. Tesser, *ibid.* **83,** 1263 (1961).
39 K. Hofmann, A. Rheiner, and W. D. Peckham, *ibid.* **75,** 6083 (1953).

40 G. W. Anderson, *ibid.*, **75,** 6081 (1953)

41 D. T. Gish and F. H. Carpenter, *ibid.* **75,** 5872 (1953).

42 K. Hofmann, W. D. Peckham, and A. Rheiner, *ibid.* **78,** 238 (1956).

43 M. Bergmann, L. Zervas, and H. Rinke, *Hoppe-Seylers Z. physiol. Chem.* **224,** 40 (1934).

44 R. Schwyzer, B. Iselin, H. Kappeler, B. Riniker, W. Rittel, and H. Zuber, *Helv. Chim. Acta* **41,** 1273 (1958).

45 R. Schwyzer, B. Iselin, H. Kappeler, B. Riniker, W. Rittel, and H. Zuber, *ibid.* **41,** 1287 (1958).

46 R. A. Boissonnas, St. Guttmann, and P-A. Jaquenoud, *ibid.*, **43,** 1349 (1960).

47 R. Schwyzer, W. Rittel, H. Kappeler, and B. Iselin, *Angew. Chem.* **72,** 915 (1960).

48 R. A. Boissonnas, St. Guttmann, R. L. Huguenin, P-A. Jaquenoud, and E. Sandrin, *Helv. Chim. Acta* **41,** 1867 (1958).

49 St. Guttmann and R. A. Boissonnas, *ibid.* **42,** 1257 (1959)

50 R. A. Boissonnas, St. Guttmann, P-A. Jaquenoud, E. Sandrin, and J-P. Waller, *ibid.* **44,** 123 (1961).

51 L. Zervas, T. T. Otani, M. Winitz, and J. P. Greenstein, *J. Am. Chem. Soc.* **81,** 2878 (1959)

52 E. Schnabel and C. H. Li, *ibid.* **82,** 4576 (1960).

53 E. Katchalski and P. Spitnik, *ibid.* **73,** 3992 (1951).

54 M. Bergmann, L. Zervas, and W. F. Ross, *J. Biol. Chem.* **111,** 245 (1935).

55 R. Roeske, F. H. C. Stewart, R. J. Stedman, and V. du Vigneaud, *J. Am. Chem. Soc.* **78,** 5883 (1956).

56 K. Hofmann, E. Stutz, G. Spühler, H. Yajima, and E. T. Schwartz, *ibid.* **82,** 3727 (1960).

57 L. A. Carpino, *ibid.* **79,** 98 (1957).

58 L. A. Carpino, *ibid.* **79,** 4427 (1957).

59 F. C. McKay and N. F. Albertson, *ibid.* **79,** 4686 (1957).

60 G. W. Anderson and A. C. McGregor, *ibid.* **79,** 6180 (1957).

61 R. Schwyzer and W. Rittel, *Helv. Chim. Acta* **44,** 159 (1961).

62 B. Bezas and L. Zervas, *J. Am. Chem. Soc.* **83,** 719 (1961).

63 E. Sondheimer and R. W. Holley, *ibid.* **79,** 3767 (1957).

64 K. Hofmann, A. Jöhl, A. E. Furlenmeier, and H. Kappeler, *ibid.* **79,** 1636 (1957).

65 R. W. Roeske, *Chem. & Ind.* **1959,** 1121.

66 G. W. Anderson and F. M. Callahan, *J. Am. Chem. Soc.* **82,** 3359 (1960).

67 E. A. Peterson and H. A. Sober, *ibid.* **78,** 751 (1956).

68 D. H. Spackman, W. H. Stein, and S. Moore, *Anal. Chem.* **30,** 1190 (1958).

69 K. Hofmann and A. Jöhl, *J. Am. Chem. Soc.* **77,** 2914 (1955).

70 K. Hofmann, T. A. Tompson, H. Yajima, E. T. Schwartz, and H. Inouye, *ibid.* **82,** 3715 (1960).

71 A. B. Lerner and J. S. McGuire, *Nature* **189,** 176 (1961).

72 K. Hofmann and S. Lande, *J. Am. Chem. Soc.*, **83,** 2286 (1961).

PAPER 4

Polyglutamic and Polyaspartic Acids: Emphasizing Hungarian Research

J. KOVACS

From the results of degradations of subtilis polypeptide (SPP) and of anthrax polypeptide (APP) it was concluded that both haptenic materials (molecular weight between 7,000 and 257,000), called native polyglutamic acid (NPG), were identical with γ-polyglutamic acid (I→II) (*1*). Since earlier investigations of Ivanovics and Bruckner (*2*) indicated that APP is composed of D-glutamic acid, the synthesis of γ-poly-D-glutamic acid was initiated by following two entirely different ideas. First, it was expected that intermolecular polyacylation of α,α′-dimethyl-γ-D-glutamyl-D-glutamic acid (key dipeptide, III) to methyl γ-poly-D-glutamate (IV) would be achieved easily by those methods which were available for the stepwise synthesis of oligopeptides. Second, the results obtained by studying the intramolecular transpeptidation reaction of polyaspartic acids and of glutamyl peptides made promising the idea of an analogous intramolecular transpeptidation of α-poly-glutamic acid into a polyglutamic acid containing predominantly γ-peptide bonds.

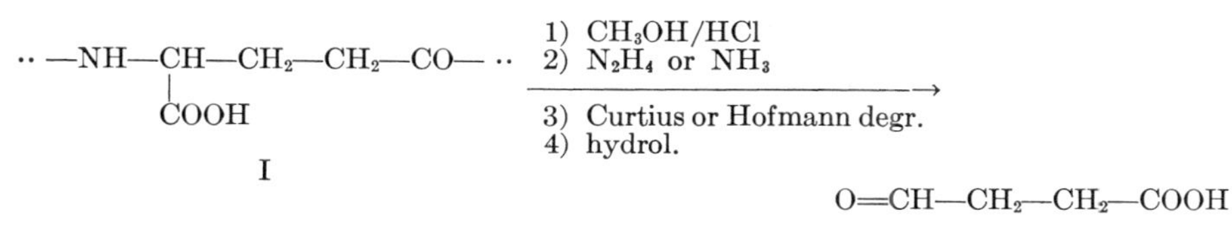

I

II

Although the former approach resulted in the first total-synthesis of a natural macromolecular haptene, the latter raised many interesting questions to be answered.

Most of these investigations are discussed in the excellent reviews of Katchalski and Sela in the *Advances in Protein Chemistry* (vols. 13 and 14) and in a more detailed form by V. Bruckner and the present author in the *Acta Chimica Acad. Sci. Hungarica* (*3*). In the present paper some of the problems which seem to be worthy of discussion are included together with results of the last few years in this field of polypeptide chemistry.

The intermolecular polyacylation of key dipeptide III (abbreviated as H_2N—**A**—COOH) to γ-polypeptide derivative IV was achieved (*3*) by (*a*) mixed anhydride method, (*b*) activated ester method, (*c*) dicyclohexyl-

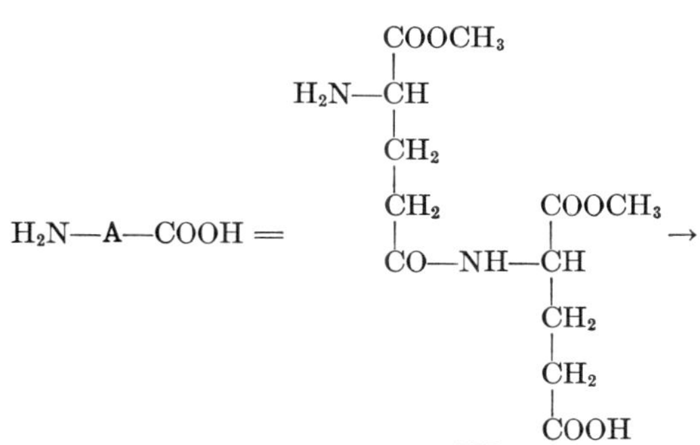

III

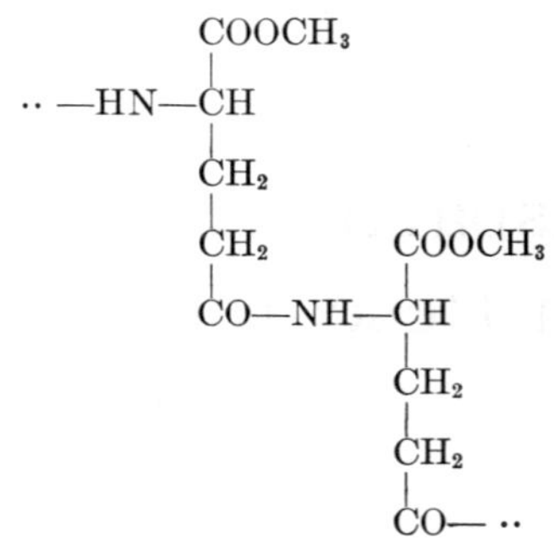

IV

Anhydride (VIII) was not investigated. However, it is reasonable to assume that the reaction of key dipeptide III under similar conditions would give *N*-carbothiophenyl derivative IX and mixed anhydride X, which upon disproportionation would give two anhydrides, XI and XII. Irreversible reaction of these anhydrides would lead to anhydride XIII. All these activated key dipeptide derivatives finally gave methyl γ-polyglutamate (IV) on heating in pyridine benzene solution. Attempts to isolate the pure intermediate compounds were unsuccessful.

$$H_2N\text{—A—COOH} + Cl\text{—}COSC_6H_5 \xrightarrow{N(C_2H_5)_3} \underset{IX}{C_6H_5\text{—S—CO—NH—A—COOH}} + \underset{X}{H_2N\text{—A—COOCOSC}_6H_5}$$

$$X \rightarrow \underset{XI}{(H_2N\text{—A—CO})_2O} + \underset{XII}{(C_6H_5\text{—SCO})_2O} \rightarrow$$

$$\underset{XIII}{C_6H_5\text{—S—CO—NH—A—}\underset{\|}{\overset{}{C}}\text{—O—}C(=O)\text{—A—NH}_2} + C_6H_5\text{—SH} + CO_2$$

carbodiimide method, (*d*) using carbothiophenylchloride. The carbothiophenyl method of Noguchi and Hayakawa gave good yield and led to some observations which deserve comment. According to the original method of Noguchi and Hayakawa (*4*), the *N*-carbothiophenyl amino acid or peptide ester was first prepared by the usual procedure. On careful acidic hydrolysis of the ester group the free *N*-carbothiophenyl derivative was obtained, which on heating in pyridine-benzene solution gave the desired polypeptide. However, as anticipated in our case, the free carboxyl group of the key dipeptide was also reacted with the acid chloride. To demonstrate, phthaloylglycine (V), as a model compound, gave with equimolecular amounts of carbothiophenylchloride (VI) and triethylamine in chloroform solution a precipitate which proved to be phthaloylglycine anhydride (VII). The following reaction was proposed.

Saponification of the methyl γ-polyglutamate (IV) obtained by the above-mentioned polymerization methods yielded γ-polyglutamic acid.

In addition to the above-discussed key dipeptide, in which both glutamyl residues were the D forms, we prepared the following dipeptide derivatives: (*a*) α,α′-dimethyl-γ-L-glutamyl-L-glutamic acid (*3*); (*b*) α,α′-dimethyl-γ-L-glutamyl-D-glutamic acid (*5*); (*c*) α,γ′-dimethyl-γ-L-glutamyl-L-glutamic acid (*6*); (*d*) α,γ′-dimethyl-γ-D-glutamyl-D-glutamic acid (*6*). Saponification of the ester groups in the polypeptides obtained by the polymerization of these dipeptides gave 1) γ-poly-L-glutamic acid, 2) γ-poly-D-glutamic acid, 3) L,D-mesoid-γ-polyglutamic acid in which L- and D-glutamyl residues alternate, 4) α,γ-poly-L-glutamic acid, and 5) α,γ-poly-D-glutamic acid. All these polyglutamic acids were used to test the specificity of the serological reaction.

$$\underset{V}{Phth{=}N\text{—}CH_2\text{—COOH}} + \underset{VI}{Cl\text{—CO—S—}C_6H_5} \xrightarrow{N(C_2H_5)_3} \underset{VII}{(Phth{=}N\text{—}CH_2\text{—CO})_2O} + \underset{VIII}{(C_6H_5\text{—S—CO})_2O}$$

Recently Bruckner and Kajtar (*7*) reported the synthesis of D,L-mesoid-γ-polyglutamic acid following the same series of reaction which were used earlier for the synthesis of L,D-mesoid-form. The serological investigation of this D,L-mesoid-form is not reported yet.

The question of racemization must be discussed in connection with the synthesis of these γ-polypeptides. Racemization does not occur during the polymerization or saponification of the ester groups. The optical purity of these γ-polypeptides was demonstrated by the optical purity of the glutamic acid obtained by total-hydrolysis. However, in connection with racemization during saponification, it is noteworthy that in contrast with α-polyglutamic acid, γ-polyglutamic acid is resistant to racemization in alkaline solution (*8*). This would be the reason why saponification of ester groups is not accompanied by racemization; one possible explanation would be that the ionized carboxyl on α-carbon atom repels the approaching hydroxyl ions and makes the ionization of α-hydrogen more difficult.

Degradation of the synthetic γ-poly-L-glutamic acid by γ-L-glutamyl peptidase was investigated by Volcani and Margalith (*9*). This enzyme hydrolyzed our preparation in 95 per cent yield to glutamic acid, indicating the 95 per cent optical purity and the presence of at least 95 per cent γ-peptide bonds. This percentage of γ-bonds shows that intramolecular $\gamma \rightarrow \alpha$ transpeptidation did not occur during the synthetic procedures. The absolute value of the specific rotation of γ-poly-L-glutamic acid, $[\alpha]_D^{20} - 23.8°$, was identical with that of APP, which was +23.5° in water solution. Identity of the specific rotations is not a proof of the optical purity because specific rotation varies with molecular weight.

The molecular weight determinations of the strongly heterodisperse synthetic γ-polypeptides are not satisfactory. Van Slyke amino-nitrogen determinations of γ-glutamyl-peptides are unreal and gave much higher values (*10*). The DNP-method was studied on γ-poly-L-glutamic acid by Waley (*11*). This method is not very precise, because reaction between a terminal glutamic acid residue can be taken to completion only with difficulty (*12*). Physical methods, such as light scattering, give much higher values; a sample of NPG with amino nitrogen value 0.14, mol. weight 10,000, gave a molecular weight of 60,000, as determined by Dr. Sedlacek (*13*). It was assumed that the real molecular weight was about six times the value obtained by van Slyke. Amino-nitrogen values (by van Slyke method) of the different synthetic γ-polyglutamic acids were between 0.15 and 2.9.

Serological investigation of the synthetic polyglutamic acids were carried out by Ivanovics (*14*). Among the different synthetic preparations only the γ-poly-D-glutamic acid precipitated the same amount of antibody as SPP from horse serum sensitized against anthrax. However, L,D-mesoid-γ-polyglutamic acid, in which L- and D-glutamyl residues alternate, gave a somewhat weaker precipitation reaction. No serological reaction was given by γ-L-polyglutamic acid and other polyglutamic acids containing α-bonds.

Recently Thorne and Leonard (*15*) have presented evidence that those subtilis polypeptides which on hydrolysis yield both L- and D-glutamic acid are mixtures of the two polypeptides, γ-poly-L and γ-poly-D, and not a single peptide containing L- and D-glutamic acids. Therefore those synthetic polypeptides, which contain L- and D-glutamyl residues, such as the mesoid-γ-polyglutamic acids cannot be considered as kinds of subtilis polypeptides, or even similar to subtilis polypeptides.

In light of this, the strong serological reaction of mesoid-γ-polyglutamic acid is remarkable because it indicates that the serological reaction is not strongly stereospecific. Distances between the repeating carboxyl groups with D-configuration of the α-carbon atoms at *p*H 7, where they are ionized and therefore the molecule uncoiled (*16*), seems to be important. Configuration of at least every second unit seems to be essential for the serological reaction. Further work is needed to evaluate this important serological reaction. For example, it would be interesting to test the precipitation reaction with a polypeptide, where every second carboxyl group is missing (poly-(γ-D-glutamyl-(γ-amino-butyric acid).

It was mentioned in the introduction that our second aim was to convert α-peptide bonds in α-polyglutamic acid into γ-peptide bonds by "intramolecular transpeptidation" reaction. The assumption that this could be done was based on two earlier observations: First, in 1952 we observed (*17*) and then studied in de-

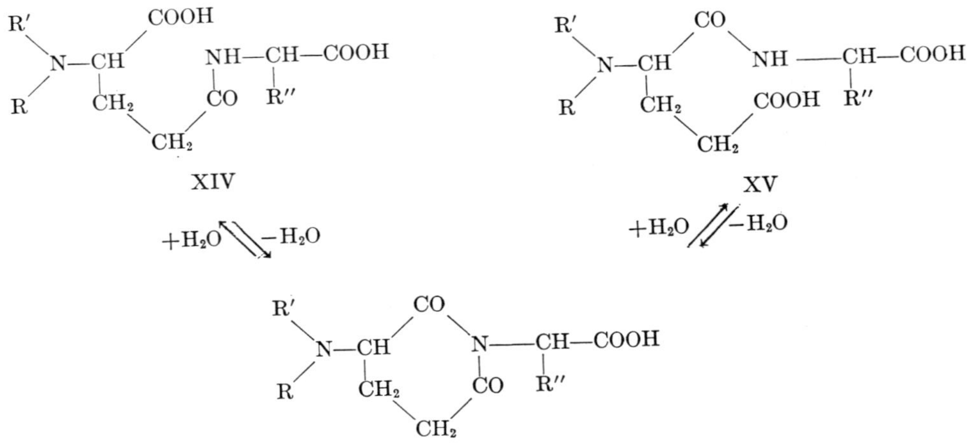

XIV XV XVI

tail (*18*) the reversible intramolecular $\alpha \rightleftharpoons \gamma$ transpeptidation of structurally isomeric α- and γ-glutamyl peptides. Both the *N*-acylated γ-glutamyl-peptides (XIV) and α-glutamyl peptides (XV) were converted under mild conditions (e.g., Ac_2O) into the same glutarimide derivative, XVI. Partial alkaline hydrolysis of XVI yielded a mixture of the starting dipeptide; however in this mixture the γ-peptide predominated (90 per cent). Second, concurrent with the degradation studies (1952) of NPG we began a study of the chemistry of polyaspartic acids—first, of those which contained omega peptide bonds. Thermal polycondensation of aspartic acid yielded anhydropolyaspartic acid (XXI); partial alkaline hydrolysis of XXI gave α,β-polyaspartic acid (XX) (*19*). It was suggested that the initial reaction in the polycondensation of aspartic acid may be the formation of either aspartic anhydride (*19*) (XIX) or dipeptide(s) (*20, 21*) (XVIII), the next product being α,β polyaspartic acid (XX) formed by polyacylation of XIX or by polycondensation of XVIII (*20, 21*). However, under the reaction conditions this material undergoes intramolecular de-

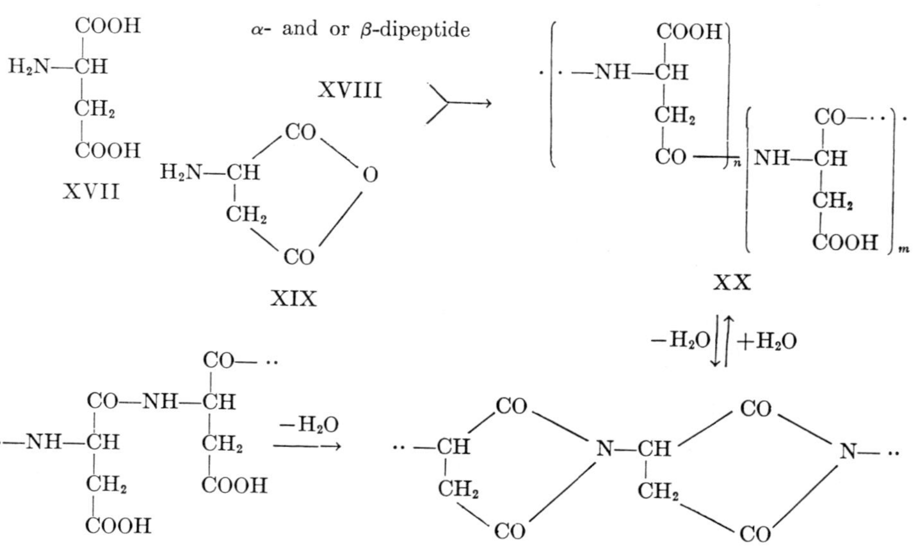

XVII XVIII XIX XX XXI XXII

hydration producing XXI, a polymeric compound of succinimide units (*20, 26, 21*). Various steps in this assumed mechanism were investigated. Aspartic anhydride (*21b*) was easily converted to XX. Recently H. Nagy carried out experiments to study reaction XVIII →XX (*22*). Comparison of the thermal polymerization of β-aspartylaspartic acid with that of aspartic acid showed that the dipeptide polymerized roughly ten times faster. In an attempt to indicate that the anhydride can be an intermediate (reaction XVII →XIX) acetylaspartic acid was heated to 145°. However instead of the expected acetylaspartic anhydride, anhydropolyaspartic acid (XXI) was formed (*21*). Acetylaspartic anhydride did not polymerize under the same conditions as above; when the temperature was raised to 200° it decomposed with the formation of acetamide. The polymerization of acetylaspartic acid is unique among the amino acids. It calls to mind the polymerization of the carbobenzoxy-asparagine carried out by Liwschitz and Zilkha at 180°, or the polymerization of *N*-carbothiophenyl aspartic acid done by Noguchi (*24*). Noguchi suggested the formation of the reactive isocyanate as an intermediate. However, here it is the β-carboxyl of acetylaspartic acid which

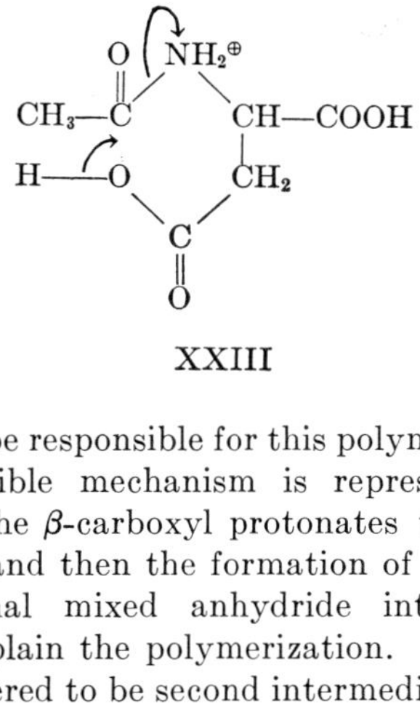

XXIII

seems to be responsible for this polymerization. One possible mechanism is represented by XXIII; the β-carboxyl protonates the amide nitrogen and then the formation of a reactive bifunctional mixed anhydride intermediate would explain the polymerization.

Considered to be second intermediate in this thermal polycondensation of aspartic acid, α,β-polyaspartic acid (XX) was converted into anhydropolyaspartic acid (XXI) under the experimental conditions of the pyrocondensation. Partial hydrolysis of XXI gave XX; in this material the ratio of α- and β-peptide bonds was 1:1.3 (*25, 21*). Also converted into XXI was α-polyaspartic acid (XXII). This on partial hydrolysis yielded XX (*26, 21*). These reactions indicate, that α-aspartyl peptide bonds are convertible into β-peptide bonds (and vice versa) within the polypeptide molecule and without breaking the peptide chain. It is interesting that the succinimide ring formation from α-aspartyl residues in a polypeptide chain is much faster than from β-aspartyl residues. This is understandable if one considers that when the α-carboxyl and the amide nitrogen approach each other for ring closure they have to take along with them the lengthy chains. On the other hand the β-carboxyl is already in position for the ring closure. The second step in the transpeptidation reaction, the opening of the imide ring, must be governed chiefly by the electrophilic character of the α- and β-carbonyl groups.

Since this intramolecular transpeptidation reaction was thought to be important in peptide chemistry, it was studied with simple model compounds (*21*). Both α- and β-acetylaspartic anilide yielded the same succinimide derivative XXIV, with acetic anhydride.

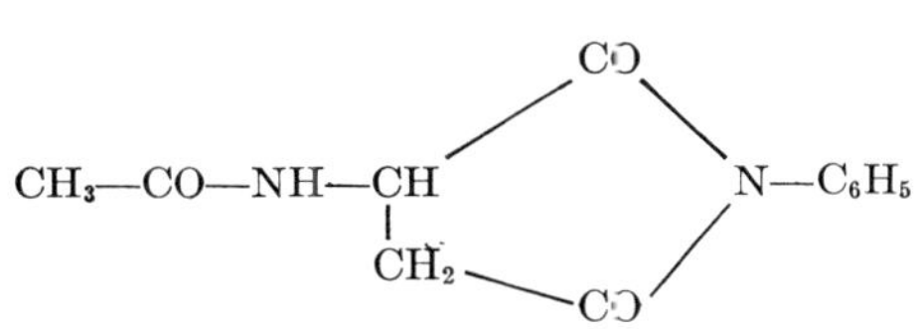

XXIV

Partial hydrolysis of XXIV gave a mixture of the starting anilides, in which the β-derivative predominated (1:8).

On the basis of observations discussed here, it should be borne in mind, that whenever an α-aspartyl or α-glutamyl residue is present in a peptide chain, imide ring formation and, through this, an intramolecular transpeptidation could occur under mild conditions. This conclusion is supported by recent observations: Abraham and Lockhart (*27*) isolated compound XXV from the acid hydrolysate of Bacitracin A; the succinimide ring was formed by the action of strong acid. Naughton, Sanger, and others (*28*) reported an interconversion of aspartyl peptides in a study of the amino-acid sequence around the reactive site of proteolytic enzymes. The esterification of α,β-polyaspartic

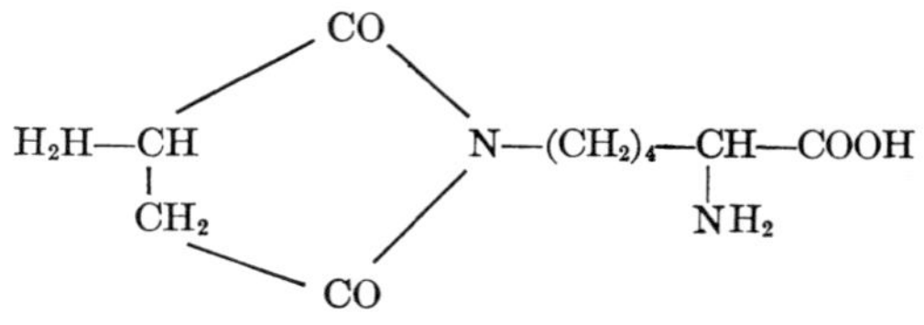

XXV

acid (*29*) with methanol-hydrogen chloride cannot be completed because of some imide ring formation. Noguchi observed imide ring formation (*24*) when the benzylester groups from benzyl α-polyaspartate were removed by hydrogen bromide. Vajda (*30*) recently investigated the fission of peptide bonds with hydrogen chloride under anhydrous conditions. Among his model compounds there were aspartyl peptides with free β-carboxyl groups. The possibility of imide ring formation probably should have been considered and the resulting presence of some water. Independently of our work, Sondheimer and Holley (*31*), Battersby and Robinson (*32*) later observed the rearrangement during alkaline hydrolysis of esters of glutamic and aspartic acid amide and peptide derivatives. Clayton and Kenner (*33*) have published other examples similar to ours in the amino-acid series.

Before the transpeptidation of α-polyglutamic acid will be discussed, it should be noted that the thermal polymerization of aspartic acid to a high molecular weight polypeptide contributed, together with other thermal reactions, to the development of Fox's thermal theory of biochemical origin of proteins (*34*). It and the thermal theory of Fox were the starting points of experiments related to the problem of the primordial origin of polypeptides. It was known long ago (*35*) that the hydrolysis of asparagine to aspartic acid in boiling water was accompanied by the formation of a noncrystalline material. This material is a mixture of aspartyl oligo- and polypeptides formed from asparagine probably by a nonenzymatic intermolecular transamidation reaction (*36*). The reaction in dilute solution at 70° gave a yield of 17 per cent as compared to 50–60 per cent at 100°. Since the formation of asparagine by the action of sunlight on paraformaldehyde and nitrate or ammonia in water solution has been reported by Bahadur (*37*), it is a logical intermediate for the formation of polypeptides under primitive conditions. We feel that the formation of peptides and polypeptides from asparagine at 70° in water solution satisfies suggested primitive conditions (*38*) better than thermal polycondensation.

The α→γ transpeptidation of α-polyglutamic acid was achieved through the partial hydrolysis of anhydroglutamic acid. On heating α-polyglutamic acid with acetic anhydride in dimethylformamide solution a polymeric material similar to anhydropolyaspartic acid separated. First a poly-glutarimide structure was suggested (*39*). However a formula with glutarimide and pyrrolidon rings (*17b*), such as XXVI, is in better agreement with the experimental results. By partial hydrolysis of anhydropolyglutamic acid we obtained a water-soluble polyglutamic acid and glutamic acid. If somewhere in the chain of α-polyglutamic acid a glutarimide residue forms, this prevents the formation of the pyrrolidon unit from the following glutamyl residue. Similarly, the pyrrolidon rings prevent the glutarimide ring formation from the preceding glutamyl unit. On the other hand, in γ-polyglutamic acid which also gave anhydropolyglutamic acid pyrrolidon residues cannot be formed. Therefore its structure can best be expressed by XXVII. The ratio of α-glutamyl to γ-glutamyl units in the polyglutamic acids obtained from α-poly-L- and -D-glutamic acids was about 1:1.5 (*17b*). These polyglutamic acids did not give the serological reaction.

On the basis of the transpeptidation of γ-esters of α-glutamyl peptides into γ-glutamyl peptides, Battersby (*32*) suggested that the α-polyglutamic acids (*40*) prepared by alkaline hydrolysis of the polyesters of α-polyglutamic acids will contain an appreciable amount of γ-linked material. However, the α-polyglutamic acid mentioned above did not exhibit properties (e.g., solubility in water) which would indicate the presence of γ-peptide bonds (*40, 17b*). Moreover, the saponification was accomplished in the presence of copper ions which form a biuret complex with the polypeptide. This saponification of the ester groups through the biuret complex was done with the purpose of preventing racemization and hydrolysis of the peptide bonds. The approach was based on the observations of Nyilasi and Kovats (*41*) that partial racemization and hydrolysis of edestine and gelatin in alkaline solution can be inhibited

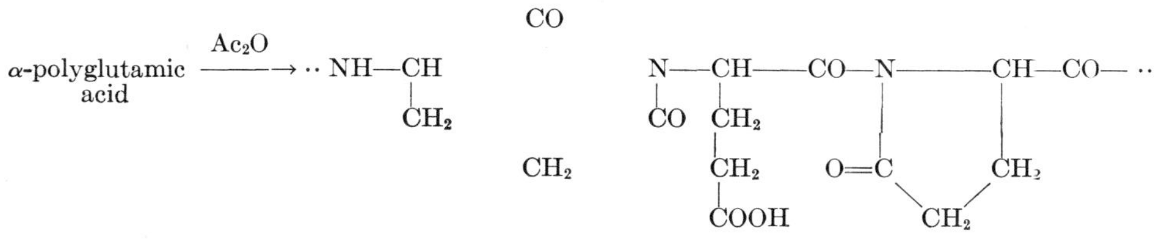

XXVI

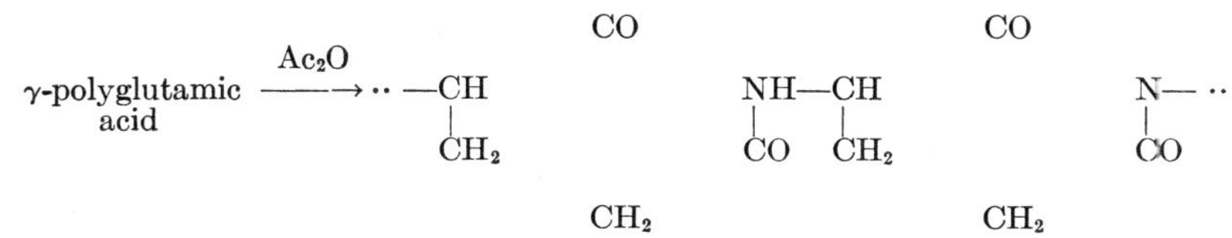

XXVII

by copper ions. Although studies of the biuret complexes of peptides appear in the literature, none are decisive as to their structure.

In the α-polyglutamic acid–biuret complex the ratio of glutamyl units and copper is 3:1 determined by colorimetric method (*40*). Searching for a structure which would explain the inhibitory action on racemization in this 3:1 complex, we proposed structure XXVIII (*17b*). This formula would at least partly explain the inhibitory action of the biuret complex on racemization by the formation and fixation of double bonds between carbonyl-carbon and nitrogen. Again, this structure would explain the inhibitory action of copper on the alkaline hydrolysis of peptide bonds.

Finally, it would explain the inhibitory action on $\alpha \rightarrow \gamma$ transpeptidation. The suggested formula of the biuret complex of the polypeptide may not be the correct one, but the significant inhibitory action on the abovementioned three reactions is definite.

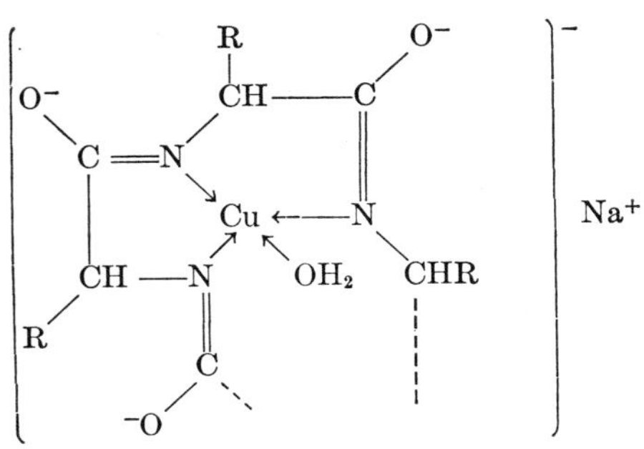

XXVIII

Recently Bruckner and others (*42*), in a careful analysis of this question on γ-methylate of benzoyl-glycyl-glycyl-α-L-glutamyl-glycine benzylamide, proved that both racemization and $\alpha \rightarrow \gamma$ transpeptidation can be prevented significantly by carrying out the alkaline hydrolysis of the ester groups in the presence of copper hydroxide.

These results suggest the use of the copper complex whenever alkaline hydrolysis of ester groups must be carried out in peptide chemistry.

Added in proof.—In a recent publication, Leonard and Thorne (*J. Bact.* **87,** 175 [1961]) reported that their effort to identify immune γ-globulins, specific to anthrax polypeptide in rabbits injected with encapsulated cells of *B. anthracis,* failed. The antisera contained basic proteins which reacted with the anthrax polypeptide when tested by the usual serological techniques. Positive reactions were also given when egg white lysozyme was substituted for antiserum. Other polyglutamic acids did not behave the same.

The present author thinks that the specificity of the precipitation reaction also indicated by Ivanovics (*14*) needs further explanation.

Acknowledgment

Most of the work reported here was conducted at the University of Budapest. Some of it was completed at St. John's University, under Grant RG 6579 from the National Institute of Health, U.S. Public Health Service. The author wishes to acknowledge the assistance of Dr. John A. Skarulis in the preparation of this manuscript. The author is also indebted to Miss Joan Bartumioli for the secretarial services in the production of this manuscript.

References

1 J. Kovacs and V. Bruckner, *Research* **5,** 194 (1952); *J. Chem. Soc.* **1952,** 4255; V. Bruckner, J. Kovacs, and H. Nagy, *ibid.* **1953,** 148; J. Kovacs, V. Bruckner, and K. Kovacs, *ibid.* **1953,** 145; V. Bruckner, J. Kovacs, and K. Kovacs, *ibid.* **1953,** 1512; V. Bruckner, J. Kovacs, I. Kandel, and G. Denes, *Acta Chim. Acad. Sci. Hung.* **7,** 223 (1955).

2 G. Ivanovics and V. Bruckner, *Naturwissenschaften* **25,** 250 (1937); *Z. Immunitätsforsch.* **90,** 304 (1937); **91,** 175 (1937).

3 V. Bruckner and J. Kovacs, *Acta Chim. Acad. Sci. Hung.* **12,** 363 (1957); V. Bruckner, M. Kajtar, J. Kovacs, H. Nagy, and J. Wein, *Tetrahedron* **2,** 211 (1958).

4 J. Noguchi and T. Hayakawa, *J. Am. Chem. Soc.* **76,** 2846 (1954).

5 V. Bruckner, J. Wein, M. Kajtar, and J. Kovacs, *Naturwissenschaften* **44,** 89 (1957).

6 V. Bruckner, M. Szekerke, and J. Kovacs, *Hoppe-Seyler's Z. physiol. Chem.* **309,** 25 (1957); *Naturwissenschaften* **42,** 179 (1955); **43,** 107 (1956); **44,** 90 (1957).

7 V. Bruckner and M. Kajtar, *Acta Chim. Acad. Sci. Hung.* **21,** 417 (1959).

8 M. Bovarnick, *J. Biol. Chem.* **145,** 415 (1942).

9 B. E. Volcani and P. Margalith, *J. Bacteriol.* **74,** 646 (1957).

10 H. Sachs and E. Brand, *J. Am. Chem. Soc.* **76,** 3601 (1954).

11 S. G. Waley, *J. Chem. Soc.* **1955,** 517.

12 A. C. Chibnall, M. W. Rees, and F. M. Richards, *Biochem. J.* **68,** 129 (1958).

13 B. Sedlacek, private communication of Mr. Sedlacek (Inst. of Organic Chemistry, Academy of Sciences, Prague).

14 G. Ivanovics, *Tetrahedron* **2,** 236 (1958).

15 C. B. Thorne and C. G. Leonard, *J. Biol. Chem.* **233,** 1109 (1958).

16 L. H. Kent, B. R. Record, and R. G. Wallis, *Phil. Trans. Roy. Soc. London* **250A,** No. 972, 1 (1957); H. Edelhoch and J. B. Bateman, *J. Am. Chem. Soc.* **79,** 6093 (1957); H. Edelhoch, R. E. Lippoldt, *Biochim. et Biophys. Acta* **45,** 205 (1960).

17 (*a*) V. Bruckner and J. Kovacs, *Magyar Tudományos Akad. Kém. Tudományok Osztályának Közleményei* **3,** 105 (1953); (*b*) *Acta Chim. Acad. Sci. Hung.* **12,** 363 (1957).

18 J. Kovacs, K. Medzichradszky, and V. Bruckner, *Naturwissenschaften* **41,** 450 (1954); *Acta Chim. Acad. Sci. Hung.* **6,** 183 (1955).

19 J. Kovacs, I. Konyves, and A. Pusztai, *Experientia* **9,** 459 (1953).

20 J. Kovacs, I. Konyves, and A. Pusztai, *Vegyipari Kutató Intézetek Közleményei* **4,** 120 (1954).

21 (*a*) J. Kovacs, H. Nagy, I. Konyves, J. Csaszar, T. Vajda, and H. Mix, *J. Org. Chem.* **26,** 1084 (1961); (*b*) H. Mix and J. Kovacs, *Naturwissenschaften* **43,** 447 (1956).

22 H. Nagy and J. Kovacs, unpublished results.

23 Y. Liwschitz and A. Zilkha, *Bull. Research Council Israel* **5A,** 309 (1956).

24 J. Noguchi, T. Saito, and M. Asai, *Nippon Kagaku Zasshi* **81,** 620 (1960).

25 J. Kovacs and I. Konyves, *Naturwissenschaften* **41,** 333 (1954).

26 J. Kovacs, I. Konyves, and J. Csaszar, *ibid.* **41,** 575 (1954).

27 I. M. Lockhart and E. P. Abraham, *Biochem. J.* **62,** 645 (1956).

28 M. A. Naughton, F. Sanger, B. S. Hartley, and D. C. Shaw, *ibid.* **77,** 149 (1960).

29 H. Nagy and J. Kovacs, unpublished results.

30 T. Vajda, *Acta Chim. Acad. Sci. Hung.* **21,** 71 (1959).

31 E. Sondheimer and R. W. Holley, *J. Am. Chem. Soc.* **76,** 2467 (1954).

32 A. R. Battersby and J. C. Robinson, *J. Chem. Soc.* **1955,** 259.

33 D. W. Clayton, G. W. Kenner, and R. C. Sheppard, *ibid.* **1956,** 371.

34 S. W. Fox, K. Harada, and H. Vegotsky, *Experientia* **15,** 81 (1959); S. W. Fox, J. E. Johnson, and A. Vegotsky, *Science* **124,** 923 (1956); S. W. Fox, *ibid.* **132,** 200 (1960).

35 F. Ehrlich and F. Lange, *Biochem. Z.* **54,** 256 (1913); E. Erlenmeyer, *ibid.* **52,** 439 (1913); C. Ravenna and G. Bosinnelli, *Atti accad. naz. Lincei* **28,** 113 (1919); *Gazz. Chim. ital.* **50,** 281 (1920); C. Ravenna and R. Nuccorini, *Gazz. chim. ital.* **60,** 140 (1930).

36 J. Kovacs and H. Nagy, *Nature* **90,** 531 (1961); unpublished results of J. Kovacs, H. Nagy, and R. Ballina.

37 K. Bahadur, *Nature* **173,** 1141 (1954).

38 S. L. Miller and H. C. Urey, *Science* **130,** 245 (1959); S. W. Fox, *ibid.* **130,** 1622 (1959); S. L. Miller and Urey, *ibid.* **130,** 1622 (1959).

39 V. Bruckner, J. Kovacs, and K. Medzihradszky, *Naturwissenschaften* **42,** 96 (1955).

40 V. Bruckner, K. Kovacs, J. Kovacs, and A. Kotai, *Experientia* **10,** 166 (1954); *Acta Chim. Acad. Sci. Hung.* **5,** 267 (1955).

41 J. Nyilasi and Z. Kovats, *Acta Chim. Acad. Sci. Hung.* **2,** 451 (1952); **3,** 273 (1953).

42 V. Bruckner, A. Kotai, and K. Kovacs, *ibid.* **21,** 427 (1959).

Discussion of Paper 4: Polyglutamic and Polyaspartic Acids

S. W. FOX

A theory which calls for substantial formation of peptide bond in aqueous solution at temperatures as high as 70° is faced with the need to demonstrate how rapid denaturation would not occur (see H. A. Barker, *J. Gen Physiol.* **17,** 21 [1933], for example). The fact that organisms are predominantly aqueous units should not be taken to signify that the macromolecules of which they are composed were first formed in aqueous environments; since these large molecules are anhydro structures, the contrary is indicated (S. W. Fox, *J. Chem. Educ.* **34,** 472 [1957]).

J. KOVACS

The polymerization of asparagine in aqueous solution represents one approach towards solving the problem of the synthesis of simple peptides and polypeptides under hypothetically primitive earth conditions. Asparagine being an amino-acid amide can be considered as an activated amino-acid derivative and its polymerization to peptides and polypeptides is then understandable. Our interest first of all was in the formation of peptide bonds in aqueous solution and not in the denaturation at 70° C. It would not be proper to discuss the problem of the denaturation of our simple peptides and polypeptides in the same environment in which they were formed.

The fact that proteins are condensation products of amino acids does not exclude that they were formed in aqueous environments. Our experiments indicated that polypeptides can be formed in aqueous solution from asparagine, which is a logical intermediate in this synthesis.

PAPER 5

The Thermal Copolymerization of α-Amino Acids

SIDNEY W. FOX, KAORU HARADA, and DUANE L. ROHLFING

The limited possibilities in production of macromolecules by thermal polymerization of unsubstituted α-amino acids have been documented (*1*). The production of a polyglycine (*2*) and of a polymer of aspartic acid (*3*), subsequently shown to be a polyimide (*4, 5*), have been the principal exceptions to this picture of futility. The difficulty has usually been explained as being one of the easy thermal decomposition of amino acids to diketopiperazines, tars, and other pyrolytic products (*1*).

Poly-α-amino acids containing a wide variety of residues have, however, been obtained by heating individual amino acids with aspartic acid, glutamic acid (*6–10*), or lysine (*11, cf. 7*) in sufficient proportion, all of the reactants being initially dry. The aspartic-acid copolymers are primarily polyimides which are easily convertible to polypeptides by mild alkaline treatment. The experiments emphasize that the results of copolymerization are not easily predicted, nor are they as limited as the results of attempted homopolymerization. The necessary conditions are thus dry heating of the otherwise thermolabile neutral amino acids in concert with sufficient proportions of the non-neutral amino acids.

Such copolymerization can be effected simultaneously so that all of the eighteen amino acids commonly found in protein are coupled within the same preparation (*10, 12*). The possibility that materials of this sort, as synthesized in the laboratory or under spontaneous conditions (*13*), could resemble proteins to a meaningful degree has faced a second conceptual difficulty. This latter is the unqualified belief that proteins themselves are unstable when heated. The destruction of catalytic activity would particularly be anticipated (*cf. 14*). Although denaturation of protein in hot water is a common observation, conditions for heating without coagulation have been known for a long time. An early report illustrating such effects is that of Barker (*15*), who found 162° to be the temperature for a half-coagulation time of 10 min. for ovalbumin initially free of measurable water. Altman and Benson have recently shown in the range of 80°–100° that "egg albumin is extraordinarily sensitive to the amount of sorbed water" (*16*). Some proteins are more readily heat-coagulable than others. Human invertase is, for instance, more thermolabile in aqueous solution than is yeast invertase. "When dry, invertase is extremely resistant to heat" (*17*). These illustrative citations indicate that denaturation or coagulation, including inactivation, is not a necessary consequence of the production of heteropoly-α-amino acids formed by heating of the dry monomers.

The data presented here and earlier indicate that one can prepare simply such panpolymers which are entirely soluble in aqueous solutions and which are largely hydrolyzable to amino acids.

Although panpolymers containing the eighteen amino acids commonly found in protein are of special interest, helpful information has been obtained from simpler copolymers which have been studied systematically (*5–9, 11*). The first such copolymer to be synthesized was formed from aspartic acid and leucine (*6, 18*). It was shown also that aspartic acid can be copolymerized with either valine or phenylalanine. The copolymerization of aspartic acid with nonhomopolymerizable α-amino acids yields polyimides as does aspartic acid (*8, 10*). As in the case of the imide type of homopolyaspartic acid, these imides can be gently hydrolyzed to peptides (*8*).

Glutamic acid, which does not form a polyimide or other polymer on heating alone, will copolymerize with other amino acids to produce mixtures of peptides (*7*). The yields are smaller in the case of thermal glutamic-acid copolymers than with aspartic acid. The reaction between aspartic acid and glutamic acid proceeds with great facility (*8*).

Each of the basic amino acids may be thermally copolymerized with other amino acids (*12*). Of these, lysine is the only one which has yet received much attention. It may also be homopolymerized thermally (*11*), and its considerable tendency to copolymerize thermally was noted early in these studies (*7*).

Copolymers of alanine and glutamic acid, alanine and aspartic acid, and alanine and lysine help to answer such questions as does the non-neutral amino-acid molecule copolymerize with more than one neutral amino-acid molecule. In each of the three copolymers mentioned, the partial hydrolyzates of the product contain the neutral amino-acid alanine in peptide fragments in which at least two alanine residues are directly linked (*12*). Evidence for this conclusion is found from chromatography of partial hydrolyzates of the polymers. Ninhydrin postive fractions obtained by paper chromatography which are not alanine give on total hydrolysis alanine only. These fragments must therefore be peptides of alanine. The result with alanine–glutamic acid copolymer is shown in Table 5.1. Chromatography of partial hydrolyzates of the lysine-alanine copolymer also yields spots which are not alanine, but which reduce only to alanine on total hydrolysis. From partial hydrolyzates of the aspartic acid–alanine copolymers, there have been obtained similar spots which reduce to alanine on total hydrolysis, but with a small proportion of aspartic acid present. When chromatographed on Amberlite IR-120 resin in the Phoenix automatic analyzer Model K-5000, the two racemate pairs of alanylalanine, free of aspartic acid, were recovered. Compositional data which support the sequential findings are recorded in Table 5.2 for the alanine–glutamic acid copolymer.

Table 5.1—Production of alanine peptides from hydrolysis of thermal polymer of alanine and glutamic acid

$$\text{3 Alanine} + \text{1 Glutamic Acid} \xrightarrow[\text{6 hr.}]{166 \pm 1^\circ}$$

$$\xrightarrow[\text{4 days}]{\text{dialysis}} \text{Poly(Ala, glu)}$$

$$\text{Poly(Ala, glu)} \xrightarrow[37^\circ\text{, 7 days}]{\text{AcOH: Conc. HCl (1:1)}}$$

8–11 spots on 2 dimensional chromatography in $BuOH:AcOH:H_2O$(4:1:1) vs. $PhOH:H_2O$(4:1) by light staining. Second fastest spot selected.

Rf's compared with *Rf* of Alanine (one dimension)

	$BuOH:HOAc:H_2O$	$PhOH:H_2O$
Spot	.37–.39	.66
Alanine	.27–.29	.59

Rf's after Total Hydrolysis

Spot	.31	.56 (tailing)
Alanine	.31	.58

Table 5.2—Analysis of copoly (alanine, glutamic acid)

	Anal. by DNP method	Anal. by analyzer
alanine	83%	79%
glutamic acid	17	21
recovery	—*	87

* Not accurately assessable, because of analytical step involving volatile solvent.

Individual amino acids have been systematically heated with aspartic acid (*12*); with glutamic acid (*7*); with aspartic acid and orthophosphoric acid; with aspartic acid and glutamic acid (*7*); and with aspartic acid, glutamic acid, and orthophosphoric acid (*12*). The propensity of an individual neutral amino acid to copolymerize tends to rank similarly in these systems.

When five amino acids are copolymerized simultaneously, there result polymers which include substantial proportions of the five amino acids (*19*).

Simultaneous copolymerization of all of the eighteen amino acids commonly found in protein give comparable results (*19*). Analyses of such polymers prepared by use of high proportions of aspartic acid and of lysine are shown in Table 5.3. Tryptophan is missing from the HCl hydrolyzates of the products, but is found in the alkaline hydrolyzates as one composition depicted in Table 5.3 demonstrates.

Because of the qualitative resemblance of the chromatograms of hydrolyzates of these two types of polymer (*cf. 19*) to a comparable one from casein, the synthetic thermal polymers are referred to as *proteinoids*, and the term is defined essentially in that manner. The type prepared with lysine is designated as a lysine proteinoid. The other is distinguished as an acid proteinoid. Inasmuch as a relatively wide variety of types can be prepared, similar appropriate designations are used in individual cases.

Table 5.3—Amino acid contents in two types of proteinoid, gm/100 gm polymer

Amino acid	Acid proteinoid*	Lysine proteinoid†
Alanine	3.0	1.8
Arginine	1.2	0.8
Aspartic acid	67.5	0.8
Cystine	0.1	0.9‡
Glutamic acid	10.7, 7.8	4.0
Glycine	2.0	1.6
Histidine	1.4	1.0
Isoleucine	1.0§	1.2§
Leucine	1.4	1.3
Lysine	2.4	37
Methionine	2.1	1.5
Phenylalanine	1.7	2.0
Proline	0.3	0.7
Serine	0.5	0.1
Threonine	0.1	0.1
Tyrosine	1.8	0.4
Tryptophan	1.0	—‖
Valine	0.9	1.0
Ammonia	—‖	1.0

* By microbial assay (*10*). Recovery 98 per cent, without correction for ash and moisture, or increase in weight due to hydrolysis.

† By amino-acid analyzer. Recovery 58 per cent without correction for ash and moisture, or increase in weight due to hydrolysis.

‡ Half-cystine is 1.8 per cent. Peak slightly displaced, as compared to standard.

§ Includes allo form.

‖ Not determined.

The acid proteinoids are found by infrared spectroscopy to have substantial proportions of imide linkage, first postulated by Kovacs and co-workers (*4*). The pertinent maxima in the acid proteinoid are at 1720 and 1780 cm^{-1}, as had been shown for polyaspartic acid (*5*). As with polyaspartic acid, this imide is entirely convertible to a peptide bond, as demonstrated by disappearance of the imide absorption when the polymer is gently treated with alkali (*5*). These shifts in spectrum were confirmed by elemental analysis of the two types of polyaspartic acid. The acid proteinoid preparations are accordingly treated with sodium bicarbonate prior to dialysis. Virtually all of the polymer dissolves as the sodium salt in the case of most such preparations. This infrared evidence also rules out the possibility of a structure in which much of the neutral amino acid is coupled in a unitary relationship with side chains of aspartic acid in a polyaspartic acid core, inasmuch as the imide type structure does not allow such aspartic-acid side chains

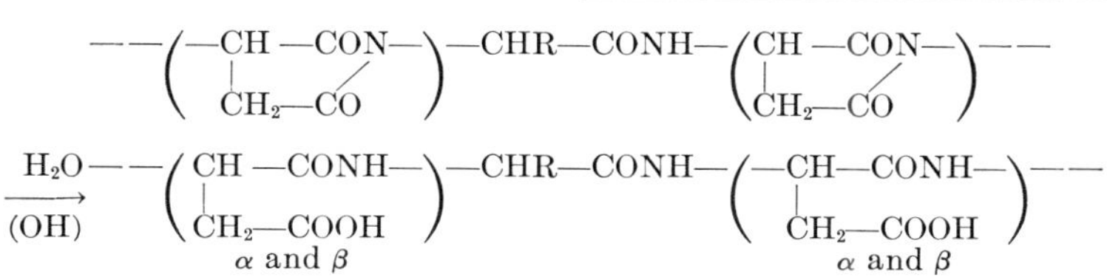

The infrared spectrograms also show absorption assignable to peptide linkages. On opening of the imide rings by treatment with sodium bicarbonate, the neutral and basic amino acids remain in the polymer. This evidence is consistent with the finding of -alanylalanyl-sequences in partial hydrolyzates of the simpler copolymers of alanine, as described earlier.

At about the time the chromatographic similarity of hydrolyzates of proteinoids to those of proteins was first observed, it was possible to compare many other properties of these two classes of material. Further testing of this relationship led to an initial detailed comparison (*10*). The fuller state of knowledge at this time is here reviewed. Data on lysine proteinoid are also first presented here.

Qualitative composition.—Seventeen of the eighteen amino acids commonly found in protein are found in HCl hydrolyzates of acid proteinoid and of lysine proteinoid. Tryptophan is found in alkaline hydrolyzates of acid proteinoid. A few ninhydrin-active spots representing traces of compounds that are also recovered in the hydrolyzates have not yet been identified.

No difficulty has been experienced in incorporating into the polymers an amino acid, *p*-fluorophenylaline, which can be accommodated by the proteosynthetic mechanism of *Lactobacillus arabinosus* (*20*).

Quantitative recovery of amino acids has not yet been achieved from any of the proteinoids. Various analyses leave 15 per cent to 40 per cent of the polymer unaccounted for, after allowance for moisture and ash.

In a series of acid proteinoids analyzed by Dr. Kenneth Woods, fourteen of the amino acids were found to be present in proportions which fall within the ranges of the contents of these amino acids in proteins (*21*). The proportions of serine and threonine are in most cases small, not rising above a few tenths of 1 per cent. In the acid proteinoids, the aspartic acid content is uncommonly high, the lowest value recorded so far being 33 per cent of the total amino acids recovered in the analysis. The aspartic acid content in lysine proteinoid is in a representative natural range, but in these cases the lysine content is high, often as much as 50 per cent.

The relationship has been studied most intensively in the case of histidine as presented in Table 5.4. These figures indicate the degree of reproducibility of four sets of syntheses by the criterion of histidine content. The table also shows that the content in the product is controlled by the content in the reaction mixture.

Non-randomness.—Earlier studies indicate a sufficient disparity between N-terminal and total compositional analyses, that some non-randomness of arrangement of residues is inferred (*10*). The relative values used were obtained by a dinitrophenylation technique which has been shown on some amino acids to be closely equivalent to results from the amino acid analyzer or by microbial assay (see section on methods).

Table 5.4—Histidine contents of several series of proteinoids

Histidine in reaction mixture (%)	Histidine in polymer gm/100 gm polymer			
	Series No.			
	I	II	III	IV
1.3	0.6	0.6	0.5	0.4
2.7	0.9	0.9	1.1	1.1
4.8	1.8	2.0	1.7	—
8.0	3.2	2.8	2.6	3.6
13.5	—	4.6	3.7	5.4

Color tests.—The classical color tests for protein (biuret, xanthoproteic, Hopkins-Cole, and Salkowski) are positive on proteinoids. The acid proteinoids react slowly, or in some cases not at all, in tests with warm ninhydrin. In this respect the lysine proteinoids, however, behave fully like protein.

Inclusion of metal ions.—As a first approach to study of the inclusion of prosthetic groups in proteinoids, metal salts have been included in reaction mixtures. An iron proteinoid, made in the presence of ferrous sulfate, was found to contain 0.06 per cent of Fe by analysis, with 0.07 per cent after ashing.

Infrared absorption spectra.—The maxima found in the acid proteinoid and in the lysine proteinoid are given in Table 5.5.

Table 5.5—Infrared absorption maxima of acid proteinoids

Maximum, cm^{-1}		
Acid proteinoid	Lysine proteinoid	Group indicated
3300	3300	(-NH-)
3080	3080	(-NH-)
1650	1650	Amide I
1550	1530	Amide II*
1720†	—	Imide
1780†	—	Imide

* Due to carbonyl radicals not found in diketopiperazines.

† These disappear when the proteinoid is subjected to alkaline treatment.

Precipitability.—The acid proteinoids have been precipitated from aqueous solutions by trichloroacetic acid, picric acid, phosphotungstic acid, phosphomolybdic acid, etc., in studies by Krampitz (*22*).

Solubility.—The solubility of the preparations varies with the conditions of synthesis. Typically, a small fraction of an acid proteinoid, following purification, is directly soluble in water. The remainder is entirely soluble in dilute sodium bicarbonate solution or phosphate buffer.

Fractions of the lysine proteinoids tend to be resinous, and soluble fractions are obtained by dissolving in mineral acid solution. Yields of the water-soluble fraction of lysine-proteinoid are typically about 10 per cent.

Response to concentrations of salt.—As the data on solubility indicate, those proteinoids which are not already soluble in water may be salted in by appropriate concentrations of mineral electrolytes. The acid and lysine proteinoids are also salted out by increased concentrations of ammonium sulfate and can be repurified by this reagent.

Mean molecular weight.—End-group assay has indicated values in the range of 3000–9000 for mean molecular weight of the acid proteinoids.

Values for lysine proteinoid and other thermal lysine polymers are higher, above 25,000 (*23*), by the approach to sedimentation-equilibrium method in the ultracentrifuge.

Retention by membranes.—The molecular weight is sufficiently high that membranes such as Visking retain the preparations. It is possible to apply the Craig escape time technique for estimation of molecular weight (*24*).

Formation of films and microspheres.—Lysine-proteinoid forms films readily. Acid proteinoids form microspherical units in the range of 1.0–3.5 μ in diameter (*25*).

Isoelectric point.—These are typically 2.8 for acid proteinoid and 8.4 for lysine proteinoid.

Proteolyzabily.—This property (*10*) has particularly been studied by Krampitz (*22*).

Nutritive quality.—This property has been reported for *Lactobacillus arabinosus* (*10*).

Catalytic activity.—The catalytic activity for the hydrolysis of *p*-nitrophenyl acetate of two of the four series of polymers reported in Table 5.4 is recorded in Fig. 5.1. It may be seen that each polymer is more active than the amount of histidine contained (*25a*). In two instances, the polymers were hydrolyzed and found to be more active than the hydrolyzates. Three proteinoids containing no histidine had catalytic activity, which other experiments

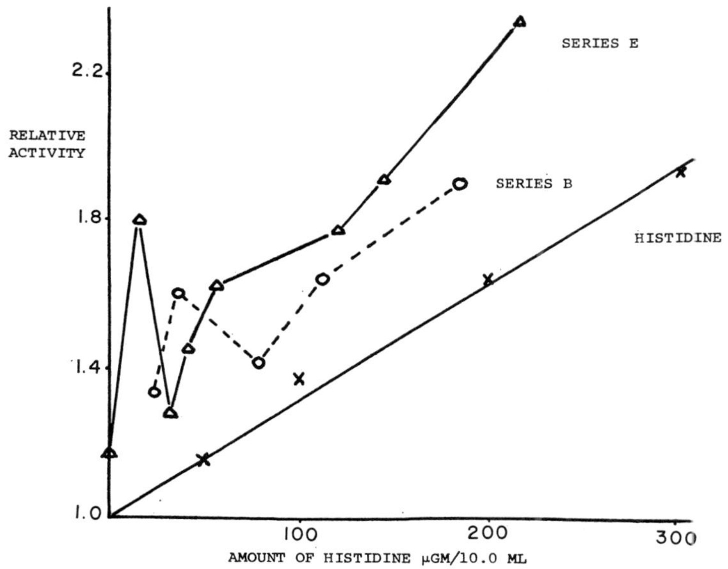

Fig. 5.1. Activity of typical proteinoids related to their histidine content. The abscissa of the lower curve represents the concentration of free histidine while those of the two uppermost curves represent the equivalent amount of histidine in polymeric form. Zero order rate constants (K) were obtained from the slopes of progress curves; relative activity is the ratio: K catalyst/K autohydrolysis. Each proteinoid was purified and employed as the sodium salt at a concentration of 4.0 mg per 10.0 ml of solution, *p*-nitrophenyl acetate at 2×10^{-3} M, in 0.067 M phosphate buffer containing 2 per cent acetone, *p*H 6.2, 31°–32° C. Reactions were followed by measuring the appearance of *p*-nitrophenol at 400 mμ in a Bausch and Lomb Spectronic 20 Colorimeter.

suggest is ascribable to other basic amino acids. In each of the four series a peak of activity relative to histidine content may be observed at about 1 per cent. The most active polymer (not presented in the figure) was found in this region in the first series, but no subsequently prepared polymer has more activity. This most active polymer has been reassayed over a period of two years with consistent results.

Although heterogeneity is suggested by N-terminal analyses (*10*) this property does not differentiate proteinoids from proteins in the natural state. Whether proteinoids can be fractionated to homogeneous preparations as easily as are proteins has not been investigated.

Among the discerned differences between protein and proteinoid, by way of summary, no proteinoid has yet been found to be 100 per cent hydrolyzable to amino acids. The possibilities of non-amino acid fractions and of difficultly hydrolyzable linkages are being investigated. The serine and threonine contents are low, and the aspartic acid or lysine content is high. The acid proteinoid is negative in warm ninhydrin tests, but the lysine proteinoid is positive.

Properties which have been less completely studied than those above are optical activity of residues within the polymer, nativity, helicity, antigenicity, branching in the molecules (*10*), and other kinds of participation in biological phenomena.

Partially optically active (about 50 per cent) glutamic acid has, however, been recovered from hydrolyzates of polymers (*12*), racemic aspartic acid has similarly been recovered (*12*), and an optically active polymer has been obtained from reaction of L-proline with DL-aspartic acid, glycine, DL-threonine, sodium dihydrogen phosphite, and "superphosphoric acid" at 85° for 5 days (*26*).

Assessment of α- and ϵ-linkages of lysine residues in lysine copolymers indicates distribution between both types (*12*).

Methods

Many of the methods used have been described earlier (*10*).

Conditions for the production of amino-acid polymers have been described earlier, with the exception of the polymer including L-proline, DL-aspartic acid, DL-threonine, and glycine, for which the conditions are given in the text. Copolymerization with free lysine was typically carried out by heating at a temperature in the range of 180° to 200° for 2–6 hours, dialyzing for several days against water, and lyophilizing the nondiffusate. For the lysine proteinoid, the reaction mixture contained 0.25 gm of an equimolar mixture of 16 amino acids (*10*), 0.25 gm of L-glutamic acid, 0.25 gm of DL-aspartic acid, and 3.0 gm of DL-lysine (Dow).

Analyses of simple copolymers were performed by a dinitrophenylation technique. The figures presented are absolute only as ratios, as in earlier papers (*8*, *9*), inasmuch as recovery values based on weight of original sample are not known. The aspartic-acid content of proteinoids determined in this way is, however, close to the value determined by microbial assay in a sample on which approximately 85 per cent recovery was registered (*10;* Tables I and VII). A similar concordance was found when copoly(alanine, glutamic acid) was analyzed by the DNP technique and on the Phoenix amino-acid analyzer (*27;* Table 5.2).

Partial hydrolysis of the alanine-lysine, alanine–glutamic acid, and alanine–aspartic acid copolymers was conducted at 37° with 12 N HCl:acetic acid (1:1) for 7 days.

Histidine was determined by the method of Macpherson (*28*). Iron was assayed by the thiocyanate method of Sandell (*29*).

Infrared analyses were conducted as before (*10*) and partly evaluated by reference to Bellamy (*30*).

Summary

Alpha-amino acids can be copolymerized by the seemingly destructive process of heating, if the mixture is dry and if minimum proportions of acidic or basic amino acids are present. These principles have been applied to including in peptide linkage all of the amino acids commonly found in protein; the products have many of the properties of protein.

New information is presented on sequences in simple copolymers, on description of thermal polymers obtained with lysine as a dominant monomer, and on catalytic activity for *p*-nitrophenyl acetate in several series of thermal polymers containing seventeen or eighteen amino acids.

Acknowledgment

From the Chemistry Department and Oceanographic Institute of the Florida State University, Tallahassee. Aided by grant numbers NSG-105-61 of the National Aeronautics and Space Administration, C-3971 of the National Institutes of Health, U. S. Public Health Service, G-4566 of the National Science Foundation, and grants from the General Foods Corporation. Part of the data on histidine and catalytic activity are from the M.S. thesis of Duane L. Rohlfing, 1960. Contribution No. 165 of the Oceanographic Institute.

References

1 *Cf.* S. W. Fox, K. Harada, and A. Vegotsky, *Experientia* **15,** 81 (1959), refs. 4–8.
2 A. B. Meggy, *J. Chem. Soc.* **1956,** 1444.
3 H. Schiff, *Chem. Ber.* **30,** 2449 (1897).
4 J. Kovacs, I. Koenyves, and A. Pusztai, *Experientia* **9,** 459 (1953).
5 A. Vegotsky, K. Harada, and S. W. Fox, *J. Am. Chem. Soc.* **80,** 3361 (1958).
6 S. W. Fox and M. Middlebrook, *Federation Proc.* **13,** 211 (1954).
7 K. Harada and S. W. Fox, *J. Am. Chem. Soc.* **80,** 2694 (1958).
8 K. Harada and S. W. Fox, *Arch. Biochem. Biophys.* **86,** 274 (1960).
9 S. W. Fox and K. Harada, *ibid.,* **86,** 281 (1960).
10 S. W. Fox and K. Harada, *J. Am. Chem. Soc.* **82,** 3745 (1960).
11 K. Harada, *Bull. Chem. Soc. Japan* **32,** 1007 (1959).
12 K. Harada and S. W. Fox, unpublished experiments.
13 S. W. Fox, *Science* **132,** 200 (1960).
14 M. H. Briggs, *Science and Culture* **26,** 160 (1960).
15 H. A. Barker, *J. Gen. Physiol.* **17,** 21 (1933).
16 R. L. Altman and S. W. Benson, *J. Am. Chem. Soc.* **82,** 3856 (1960).
17 C. Neuberg and I. Mandl, in J. B. Sumner and K. Myrbäck, "The Enzymes," Academic Press, Inc., N. Y., 1950, Vol. I, Part 1, p. 536.
18 G. D. Maier, M.S. thesis, Iowa State University (1956).
19 S. W. Fox and K. Harada, *Science* **128,** 1214 (1958).
20 R. S. Baker, J. E. Johnson, and S. W. Fox, *Biochim. et Biophys. Acta* **28,** 318 (1958).
21 K. Woods, K. Harada, A. Vegotsky, and S. W. Fox, unpublished analyses.
22 G. Krampitz, *Naturwissenschaften* **46,** 558 (1959).
23 C. Genaux and S. W. Fox, unpublished experiments.
24 L. C. Craig, T. P. King, and A. Stracher, *J. Am. Chem. Soc.* **79,** 3729 (1957).
25 S. W. Fox, K. Harada, and J. Kendrick, *Science* **129,** 1221 (1959).
25a M. L. Bender, *Chem. Revs.* **60,** 78 (1960); E. Katchalski *et al., Arch. Biochem. Biophys.* **88,** 361 (1960).
26 P. D. Hoagland and S. W. Fox, unpublished experiments.
27 D. H. Spackman, W. H. Stein, and S. Moore, *Anal. Chem.* **30,** 1190 (1958); K. R. Woods and R. L. Engle, Jr., *Ann. N. Y. Acad. Sci.* **87,** 764 (1960).
28 H. T. Macpherson, *Biochem. J.* **36,** 59 (1942).
29 E. B. Sandell, "Colorimetric Determination of Traces of Metals," Interscience Publishers, New York, 1959, 3rd ed., p. 553.
30 L. J. Bellamy, "The Infra-red Spectra of Complex Molecules," Methuen and Co., Ltd., London, 1958, pp. 223–30.

Discussion of Paper 5: Thermal Copolymerization of Amino Acids

J. KOVACS

According to our experience anhydropolyaspartic acid reacts very easily with amino compounds. Therefore, it is a possibility that during the formation of acidic proteinoids aspartic acid will polymerize first and then the amino groups of other amino acids will open the imide rings. This would give branched polypeptides. The ratio of aspartic acid to other amino acids in the acidic proteinoids is high. Therefore, the presence of imide rings (IR evidence) does not rule out the above-mentioned structure, where an appreciable amount of the basic and neutral amino acids would be in the side chain. Quantitative determination of the imide rings would give a solution. Have you considered making such determinations?

S. W. FOX

Although the accumulated evidence (particularly from partial hydrolyzates, and from IR, etc.) indicates a structure including neutral amino acids in adjacent positions, such evidence certainly does not rule out appreciable proportions of single nonacidic amino acids attached to side chains of dicarboxylic amino-acid residues in proteinoids. The proportion of this type of postulated peripheral structure must also be minor in the glutamic acid–alanine copolymer. Also, Dr. Kovacs' published experience with amino compounds and anhydropolyaspartic acid is under alkaline conditions, involving, for example, excess aniline, whereas we are discussing polymers prepared under acidic conditions. Dr. Kovacs has shown, as have we, that the imide structure is relatively quite stable in acidic and unstable in alkaline media.

Determination of quantitative proportions of imide linkage is indeed desirable, and we are testing various methods.

We agree with the comments on the need for systematic studies of polymers containing as many as eighteen amino acids. In fact, much systematic study of polymers from two, three, and five amino acids is already recorded in our paper and in others cited in the bibliography (refs. *7–11;* also *20*). It is also possible to be systematic at either simple or complex levels. Studies of both types are continuing in the belief that work at each level aids the other. By copolymerizing two amino acids thermally, new vistas for research have appeared, even though the statements in the literature ruled out thermal polymerization of amino acids in general. With eighteen or seventeen amino acids in the same panpolymer, there has resulted such new information as:

1. The combined amino acids are recoverable to a major degree.
2. The panpolymers are generally more active catalytically than polymers containing few amino acids.
3. Amino acids other than histidine contribute to the catalytic activity of the panpolymers (Fig. 5.1, Series E).
4. The catalytically active panpolymers are thermally inactivatable *in aqueous solution.* One suggested interpretation is that subtle intramolecular structures exist in such complex polymers.
5. Nutritional utilizability of proteinoids by organisms having multiple amino-acid requirements could not have been demonstrated with polymer preparations that are much less heterocompositional.
6. From such studies has emerged the realization that materials of a degree of complexity comparable to that of proteins could have arisen by simple processes on the primitive earth.

If the interest in poly-α-amino acids is truly in their employment as protein models, I believe we are obliged to focus attention upon those which compositionally most closely resemble proteins, now that it has been shown that such are obtainable in a reproducible way by polymerization of amino acids.

In response to a question from Dr. Sela, our paper includes studies of sequence in partial hydrolyzates of simple polymers, comparable studies on proteinoids are under way, and we have published comparisons of total and N-terminal compositions in proteinoids and simple polyamino acids (*J. Am. Chem. Soc.* **82,** 3745 [1960]).

In response to questions from Drs. Kovacs and Durrum, thermal copolymerization has occurred at temperatures as low as 70° by adding to the dry amino acids proportions of polyphosphoric acid.

PAPER 6

On the Copolymerization of Amino Acids and the Proteolyzability of Proteinoids

GOTTFRIED KRAMPITZ

Heating of amino acids results ordinarily in products such as diketopiperazines, tars, and other pyrolytic matter (*1*). In the presence of large proportions of aspartic acid or glutamic acid it is possible to prepare copolymeric peptides thermally from amino-acid mixtures (*2*). In such experiments anhydropolymers can be obtained consisting of all of the eighteen amino acids usually present in proteins. The properties of those preparations are similar to those of protein (*3*).

The conditions employed are a temperature range of 160° to 210° (140° to 210° in the presence of orthophosphoric acid) or temperatures below 100° if polyphosphoric acid is employed (*4*). After hydrolysis of proteinoids (polymers containing all common amino acids) by hydrochloric acid, usually the same amino-acid pattern can be observed as in casein (*2*).

A main question is the susceptibility of proteinoids to proteolysis. Fox reported recently that proteinoids replace peptones in the broth cultures of *Lactobacillus arabinosus*, a fact which indicates that the linkages are susceptible to the attack of bacterial proteases (*2*). Furthermore, Fox has reported that proteinoids could be attacked by proteases other than bacterial (*3, 5*). In this laboratory it could be observed that pepsin, trypsin, and papain can hydrolyze synthetic proteinoids (*6*). This report deals with the action of proteolytic enzymes on proteinoids.

Methods

Reactants.—The proportions of all sixteen amino acids (Table 6.1) exclusive of the dicarboxylic amino acids were the same in all experiments. Sample materials were intimately ground to a fine powder. The mixture A-1 consisted of 2 parts of aspartic acid, 2 parts of glutamic acid, and 1 part of the mixture of neutral and basic amino acids. The mixture A-2 consisted of 4 parts of aspartic acid, 4 parts of glutamic acid, and 1 part of the neutral and basic amino acids.

These described mixtures of the types A-1 or A-2 were heated in closed vessels under an atmosphere of nitrogen at a temperature of 170° for three hours (*3*). The molten mixture showed an amber color. After the thermal procedure the vitreous material was rubbed vigorously with water, and after it had stood overnight the solid was separated by centrifugation and washed with water, cold methanol, and water again. Following this the preparations were dissolved in sodium bicarbonate solution and dialyzed in cellulose tubings (Visking 27/32 inch) in an agitated water bath for four days.

Tests of protein properties.—Precipitation could be achieved with the following reagents:

Table 6.1

1.00 mol each of:	
DL-alanine*	L(+) arginine†
DL-isoleucine*	L(+) cystine*
DL-threonine*	L(−) histidine†
DL-valine*	L− proline‡
DL-methionine*	L− leucine‡
DL-phenylalanine*	L− tyrosine‡
DL-serine*	glycine†
DL-tryptophan*	lysine-HCl†

These samples were obtained from:
* Degussa, Frankfurt (Main)
† Pharmacell, Aschaffenburger Zellstoffwerke, Raubling
‡ Merck, Darmstadt

We are greatly indebted to Dr. A. Becher, Degussa, Frankfurt (Main) for samples of amino acids.

trichloracetic acid	phosphotungstic acid
picric acid	phosphomolybdic acid
sulfosalicylic acid	sublimate solution
perchloric acid	zinc hydroxide
acetic acid	hot methanol

Color tests which are positive for protein include the biuret, xanthoproteic, Hopkins-Cole, Salkowski, and a test with azocarmin B. These were performed with aliquots of each sample of the polymers.

Amino-acid assay of proteinoid.—Qualitative amino-acid composition was checked by paper chromatography (*7*). The quantitative amino-acid composition was determined by the Moore and Stein ion-exchange technique using automatic recording equipment (*8, 9, 10*). The excess of dicarboxylic amino acids was removed from the proteinoid hydrolysates by Dowex 50×4 (*11*); these amino acids were determined separately; otherwise the high concentration of both amino acids would harm the Teflon coil of the amino-acid analyzer.

Determination of N-terminal amino acids.—The determination of N-terminal amino acids was performed by addition of an excess of 4 per cent alcoholic 2,4-dinitrofluorobenzene solution of aliquots of polymers in sodium bicarbonate solution. After three hours standing the excess of 2,4-dinitrofluorobenzene was removed with diethyl ether. The insoluble residue was washed until no yellow color could be extracted. The subsequent fission with formic acid (90 per cent), anhydrous acetic acid, perchloric acid (60 per cent; 90:5.5:60) was performed overnight in a water bath under reflux. The DNP-amino acids insoluble in water were extracted with ether. The residue was taken to dryness in a rotary evaporator. The chromatograms ran on Ederol in the systems: toluene, ethylenechlorhydrin, pyridine, 0.8 N ammonium hydroxide (5:3:1.5:3) and in 1 M NaH_2PO_4+0.5 M Na_2HPO_4 (1.5 M phosphate buffer solution) (*7, 12*).

Proteolyzability.—For the enzymic splitting of the proteinoids we employed the following method. Purified enzymes were obtained from Merck, Darmstadt, and Schwab, Munich. The enzyme preparations were dissolved as described below:

I. Papain

a) 200 mg papain, 100 mg Na_2S dissolved in 100 ml Na citrate-phosphate buffer *p*H 5

Composition of Na citrate-phosphate buffer solution:
48.5 ml 0.1 M citric-acid solution
51.5 ml 0.2 M Na_2HPO_4 solution

b) 200 mg papain, 100 mg Na_2S dissolved in 100 ml Na citrate-phosphate buffer *p*H 7.5

Composition of Na citrate-phosphate buffer *p*H 7.5:
9.2 ml 0.1 M citric-acid solution
90.8 ml 0.2 M Na_2HPO_4 solution

II. Pepsin

200 mg pepsin in 100 ml Na citrate buffer *p*H 2.2

This buffer solution was prepared according to Bailey, Moore, and Stein (*13*):
105 gm citric acid
42 gm NaOH
80 ml conc. HCl were made up to a volume of 5 liters.

III. Trypsin

200 mg trypsin in 100 ml Na phosphate buffer *p*H 7.0

This buffer was prepared according to Hirs, Moore, and Stein (*14*):
8.28 gm NaH_2PO_4
19.88 gm Na_2HPO_4 were made up to a volume of 1 liter.

Other proteolytic enzymes such as peptidase or erepsin were not used in this study because of their high autolysis rate. Two hundred mg of purified (dialyzed) polymer was dissolved in the buffer solutions according to the enzymes in question. Twenty ml of the enzyme solution was added. Enzymic hydrolysis was performed at 38°. Controls consisting of polymers dissolved in the appropriate buffer solution but not containing enzymes were employed. No appreciable autodigestion of the polymers could be observed in the blanks.

Aliquots of 0.5 ml of the fermentation samples were taken at the beginning, after 30 minutes, after one hour, and after two hours of fermentation. To each aliquot one drop of trichloracetic acid (30 per cent) was added for inactivation of the enzymes.

The course of the enzymic hydrolysis was controlled by means of the ninhydrin reaction. The following ninhydrin reagent was employed because of the phosphate-containing buffer solutions:

Solution 1 1 gm ascorbic acid
1 gm ninhydrin dissolved in 160 ml

Methyl Cellosolve
Solution 2 20 gm ninhydrin dissolved in 1 liter Methyl Cellosolve
Solution 3 4 N Na acetate buffer, *p*H 5.5 (*15*)

Immediately before use these three solutions were mixed in the stated sequence employing the volume ratio 1:2:1. One ml of the final solution was sufficient for the maximum color development. The samples were shaken vigorously and heated in a boiling water bath for 15 minutes. After cooling under running tap water, the samples were diluted with 15 ml of a mixture of propanol and water (1:1). Each sample was shaken for 30 seconds. The optical density was measured in a Beckman spectrophotometer B (wave length 570 mμ) (*16*).

Hydrolysis of proteinoid.—Hydrolysis of proteinoid was performed according to Dustin *et al.* (*18*).

Measurements of radioactivity.—Radioactive hydrolysates of proteinoid were chromatographed on Dowex 50×12 columns (*8*). The radioactivity was measured by a plastic scintillator spiral flow cell (*19*).

Experimental and results

Series A.—All the preparations of proteinoid series A-1 (proportions given in preceding section on methods) could be precipitated by the stated reagents. All color reactions were protein-positive. On paper chromatograms all amino acids except tryptophan could be detected after hydrolysis with HCl. The destruction of tryptophan by HCl hydrolysis is well known.

The amino-acid composition of one preparation of proteinoid series A-1 is found in Table 6.2. The results are in good agreement with microbial assays described by Fox and coworkers (*3*). The total of all amino acids found after hydrolysis is about 100 per cent without correction for ash and moisture or increase in weight due to hydrolysis.

The N-terminal amino acids in series A-1 were aspartic acid, glutamic acid, and traces of neutral and basic amino acids. The quantitative results of the enzymic hydrolysis showed some variation between experiments.

The amber melt of series A-2 was purified as stated above. When these preparations were submitted to the above tests, all samples reacted positively and they could be precipitated by the reagents described.

Table 6.2—Amino-acid composition of a proteinoid series A-1*

Threonine	0.1	Leucine	1.2
Serine	0.6	Tyrosine	1.4
Proline	0.2	Phenylalanine	1.3
Glycine	1.5	Lysine	2.8
Alanine	3.2	Histidine	1.8
Cystine†	0.2	Arginine	0.8
Valine	1.2	Tryptophan‡	1.1
Methionine	3.0	Glutamic acid	11.5
Isoleucine	0.8	Aspartic acid	70.3

* The values are expressed as grams of amino acid per 100 grams proteinoid without correction for ash and moisture, or increase in weight due to hydrolysis.

† Determination of cystine as cysteic acid.

‡ Determination of tryptophan after alkaline hydrolysis.

The same amino-acid pattern after HCl hydrolysis was observed as for A-1 and the N-terminal amino acids were the same. In the quantitative composition of proteinoids from series A-2 were found some differences from series A-1. In Table 6.3 the amino-acid composition of preparations of proteinoids series A-2 is demonstrated.

The quantitative results of the hydrolysis by proteolytic enzymes were variable, although some preparations were oxidized by performic acid in order to improve the proteolyzability according to the findings of Hirs (*17*) that ribonuclease was more susceptible to trypsin after treatment with performic acid. The reasons for those irregular results could be due to heterogenous mixing of the amino acids or to the changing distribution of the components.

Series B.—Therefore the reaction mixtures were dissolved in 5 per cent H_2SO_3. Subsequently the solvent was evaporated and the residue was used for synthesis. This type of

Table 6.3—Amino-acid composition of proteinoid series A-2*

Threonine	0.1	Isoleucine	0.9
Serine	0.4	Leucine	1.1
Proline	0.3	Tyrosine	1.0
Glycine	1.0	Phenylalanine	1.1
Alanine	2.8	Lysine	2.4
Cystine†	0.1	Histidine	1.2
Valine	1.4	Arginine	0.8
Methionine	2.6	Tryptophan‡	0.8

* The values are expressed as grams of amino acid per 100 grams proteinoid without correction for ash and moisture, or increase in weight due to hydrolysis.

† Determination of cystine as cysteic acid.

‡ Determination of tryptophan after alkaline hydrolysis

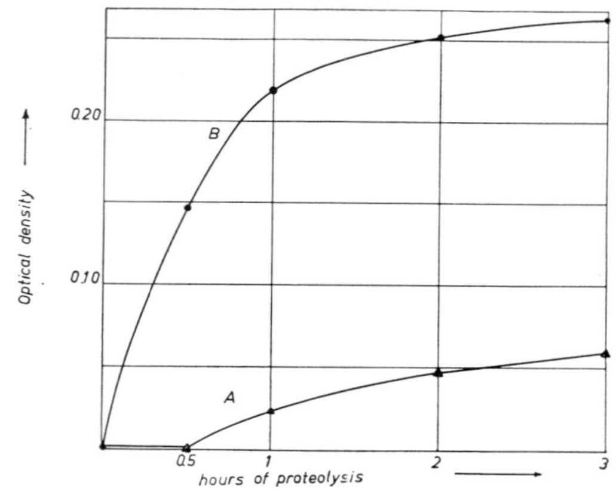

Fig. 6.1. Tryptic hydrolysis of proteinoid series D. *Curve A:* nonoxidized preparation. *Curve B:* performic acid oxidized sample.

preparation was marked as series B. The weight ratio was the same as stated for A-2.

The reaction products after the heating period looked light yellow instead of amber in series A-1 and A-2. The color and precipitation tests were positive. The amino-acid composition did not differ remarkably from these of the preparations of series A-2. The N-terminal amino acids were the same as in A-1.

To the contrary absolutely no enzymic hydrolysis could be observed in the products of series B. We suppose that either few or no peptide bonds were formed or that in this type of synthesis secondary structures are produced which mask all such bonds.

Series C.—In the case that only few peptide linkages had been formed in series B we hoped that more purified preparations would result in better enzymic hydrolysis. We expected better results by precipitating dialyzed proteinoids. Precipitations were carried out with the reagents described in the preceding section on methods. The precipitated material was marked as series C, but none of the precipitation reagents had a positive effect on enzymic hydrolysis. No enzymic hydrolysis could be observed even when this material was oxidized by performic acid.

Series D.—Weight ratio, synthesis, and purification were the same as described for series B. The polymers were not precipitated as in series C, but an additional step was introduced in this procedure by treating the proteinoids (oxidized or not oxidized) with diluted HCl. In a typical experiment 500 mg of proteinoid were placed in 20 ml of 0.1 N HCl. Other HCl concentrations and time intervals of 6 and 24 hours were employed. After that the solids were washed with water until the material was free of chloride. Subsequently these preparations were carefully dried over P_2O_5 and subjected to enzymic hydrolysis. This was carried out with the enzymes and methods described in the preceding section on methods.

A regular and reproducible attack of proteinoids by all enzymes employed was observed.

Figure 6.1 indicates that performic-acid treatment results in a higher hydrolysis rate as evidently peptide bonds in this proteinoid become more susceptible to the action of trypsin.

In further experiments it was checked at which HCl concentration and at which time of

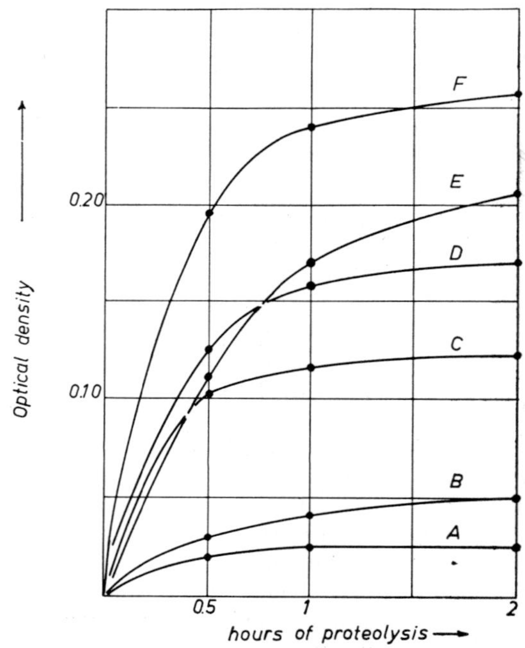

Fig. 6.2. Influence of HCl concentration and pretreatment time on tryptic hydrolysis rate of oxidized amino-acid copolymers series D.

	HCl concentration	*Time of HCl treatment*
Curve A	1.0 N HCl	6 hours
Curve B	0.5 N HCl	6 hours
Curve C	1.0 N HCl	24 hours
Curve D	0.5 N HCl	24 hours
Curve E	0.1 N HCl	6 hours
Curve F	0.1 N HCl	24 hours

treatment with HCl the best results could be obtained in enzymic hydrolysis rate. The most satisfactory conditions found were those in which the proteinoids were treated with 0.1 N HCl for 24 hours at room temperature (Figure 6.2).

Additional experiments were performed in testing hydrolysis rate of proteinoids employing pepsin (Figure 6.3).

The same tendency can be observed here as was demonstrated for the tryptic hydrolysis. Evidently the proteinoids were made more susceptible to enzyme action by performic-acid pretreatment. The rate of hydrolysis of oxidized samples is higher than for nonoxidized products. Oxidized samples could not be dissolved in the buffer solution *p*H 7.50; therefore proteolysis by papain could not be performed. In the case of papain hydrolysis of proteinoids series D (Figure 6.4) the assumption is made that performic-acid treatment of proteinoids in some way renders certain peptide linkages susceptible to the action of the proteolytic enzymes employed in these experiments.

Incorporation of radioactivity.—An attempt was made to incorporate labeled amino acids into proteinoids using information from the experiments described above. *Torula utilis* was inoculated with ^{35}S-sulfate and hydrolyzed in sealed tubes under nitrogen to avoid oxidation of thioamino acids. Four parts of aspartic acid, four parts of glutamic acid, and one part of carefully dried yeast hydrolysate were dissolved in 5 per cent H_2SO_3. After solution of the components the solvent was evaporated as described before and the residue was heated under a nitrogen atmosphere. After cooling, the vitreous mass was treated as in series D. The purified preparation was hydrolyzed with 6 N HCl after pretreatment with diluted hydrochloric acid.

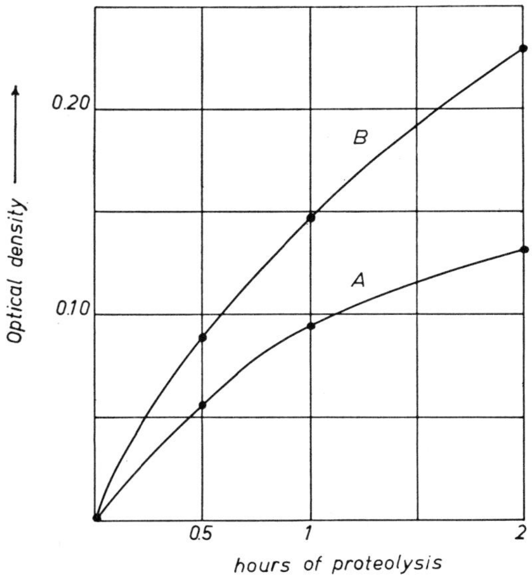

Fig. 6.3. Hydrolysis of proteinoid series D. by pepsin. *Curve A:* nonoxidized sample. *Curve B:* performic acid oxidized preparation.

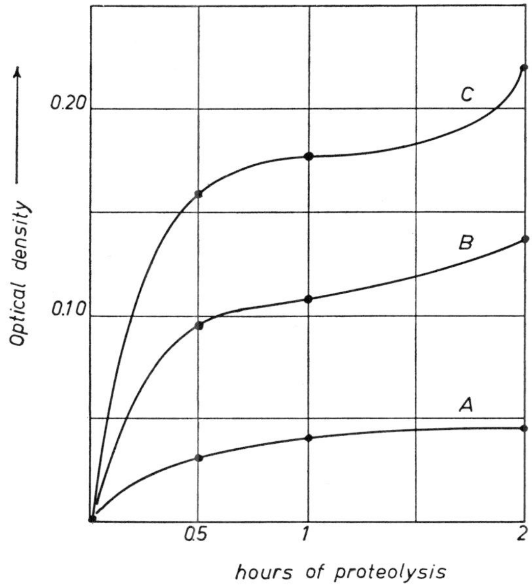

Fig. 6.4. Hydrolysis of proteinoid series D by papain. *Curve A:* nonoxidized sample; hydrolysis at *p*H 5.00. *Curve B:* performic acid oxidized sample; hydrolysis at *p*H 5.00. *Curve C:* nonoxidized sample; hydrolysis at *p*H 7.50.

The diagram of the chromatographed sample showed radioactivity in the position of the cysteic acid (sulfate?), traces of activity in the position of cystine, and more activity in the position of methionine. Another radioactive component was eluted by 0.2 N NaOH. This component was present in hydrolysates of *Torula utilis* also (*20*).

The latter results indicate that it is possible to copolymerize hydrolysates of yeast with dicarboxylic amino acids and that ^{35}-S-labeled thioamino acids can be found in hydrolysates of the synthesized proteinoid.

Table 6.4

2:2:1 mixture of amino acids
↓ 170°, 3 hours
crude polymer
↓ washed with water, cold methanol, water; dissolved in sodium bicarbonate solution; dialysed for four days
A-1 → enymic hydrolysis*

4:4:1 mixture of amino acids
↓ 170°, 3 hours
crude polymer
↓ washed with water, cold methanol, water; dissolved in sodium bicarbonate solution; dialysed for four days
A-2 → enzymic hydrolysis*

4:4:1 mixture of amino acids dissolved in 5% H_2SO_3
↓
solution of 18 amino acids; evaporation of H_2SO_3
↓
homogenous solid mixture heated at 170°, 3 hours
↓
crude polymer
↓ washed with water, methanol, water; dissolved in sodium bicarbonate solution; dialysed for four days
polymer series B → enzymic hydrolysis†
↙ precipitated with trichloracetic acid etc. ↓ series C→enz. hydrol.‡ (↙ oxidized; ↘ not oxidized)
↘ treated with dilut. HCl, washed free of chloride ↓ series D→enz. hydrol.‡ (↙ oxidized; ↘ not oxidized)

* Results variable.
† No enzymic hydrolysis possible.
‡ Hydrolyzed by proteolytic enzymes.

Discussion

It was stated by Fox that broth cultures of *Lactobacillus arabinosus* showed a somewhat reduced rate of growth when proteinoids replace peptones (*2*). Thus, proteinoids are attacked by bacterial proteases. In further investigations Fox and co-workers (*3*) describe a proteolysis by pepsin and chymotrypsin. Compared with the proteolysis of casein or of performic acid-oxidized proteinoids the enzymic hydrolysis of proteinoids is low.

On the other hand, it was reported by Hirs (*17*) that the hydrolysis rate of ribonuclease was low, but when the protein was treated with performic acid a severalfold higher hydrolysis rate could be observed because of splitting of disulfide bridges.

We tried a procedure of rendering proteinoids more susceptible for proteolysis. In Table 6.4 the steps of synthesis and treatments are summarized.

An important feature of this method is the treatment of proteinoids with performic acid prior to enzymic hydrolysis. Treatment of proteins with performic acid leads to dissolving disulfide bridges along a peptide chain and between peptide chains. Perhaps splitting of disulfide bridges, if present, occurs in proteinoids too, although only very few can be present. Other linkages may also be cleaved. Possibly this is the reason that oxydized proteinoids show an improved proteolyzability.

A second reason for an improved proteolyzability of proteinoids series D is the treatment with diluted hydrochloric acid, as the results indicate. It is assumed that diluted HCl removes perhaps secondary structures which may mask peptide bonds attacked normally by proteolytic enzymes. This would be a plausible hypothesis, but it is not proved by the experiments reported.

Our object, however, was to assess the possibility of splitting proteinoids with proteolytic enzymes of the digestive tract. Perhaps proteinoids can be utilized by animals, but this presumes proteolyzability of these products.

In experiments A-1 to D we worked with mixtures of purified amino acids which were vis-à-vis in a molar ratio. No other material than amino acids was employed.

In the last experiments described hydrolysates of yeast were used for synthesis and such hydrolysates usually contain components other than amino acids. The individual neutral and basic amino acids were not brought to an equal ratio, but nevertheless copolymerization occurred. The results show that methionine and some cystine were truly fixed in the structure of the proteinoid.

Summary

By use of an excess of dicarboxylic amino acids, the dry eighteen amino acids usually present in proteins, as well as hydrolysates of *Torula utilis*, can be copolymerized at temperatures of 170°.

These products are slowly attacked by proteolytic enzymes. Dissolving the components in H_2SO_3 before thermal copolymerization and treatment of the proteinoids with performic acid and dilute HCl results in improved proteolyzability.

References

1 K. Harada and S. Fox, *Arch. Biochem. Biophys.* **86,** 274 (1960).
2 S. Fox, K. Harada, and A. Vegotsky, *Experientia* **15,** 81 (1959).
3 S. Fox and K. Harada, *J. Am. Chem. Soc.* **82,** 3745 (1960).
4 137th ACS National Meeting Biological Chemistry, *Chem. Eng. News* **38,** 68 (1960).
5 S. Fox, *Science* **132,** 200 (1960).
6 G. Krampitz, *Naturwissenschaften* **46,** 558 (1959).
7 F. Cramer, "Papierchromatographie," Verlag Chemie, Weinheim, 4th ed., 1954..
8 S. Moore, D. H. Spackman, and W. H. Stein, *Anal. Chem.* **30,** 1185 (1958).
9 D. H. Spackman, W. H. Stein, and S. Moore, *ibid.* **30,** 1190 (1958).
10 G. Krampitz, *Z. Tierphys. Tierern. Futtermkde.* **15,** 40 (1960).
11 C. H. W. Hirs, S. Moore, and W. H. Stein, *J. Am. Chem. Soc.* **76,** 6063 (1954).
12 A. L. Levy, *Nature* **174,** 126 (1954).
13 J. Bailey, S. Moore, and W. H. Stein, *J. Biol. Chem.* **221,** 143 (1956).
14 C. H. W. Hirs, S. Moore, and W. H. Stein, *ibid.* **219,** 623 (1956).
15 S. Moore and W. H. Stein. *ibid.* **211,** 907 (1954).
16 R. Müller and G. Krampitz, *Z. Tierphys. Tierern. Futtermkde.* **11,** 227 (1956).
17 C. H. W. Hirs, *J. Biol. Chem.* **219,** 611 (1956).
18 J. P. Dustin, C. Czajkowska, S. Moore, and E. J. Bigwood, *Anal. Chim. Acta,* **9,** 256 (1953).
19. F. Knappen, *Landwirtsch. Forsch.* (in press).
20. F. Knappen and G. Krampitz, *Nature* (in press).

Discussion of Paper 6: Copolymerization of Amino Acids

S. W. FOX

Although Drs. Bamford and Katchalski state that polymerization of the NCA's does not give reproducible products, the same situation does not apply to thermal copolymerization and we have seen fit to collect many such data.

For proteinoids, we have a study of replication of histidine content in Table 5.3 of our paper (Paper 5). Dr. Krampitz reports with proteinoid made and analyzed on the automatic machine in Bonn compositional figures very much like our own, as determined by microbial assay. Although the absolute values are subject to correction, regular trends in N-terminal and total composition are observed in Tables VI, VII, and VIII, and IX of proteinoids described in *J. Am. Chem. Soc.* **82,** 3745 (1960).

While the thermal copolymerization will not occur in the presence of much excess water, it is not seriously disturbed by traces of water as, is the NCA polymerization, except when conducted in properly buffered solvent. The reactions are rugged, yet respond to changes in conditions of reactions.

Relative to Dr. Katchalski's remarks on the difficulty of copolymerizing in single macromolecules the NCA's of glutamic acid and proline, we have made a similar observation. Although other amino acids copolymerize easily under thermal conditions with glutamic acid, proline is unusual in not doing so.

PART II

Kinetics and Mechanism of the Polymerization Reaction

PAPER 7

The Polymerization of α-Amino Acid N-Carbonic Anhydrides

C. H. BAMFORD and H. BLOCK

Although α-amino acid N-carbonic anhydrides (NCA's) are the monomers most extensively used for the synthesis of high-molecular-weight polypeptides, no comprehensive and generally accepted scheme for their polymerization has been advanced. A large variety of bases, including primary, secondary and tertiary amines, salts of weak acids, alkali and alkoxide ions, can initiate polymerization (*1, 2*). It has been the custom to divide initiators into bases with labile protons, such as primary and secondary amines, and aprotic bases, such as tertiary bases, and to consider separately the reactions of each type. Such a division raises two important questions: first, why do not primary and secondary bases follow, at least in part, the tertiary-base reaction path; and secondly, what evidence is there that, in fact, they do not. These considerations emphasize the need to consider the reactions of all types of initiators in a unified scheme. For some time we have been aware of the necessity for reappraising the experimental evidence relating to the mechanism of polymerization, and while collating the evidence we have been led to test experimentally predictions previously unsupported. Our reasoning and attempts to devise a unified scheme are given in the following pages.

The kinetic behavior during polymerization is not always wholly determined by purely chemical factors, such as those referred to above, and phenomena which appear to be essentially physical in their origin must also be considered. Among these are the chain effect (*3*) and the conformational effects first described by Doty, Blout and their colleagues (*4, 5, 6*), and we conclude the paper by a discussion of these topics.

The chemistry of the polymerization reactions

INITIATORS WITH LABILE PROTONS

The reaction of primary and secondary amines with a NCA is generally thought to involve attack on the 5-carbonyl group of the ring:

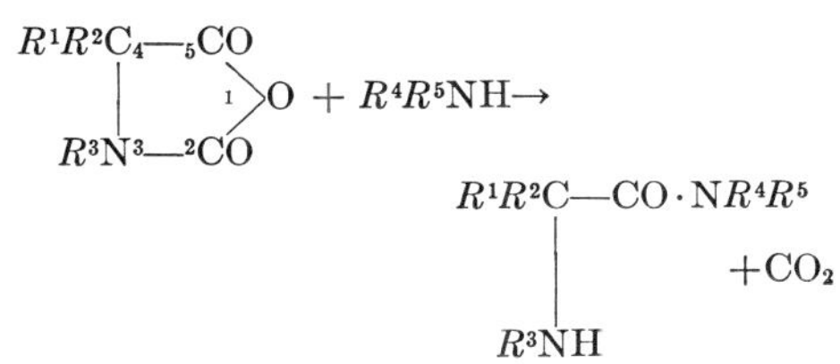

(1)

Repetition of (1), which is the "normal" propagation step, results in the formation of polymer. The occurrence of (1) is indicated by the production of substituted amino-acid amides in good yield when the NCA is treated with an excess of base. In many cases this reaction is undoubtedly the major one taking place during polymerization initiated by primary and secondary bases, as shown by the first-order dependence of the rate on the monomer and initiator concentrations (*7, 8*). If (1) were the exclusive reaction the number-average degree of polymerization $\overline{P}_n$ would be given by

$$\overline{P}_n = \frac{[M]_o}{[I]_o} \qquad (2)$$

where $[M]_o$, $[I]_o$ are the initial monomer and base concentrations respectively, and further, if the initiation and propagation rate constants were identical the molecular weight distribution would be very narrow and follow a Poisson law (*9a, b*). Fessler and Ogston (*10*) and

Pope, Weakley, and Williams (*11*) have shown that for polysarcosine ($R^1=R^2=H$, $R^3=CH_3$) the observed distribution is not inconsistent with that expected.

However, the simple relation (2) does not hold in all cases. Blout and Karlson (*12*), when studying the polymerization of γ-benzyl-L-glutamate NCA, found that the intrinsic viscosities of polymers prepared by *n*-hexylamine and diethylamine initiation were much less dependent on $[I]_o$, and higher than expected for degrees of polymerization given by (2). These findings resemble to some extent those with aprotic initiation (*13, 14*). We have recently demonstrated (*15*) that the characteristics of aprotic (tertiary base) polymerization are shown if the initiator is a secondary base which for steric reasons cannot add readily to 5-CO. Di-isopropylamine is such a base; the rate of polymerization of γ-ethyl-L-glutamate NCA initiated by this base in *N,N*-dimethylformamide solution at 25° C. is shown in Fig. 7.1. It will be seen that the reaction is much faster than the polymerization initiated by *n*-hexylamine. Further, the di-isopropylamine-initiated polymerization gives much higher polymers; the intrinsic viscosity (in dichloracetic :trifluoracetic acid mixture, 9:1) for the conditions shown in Fig. 7.1 was 6.12 b.mole^{-1}l. ($\overline{P}_n$ in the range 150–250), while for *n*-hexylamine initiation the corresponding value was 1.46 b.mole^{-1}l. ($\overline{P}_n$ in the range 60–100). From Fig. 7.1 it is clear that the high degrees of polymerization cannot be attributed to slow initiation by di-isopropylamine followed by a more rapid but normal propagation reaction. The polymerization of sarcosine NCA initiated by di-isopropylamine (Fig. 7.1) starts slowly and accelerates, indicating that the first step—the addition of the initiator to 5-CO—is relatively slow. As will be shown later, the tertiary base type of initiation does not occur with 3-substituted NCA's; clearly the rapid reaction in the case of γ-ethyl-L-glutamic acid NCA does not mainly involve addition to 5-CO.

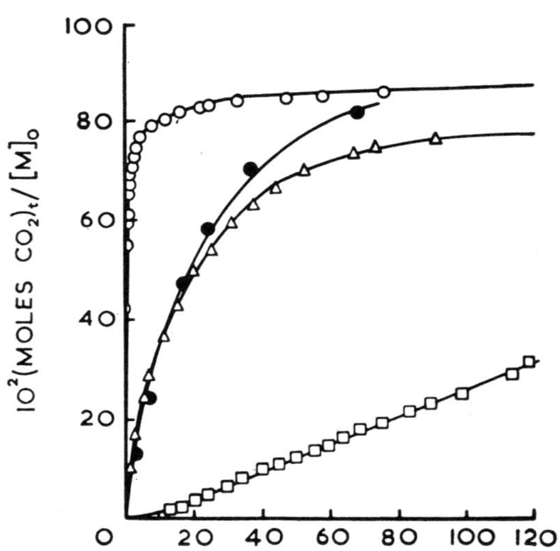

Fig. 7.1. Conversion-time curves for polymerization of NCA's (0.224 mole 1.$^{-1}$) with amines (0.015 mole 1.$^{-1}$) at 25° C. in *N,N*-dimethylformamide.

- ○ γ-Ethyl-L-glutamate NCA—di-isopropylamine.
- △ γ-Ethyl-L-glutamate NCA—*n*-hexylamine.
- □ Sarcosine NCA—di-isopropylamine.
- ● Sarcosine NCA—*n*-hexylamine.

Bearing in mind the fact that primary and secondary amines do not necessarily react *in toto* by addition to the 5-CO of the NCA ring unless the ring is 3-substituted we can, nevertheless, write one step of that part of the polymerization which follows such a reaction as shown below.

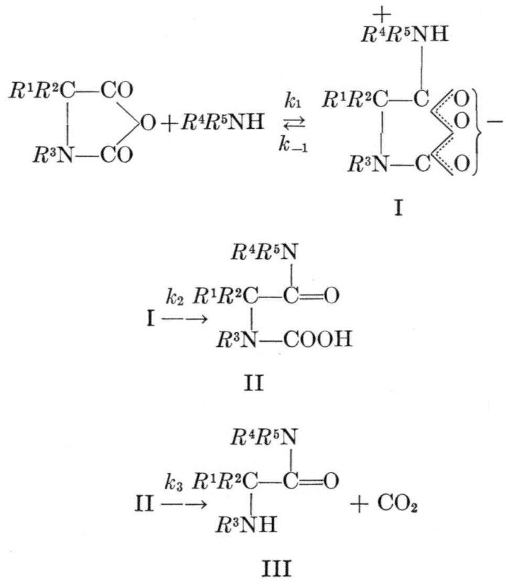

(3)

In (I) the labile proton on the amide nitrogen may tautomerize between any one of the oxygen atoms or the 3-N atom. Normally the decarboxylation of the carbamic acid (II) is not a rate-determining step as has been shown by the absence of an isotope effect when C^{12} is replaced by C^{13} (*16*) and by measurements of the half-life of the carbamic acid derived from sarcosine dimethylamide (*8*). The investigations of Heyns and Schultze (*17*) on the isotope effect associated with the endocyclic oxygen atom showed that either k_1 or k_2 in (3) can be rate-determining, and that in some cases k_1, k_2 are of comparable magnitude. These conclusions had already been drawn from kinetic observations (*8*). In the polymerization of sarcosine NCA the pre-equilibrium in (3) is established, as shown by the negative temperature coefficient (*9b*); with DL-phenylalanine- and DL-leucine NCA's the temperature coefficient is positive and in these cases k_1 may be rate-determining (*8*).

The polymerization of sarcosine NCA differs from that of some other NCA's in that it is catalyzed by carbon dioxide and by α-picolinic acid (*8*). Inspection of the kinetic features of these reactions shows that the catalytic species are the carbamate and α-picolinate ions respectively. There are two steps in (3) which may be catalyzed. The catalytic ion may add more readily than the propagating base to 5-CO (*cf.* equation [7c]) forming an open-chain anhydride which then acylates the propagating secondary amine,* or the ion, functioning as a base, removes a proton from (I) thereby facilitating ring-opening. In the former case, the unusual behavior of sarcosine NCA must be attributed to the greater ease of formation of the open anhydride by the sterically unhindered ion compared to the formation of (I) with the propagating secondary base. In the latter case, the different behavior of sarcosine and other NCA's (e.g., DL-phenylalanine NCA) arises from the fact that (as shown by the negative temperature coefficient) the pre-equilibrium is established with sarcosine NCA, while with DL-phenylalanine NCA the forward reaction (k_1) would be rate determining.

The mechanisms so far discussed all involve attack by the base on 5-CO. This is not the only carbonyl addition possible; there is evidence (*2*) that some attack on 2-CO to form urea derivatives may also occur. A reaction of this type is a termination process; since it has a comparatively high activation energy, it is particularly important at high temperatures, or when high base concentrations are used.

Aprotic initiators

Tertiary bases.—The reactions shown in (3) clearly cannot occur if the initiator does not have a proton which can be transferred to form the propagating peptide amine end-group. Tertiary amines are an important class of aprotic initiators which we have examined in some detail. One of the most important and controversial questions relating to this type of initiation is whether only NCA's unsubstituted in the 3 position can react with tertiary bases (*2, 13, 18*). We have long maintained the view (*1*) that such substitution precludes reaction and more recent investigations (*19*) on sarcosine and L-proline NCA's confirm this. The sensitivity of the experiments is brought out by comparing the rate curves for these NCA's with that for γ-ethyl-L-glutamic acid NCA (Fig. 7.2); although the base concentration in the latter case is less than 10 per cent of that used with sarcosine and L-proline NCA's, the initial rate of reaction is at least thirty times greater. For comparison, the reactivities of the NCA's towards primary bases are in the order L-proline > γ-ethyl-L-glutamate ∼ sarcosine (in *N,N*-dimethylformamide solution). We noted that these NCA's, and L-proline NCA in particular, are very sensitive to trace impurities, and reproducible results could only be obtained after very rigorous purification. This fact may well account for the many and various reports in the literature of the polymerization of such NCA's initiated by tertiary bases (see e.g., ref. [2]). Investigations into other systems with aprotic base initiation and *N*-substituted NCA's must obviously be carried out under similar rigorous conditions and in the absence of proton sources.

We have found that addition of 3-methyl hydantoin† to sarcosine NCA in the presence of a tertiary base results in reaction (*19*), although in the absence of the base no reaction is observed (Fig. 7.2). This result closely parallels that previously reported relating to initia-

* Note that such acylation is impossible in scheme (7) because no free amine is present.

† For numbering of atoms in the hydantoin ring see equation (4a).

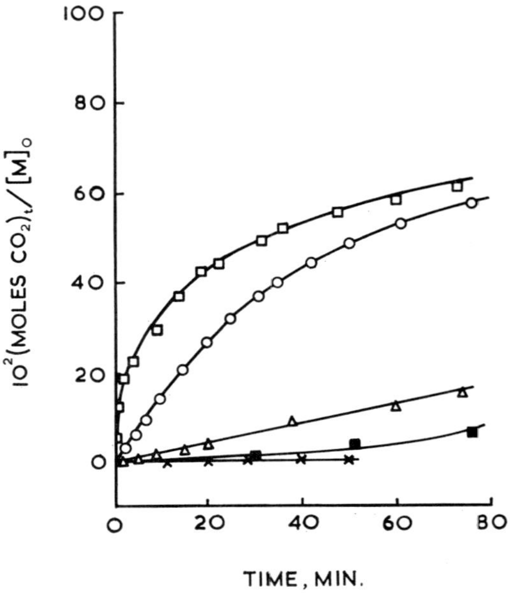

Fig. 7.2. Conversion-time curves for reactions of NCA's (M), with tri-n-butylamine (B) in N,N-dimethylformamide at 25° C.

□ γ-Ethyl-L-glutamate NCA ($[M]_o=0.1$ mole $l.^{-1}$, $[B]=0.0126$ mole $l.^{-1}$).
○ Sarcosine NCA ($[M]_o=0.332$ mole $l.^{-1}$, $[B]=0.042$ mole $l.^{-1}$) +3-methyl hydantoin (0.178 mole $l.^{-1}$).
△ L-Proline NCA ($[M]_o=0.291$ mole $l.^{-1}$, $[B]=0.129$ mole $l.^{-1}$).
◪ Sarcosine NCA ($[M]_o=0.289$ mole $l.^{-1}$, $[B]=0.262$ mole $l.^{-1}$).
× Sarcosine NCA ($[M]_o=0.323$ mole $l.^{-1}$, $[B]=0$ mole $l.^{-1}$), +3-methyl hydantoin (0.175 mole $l.^{-1}$).

tion by lithium chloride (*20*). These observations, together with the finding that 1,3-dimethyl hydantoin is ineffective (*21*), strongly suggest that the proton attached to 1-N is involved in initiation. The compound 3-methyl hydantoin was chosen because of its structural resemblances to the unsubsituted NCA ring, and its mode of reaction is likely to be similar. It is not unique in its ability to induce the tertiary base-initiated polymerization of sarcosine NCA, as will be seen from Table 7.1, but it is the most effective of such substances so far examined.

The 3-substituted hydantoin residue becomes incorporated into the polymer chain during the reaction. This has been shown by the use of 3-(9-methylene anthracene) hydantoin which has an ultraviolet absorption band with a maximum near 3650 Å conveniently situated for identification. Polysarcosine prepared with the aid of this compound and tri-n-butylamine in N,N-dimethylformamide solution exhibits the characteristic absorption peaks.

The use of N-phenylglycine NCA, which does not polymerize readily (*22*), has enabled us to determine the nature of the initiation step in these systems. Reaction of the NCA with 3-methyl hydantoin in the presence of tertiary bases results in the formation of 3-methyl-1-(N-phenylglycyl) hydantoin, which has been isolated and shown to have the structure (IV) (*19*).

Table 7.1—Relative reactivities of various compounds (X) in inducing the tri-n-butylamine (B)-initiated polymerization of sarcosine NCA (M) in N,N-dimethylformamide at 25° C.

Compound	$\frac{10^3}{[B][M][X]}\frac{d[CO_2]}{dt}$ $l.^2$ mole^{-2} sec.$^{-1}$	$\frac{10^4}{[B][M]}\frac{d[CO_2]}{dt}$ l. mole^{-1} sec.$^{-1}$
α-pyrrolidone	1.4	1.8
succinimide	0.8	1.6
N-carbomethoxyglycine ethyl ester	0.8	1.3
N,N'-dimethyl N-acetyl urea	1.3	2.4
oxalic acid	0.2	0.2
methyl dihydrouracil	7.8	11
hydantoin	11	20
3-(9-methylene anthracene) hydantoin	23	38
3-methyl hydantoin	47	70
none	—	2.2

* For numbering of atoms in the hydantoin ring see equation (4a).

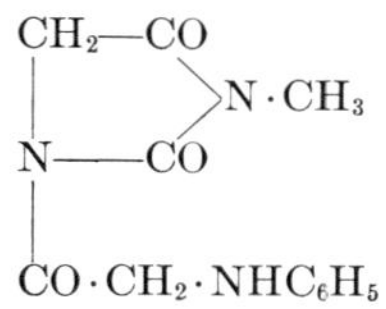

IV

From the analogous system lithium chloride, 3-methyl hydantoin, *N*-phenyl-glycine NCA (IV) has also been isolated, thus emphasizing the close similarity, already mentioned, between these examples of aprotic initiation.

There appear to be only two alternative initiation paths leading to (IV) and these are (4a), (4d) in the scheme below, in which *B* represents the base, i.e., Cl^- or NR_3. On the basis

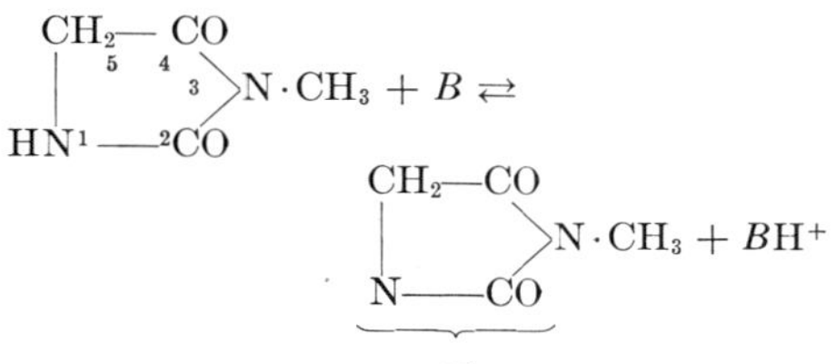

(4a)

V

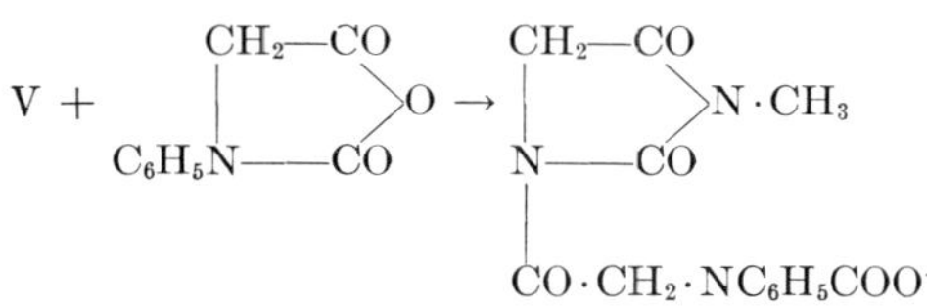

(4b)

VI

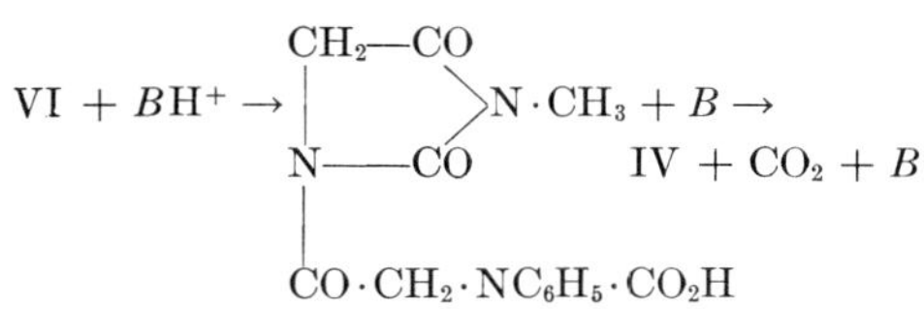

(4c)

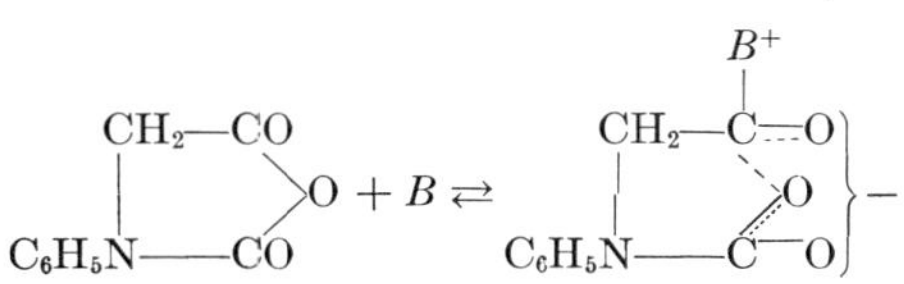

(4d)

VII

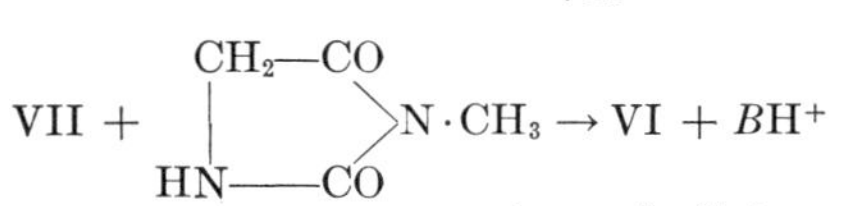

(4e)

etc., as in (4c)

of evidence relating to the reactions of 3-unsubstituted NCA's (given below) we prefer

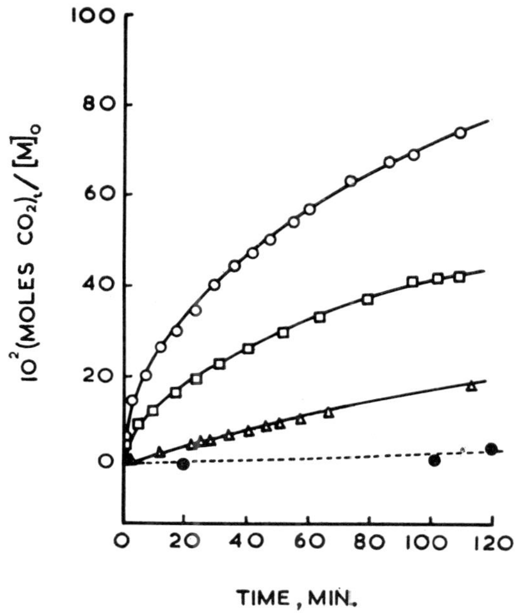

Fig. 7.3. Conversion-time curves for the polymerization of γ-ethyl-L-glutamate NCA (0.2 mole $1.^{-1}$) in *N, N*-dimethylformamide by pyridine and homologues (2.0 mole $1.^{-1}$) at 25° C.

○ 2,6-Lutidine.
□ α-Picoline.
△ Pyridine.

The broken curve shows a comparative experiment with sarcosine NCA (0.2 mole $1.^{-1}$) and 2,6-histidine (2.0 mole $1.^{-1}$).

(4a), although this does not exclude the possibility that the equilibrium in (4d) is established. That the initial ionization in (4a) is small is indicated by the observation that no measurable increase in the electrical conductivity of solutions of 3-methyl hydantoin in *N,N*-dimethylformamide is produced by adding equimolar amounts of triethylamine. This finding is consistent with the literature (*23*). Further, there is no observable change in the ultraviolet absorption spectrum of tri-*n*-butylamine in *N,N*-dimethyl-formamide solution on the addition of 3-methyl hydantoin.

The experiments with 3-substituted hydantoins strongly suggest that a 3-unsubstituted NCA is polymerized by aprotic bases in an analogous mannner; that is, one molecule of the NCA provides a proton for the formation of an amine end-group from a second NCA molecule

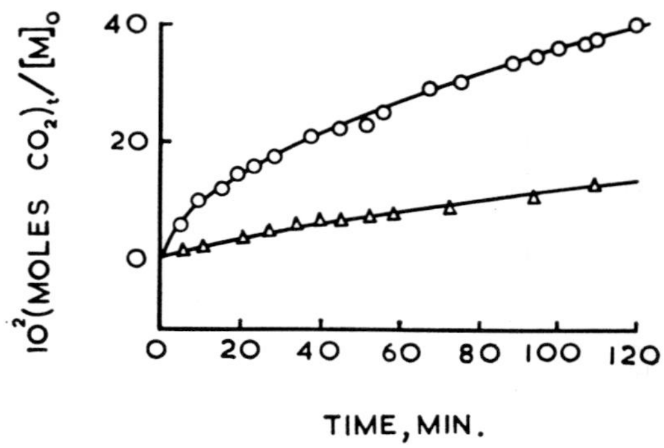

Fig. 7.4. Conversion-time curves for the polymerization of DL-phenylalanine NCA (0.2 mole l.$^{-1}$) in nitrobenzene catalyzed by pyridine and 2,6-lutidine (2.0 mole l.$^{-1}$) at 25° C.

△ Pyridine.
○ 2,6-Lutidine.

and is itself incorporated in the chain. Two mechanisms of this type may be proposed, analogous to those given above for the system 3-methyl hydantoin, *N*-phenylglycine NCA, base, depending on whether the base initially ionizes the 3-unsubstituted NCA or adds to 5-CO. We have performed diagnostic experiments similar to those carried out by Gold and Jefferson (*24*) in studying the base-catalysed hydrolysis of carboxylic acid anhydrides. The method involves using the bases pyridine, α-picoline, and 2,6-lutidine; if ionization is involved their activities should be in the order of base strengths—i.e., in the order given—whereas association with a carbonyl group would lead to the reverse order because of the increased steric shielding of the basic nitrogen by the α-substituents. The rates of initiation of the polymerization at 25° C. of γ-ethyl-L-glutamate NCA in *N*,*N*-dimethylformamide and DL-phenylalanine NCA in nitrobenzene by these bases (highly purified) have been found to be in the order 2,6-lutidine $>\alpha$-picoline $>$pyridine (Figs. 7.3 and 7.4). Clearly ionization, and not addition to a carbonyl group, is operative.

In view of the results described above we believe that the mechanism of the polymerization of NCA's, initiated by tertiary bases (B) is that shown in (5a)–(5h) (*cf.* refs. *13, 25*).

INITIATION

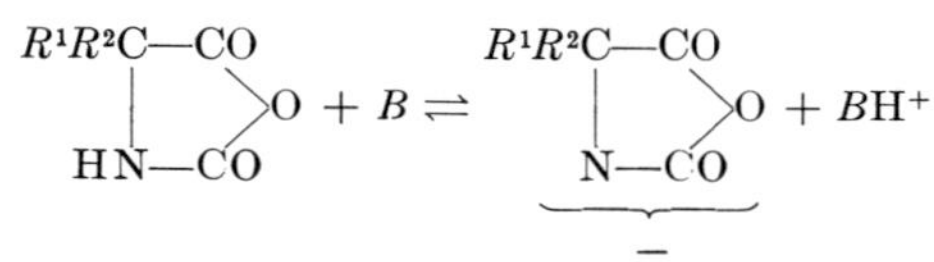

VIII

(5a)

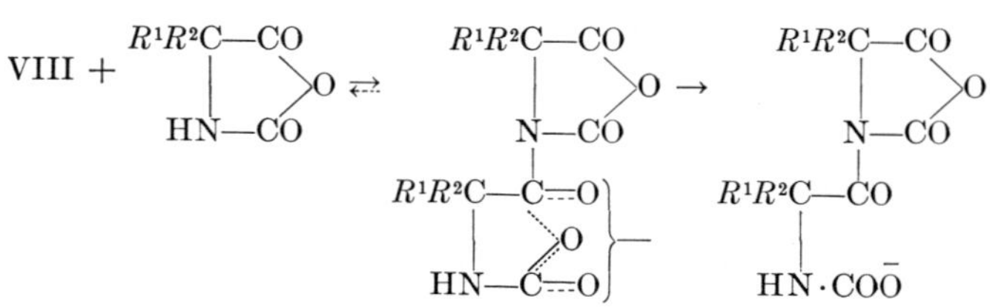

IX X

(5b)

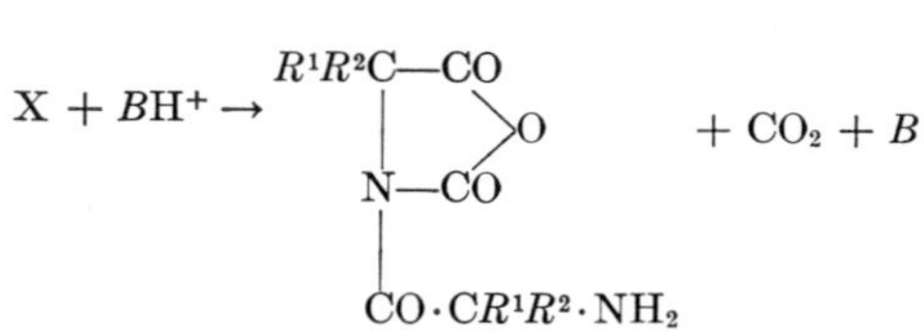

XI_1

(5c)

PROPAGATION

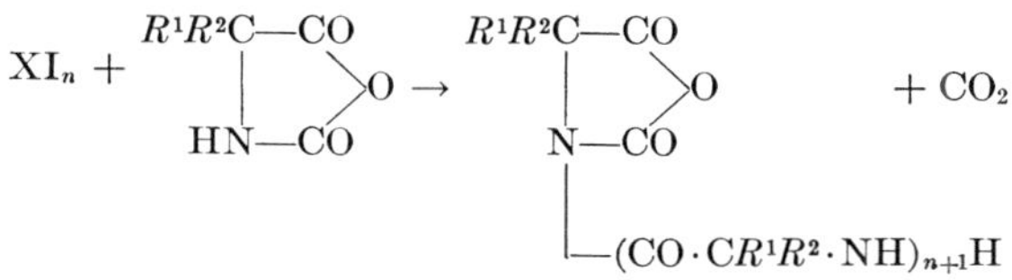

XI_{n+1} (5d)

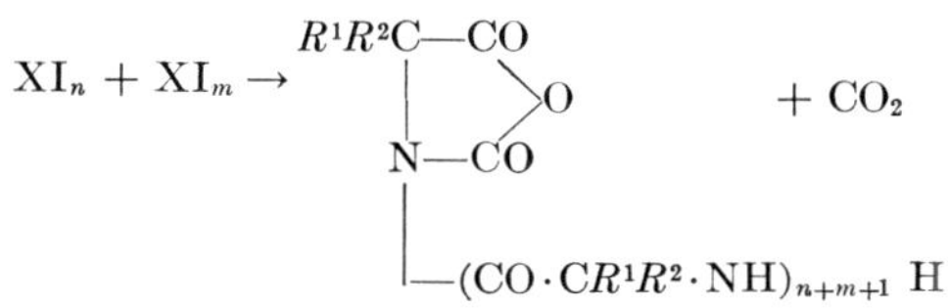

(5e)

$$XI_n + VIII + BH^+ \rightarrow XI_{n+1} + CO_2 + B \quad (5f)$$

TERMINATION

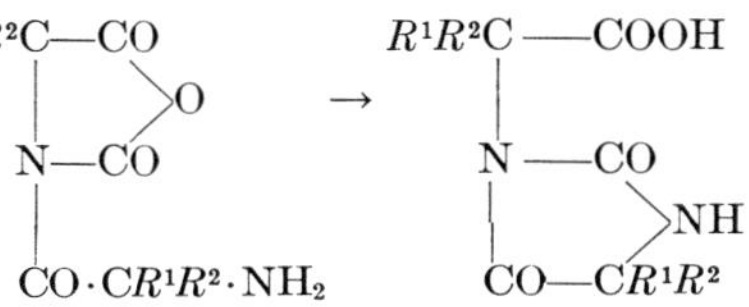

XI_1 $\qquad$ XII (5g)

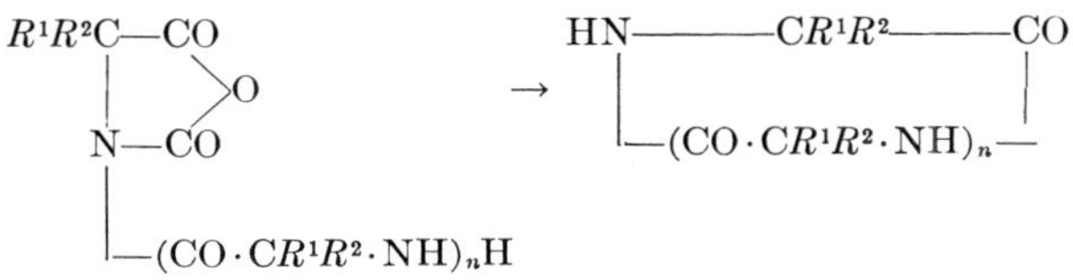

XI_n $\qquad$ XIII (5h)

The formation of the bifunctional intermediate (XI_1) follows the steps already discussed. We use the symbol (XI_n) to describe a similar bifunctional intermediate containing n amino acid residues in addition to its terminal NCA group. The propagation reactions (5d), (5e) represent the attack of the primary amine end-group of (XI_n), according to equation (3), on the 5-CO of the monomer and the anhydride end of the intermediate (XI_m), respectively. The former is the "normal" propagation, the latter a coupling of bifunctional molecules. Reaction (5f) is the growth of an intermediate (XI_n) by addition of ion (VIII) to the 5-CO of its terminal anhydride group. The termination reactions (5g) (leading to substituted hydantoin 3-acetic acids [XII]) and (5h) (leading to cyclic peptides) are not necessarily the only chain-ending steps, but there is little evidence as to the nature of other termination processes.

This reaction scheme is consistent with the following observations. First, both hydantoin 3-acetic acid derivatives and cyclic peptides can be major products (*20, 25*). Second, as expected, no cyclic dimers (diketopiperazines) are formed, since cyclization of (XI_1) involving 5-CO is sterically impossible (*20*). Third, the degrees of polymerization are generally higher than the $[M]_o/[I]_o$ values and, for given $[M]_o$, less dependent on $[I]_o$ than is this ratio (*12, 13, 26*). Fourth, high dilution at a given $[M]_o/[I]_o$ increases the yield of substituted hydantoin 3-acetic acids (*13*).

A general explicit solution of the rate equations for the above scheme, giving the rate of evolution of carbon dioxide as a function of time, is too complicated to be useful, although an approximate treatment suggests that satisfactory agreement with the rather meager kinetic data available may be obtained (*13*).

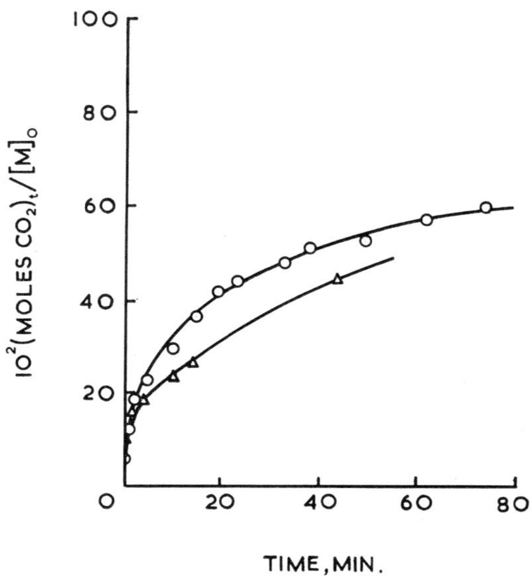

Fig. 7.5. Conversion-time curves for polymerization of γ-ethyl-L-glutamate NCA (*M*) with amines (*I*) in *N,N*-dimethylformamide solution at 25° C.

○ I = tri-*n*-butylamine, 0.0126 mole $1.^{-1}$; M = 0.1 mole $1.^{-1}$

△ I = *n*-hexylamine, 0.0075 mole $1.^{-1}$; M = 0.075 mole $1.^{-1}$

We believe that it would be useful to measure the concentration of monomer, independently of the carbon dioxide evolution, as a function of time; this would enable direct estimation of the total concentration of the species (XI_n) to be made. We are exploring the possibility of using NMR spectroscopy for this purpose.

In comparing schemes (3) and (5) we may note that for NCA's which are not 3-substituted initial attack by primary and secondary bases may follow either scheme. However, they do not do so with equal facility. For all types of base the degree of ionization in (5a) is directly paralleled by the base strength, while for unhindered primary and secondary bases the association with 5-CO shown in (3) would be expected to be less sensitive to base strength. This arises because the proton transfer from nitrogen to oxygen greatly facilitates association, and its ease is an inverse function of the base strength. Thus, steric questions apart, the weaker the base the greater is the relative importance of association compared to ionization.

Steric considerations must be superimposed, since the association in (3) is very sensitive to the environment of the amino nitrogen while the ionization is not. We can thus understand the behavior of initiators such as *n*-hexylamine and di-isopropylamine mentioned above. It is important to note that the initial rates of reaction of such bases as tri-*n*-butylamine and *n*-hexylamine are not very different (Fig. 7.5) and follow the order of their base strengths. Both are strong bases. Presumably *n*-hexylamine reacts initially in part according to scheme (5); the weaker peptide bases formed by reaction (3) cause much less extensive ionization, so that the reaction decelerates and subsequently mainly follows route (3). With tri-*n*-butylamine initiation the high rate of reaction is maintained, since the tertiary base (unlike the primary) is not consumed. It has been argued that the high initial rate obtained with *n*-hexylamine is solely due to the fact that its reactivity according to (3) is greater than that of the propagating base (*1*). This contention does not explain the continuance of the high rate well beyond the stage of conversion of the initiator into the amino-acid hexylamide (Fig. 7.5), nor the formation of polymers with $\overline{P}_n > [M]_o/[I]_o$.

As explained above, part of the evidence for the scheme (5) is derived from the use of hydantoin derivatives as models for the NCA molecule. It is pertinent to inquire at this point how far the analogy is valid; in particular, the hydantoin ring is considerably more stable than the NCA ring so that comparison taken beyond the initial process may be misleading. For reasons already given there can be no doubt that (5a) correctly represents the initial attack of the base; as an alternative to (5b) we may consider reactions based on the opening of the ring of (VIII):

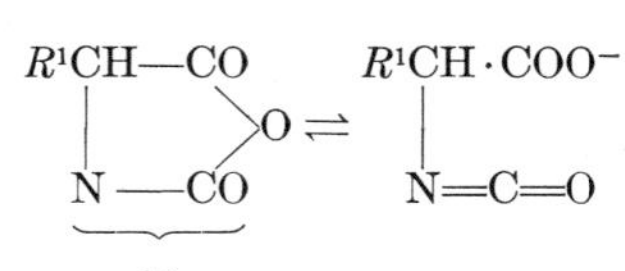

(6a)

VIII XIV

The existence of an equilibrium of this type is not inconsistent with views prevalent in the literature for many years (*27*). Kopple (*28*) has suggested that the tertiary base-catalyzed poly-

$$\mathrm{XIV} + B\mathrm{H}^{+} \rightleftharpoons \begin{array}{c} R^{1}\mathrm{CH}\cdot\mathrm{COOH} \\ | \\ \mathrm{N{=}C{=}O} \end{array} + B \tag{6b}$$

XV

$$\mathrm{XIV} + \mathrm{XV} \rightarrow \mathrm{OCN}\cdot\mathrm{CH}R^{1}\cdot\mathrm{CO}\cdot\mathrm{O}\cdot\mathrm{OC}\cdot\mathrm{NH}\cdot\mathrm{CH}R^{1}\mathrm{COO}^{-} \tag{6c}$$

XVI

$$\mathrm{XVI} \rightarrow \mathrm{OCN}\cdot\mathrm{CH}R^{1}\cdot\mathrm{CO}\cdot\mathrm{NH}\cdot\mathrm{CH}R^{1}\cdot\mathrm{COO}^{-} + \mathrm{CO_2} \tag{6d}$$

XVII

XVII →

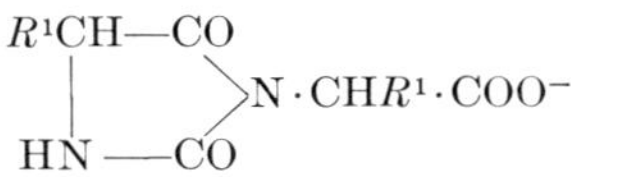

(6e)

XII(R^2 = H)

merization may occur through the intermediate XIV, according to the reactions (6b)–(6e). The bifunctional intermediate (XVII) is supposed to grow by reaction of carboxyl with isocyanate—i.e., by addition of (XIV) or (XV) or of larger fragments similar to itself—and to form hydantoin 3-acetic acid derivatives by reaction (6e). These may enter into coupling reactions, or cyclization of the terminal portions of large fragments of type (XVII) may occur, both processes giving peptides with terminal hydantoin residues. It is considered that (XVII) cannot yield the diketopiperazine on account of the improbability of forming an eight-membered ring. This scheme is similar in many respects to that in (5), but we consider it less satisfactory for several reasons. It seems doubtful to us (from analogies in the literature [*29, 30*]) whether the coupling reaction between carboxyl and isocyante groups could, in fact, be sufficiently rapid to account for the high rates of some tertiary base-initiated polymerizations. Reaction (6e) would not be expected to occur at all readily, and hardly seems capable of accounting for the high yields of hydantoin 3-acetic acid which have been reported to arise from glycine NCA (*25*). We do not know of any evidence in favor of hydantoin chain-terminations; in our experience the total acid formed is not significantly different from the yield of free hydantoin 3-acetic acid derivatives. We have recently studied the polymerization of γ-ethyl-L-glutamate NCA initiated by equimolar concentrations of n-hexylamine and tri-n-butylamine. The initial rate of carbon dioxide evolution is greater than that with the tertiary base alone. If (6a) were rate-determining, one would expect preferential reaction of isocyanate with primary amine groups (*30*) resulting in initial retardation in carbon dioxide evolution. Therefore, if this series of reactions occurs, (6c) must be the rate-determining step. As will be shown later, this cannot be the case when the initiator is the anion of a weak carboxylic acid; hence if the mechanism (6) holds for initiation by tertiary bases, the reactions must differ from those initiated by such salts.

Other aprotic initiators.—Initiators of this type which have been studied include salts of weak acids ([*20, 25, 31*]; e.g., halides in N,N-dimethylformamide solution), sodium methoxide (*12, 14, 32, 33*), sodium borohydride (*12*), and sodium triphenylmethyl (*14*).

The reactions initiated by alkali metal halides in N,N-dimethylformamide solution resemble in all respects the corresponding tertiary base reactions. As already mentioned, reaction is confined to 3-unsubstituted NCA's (*20, 25*); with other NCA's reaction can be induced by 3-methyl hydantoin (*20, 25*), and the use of N-phenylglycine NCA in these circumstances yields the compound (IV) (*19*). In a kinetic study of the system sarcosine NCA, 3-methyl hydantoin, lithium chloride in N,N-dimethylformamide solution (*20*) satisfactory agreement with observations of the rate of carbon dioxide evolution and secondary (propagating) base formation has been obtained using a kinetic scheme essentially similar to (4a, 4b, 4c). Although complex, the rate equation contains only two adjustable parameters, which were evaluated from the yield of secondary base; the fit with the rate data is therefore evidence for the correctness of the reaction scheme. However, fractionation studies on the polymer suggest that the molecular weight distribution is narrower than that calculated (*11*). The reason for this is not clear; it is difficult to see how the polymer resulting from an autocatalytic reaction such as this could have a distribution as narrow as that suggested.

With alkali metal halides (in the absence of 3-methyl hydantoin) 3-unsubstituted NCA's react rapidly, the major products being cyclic peptides and hydantoin 3-acetic acid derivatives. The relative yield of the latter decreases as the initiator concentration increases (*20*). A similar observation has been made with tertiary-base initiation and follows readily from scheme (5), as already mentioned. No diketopiperazines have been detected. From the reaction between glycine NCA and lithium chloride in *N*,*N*-dimethylformamide, cyclic hexaglycine has been isolated.

Salts of carboxylic acids were shown to be initiators of 3-unsubstituted NCA's by Ballard and Bamford (*31*). The features characteristic of tertiary-base initiation again appear in these reactions; thus they are often comparatively rapid and may lead to high polymers together with some hydantoin 3-acetic acid derivatives. The yield of hydantoin 3-acetic acid from glycine NCA decreases with increasing initiator concentration as described in earlier instances. Moreover, at least under the conditions investigated (*N*-methylformamide solution at 25° C.), 3-substitution of the NCA very greatly reduces the rate of reaction. This latter observation would appear to exclude an anhydride-exchange mechanism, e.g., of the types (7a)–(7d), as a major reaction.

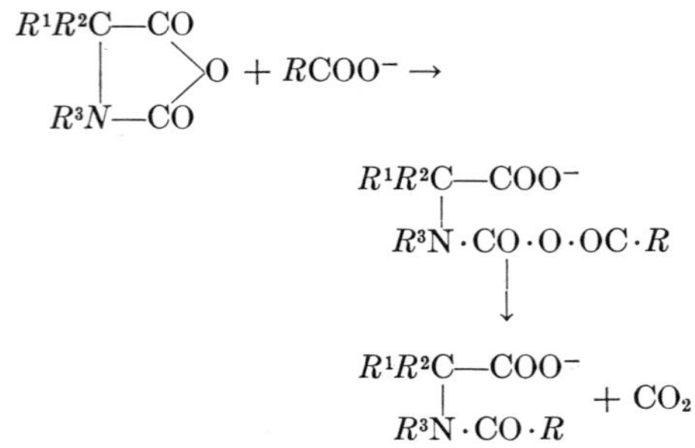

(7a)

XVIII

Further, the mechanism of equations (6a)–(6e) hardly seems applicable in this case, since the (monofunctional) carboxylate ions of the initiator would be exepcted to compete in the coupling reactions of the bifunctional intermediates (e.g., [6c]) and prevent the formation of high polymers. Such competition is likely to be severe, since the initiating ion is expected to be a stronger base than the ion (XIV), and is probably generally present in higher concentration. Reactions between the initiating ion and the isocyante group of (XIV) or (XV) will lead to a monofunctional N-acylated amino

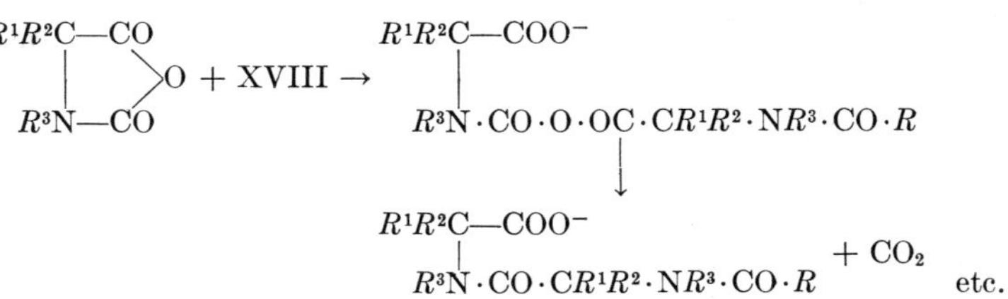

(7b)

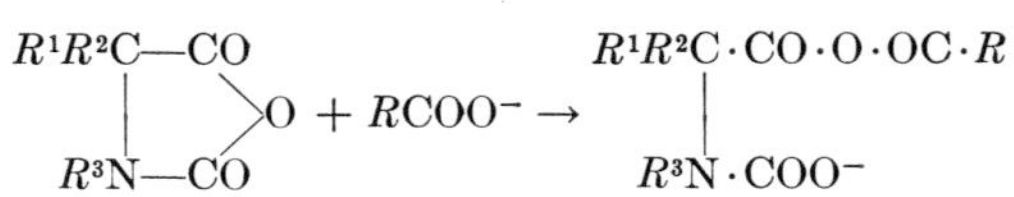

(7c)

XIX

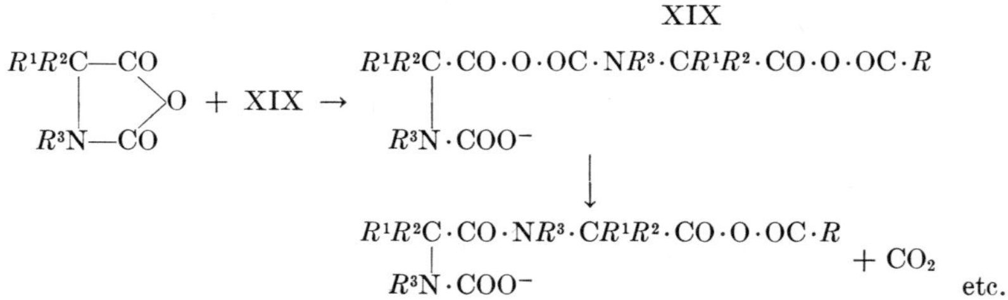

(7d)

acid (or ion) which will be the major propagating species, and result in a degree of polymerization approximating that given by (2). The polymers obtained in practice have degrees of polymerization much greater than this (*31*).

The evidence we have presented suggests very strongly that the tertiary base- and salt-initiated reactions of NCA's follow the same route. This, we believe is best represented by scheme (5).

Doty and Blout and their colleagues (*12*, *14*, *32*) have shown that high-molecular-weight polypeptides can often be prepared by the use of initiators which are strong bases—e.g., sodium hydroxide, sodium methoxide—and Idelson and Blout (*14*) have studied in detail the polymerization of γ-benzyl-L-glutamate NCA initiated by sodium methoxide. The reaction is autocatalytic, and the resulting polymers have degrees of polymerization much larger than the ratio $[M]_o/[I]_o$. Idelson and Blout point out that sodium methoxide is known to react rapidly with the NCA to yield two products (*33*):

$$\begin{array}{c} R^1CH\cdot COO^-Na^+ \\ | \\ HN\cdot CO\cdot OCH_3 \end{array} \quad \text{and} \quad \begin{array}{c} R^1CH\cdot CO\cdot OCH_3 \\ | \\ HN\cdot COO^-Na^+ \end{array}$$

They suggest that these enter into anhydride-exchange reactions essentially similar to those shown in (7a)–(7d). Such a mechanism is consistent with these authors' view that the reactions can proceed with 3-substituted NCA's; however, this contention appears to be based entirely on work with L-proline NCA, details of which have not been published. On account of the high reactivity of L-proline NCA already referred to, and the misleading reports in the literature about its behavior with tertiary bases, it is, in our opinion, very desirable that the observations should be extended to other 3-substituted NCA's. The stated reactivity of 3-substituted NCA's is one reason advanced by Idelson and Blout (*14*) for rejecting the mechanism of scheme (5) for strong-base initiated polymerizations. The other reasons are that "no amine groups can be formed in triphenylsodium-initiated polymerizations" and "the postulated addition to both 'amine' ends and 'NCA' ends of growing chains would lead to a broader molecular weight distribution than has been observed." The present authors cannot agree with either statement; it is not obvious to them why amine groups cannot be formed, since presumably the initiator would be destroyed rapidly, and further, discussion of molecular weights and their distributions presupposes a complete knowledge of the termination mechanism, which is not available.

Physical features in polymerization reactions

In the introduction we pointed out that physical, as well as chemical, features may have an important influence on the kinetics of polymerization of NCA's. Effects such as chain conformation and adsorption of monomer on the growing polypeptide chains may, individually or in combination, greatly complicate the kinetic picture. We believe that the study of these phenomena is only in its infancy and explanations advanced at this stage are unlikely to be complete.

The chain effect

The polymerization of some DL-α-amino acid NCA's with polysarcosine as initiator exhibits a number of unexpected features (*3*). First, the initial rate of reaction is a function of the degree of polymerization of the initiating base, increasing with $\overline{P}_n$ and approaching a limit at $\overline{P}_n = 20$, approximately. Second, the enhanced rate is high, being up to twenty times greater than that for sarcosine dimethylamide initiation. Third, the reaction is self-inhibiting. Fourth, it is also inhibited by the addition of a polymer of the α-amino acid whose NCA is being polymerized, and by the addition of *N*-acetyl-polysarcosine dimethylamide. Fifth, for the occurrence of the chain-effect it is necessary that the initiating base should be attached to the polysarcosine chain; *N*-acetyl-polysarcosine together with an equivalent of low-molecular-weight base shows no such behavior.

Intrinsic viscosity measurements (*3*) on polymers isolated after these reactions showed that they are true block copolymers, their viscosities being identical with those of copolymers prepared by polymerizing sarcosine NCA with the corresponding preformed poly-α-amino acid derivatives as initiators. (The latter reaction does not show a chain-effect.)

The only satisfactory explanation (*3*) of

these findings appears to be that the monomer is adsorbed on the polysarcosine chain with the result that the former is brought into closer proximity with the base group of the latter, resulting in a higher local concentration of monomer. The monomer is also possibly activated by adsorption. As observed (*3*), at high monomer concentration the rate would be expected to become zero order in monomer. Introduction of dead polymer (e.g., *N*-acetyl-polysarcosine dimethylamide), which itself adsorbs monomer, reduces the adsorption on the growing chains and thereby reduces the reaction rate. Association can also occur between polysarcosine and added poly-α-amino acids (with unsubstituted NH groups); hence addition of such polymers also reduces the monomer adsorption and the rate. The self-inhibition observed in this type of reaction may arise from this cause, or from the increased separation of the growing end from the monomer adsorbed on the polysarcosine.

The kinetic scheme used in the analysis of the experimental results (*3*) is shown below in (8a) and (8b), where M represents the monomer, S_i the sarcosine unit of the chain in position i from the terminal base, E_i the adsorbed monomer on this site, and X_n a polymer chain containing n amino-acid residues. The rate of reaction is given by (9), where $[E] = \sum_i [E_i]$.

$$M + S_i \overset{K}{\rightleftharpoons} E_i \tag{8a}$$

$$E_i(+X_n) \overset{k_i}{\rightarrow} CO_2 + S_i(+X_{n+1}) \tag{8b}$$

$$\frac{d[CO_2]}{dt} = [E] \sum_1^n k_i/n = \bar{k}_n[E] \tag{9}$$

Results allow the determination of k_i as a function of i and show that k_i has a sharp maximum at $i = 7$. This value may be taken as the optimum length of the polysarcosine chain for coiling back on itself.

The type of phenomena described and the kinetic scheme used show close analogies with those appertaining to enzymic reactions. Thus (8a) and (8b) resemble closely the familiar Michaelis-Menten scheme (*34*).

Conformational effects

There have been a number of reports in the literature (*4, 5, 6, 25*), that, during the primary-base initiated polymerization of single enantiomorphs of NCA's, increases in reaction rate set in rather sharply when the growing chains reach a mean size of 7–12 units. Doty and Lundberg (*4*) suggested that, since at this stage a change from random-coil to α-helical conformation takes place (*36*), the higher rate of propagation is in some way connected with the adoption of the α-helical form. Their explanation (*6*) is that the change in rate is connected with a change in the entropy of activation resulting from "an increased accommodation provided for the anhydride molecule at the end of a helix." An increase in entropy of activation could be due to adsorption of the NCA on the end of the helix so that it becomes suitably disposed for reaction (*37*) or, alternatively, to a greater collision frequency resulting from reduced obstruction by the helix compared to the random coil. Buyle and Hargitay (*36*) have suggested an alternative explanation also based on the random coil–helix transition. These authors consider that a carbamate ion at the end of an α-helix is stabilized by hydrogen bonding and is responsible for a more rapid propagation. Such a view seems untenable for several reasons, e.g., the observed absence of catalysis by carbon dioxide.

In studies in our laboratories (*38, 39*) we have not been able to obtain reproducible results with the system most extensively studied by Doty, Blout, and their colleagues (γ-benzyl-L-glutamate NCA in dioxan) since gel formation frequently occurred during reaction. It is known that gelation or precipitation can give rise to enhanced rates (*40*) even in systems from which α-helical conformations are excluded. In some systems showing the "two-stage propagation" (e.g., L-leucine NCA in nitrobenzene) precipitation is known to take place (*6*). We have recently observed small effects of the type described by Doty and Lundberg (*4*) during the polymerization by preformed polymer of γ-ethyl-L-glutamate NCA in *N,N*-dimethylformamide solution, which is undoubtedly homogeneous (*41*). As will be seen from Fig. 7.6, the change in the velocity coefficient during the addition of 25 units to a 5-mer is small and the coefficient varies smoothly between the values for addition of 5 units to a 5-mer and 25-mer. These effects may well be manifestations of the conformational phenomena mentioned by Lundberg and Doty (*6*).

STEREOISOMERIC EFFECTS

Doty and Lundberg (*4*) first reported that initiation by a preformed polymer ($\overline{P}_n = 20$) is stereospecific in the sense that a polymer consisting entirely of L-residues initiates reaction with an L-NCA much more rapidly than with a D-NCA and vice versa. They also investigated the rate of primary base-initiated polymerization as a function of the ratio D:L for γ-benzyl-glutamate NCA in *N,N*-dimethylformamide, and found that the rate coefficient passes through a minimum for the racemic mixture. Subsequently, similar observations have been made with other types of initiators, e.g., sodium methoxide (*14*) and tertiary bases (*42*). We have found that with initiation by preformed polymer the optical specificity is much greater when the initiator is a 15-mer than when it is a 3-mer (*39*). This is a further indication that the specificity is connected with the α-helical conformation, as originally proposed by Lundberg and Doty (*6*). The addition of lithium perchlorate to the reaction mixtures eliminates the optical specificity, both enantiomorphs reacting with a preformed 15-mer at approximately the rates expected for a preformed 3-mer (*39*). Lithium perchlorate, by preferential bonding, would be expected to reduce any hydrogen-bonding of the NCA to the helix; consequently this result shows that the specificity may, in part, be connected with adsorption. Preliminary experiments have been carried out in which the reactions of the individual enantiomorphic γ-benzyl glutamate NCA's with preformed polymers in *N,N*-dimethylformamide solution at 0° C. have been studied. The polymers used were poly-γ-benzyl-L-glutamate ($\overline{P}_n = 15$), and a similar polymer to which a terminal unit of γ-benzyl-D-glutamate had been added statistically. The initiator carrying one D-residue was found to react comparatively slowly with *both* enantiomorphic NCA's, the addition being about 6–8 times slower than the straightforward L-L reaction.

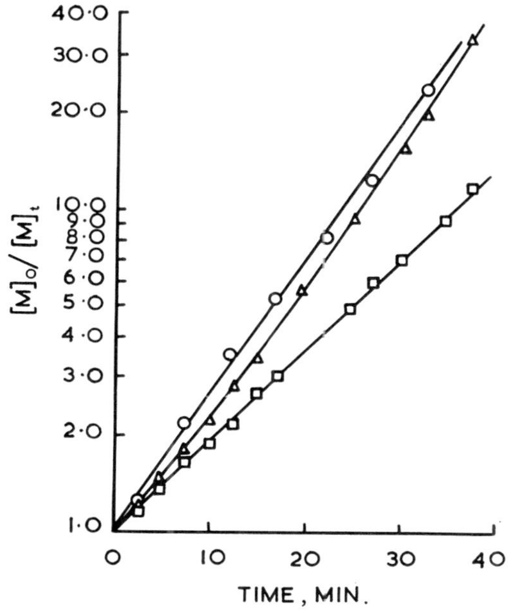

Fig. 7.6. Effect of chain length on the polymerization of γ-ethyl-L-glutamate NCA, with preformed poly-γ-ethyl-L-glutamate (1.5×10^{-2} mole $1.^{-1}$) in *N,N*-dimethylformamide at 15° C.

○ Initiator $\overline{P}_n = 25$, $[M]_o/[I]_o = 5$.
△ Initiator $\overline{P}_n = 5$, $[M]_o/[I]_o = 25$.
□ Initiator $\overline{P}_n = 5$, $[M]_o/[I]_o = 5$.

Models show that the complexing of the NCA's with the non-hydrogen bonded >*N*H group of an α-helix is greatly reduced if the terminal residue (at the growing end) has the "wrong" conformation. This is not surprising. According to the considerations advanced by Huggins (*43*) the side chain of the "wrong" residue makes bad contacts with the body of the helix; these can, however, be relieved by rotation of the residue about the *C′*-α*C* bond, but, as a result, sterically favorable contact between the adsorbed NCA and the terminal base group becomes less likely. The observation that a single wrong residue on a helix greatly reduces the rate of addition of *both* enantiomorphs clarifies the finding of Lundberg and Doty (*6*) that the rate of reaction in D,L mixtures is lower than that calculated on the basis of absolute preference of one type of chain for its own enantiomorph. The addition of the first D-residue to an α-helix composed entirely of L-residues was also found to be about six times slower than the L-L addition under the same conditions, in agreement with Lundberg and Doty (*4, 6*). This is similarly understandable in terms of Huggins' postulate (*43*) if the transition complex has a configuration close to that of the final state.

In our experience, the differences in rate between the addition of D- and L-residues to a helix of either type are more marked than the kinetic changes associated with the random coil–helix transition in pure enantiomorphs (Fig. 7.6). A similar conclusion may be drawn from the work of Lundberg and Doty (using solutions in *N,N*-dimethylformamide; ref. *6*). The reasons for this are at present obscure. It may be that owing to unelucidated steric features a helical chain can only continue to grow by reaction with adsorbed molecules as described. The situation can only be clarified by further work; in our opinion studies with preformed polymers, in which the isomeric terminal sequences are uniquely determined by synthesis, would be helpful.

Acknowledgment

We wish to thank Dr. A. C. P. Pugh for useful discussion and for his willingness to allow us to quote some of his unpublished results, and also Mr. D. R. Boreham, Mrs. D. Kapur, and Mrs. J. Watson for assistance with the experimental work. We are indebted to Miss D. Eldridge for secretarial services in the production of this manuscript.

References

1 Bamford, Elliott, and Hanby, "Synthetic Polypeptides," Academic Press, Inc., New York, 1956, Chapter III.
2 Katchalski and Sela, *Advances in Protein Chem.* **13,** 243 (1958).
3 Ballard and Bamford, *Proc. Roy. Soc.* **A236,** 384 (1956).
4 Doty and Lundberg, *J. Am. Chem. Soc.* **78,** 4810 (1956).
5 Idelson and Blout, *ibid.* **79,** 3948 (1957).
6 Lundberg and Doty, *ibid.* **79,** 3961 (1957).
7 Breitenbach and Allinger, *Monatsh. Chem.* **84,** 1103 (1953).
8 Ballard and Bamford, *Proc. Roy. Soc.* **A223,** 495 (1954).
9 (*a*) Flory, *J. Am. Chem. Soc.* **62,** 1561 (1940); (*b*) Waley and Watson, *Proc. Roy. Soc.* **A199,** 499 (1949).
10 Fessler and Ogston, *Trans. Faraday Soc.* **47,** 667 (1951).
11 Pope, Weakley, and Williams, *J. Chem. Soc.* **1959,** 3442.
12 Blout and Karlson, *J. Am. Chem. Soc.* **78,** 941 (1956).
13 Ballard and Bamford, *J. Chem. Soc.* **1956,** 381.
14 Idelson and Blout, *J. Am. Chem. Soc.* **80,** 2387 (1958).
15 Banford and Block, *J. Chem. Soc.* **1961,** 4989, 4992.
16 Heyns, Schultze, and Brockmann, *Ann. Chem.* **611,** 33 (1958)
17 Heyns and Schultze, *ibid.* **611,** 40 (1958).
18 Bilek, Derkosch, Michl, and Wessely, *Monatsh. Chem.* **84,** 717 (1953).
19 Bamford, Block, and Pugh, *J. Chem. Soc.* **1961,** 2057.
20 Ballard, Bamford, and Weymouth, *Proc. Roy. Soc.* **A227,** 155 (1955).
21 Bamford, *Chem. Spec. Publ.* No. 2, 47 (1955).
22 Fuchs, *Ber. deut. chem. Ges.* **55,** 2943 (1922).
23 Ware, *Chem. Revs.* **46,** 403 (1950).
24 Gold and Jefferson, *J. Chem. Soc.* **1953,** 1409, 1416.
25 Ballard, Bamford, and Weymouth, *Nature* **174,** 173 (1954).
26 Doty, Bradbury, and Holtzer, *J. Am. Chem. Soc.* **78;** 947 (1956).
27 See, e.g., Hurd, Buess, and Bauer, *J. Org. Chem.* **17.1,** 865 (1952).
28 Kopple, *J. Am. Chem. Soc.* **79,** 6442 (1957).
29 Goldschmidt and Wick, *Ann. Chem.* **575,** 217 (1952).
30 Saunders and Slocombe, *Chem. Revs.* **43,** 203 (1948).
31 Ballard and Bamford, *Chem. Soc. Spec. Publ.* No. 2, 25 (1955).
32 Blout, Karlson, Doty and Hargitay, *J. Am. Chem. Soc.* **76,** 4492 (1954).
33 Berger, Sela, and Katchalski, *Anal. Chem.* **25,** 1554 (1953).
34 Michaelis and Menten, *Biochem. Z.* **49,** 333 (1913).
35 Mitchell, Woodward, and Doty, *J. Am. Chem. Soc.* **79,** 3955 (1957).
36 Buyle and Hargitay, Symposium über Makromolecüle in Wiesbaden, Sektion V A1 (1959).
37 Weingarten, *J. Am. Chem. Soc.* **80,** 352 (1958).
38 Ballard and Bamford, *ibid.*, **79,** 2336 (1957).
39 Ballard, Bamford, and Elliott, *Makromol. Chem.* **35,** 222 (1960).
40 Ballard and Bamford, *J. Chem. Soc.* **1959,** 1039.
41 To be published.
42 Watson, personal communication.
43 Huggins, *J. Am. Chem. Soc.* **74,** 3963 (1952).

Discussion of Paper 7: Polymerization of *N*-Carboxy Anhydrides

E. R. BLOUT and M. IDELSON

In discussing the reaction mechanism for the polymerization of amino-acid *N*-carboxy anhydrides under various conditions we must be aware that the initiating conditions appear to determine to a considerable extent the nature of the product. For example, in previous work Dr. Bamford has shown that initiation with lithium halides leads to low molecular weight products such as cyclohexaglycine and hydantoin-3-acetic acid derivatives. On the other hand, we have shown that strong alkali initiation, e.g., sodium methoxide, leads to a high yield of very high molecular weight polypeptides. We have been unable to isolate any by-products from this type of polymerization. Also, when lithium bromide is added to a sodium methoxide–initiated polymerization, the reaction slows down and lower molecular weight polypeptides are produced. It would, therefore, appear that the reactions of *N*-carboxy anhydrides with lithium halides, on the one hand, and of *N*-carboxy anhydrides with strong alkalies lead to different products; hence one may conclude that they are different reactions proceeding by different mechanisms.

We must comment now on Drs. Bamford's and Block's disagreement with our published statements (*a*) that "no amine groups can be formed in triphenylmethylsodium-initiated polymerizations" and (*b*) that the mechanism of Bamford *et al.* which involves "the postulated addition to both 'amine' ends and 'NCA' ends of growing chains would lead to a broader molecular weight distribution than has been observed." Reaction scheme (5) outlined in the paper above leads to an entity bearing both amine and NCA groups in the ends. If we assume that steps (5a and b) are plausible (reaction [5a] has been studied by Berger, Sela, and Katchalski [ref. *33*] and by Blout and Karlson [ref. *12*]), then step (5c) requires that the sodium carbamate salt abstracts a proton from the protonated base. When the base is a strong one, e.g., triphenylmethyl anion, the Brönsted acid derived from it is very weak—much weaker than carbamic acids. Therefore, step (5c) will not proceed and amine groups cannot be formed by this route.

We now consider the question of molecular weight distribution. A polymerization of the type illustrated by scheme (5) is similar to condensation polymerizations which are known to produce broad molecular weight distributions. In contrast to this, Doty and Yang have observed that amino-acid *N*-carboxyanhydride polymerizations when initiated by sodium methoxide have narrow Poisson molecular weight distributions typical of addition polymerizations involving the opening of a ring (see P. J. Flory, "Principles of Polymer Chemistry," Cornell University Press, Ithaca, N. Y., 1953, p. 336, *et seq.*).

SETSUJI SAKURAI

In the polymerization of *N*-carboxyanhydrides of α-amino acids the molecular weight of the polypeptides obtained depends upon the characteristic of the polymerization medium, even when the same initiator is applied. Further, we have observed that in some cases an initiator which has little or no initiating ability in one solvent is effective in another solvent. Thus, for a discussion of polymerization mechanism consideration should be taken of the concerting effect of both the solvent and the initiator.

It was found that the polymerization initiated heterogeneously, namely, in a suspension of crystals of γ-methyl-L-glutamate-*N*-carboxylic anhydride (L-MG-NCA) in chloroform (or methylene dichloride), proceeded apparently much more rapidly than expected from its solubility in this solvent. The solubility of the NCA was estimated at only 2 per cent in chloroform. For example, the diethylamine-initiated polymerization of this type (at an A/I of 500 and 10 per cent NCA concentration) completed within 1.5 hours and produced

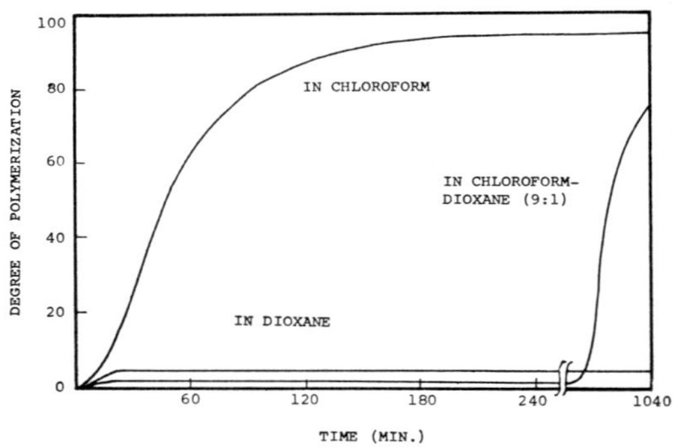

Fig. 7.7. Conversion-time curves for polymerization of γ-methyl-L-glutamate-NCA initiated with acetic acid ($A/I = 1.3$) in solution.

a polypeptide with a DP of 1500, while in the polymerization initiated homogeneously in dioxane solution more than 60 hours was required for the completion and the DP found was 800. Moreover, that as a polymerization medium for L-MG-NCA chloroform (as well as methylene dichloride) is extraordinary is clearly indicated by the polymerization experiment using acetic acid as initiator. Acetic acid, which has little initiating ability, was found to be able to cause the fast polymerization in solution as well as in suspension in chloroform. The typical results are shown in Fig. 7.7, from which it is seen that the difference with dioxane is very remarkable. Also indicated is that even an addition of 10 per cent dioxane retards markedly the polymerization. Because L-MG-NCA is probably more soluble in chloroform containing poly-γ-methyl-L-glutamate (L-PMG) than the original medium, the unexpectedly rapid disappearance of the crystals of NCA may be connected with the increase in its concentration in the proximity of the growing polypeptide chain.

K. D. KOPPLE

Dr. Bramford has presented a number of convincing arguments against isocyanates and mixed carbamic-carboxylic anhydrides as intermediates in aprotic base-catalyzed NCA polymerizations. The strongest argument is that such a path is too slow to account for observed propagation rates. For the reaction of ethyl isocyanate with phenylacetic acid in 1,2-dimethoxyethane at 24°, rate constants of 1.8×10^{-3} M^{-1} sec^{-1} for anhydride formation and 2.4×10^{-5} sec^{-1} for decomposition of anhydride to amine have been determined (K. D. Kopple and R. A. Thursack, unpublished data). The latter reaction is catalyzed by triethylamine, but even with the most foolishly optimistic assumptions about isocyanato acid or mixed anhydride concentrations in a polymerizing mixture, reaction velocities calculated using these constants are at least a hundredfold lower than those actually observed.

If the amide ions are indeed the propagating species in these polymerizations, perhaps one might worry about the occasional misstep in which amide ion derived from a peptide bond (or a carbobenzoxyamino link) attacks anhydride. Table 7.1 suggests low reactivity of amides or urethans for this process, but at later stages of polymerization (high amide, low NCA concentrations) it may be competitive. The result of this reaction would be a branched-chain with a diacylamide link at the branch point. If the diacylamide is cleaved during polymerization or work-up, a broadening of molecular weight distribution, rather than branching, would result.

PAPER 8

Aqueous Polymerization of *N*-Carboxy–α-Amino Acid Anhydrides

THOMAS K. MIWA and MARK A. STAHMANN

N-carboxy-amino acid anhydrides represent a class of high-energy amino-acid derivatives which polymerize to form polypeptides. It was known since the first synthesis in 1906 (*13*) that water in small quantities catalyzed the polymerization of these anhydrides; but in aqueous solutions, it was believed that these were hydrolyzed to their parent free amino acids with simultaneous evolution of CO_2 (*12*).

In 1953, Becker *et al.* (*5*) reported the polymerization of several of these anhydrides in buffered aqueous solutions. The effect of various buffer systems on the polymerization of *N*-carboxy-glycine anhydride was also reported. In a sodium phosphate buffer at *p*H 7.4 containing 1 per cent glycine anhydride, as much as 95 per cent yield of polymer was obtained; while in a sodium cacodylate buffer at *p*H 6.1 containing 2 per cent glycine anhydride, only 23 per cent polymerized. Later Tsuyuki *et al.* (*19*) reported that in sodium bicarbonate essentially 100 per cent yields for *N*-carboxy-L-leucine anhydride polymerizations were obtained. It was evident that the anion played an important role in the course of the polymerization of *N*-carboxy–amino acid anhydrides in aqueous systems.

This paper presents a further study on the effect of ions on the polymerization of *N*-carboxy–amino acid anhydrides. By using various ions at different concentrations and determining their effects on the turbidity formation, the extent of polymerization, and the rate of uptake of the *N*-carboxy–amino acid anhydride, it was found that the concentration as well as the nature of the various ions had a pronounced effect on the extent and rate of polymerization. The degree of polymerization increased with increase in concentration of the ion until an optimum concentration was reached, after which polymerization decreased with further increase in ionic concentration. In most cases the rate of uptake of the *N*-carboxy–amino acid anhydride increased with increase in the concentration of the ion. Exceptions were the chlorides of sodium, lithium, and hydrogen, which showed decreases at higher concentrations.

The results obtained were explained by the postulation of a polymerization mechanism involving a mixed anhydride intermediate of the anion and the *N*-carboxy–amino acid. An experiment was carried out whereby a by-product, benzoyl-L-leucine, predicted by this mechanism was isolated. Under the conditions employed, the by-product could have resulted from the formation of such mixed anhydride.

The rate of polymerization was measured by observing turbidity formation. After completion of the reaction, the fraction of anhydride that was hydrolyzed was determined colorimetrically with ninhydrin by comparison with a standard containing the free amino acid at the same concentration as the anhydride. The rest of the anhydride was assumed to have undergone polymerization.

Preliminary studies were carried out to determine the best concentration and type of the *N*-carboxy–amino acid anhydride with which to study the effect of ions on the polymerization of these anhydrides in aqueous systems. *N*-carboxy-L-leucine anhydride at 0.01 M concentration was found to be most suitable.

The anions studied included phosphate, bicarbonate, cacodylate, and chloride; the cations were sodium, lithium, and calcium. The common cation for the anions was sodium, while the common anion for the cations was chloride. Ammonia, which is used as an initiator for polypeptide syntheses, L-leucine, the hydrolysis product of the anhydride, and nucleic acids were also included. Hydrochloric acid and sodium hydroxide were used to study the effect of extremely low and high *p*H's.

Experimental

POLYMERIZATION OF *N*-CARBOXY-L-LEUCINE ANHYDRIDE: TURBIDITY AND ANHYDRIDE DISAPPEARANCE MEASUREMENTS

Each aqueous ionic solution was adjusted so that the desired final ionic concentration could be obtained when 3 ml of a dioxane solution of L-leucine anhydride was added to 12 ml of the aqueous solvent. The dioxane solution of anhydride was forced rapidly into the aqueous ionic solvent, resulting in a 0.0100 M anhydride solution containing 20 per cent dioxane by volume. Since the anhydride was only slightly soluble in water, it was necessary to use a concentrated dioxane solution which was diluted fivefold upon addition to the aqueous systems. However, the completely hydrolyzed product (free amino acid) was soluble at this concentration and gave no turbidity. After adding the anhydride to the ionic solution, the tube was swirled and shaken for 15 seconds, then placed in the spectrophotometer and allowed to stand for 10 seconds. Then the first 1 ml aliquot was taken and added to the hydroxylamine reagent for anhydride concentration determination, while simultaneously a turbidity reading at 540 mμ was taken at 30 seconds after the mixing. Aliquots for anhydride concentration determination and the turbidity readings were taken subsequently at given intervals. The reaction tube was never shaken, except for the initial vigorous shaking. All experiments were carried out at room temperature (*ca.* 23° C.).

DETERMINATION OF *N*-CARBOXY–AMINO ACID ANHYDRIDE CONCENTRATION: A GENERAL METHOD

The *N*-carboxy–amino acid anhydride was added to 4.00 ml of aqueous 2.0 M hydroxylamine hydrochloride solution (*p*H 3.4) and mechanically shaken for exactly 15 minutes. A colorless hydroxamic acid hydrochloride was produced, which upon addition of a large excess of hydrochloric acid (2.00 ml of 4 N HCl) and ferric ion (2.00 ml of 0.37 M $FeCl_3$) formed a brown ferric hydroxamate complex. The final volume was adjusted to 10.00 ml by adding water and the optical density of the solution measured at 540 mμ. Interfering polymers were removed by centrifugation and filtration.

The purity of the anhydride preparation was determined by measuring the CO_2 gas evolved in a modified Van Slyke volumetric (blood volume) apparatus.

DETERMINATION OF EXTENT OF POLYMERIZATION

The percentage of hydrolysis of the anhydride for each reaction was determined by the colorimetric ninhydrin method (*19*), using free L-leucine as standard. The percentage of polymerization was calculated from the relationship:

$$\%\ \text{POLYMERIZATION} = \left(1 - \frac{\text{Optical density for sample}}{\text{Optical density for 0.01 M } L\text{-leucine}}\right) \times 100.$$

DETERMINATION OF CHAIN LENGTH OF SOLUBLE L-LEUCINE POLYMER

Total nitrogen was determined by the Kjeldahl method and amino nitrogen by the nitrous acid–Van Slyke volumetric method. The chain length was calculated by dividing the concentration of total nitrogen by the concentration of amino nitrogen.

POLYMERIZATION OF *N*-CARBOXY-L-LEUCINE ANHYDRIDE IN NEUTRAL SODIUM BENZOATE

The sodium benzoate buffer was prepared by dissolving 153 mg (1.25 mmoles) of benzoic acid and 50.0 mg (1.25 mmoles) of sodium hydroxide in 100 ml of solution. The *p*H of the solution was 6.9 (23° C.). The *N*-carboxy-L-leucine anhydride (0.500 mmole) was dissolved in 10.0 ml of dioxane and to this was added 40.0 ml of the neutral sodium benzoate solution, resulting in a 0.010 M solution of anhydride and sodium benzoate. After the system was shaken for one hour, the mixture was acidified and shaken with two 50 ml portions of diethyl ether to extract the benzoic acid and benzoyl-leucine. The ether extract was concentrated to near dryness and dissolved again in diethyl ether. This solution (10.0 ml total) was analyzed for benzoic acid and benzoyl-leucine by silica gel column chromatography and ultraviolet absorption spectrophotometry. The silicic acid was pretreated with 0.5 N sulfuric acid and packed in a 0.9×40 cm glass column. The

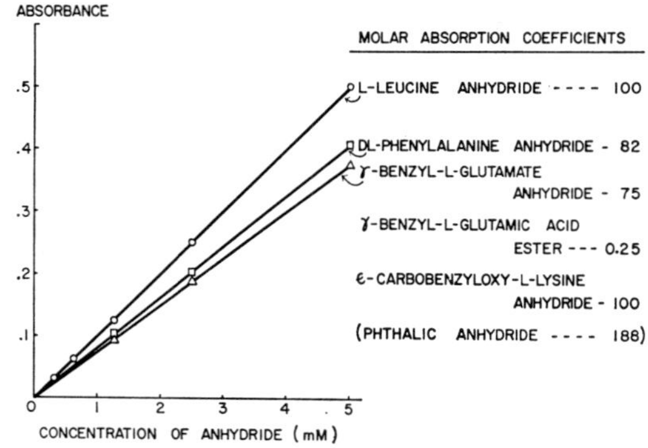

Fig. 8.1. Ferric hydroxamate complexes of *N*-carboxy-α-amino acid anhydrides (hydroxylamine solution *p*H 3.4; 540 mμ).

sample was eluted with *n*-butanol-0.5 N aqueous H_2SO_4—glacial acetic acid (13:2:3, *v/v/v*), 0.10 ml of each 5 ml eluate diluted to 3.0 ml with ether, and quantitatively analyzed with a Cary automatic spectrophotometer.

Results and discussion

Stability of *N*-carboxy–amino acid anhydrides in dioxane and choice of l-leucine anhydride for polymerization measurements

The linear relationship between the absorbance of the ferric hydroxamate complexes from the *N*-carboxy–amino acid anhydrides and the concentration of the anhydrides is shown in Fig. 8.1.

Among the four *N*-carboxy–amino acid anhydrides tested, L-leucine anhydride was found to be the most stable at low concentrations. As shown in Table 8.1, at 0.025 M no change in

Table 8.1—Stability of *N*-carboxy–amino acid anhydrides in dioxane (% remaining after 8 days)

Anhydride	0.1 M	0.05 M	0.025 M
L-leucine	10%	40%	100%
DL-phenylalanine	24	65	85
γ-benzyl-L-glutamic	22	44	98
ε-carbobenzoxy-L-lysine			83

concentration of L-leucine anhydride was observed after it had stood for eight days at room temperature. However at higher concentrations, such as 0.1 M, L-leucine disappeared most rapidly.

The rate of disappearance of *N*-carboxy-L-leucine anhydride in dioxane at various concentrations is shown in Fig. 8.2. No initiator was added and no shaking or stirring was applied. The dotted lines indicate the changes occurring in anhydride concentration. The fact that the 0.05 M solution was stable for three days made possible, for kinetic studies, the preparation of an anhydride stock solution at this concentration. This stock solution was diluted fivefold into the aqueous ionic solutions for polymerization measurements. Since L-leucine anhydride was poorly soluble in aqueous solutions, it was necessary to dissolve it in dioxane before addition to the aqueous solvents. A low anhydride concentration was favored because, when 100 per cent hydrolysis had occurred, the solution remained clear with no measurable turbidance. Also when the anhydride concentration in the aqueous ionic solutions was greater than 0.01 M, the rate of polymerization and anhydride disappearance was too fast to measure and the differences among the various ionic solutions as well as the effect of variations in ionic concentration could not be shown to full advantage. Therefore, all measurements were made at 0.0100 M anhydride concentration in aqueous solutions containing 20 per cent by volume of dioxane.

The soluble fraction of a polyleucine reaction mixture which had shown a negative ninhydrin reaction was analyzed for chain length by determination of the total nitrogen and the amino nitrogen. The measured value of 2.6 meant that peptides larger than about four residues had precipitated from the system. This would also

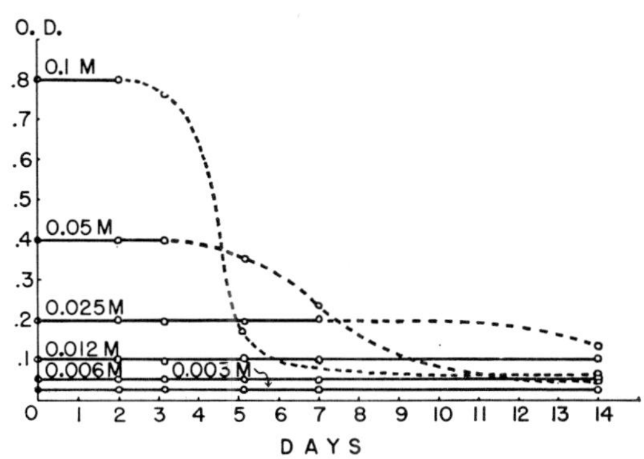

Fig. 8.2. Reaction of *N*-carboxy-L-leucine anhydride in purified dioxane. (No initiator added; change in anhydride concentration observed.)

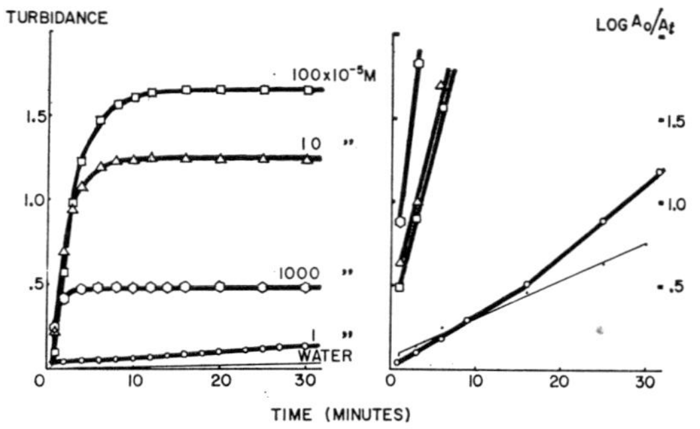

Fig. 8.3. Polymerization in aqueous sodium phosphate (0.01 M *N*-carboxy-L-leucine anhydride).

mean that a significant extent of polymerization could occur without any turbidity formation, as was actually observed in some of the experiments carried out.

Effect of ions on the polymerization of *N*-carboxy-L-leucine anhydride

Sodium phosphate.—The effect of sodium phosphate at four different concentrations is shown in Fig. 8.3. At the left is shown the rate of polymerization as a plot of turbidance against time. The bottom thin line is for water without any added ions. Turbidance increased with the increase in concentration of phosphate until an optimum concentration was reached, after which the turbidance decreased until no turbidity was observed at concentrations of 0.1 M or more. Although the experiment was run at twelve different concentrations of the phosphate, only four were plotted to avoid crowding the figure. The plot of turbidance versus concentration is given and discussed in a later section.

On the right half of Fig. 8.3, the rate of anhydride disappearance is plotted as the logarithm of the ratio of the initial concentration A_o, to anhydride concentration at the given time A_t, against time (*11*). The greater the slope, the faster the anhydride disappears. The rate constants for the curves are given in a later section. At the lowest concentration of phosphate, a slight secondary rate increase was observed.

Sodium bicarbonate and sodium cacodylate.—Similar plots for sodium bicarbonate and sodium cacodylate are given in Figs. 8.4 and 8.5. The rates of anhydride disappearance increased with increase in ionic concentrations. It is interesting to note that at 10×10^{-5} M concentration, cacodylate produced a greater amount of turbidity than bicarbonate, while at 1000×10^{-5} M (the same as that of the anhydride) cacodylate produced very little turbidity and bicarbonate gave a yield even greater than phosphate.

Comparison of various ions at the same molar concentration.—The concentration 0.001 M was chosen for comparing the various ions since the preceding three anions gave maximum turbidity at about this concentration. If these ions functioned as initiators which remained attached to the growing chains, then theoretically the average chain length would be 10. If the concentration of the ions and the anhydride were both 0.01 M, there would be no polymerization, provided the initiation was much faster than propagation. Such, however, was not the case.

As seen in Fig. 8.6, sodium phosphate gave

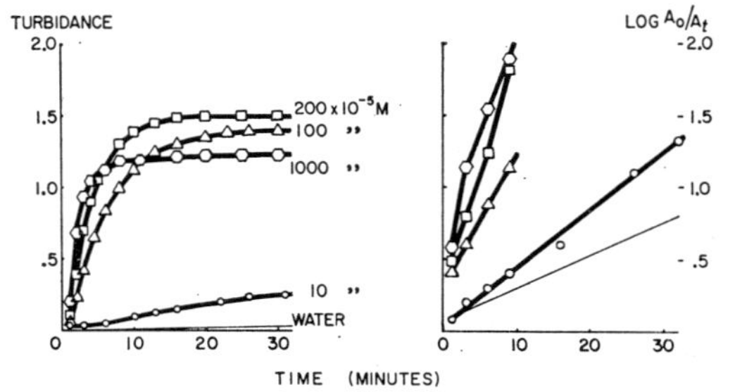

Fig. 8.4. Polymerization in aqueous sodium bicarbonate (0.01 M *N*-carboxy-L-leucine anhydride).

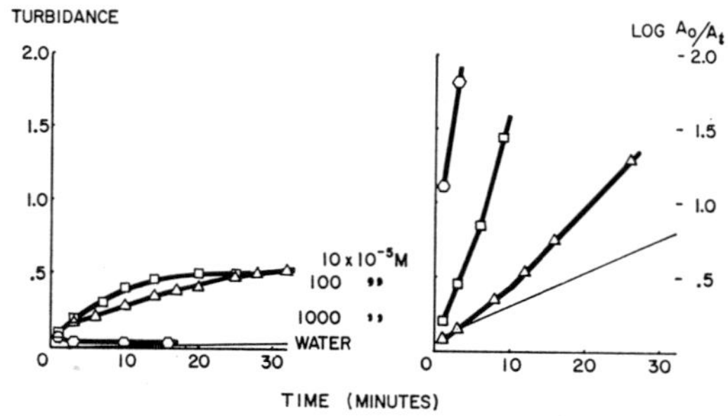

Fig. 8.5. Polymerization in aqueous sodium cacodylate (0.01 M *N*-carboxy-L-leucine anhydride).

the greatest amount of turbidity, sodium bicarbonate slightly less, sodium cacodylate only a moderate amount, and sodium chloride none at all. The broken line represents water without added ions. Ammonia produced only a small amount of turbidity and L-leucine none at all. This may be due to the reduction in activity of the amino group by protonation. Lithium chloride produced about the same amount as sodium cacodylate but at a slower rate. Calcium chloride was intermediate between sodium phosphate and sodium chloride. Figure 8.6 shows the striking differences among the anions having the common sodium cation.

Figure 8.7 shows the relative rates of anhydride disappearance in these same ionic solutions. Phosphate and cacodylate were very rapid, bicarbonate slightly less, and sodium chloride even slower than water alone. Calcium chloride was intermediate, and lithium chloride very slow with a secondary decrease in its rate.

Cations.—The effect of changing the concentration of cations on the polymerization of anhydride is shown in Fig. 8.8, chloride being the common anion. At the left is shown the maximum turbidance obtained at the various concentrations of the ions. Sodium chloride failed to produce any turbidity at the range of 0.1 M to 0.001 M. Maximum turbidity formation with lithium chloride occurred at 0.01 M, the same as the anhydride concentration, while with calcium it was 0.001 M. At the right is shown the

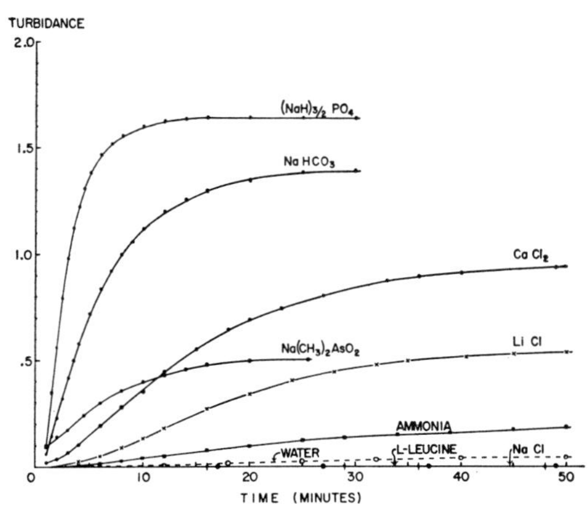

Fig. 8.6. Polymerization in 0.001 M aqueous ionic solutions (0.01 M *N*-carboxy-L-leucine anhydride).

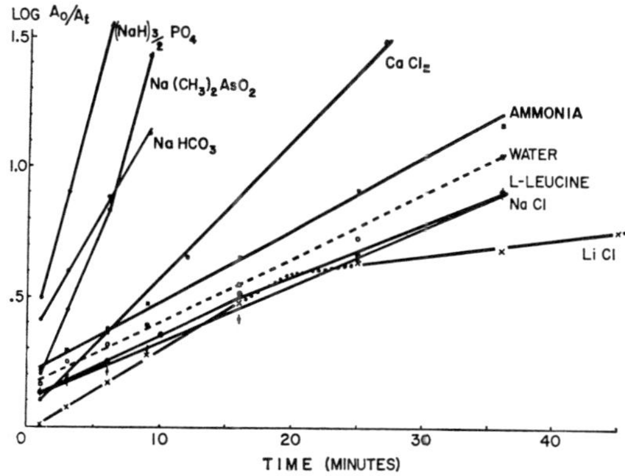

Fig. 8.7. Polymerization in 0.001 M aqueous ionic solutions (0.01 M *N*-carboxy-L-leucine anhydride).

extent of polymerization calculated from the percentage hydrolyzed. All three chlorides produced maximum polymerization at a concentration equal to that of the anhydride.

Variation in concentration of the anions.—The effect of variation in concentration of the various types of anions on the turbidity formation by the anhydride is shown in Fig. 8.9. Phosphate, bicarbonate, and cacodylate showed maximum turbidity production at around 0.001 M or one-tenth that of the anhydride. The deoxyribonucleic and ribonucleic acids had their maximum at around 0.01 "N." The concentration of DNA and RNA is expressed as equivalents of nucleotide per liter. For convenience, the equivalent weight was assumed to be 400. At ionic concentrations higher than that of the anhydride, the turbidance decreased rapidly in every case.

The effect of these anions on the extent of

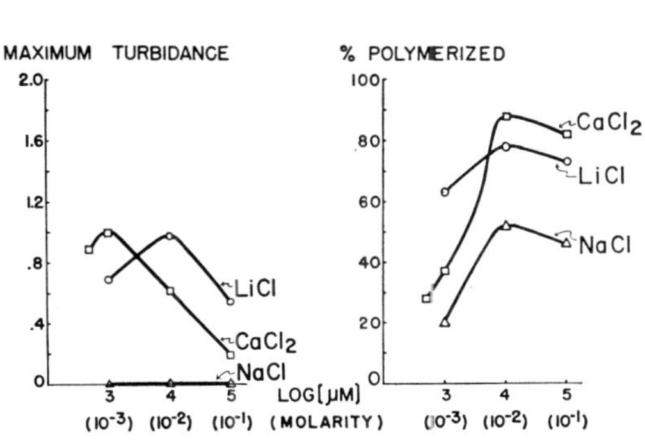

Fig. 8.8. Effect of cations at various concentrations on the polymerization of 0.01 M *N*-carboxy-L-leucine anhydride.

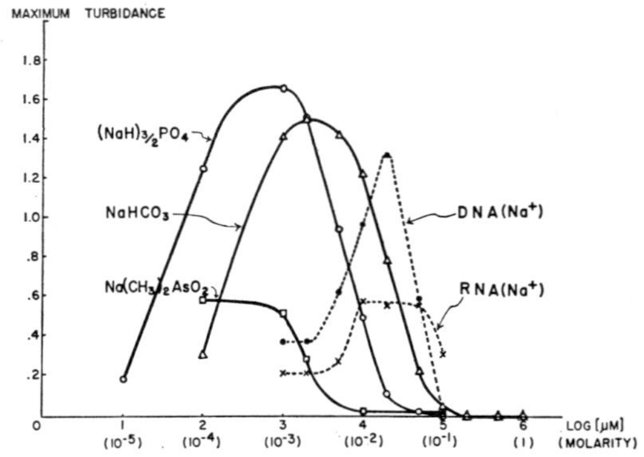

Fig. 8.9 Effect of type and concentration of anions on the turbidity formation from 0.01 M *N*-carboxy-L-leucine anhydride.

polymerization is shown in Fig. 8.10. At very low ionic concentrations phosphate was most effective, while at higher concentrations bicarbonate was as effective as phosphate. (Polymerizations greater than 95 per cent could be considered complete since the short-chain polymers may have contributed about 5 per cent coloration in the ninhydrin tests.) It is interesting to note than DNA produced maximum polymerization at a concentration equal to that of the anhydride. This may suggest that the anhydride molecules became attached to the phosphates of the polynucleotides and a favorable alignment resulted which could readily react to form the polypeptide. The mixed anhydride mechanism, which is postulated in a later section, may provide an explanation for these reactions. When the concentration of the

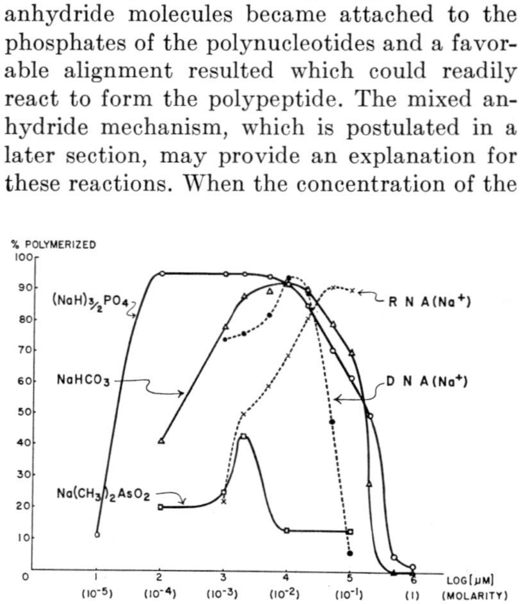

Fig. 8.10. Effect of type and concentration of anions on the extent of polymerization of 0.01 M *N*-carboxy-L-leucine anhydride.

nucleotide phosphate was greater than that of the anhydride, gaps between the anhydride molecules may have prevented peptide formation, thus increasing the percentage of hydrolysis. The rapid drop in percentage of polymerization at high concentrations of DNA give strong support to this view. If the DNA phosphate were to act similarly to inorganic phosphate, the curve obtained would have resembled that of inorganic phosphate and would have a relatively broad peak.

These studies show that neutral sodium phosphate has the widest concentration range for effective aqueous polymerization of *N*-carboxy-L-leucine anhydride.

Acid and base.—The effect of high concentrations of acid and base on the extent of polymerization is shown in Fig. 8.11. Between *p*H's 0 and 3, HCl produced no turbidity, although at 3 (which is one-tenth the concentration of the anhydride) about 35 per cent of the anhydride had formed peptide bonds. The rate of anhydride disappearance in acid solutions was extremely slow especially at the higher acid concentrations. This was not surprising since, during synthesis of the anhydride, HCl is liberated in the cyclization step.

In contrast, sodium hydroxide produced a large amount of turbidity and 93 per cent polymerization at 0.001 M (or *p*OH 3). The rate of anhydride disappearance for sodium hydroxide was greater than any of the other ions tested. Therefore, with proper adjustment of the ratio of anhydride to alkali base, polypeptides may be synthesized in aqueous alkaline systems.

Rates of anhydride disappearance.—A set of curves showing the rate of anhydride disappearance at various concentrations of sodium

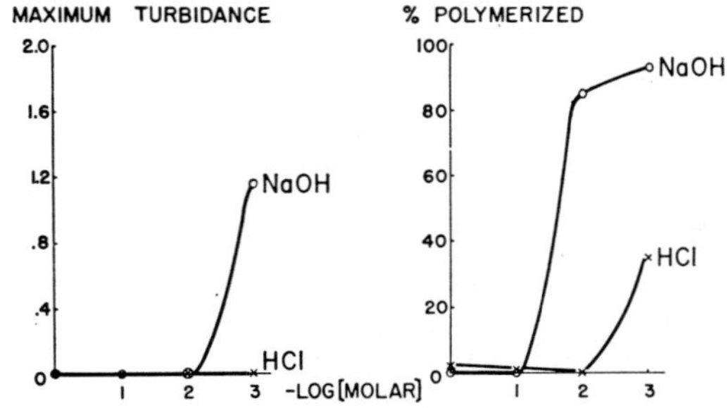

Fig. 8.11. Effect of HCl and NaOH at various concentrations on the polymerization of 0.01 M *N*-carboxy-L-leucine anhydride.

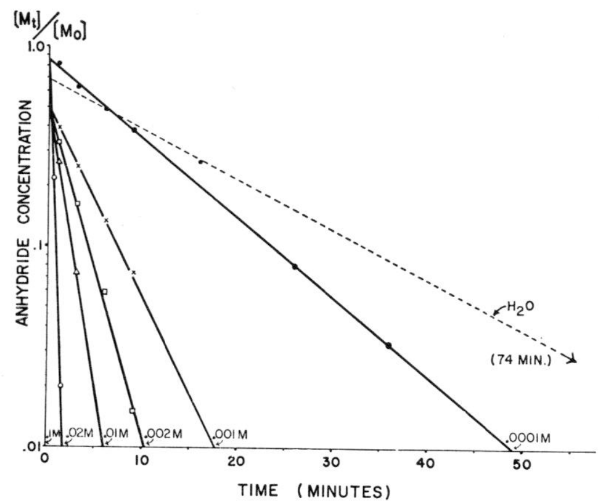

Fig. 8.12. Rate of disappearance of *N*-carboxy-L-leucine anhydride in various concentrations of sodium bicarbonate solutions; $[M_o] = 0.01$ M.

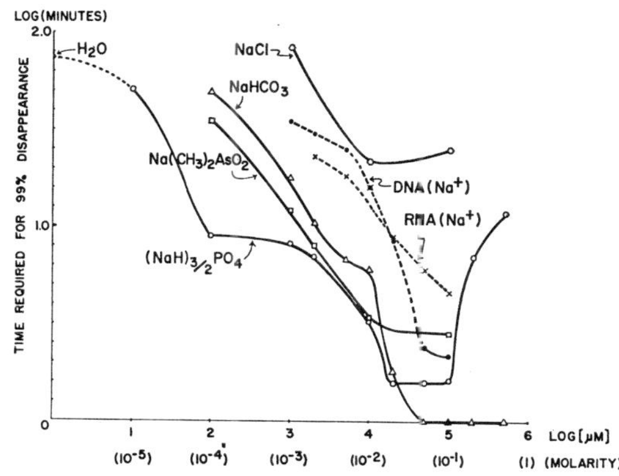

Fig. 8.13. Effect of type and concentration of anions on the rate of disappearance of 0.01 M *N*-carboxy-L-leucine anhydride.

bicarbonate is shown in Fig. 8.12. The fraction of anhydride remaining at a given time is plotted on a semi-log scale against time. On the ordinate, 0.1 represents 90 per cent and 0.01 means 99 per cent disappearance. The dotted line is for water without added ions. The concentration of bicarbonate decreases from left to right and so does the rate of anhydride disappearance.

The time required for 99 per cent disappearance of the anhydride in solutions of the various types of anions was plotted on semi-log scale against the logarithm of the concentration of the ions, as shown in Fig. 8.13. All salts had sodium as the common cation. Sodium hydroxide which does not appear in the figure had the shortest reaction times, being zero between 1 M and 0.01 M and less than 0.5 on the ordinate at 0.0001 M. Sodium chloride had the slowest rate of anhydride disappearance. At extremely high ionic concentrations of phosphate, the rate of anhydride disappearance reversed to a slower rate. This reversal is explained later in the section discussing the mechanism of reaction.

Anhydride disappearance rate constants.—The bimolecular rate constants for these reaction systems are listed in Table 8.2. These were calculated in the conventional manner for first-order reactions (*8*). Although the concentrations of both the anhydride and the ion are in-

Table 8.2—Anhydride disappearance rate constants for the polymerization of 0.01 M *N*-carboxy-L-leucine anhydride in the various aqueous ionic solutions (k = liter/mole/sec)

Molarity =	.1	.01	.001	.0001	10^{-5}
NaOH	≥.67	≥6.7	20	70	
$(NaH)_{3/2}PO_4$	.125	.78	3.50	30	53 [71]*
$NaHCO_3$	≥.67	.46	1.57	6.6	
$Na(CH_3)_2AsO_2$	.075 – .17	.59	2.1	6.7 [8.9]	
$CaCl_2$	.015 (.0037)†	.16 (.024)	.86		
LiCl	.0090 (.0023)	.23 (.095)	.52 (.20)		
NaCl	.0062 (.0020)	.040 (.014)	.37		
HCl	.0027 (.00073)	.040 (.024)	.35 (.071)		
L-leucine			.38		
$NH3 \cdot H_2O$			.50		
DNA (Na^+)	<.27	.43	2.0		
RNA (Na^+)	.052	.125	.87		

* Brackets indicate secondary rate increase.
† Parentheses indicate secondary rate decrease.

volved, the effective concentration of the ion (or initiator) remains constant and therefore the reaction behaves as one in the first order. A discussion on the applicability of this assumption has been given by Idelson *et al.* (*11*). The straight lines we have obtained by such plots suggest the existence of first-order reactions in these aqueous ionic solutions.

The rate constants varied continuously with change in concentration of the ions. In every case the constant decreased with increase in concentration of the ion. This may be due to the gradual approach of the apparent or effective initiator concentration of the ion to the true initiator concentration. At low initiator concentrations, the effective concentrations of the ion appeared to be greater than the true initiator concentrations.

Secondary increases and decreases in disappearance rates are listed in the table in brackets and parentheses, respectively. Except for sodium bicarbonate and L-leucine, all salts were at neutral *p*H. Sodium bicarbonate ranged from *p*H 8.6 at the highest concentration to 7.4 at the lowest. L-leucine at 0.001 M was *p*H 6.2. At the higher concentrations of sodium hydroxide, bicarbonate, cacodylate, and DNA, exact measurements were not possible because of the completion of the reaction before or immediately after the initial determination of the anhydride concentration. Sodium hydroxide gave, by far, the highest disappearance rate while sodium chloride and hydrochloric acid gave the lowest.

The kinetic study of *N*-carboxy–amino acid anhydrides in aqueous solutions reported by Bartlett *et al.* (*4*) was made at very high ionic concentrations. The only condition employed by these workers which could be compared was their reaction of 0.02 M *N*-carboxy–α-amino isobutyric acid anhydride in 0.1 M HCl. Their reported value for the rate constant, 2.4×10^{-3}, is close to the value 2.7×10^{-3} obtained here for 0.01 M *N*-carboxy-L-leucine anhydride in 0.1 M HCl.

The values reported by other workers (*7, 11, 3, 14*) for polymerizations in organic media were obtained with higher anhydride concentrations; but in general, the reaction rates in aqueous ionic systems were about a hundredfold to a thousandfold greater than those in organic systems. Blout *et al.* (*7*) have found that strong-base initiated polymerizations in organic solvents give rate constants about a hundred times that for amine-initiated polymerizations.

No attempt will be made to theorize on the secondary rate increase or decrease at this time because of lack of experimental data. However, it could be pointed out that the secondary rate increases occurred only at very low concentrations of the anions which gave the higher rate constants, and that the secondary rate decreases occurred only with the chlorides over a wide (a hundredfold) range of ionic concentrations. These chlorides gave the lowest rate constants.

Mechanism of polymerization of *N*-carboxy-L-leucine anhydride in aqueous systems

Postulation of a mixed anhydride intermediate.—The pronounced effect of ions, especially the anions, on the rate of anhydride consumption and extent of polymerization has led us to postulate a mechanism of polymerization involving a mixed anhydride intermediate of the particular anion and the *N*-carboxy–amino acid as illustrated in Fig. 8.14.

The initial step of polymerization is the addition of the anion to the 5-carbonyl of the oxazolidine ring as follows:

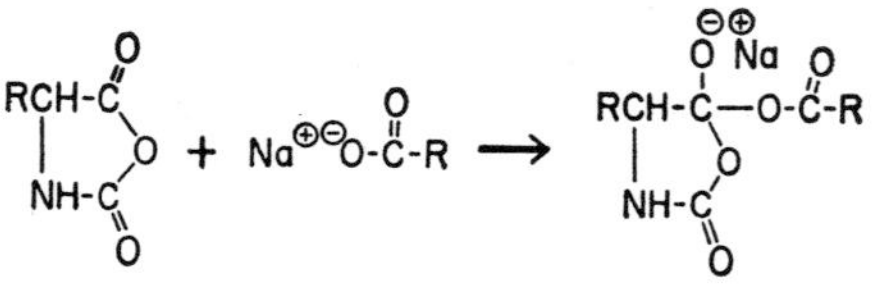

The return of the alcoholate to the carbonyl and the simultaneous opening of the ring produces the N-carboxylate of the α-carboxy-mixed anhydride which is shown as (II) in Fig. 8.14. The benzoate ion is shown since it was employed in an experiment to provide evidence for the mixed anhydride intermediate. The existence of the N-carboxylate has been reported by previous investigators (*1, 18*) but the mixed anhydride has never been previously

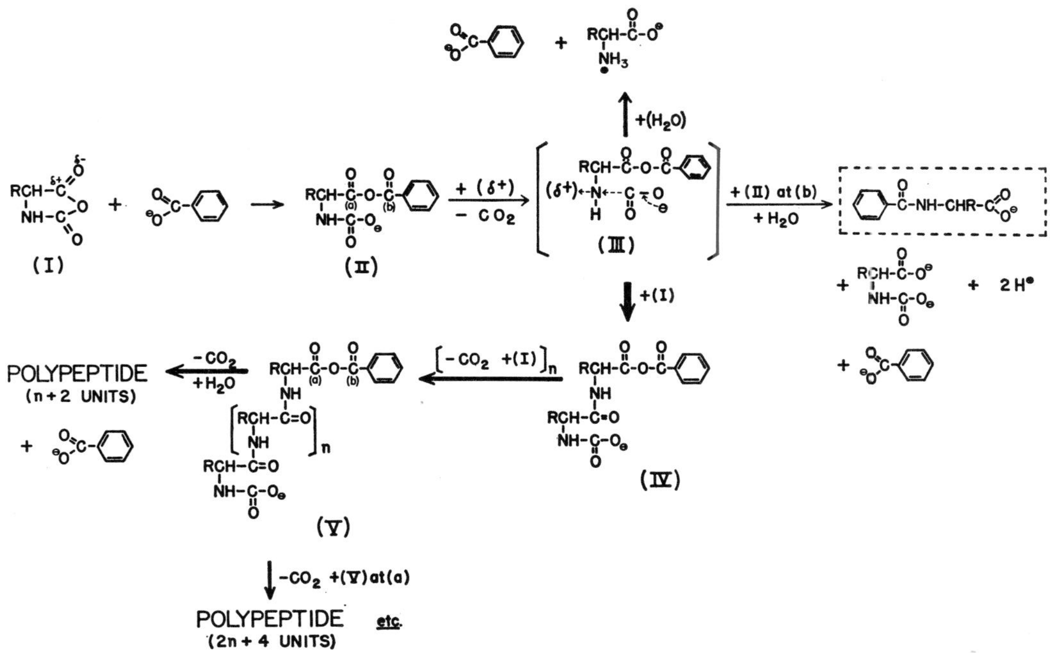

Fig. 8.14. Proposed mechanism for the polymerization of *N*-carboxy-L-leucine anhydride in neutral aqueous sodium benzoate solutions.

suggested. This initiation reaction is very rapid in the cases of phosphate, cacodylate, and bicarbonate, but very slow with chlorides. The bulkiness in ionic radius of the chloride may be the hindering cause.

Formation of polypeptides.—Upon approach of a positive center toward the α-nitrogen, the N-carboxylate decomposes as the bond between the positive center and the nitrogen is formed. This is shown as (III) in Fig. 8.14. If the positive center is the 5-carbonyl of an anhydride molecule (I), a peptide bond is formed as is shown by the downward arrow. Repetition of this step leads to the formation of a polypeptide molecule. The polypeptide intermediate (V) is still activated at both the amino and carboxyl ends and may interact intramolecularly to form a cyclic polypeptide or intermolecularly to form a polypeptide much longer than calculated initially. In less polar solvents where the N-carboxylates are unstable, (*a*) the return of the alcoholate to the carbonyl, (*b*) opening of the ring to form the N-carboxylate, (*c*) decarboxylation, and (*d*) formation of the (δ+) N bond occur simultaneously. (II), (III), and (IV) would then be represented as:

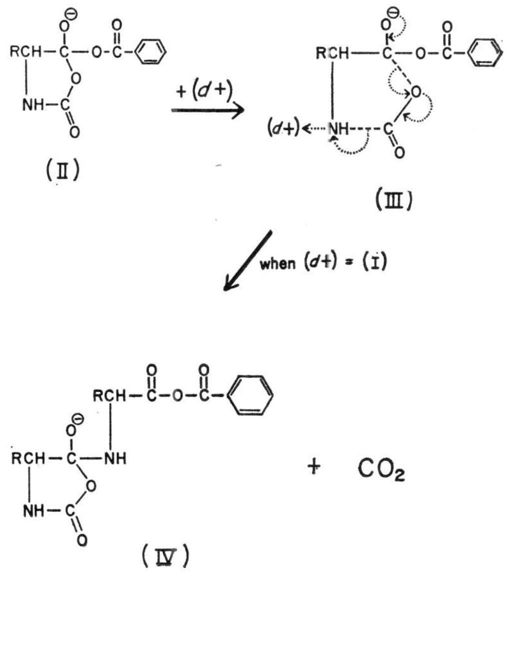

Initiators that liberate protons may retard this push-pull reaction by covalent bond formation with the alcoholate anion.

In organic solvents, when an alkoxide of an alkali is used as initiator, the ester (which may be considered an anhydride of a carboxylic acid and a very weak alcoholic acid) could interact with the N-carboxylate to form a polypeptide of chain length greater than calculated by the anhydride to initiator ratio. This may explain the results reported by Blout *et al.* (*6*). An intramolecular ester-amide interchange releasing an alcohol and producing a cyclic polypeptide has been observed by Lundberg *et al.* (*14*). They have observed that when the reaction time for polymerization is prolonged, this cyclization reaction becomes pronounced, while at shorter times cyclization is negligible. Similar observations were reported by Hanby *et al.* (*9*).

Now if (IV) were to cyclize, a diketopiperazine derivative will be obtained (*18*). However, due to steric hindrance this reaction would proceed with great difficulty as was found to be true even in organic solvents (*13, 18*).

When the concentration of the anhydride (I) is greater than that of the anion, the intermediate (II) could react with the excess anhydride (I) to form peptides. However, if the concentrations of both the anhydride and the anion were about equal and the initiation step were rapid, the predominating species would be the intermediate (II). If this intermediate anhydride is unstable, hydrolysis would occur. If it were stable, then it could interact with another intermediate anhydride to form a peptide. The decarboxylating nitrogen of one could form a bond with the α-carbonyl at site (*a*) of the mixed anhydride of the second. This may explain the short-chain soluble peptides obtained at higher concentrations of phosphate and bicarbonate.

The mixed anhydride of L-leucine and phosphate appeared to be the most stable. When the phosphate concentration was increased twentyfold to fiftyfold in excess of the anhydride, the rate of the anhydride disappearance, as determined by the ferric hydroxamate method, decreased drastically as pointed out during the discussion of Fig. 8.13. The high concentration of phosphate had retarded the hydrolysis reaction (upward reaction in Fig. 8.14) in accordance with the law of mass action, and since no anhydride (I) was available for peptide formation, the mixed anhydride (II) was stabilized and produced positive ferric hydroxamate tests for an extended period.

Formation of urea and hydantoin derivatives.—If the intermediates (II), (IV), or (V) were to react with the 2-carbonyl of (I), a urea derivative would be formed. The urea derivative could readily cyclize to form a hydantoin. Since the α-carboxyl of the incoming mixed anhydride (II) would still be activated, the cyclization proceeds with greater ease than if it were the free carboxylate. It is not surprising, therefore, that in providing evidence for the formation of urea derivatives from *N*-carboxy–amino acid anhydrides, various authors (*16, 17, 18*) isolated derivatives of hydantoin 3-acetic acid corresponding to the amino-acid anhydrides used.

Hydrolysis.—Conversion of the *N*-carboxy-L-leucine anhydride to the parent amino acid results when the anhydride bond of the intermediate (II) is hydrolyzed and the positive center ($\delta+$) approaching the N-carboxylate is a proton. When the concentration of the hydrogen ion was equal to or greater than that of the anhydride, the positive centers were predominately hydrogen ions; therefore, very little or no polymerization occurred, as shown in Fig. 8.11.

Evidence for the mechanism of polymerization involving a mixed anhydride.—When the concentrations of the *N*-carboxy–amino acid anhydride (I) and the anion were about the same and the mixed anhydride (II) predominated, (II) could interact, as discussed previously, at site (*a*) of the mixed anhydride to form a peptide bond. It could also interact at site (*b*), the carboxyl of the anion, to form the amide of the anion.

A reaction was carried out using neutral aqueous sodium benzoate and *N*-carboxy-L-leucine anhydride at equal concentrations in a manner as described in the experimental section. The product benzoyl-leucine, which was isolated by silica gel column chromatography, could have formed by means of the mixed anhydride intermediate as discussed above. In Fig. 8.14, the benzoyl-leucine is shown at the right enclosed in broken lines.

The absorption curves for benzoyl-leucine eluted from the columns, containing known benzoyl-leucine in one case and the reaction mixture in another, were almost identical. Benzoyl-leucine was well separated from benzoic acid. Benzoic acid appeared in fractions No. 2, 3, and 4, the peak being at No. 3. Benzoyl-leucine in the standard appeared in fraction No. 20, but in the sample from the reaction mixture there was a broad peak starting from No. 18 and leveling off at No. 29.

By comparing the optical density units for benzoic acid and benzoyl-leucine, it was found that 7 per cent of the benzoate was converted to benzoyl-leucine. Also, since the benzoic acid and benzoyl-leucine concentration in the soluble fraction accounted for only 90 per cent of the original benzoate, it suggested that some benzoyl-polyleucine was formed. The precipitate from the polymerization reaction of *N*-carboxy-L-leucine anhydride and neutral sodium benzoate was thoroughly washed to remove all of the free benzoic acid, the ether washings checked spectrophotometrically. Then upon hydrolysis of the precipitate, benzoic acid was detected spectrophotometrically, accounting for 1 per cent of the original benzoate. Further evidence for the presence of benzoyl-leucine was obtained when the ether extract from the acidified original reaction mixture was analyzed for leucine before and after acid hydrolysis. The ether extract before hydrolysis gave a negative ninhydrin test, but upon hydrolysis gave a dark blue color.

Summary

Investigation of the effect of ions on polymerization of *N*-carboxy-leucine anhydride in aqueous systems was carried out by allowing the polymerization to occur in the presence of various ions at different concentrations and determining their effects on turbidity formation, extent of polymerization, and rate of uptake of the *N*-carboxy–amino acid anhydride. It was found that the concentration as well as the nature of the various ions had a pronounced effect on the extent and rate of polymerization. The extent of polymerization increased with increase in concentration of the ion until an optimum concentration of the ion was reached, after which polymerization decreased with increase in ionic concentration. The rate of uptake of the *N*-carboxy–amino acid anhydride increased with increase in the concentration of the ion.

Pronounced differences among the ionic solutions were observed in (1) the extent of polymerization, determined at the end of the reaction as the fraction not hydrolyzed to the parent amino acid (by colorimetric ninhydrin); (2) the maximum turbidity produced at each concentration of the ions; and (3) the rate of anhydride disappearance, determined by an anhydride analysis based on a colorimetric ferric hydroxamate method developed for these studies. For example, sodium phosphate and sodium bicarbonate produced very high yields of leucine polymers at very rapid rates over a wide range of ionic concentrations, but sodium chloride produced very low yields at very slow rates.

The effect of high concentrations of acid and base was also studied. Sodium hydroxide, at concentrations equal to or less than that of the anhydride, was very effective in causing fast polymerization, while hydrochloric acid, at concentrations equal to or greater than that of the anhydride, inhibited the rate and extent of polymerization markedly.

The nucleic acids, deoxyribonucleic and ribonucleic, produced maximum amounts of polymers when the equivalent concentration of the nucleotide phosphate was equal to the molar concentration of the *N*-carboxy–amino acid anhydride.

Propagation rate constants for these pseudo-first-order reactions were determined and the results showed sodium hydroxide solutions to have the highest rate constants and hydrochloric acid solutions the lowest.

These effects of the ions, especially the anions, were explained by postulation of a mechanism of polymerization involving a mixed anhydride intermediate of the particular anion and the *N*-carboxy–amino acid. An experiment was carried out whereby a by-product predicted by this mechanism was isolated, neutral benzoate being used as the anion. The by-product of the polymerization, benzoyl-leucine, which was isolated by silica gel column chromatography, seems to have been formed by means of a mixed anhydride of the benzoate anion and *N*-carboxy-leucine.

Acknowledgment

This work was supported by research grants (No. E-101) from the National Microbiological Institute of the National Institutes of Health, U.S. Public Health Service.

References

1 J. L. Bailey, *J. Chem. Soc.* **1950,** 3461.
2 D. G. H. Ballard, C. H. Bamford, and F. J. Weymouth, *Nature* **174,** 173 (1954).
3 C. H. Bamford, A. Elliott, and W. E. Hanby, "Synthetic Polypeptides," Academic Press, Inc., New York, 1956.
4 P. D. Bartlett and R. H. Jones, *J. Am. Chem. Soc.* **79,** 2153, 2159 (1957).
5 R. R. Becker and M. A. Stahmann, *J. Biol. Chem.* **204,** 737 (1953).
6 E. R. Blout, R. H. Karlson, P. Doty, and B. Hargitay, *J. Am. Chem. Soc.* **76,** 4492 (1954).
7 E. R. Blout and M. Idelson, *ibid.* **78,** 3857 (1956).
8 F. Daniels, "Outlines of Physical Chemistry," John Wiley and Sons, Inc., New York, 1954, p. 345.
9 W. E. Hanby, S. G. Waley, and J. Watson, *J. Chem. Soc.* **1950,** 3239.
10 A. Hiller, J. Plazin, and D. D. Van Slyke, *J. Biol. Chem.* **176,** 1401 (1948).
11 M. Idelson and E. R. Blout, *J. Am. Chem. Soc.* **79,** 3948 (1957).
12 E. Katchalski, *Advances in Protein Chem.* **6,** 123 (1951).
13 H. Leuchs, *Ber. der Deut. Gesellschaft.* **39,** 857 (1906).
14 R. D. Lundberg and P. Doty *J. Am. Chem. Soc.* **79,** 3961 (1957).
15 S. Moore and W. W. Stein, *J. Biol. Chem.* **176,** 367 (1948).
16 M. Sela and A. Berger, *J. Am. Chem. Soc.* **75,** 6350 (1953).
17 M. Sela and A. Berger, *ibid.* **77,** 1893 (1955).
18 L. A. Sluyterman and B. Labruyere, *Rec. trav. chim.* **73,** 347 (1954).
19 H. Tsuyuki, H. VanKley, and M. A. Stahmann, *J. Am. Chem. Soc.* **78,** 764 (1956).

Discussion of Paper 8: Aqueous Polymerization of *N*-Carboxy Anhydrides

C. H. BAMFORD

Dr. Miwa and Professor Stahmann postulate addition of carboxylate ions to the 5-CO group of the NCA in the primary step of their mechanism, whereas Dr. Block and I suggest in our paper that similar ions react by abstracting a proton from the NH < group. These views are not necessarily inconsistent, since the reactions were carried out under different conditions—those of Miwa and Stahmann in aqueous solution, ours in organic media. The carboxylate ion is a weaker base in aqueous media and, further, reactions involving the ionized form of the NCA would hardly be expected in dilute aqueous solution, owing to the high availability of protons.

The propagation reaction between carbamate ions and NCA suggested by Miwa and Stahmann does not seem to occur in nonaqueous media; otherwise the propagation reaction would be catalyzed by carbon dioxide. This is not observed. (The catalysis found in the case of sarcosine NCA is common to weak acids in general, not specific to carbon dioxide.)

M. A. STAHMANN

The mechanism of the polymerization reaction for aqueous *N*-carboxy–amino acid anhydride polymerizations in the presence of sodium benzoate must explain the activation of the carboxyl group of the benzoate, for we isolated benzoyl leucine. We have suggested that a mixed anhydride intermediate may explain its formation. Bamford suggests that in salt-initiated reactions the function of the initiator is first to remove a proton from the nitrogen at the three position of the oxazolidine ring and subsequently to transfer this proton to the carbamate ion formed during initiation to permit decarboxylation. The initiator is thus regenerated prior to the propagation.

As shown in Figure 8.15 we have followed the disappearance of one initiator, sodium *p*-nitrophenolate, during polymerization of γ-benzyl-L-glutamate *N*-carboxy anhydride in dimethyl formamide at an anhydride to initiator ratio of 60. By measurement of the change in optical density at 428 mμ it was shown that 91 per cent of the yellow color disappeared within 4 minutes and that less than 1 per cent of the color remained after 30 minutes. Since carbon dioxide did not discolor solutions of sodium *p*-nitrophenolate in dimethyl formamide, we concluded that this initiating ion was removed by the reaction and was not regenerated but rather must have been incorporated into the growing polymer. By precipitation of the polymer with water followed by washing, a product was obtained that liberated *p*-nitrophenolate ion when heated in alkaline solution. These results clearly indicate that the initiating species, the *p*-nitrophenolate anion, was not regenerated during the polymerization reaction and was incorporated into the polymer. The appearance of the nitrophenolate color on heating suggests

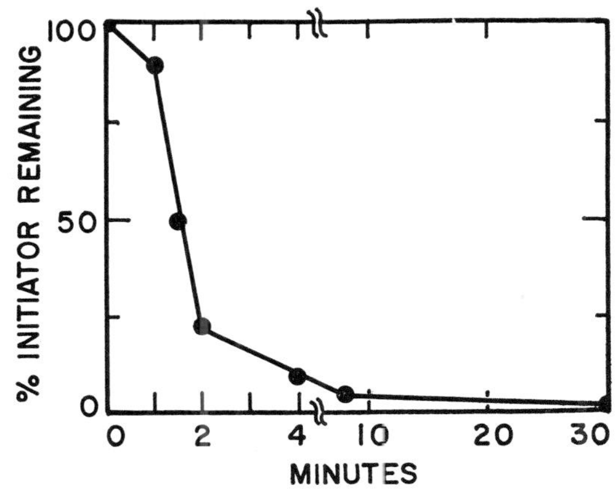

Fig. 8.15. Disappearance of *p*-nitrophenolate anion during polymerization of γ-benzyl-L-glutamate *N*-carboxy anhydride. The decrease in optical density at 428 mμ was measured in a reaction mixture containing 1 per cent anhydride in dimethyl formamide and sodium *p*-nitrophenolate an an anhydride to initiator ratio of 60.

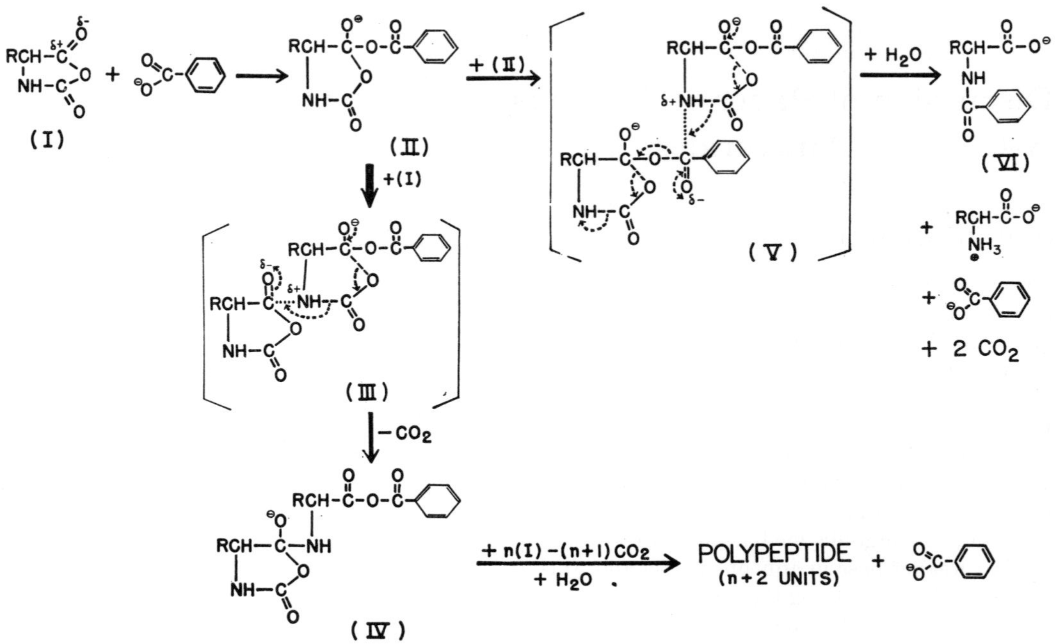

Fig. 8.16. Modified mechanism for the polymerization of *N*-carboxy–amino acid anhydrides to illustrate the electron shifts and simultaneous acylation with the elimination of carbon dioxide to facilitate the polymerization reaction.

that this anion was linked to the carboxyl group as an active ester. This is analogous to the formation of an anhydride at the carboxyl which was suggested for the benzoate anion.

Figure 8.16 is a drawing of a modification of the polymerization mechanism in aqueous solutions in which the mixed anhydride II undergoes an electron shift that activates the nitrogen for subsequent acylation. The acylation reactions during the polymerization as shown below the heavy arrow are very fast because they are facilitated by the electron shifts shown in III and the simultaneous elimination of carbon dioxide during the formation of the new bond to the nitrogen. Thus, in this scheme the formation of carbon dioxide is not due to the decarboxylation of an *N*-carboxy acid but an elimination of carbon dioxide through a push-pull mechanism during the acylation. It should be insensitive to the carbon dioxide pressure because the formation of carbon dioxide is not reversible but rather is an important part of the driving force for the acylation reaction. This scheme would explain the formation of benzoyl leucine by the acylation of II by the mixed anhydride as shown in the intermediate V. It would also allow for the formation of cyclic peptides by an intramolecular acylation.

PAPER 9

Chromatographic Analysis of Polylysine with Carboxymethyl Cellulose

JOHN W. STEWART and MARK A. STAHMANN

The fractionation of a polyamino-acid preparation into its individual molecular components constitutes a challenging isolation problem. Solution of this problem should permit the accumulation of data on the distribution of the molecular species in the polymerization products; such data would be useful in the study of the polymerization mechanisms. Complete separation of members of a homologous series of polyamino acids may be practicable only in the region of relatively low molecular weights. Species of low molecular weight occur in many preparations of poly-γ-benzyl-L-glutamate (*1–4*), and investigation as to both the structures and molecular weight distributions of such species should be of interest. Theoretical analyses have predicted molecular weight distributions in the low molecular weight region as a function of the time of polymerization (*5, 6*), and a test of these predictions might be made.

Fractionation of homologous series of certain polyions has been described previously. Paper chromatography was used to separate the seven lowest members of the oligolysine series, namely, lysine through heptalysine; higher members were not completely resolved (*7*). A technique more promising than paper chromatography became available in the development by Peterson and Sober (*8*) of the substituted cellulose ion exchangers. The aminoalkyl substituted cellulose exchangers DEAE-cellulose and ECTEOLA-cellulose were employed by Tener *et al.* (*9*) for fractionation of mixtures containing synthetic low molecular weight oligonucleotides. DEAE-cellulose was used also by Staehelin *et al.* (*10*) for separation of the six lowest members of the polyadenylic acid series. These results led us to investigate the application of gradient elution chromatography on carboxymethyl cellulose to the fractionation of polylysine.

This investigation commenced with efforts to determine the relationship between the order of elution from carboxymethyl cellulose columns and the molecular weight of members of a homologous series of lysine oligopeptides. Lysine oligopeptides emerged in order of their increasing molecular weight. In fact, the logarithm of the number of lysine residues per molecule of peptide in the eluate was directly proportional to eluent molarity when oligomer mixtures were developed with linear gradients of sodium chloride. Some effects of changing the chromatographic conditions on both the resolution and the elution positions of homologs were studied. Exponential gradients, formed with a constant volume mixer, were superior to linear gradients for fractionation of a homologous series of peptides. True chromatography (a condition in which instantaneous R_f values are always much less than 1) through 40 cm columns of carboxymethyl cellulose was conveniently achieved with lysine oligopeptides as large as the 15-mer. Finally, a few poly-L-lysine preparations were analyzed for their polydispersity. The preparations contained polypeptides with a bimodal distribution of molecular weights when initiation of polymerization of ε-carbobenzoxy-L-lysine *N*-carboxy anhydride in dioxane solution was performed with ammonia. The lower molecular weight population consisted of more than one homologous series of lysine oligopeptides.

Gradient elution chromatography of lysine oligopeptides

Molecular weight dependence of elution order. —A sample whose fractionation would lead most readily to a correlation of molecular weight and elution position should contain a single homologous series of polypeptides for which there exist straightforward means of identification. A partial aqueous acid hydrolysate of a high molecular weight polylysine should provide such a sample. The acid hydrolysis of polylysine proceeds by random cleavage (*11*) and yields therefore at intermediate stages a homologous series of low molecular weight peptides. The partial acid hydrolysis of a polymer initiated with methoxide ion at a high molar ratio of *N*-carboxy anhydride to initiator should yield a homologous series of peptides terminating with an amino and a carboxyl group. Only minor contamination with the peptide esters would be expected. Since the products of the action of trypsin on polylysine have been identified as dilysine, trilysine, tetralysine, and pentalysine (*7*), comparison with known lysine and trypsin hydrolysates should permit ready identification of the lower members of the oligolysine series in the acid hydrolysate. Accordingly, a partial acid hydrolysate of polylysine obtained by initiation of ε-carbobenzoxy-L-lysine *N*-carboxy anhydride with sodium methoxide in dioxane at a molar ratio of anhydride to initiator (A/I) of 320 was selected for fractionation.

Development of the polylysine partial acid

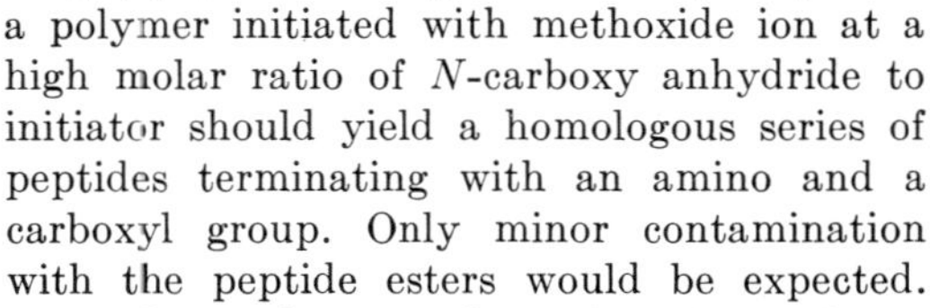

Fig. 9.1. Linear gradient elution chromatogram of a partial acid hydrolysate of polylysine.

Eluate from the column passed through a flow-through cuvette in a Zeiss PM Q II spectrophotometer; transmittancy at 220 mμ was recorded continuously against time with a self-balancing potentiometric recorder. Optical density was adjusted to zero after the breakthrough peak, just before elution of the small peak, number 1. Thirty mg of polylysine hydrolysate at *p*H 6.6 was applied to a 0.9×40 cm column of 4 gm of carboxymethyl cellulose, 0.7 meq carboxyl groups per gram, Na^+ form. The hydrolysate was prepared by a 72-hour reaction of high molecular weight polylysine with 6 N HCl at room temperature. The polylysine preparation was that designated in Fig. 9.9 as $NaOCH_3$ A/I 320. For development a linear gradient from 0 to 1.012 M NaCl over 850 ml was pumped at 1.0 ml per min.

hydrolysate on carboxymethyl cellulose columns with linear gradients of sodium chloride in water gave elution patterns like that shown in Fig. 9.1. The eluate contained a sequence of peaks which became progressively closer as the molarity of sodium chloride increased. The first six or seven peaks following the breakthrough peak were completely separated; subsequent peaks overlapped. The optical density of the valleys rose progressively until at the twenty-second peak distinction between peaks and valleys disappeared.

Rechromatography of the first, second, and tenth peaks yielded single symmetrical peaks at their original elution position; hence reversible aggregation had not occurred and reproducible fractionation was achieved. Identification in this chromatogram of the five lowest members of the oligolysine homologous series was undertaken by comparative chromatography with a trypsin hydrolysate of polylysine and with known synthetic samples. When the trypsin hydrolysate and the partial acid hydrolysate were chromatographed in turn on the same column under identical elution conditions, the elution positions of the first four (and largest) peaks of the trypsin hydrolysate overlapped those of the second, third, fourth, and fifth peaks of the acid hydrolysate. Co-chromatography of the first peak isolated from the acid hydrolysate with known lysine and of the second peak with synthetic dilysine gave in each case a single symmetrical peak at the original position of the isolated peak. Thus comparative column chromatography with known peptides led to the identification of the first five peaks eluted from the partial acid hydrolysate as the five lowest members of the homologous series of lysyl peptides: lysine, dilysine, trilysine, tetralysine, and pentalysine. From this data we conclude that the sixth peak was hexalysine, the seventh was heptalysine, and that the *N*th peak corresponded to an oligolysine molecule consisting of *N* residues of lysine. Thus, the peak number corresponded to the number of lysyl residues per molecule. Molecular weight was a monotonic increasing function of the concentration of sodium chloride required for elution.

The relationship between molecular weight and the concentration of sodium chloride required for elution could be defined more specifically. The logarithm of peak number for each of the twenty-two peaks in Fig. 9.1 was plotted against the corresponding sodium chloride molarity, calculated in the column at the time of elution. The graph of the logarithm of the peak number against molarity is shown in Fig. 9.2. The nineteen points above the first three, which correspond to lysine, dilysine, and trilysine, fall along a straight line. Deviations of the first three peaks toward higher values of sodium chloride molarity required for elution may be due to the decrease in degree of ionization of the α-amino group associated with increased molecular weight. The pK_a of the α-amino group of lysine is 8.95 (*12*), of dilysine 7.56, of trilysine 7.26, of tetralysine 7.15, and of pentalysine 7.07 (*13*). In aqueous nonbuffered solution at the pH of distilled water, about 6.5, the percentage of each of these in the ionized form would be respectively: 100, 92, 85, 82, and 79 per cent.

The equation which describes the straight line portion of Fig. 9.2 may be given by:

$$\text{Peak Number} = ae^{bC} \quad (1)$$

where a is the ordinate intercept, b is the slope of the straight line, C is the concentration of sodium chloride required for elution of each peak, and e is the base of the natural logarithms. This equation describes the exponential relationship between the peak number or number of lysyl residues per molecule and the elution molarity. The equation accurately described the elution behavior for the fourth through the twenty-second peak from an acid hydrolysate of polylysine developed with a linear gradient. Such an exponential relationship resulted from several elution conditions as Fig. 9.3 shows. We found a similar exponential relationship between the number of glutamic acid residues or of thymidine residues per molecule and the elution molarity when the recently published data of Miller on the DEAE-cellulose chromatography of polyglutamic acid (*14*) or of Khorana and Vizsolyi on DEAE-cellulose chromatography of thymidine polynucleotides (*15*) was plotted in the same way.

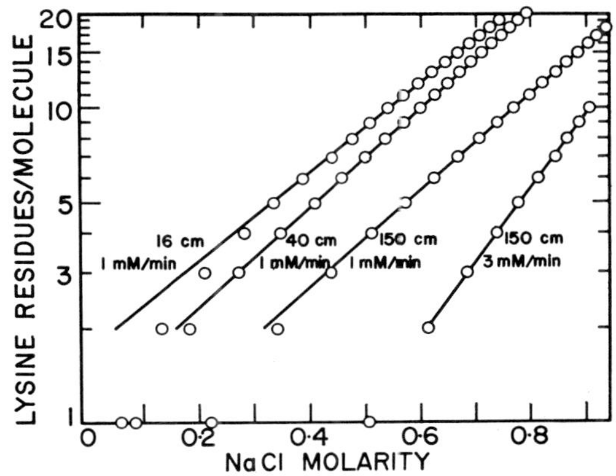

Fig. 9.3. Effect of column length and gradient rate on the relationship between the number of lysine residues per molecule (log scale) and the eluent molarity at elution for linear gradient elution development of oligolysines.

Flow rate was 1.0 ml/min. The 16 cm column contained 1.5 gm of carboxymethyl cellulose, Na^+, 0.7 meq/gm, with a holdup volume of 10.2 ml. The 40 cm column contained 4.0 gm carboxymethyl cellulose, Na^+, 0.7 meq/gm with a holdup volume of 26 ml. The 150 cm column of 0.7 meq/gm carboxymethyl cellulose, Na^+, had a holdup volume of 92 ml.

We have shown that an exponential relationship exists between the number of charged residues per molecule and the concentration of salt required to elute members of homologous series of three polyelectrolytes from two types of substituted cellulose exchangers with linear gradients, and we believe that this relationship can be used both in the design of experiments for separation of polyelectrolyte homologs and in the estimation of the degree of polymerization of members of homologous series. We suggest that application of this observation should yield new information on the molecular species present in preparations of polyamino acids. These findings should also contribute to a better understanding of the mechanism of gradient elution chromatography on modified cellulose exchangers.

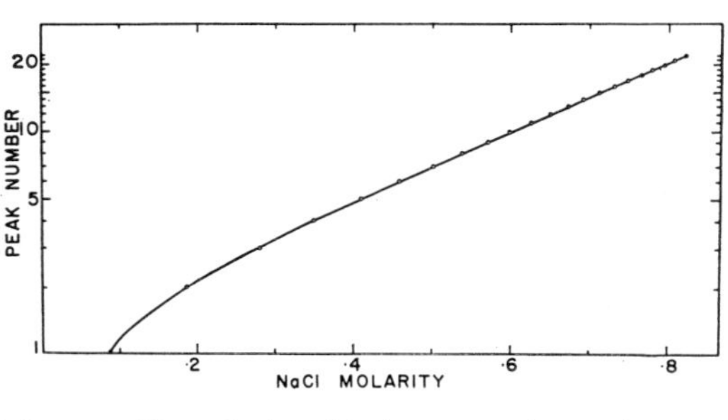

Fig. 9.2. The relationship between the peak number (log scale) from Fig. 9.1 and the sodium chloride molarity required for elution.

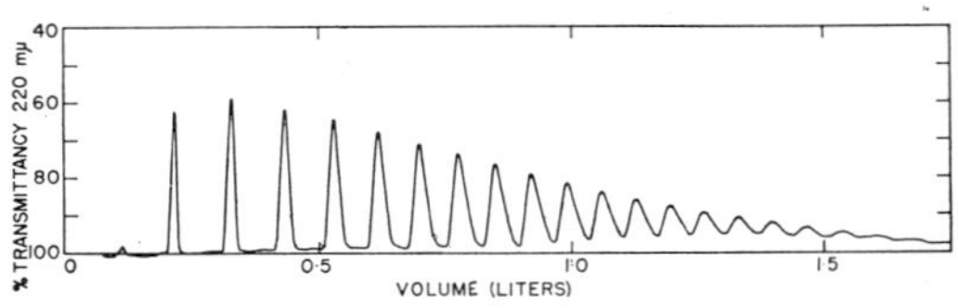

Fig. 9.4. Convex gradient chromatogram of a partial acid hydrolysate of polylysine.

Twenty mg of the hydrolysate described under Fig. 9.1 was applied to a 0.9×40 cm column of 4 gm of carboxymethyl cellulose, 0.7 meq carboxyl groups per gram, Na^+ form. The mixing flask was filled initially with 1000 ml of water. The limit flask contained 0.823 M NaCl. Flow rate was 2.0 ml per min.

Convex gradients: effects of variables on resolution and on elution position.—Members of a homologous series of oligolysine molecules were eluted progressively closer to each other as the molecular weight increased when they were developed with a linear gradient of sodium chloride; resolution progressively decreased. If gradient conditions could be found that would separate the homologs by a constant eluate volume, then the decrease in resolution with increase in molecular weight would be less than that resulting from linear gradient development; it would be limited only by the rate of increase in width of peaks. To achieve such conditions a convex gradient, one in which a continuous reduction in the gradient slope occurs, would be required. Convex gradients are generated by the constant volume type of mixer. Such a mixer consists of two reservoirs, one containing solution of limit concentration and the other, the mixing flask, containing a constant volume of solution of continuously increasing concentration. As liquid is fed to the column from the mixing flask, a volume of liquid equal to that removed is replaced from the reservoir containing solution of limit concentration. The concentration in the solution emerging from the mixing flask increases according to the equation:

$$C = C_2 - De^{-k} \tag{2}$$

where C is the concentration of the solution being withdrawn, C_2 that of the limit solution, D is the difference between the initial concentration of the solution in the limit reservoir and that in the mixing flask, e is the base of the natural logarithms, and k is the ratio of the volume removed up to that point to the fixed volume of solution in the mixing flask (*16*).

The chromatogram shown in Fig. 9.4 was developed with a convex gradient of sodium chloride concentration in which the average rate of increase over the range represented was 0.7 mM per min. In this chromatogram the first ten peaks were separated completely; the next twelve additional peaks were resolved sufficiently as to permit only little rise in the optical density of the valleys. In contrast to this, chromatography of the same hydrolysate with a linear gradient of 1.2 mM per min increase in sodium chloride concentration as is shown in Fig. 9.1 led to complete separation of about six or seven peaks and to a subsequent rapid rise in the optical density of valleys. Thus, significant improvement in the resolution of higher molecular weight homologs relative to that of the lower ones occurred when the partial acid hydrolysate was developed with a convex gradient that eluted peaks at intervals of approximately constant volume.

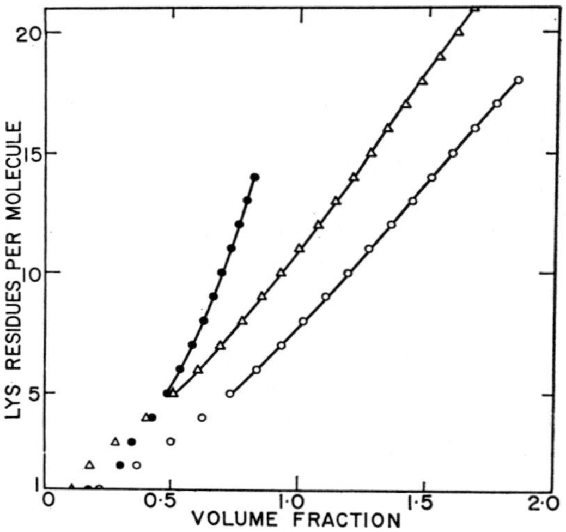

Fig. 9.5. Effects of variations in mixer volume and limit concentration.

Twenty mg of a partial acid hydrolysate of polylysine was applied to a 0.9×40 cm column of 4 gm carboxymethyl cellulose, 0.7 meq carboxyl groups per gm, Na^+ form, and was developed with a convex gradient at 2 ml/min flow rate. *Triangles:* 1000 ml H_2O in mixing flask, 0.823 M NaCl in limit flask. *Open circles:* 370 ml H_2O in mixing flask, 0.823 M NaCl in limit flask. *Closed circles:* 370 ml H_2O in mixing flask, 1.372 M NaCl in limit flask. Volume is expressed as multiples of the liquid volume in the mixing flask.

Some additional features of exponential gradient elution chromatograms of lysine oligopeptides were observed. An increase in the gradient rate over that of Fig. 9.4, produced either by increasing the concentration of limit solution from 0.82 to 1.37 M or by decreasing the mixer volume from 1000 to 370 ml, caused a considerable decrease in resolution. Elution positions were affected by variations in the mixer volume and the limit concentration as is shown in Fig. 9.5. Volume is expressed as the fraction of the volume of liquid in the mixer to facilitate plotting data from three different experiments. The volumes corresponding to the peaks in Fig. 9.4 are plotted as triangles against the number of lysine residues per molecule. The number of lysine residues per molecule in the eluate was nearly linearly related to eluent volume. An increase in the gradient rate, produced by decreasing the mixer volume from 1000 to 370 ml, led to the data plotted as open circles. In both of these chromatograms, eluent concentrations were identical at identical volume fractions since the limit concentration and flow rate of both were identical. An increase in the gradient rate due to a reduction of mixer volume led to a translation of the relationship between molecular size and eluent concentration to higher values of concentration. Changing the concentration of the limit solution from 0.82 M (open circles) to 1.37 M (solid circles) caused an increase in the slope of the relationship between molecular size and volume fraction. In these two chromatograms, flow times were identical at identical volume fractions since the mixer volume and flow rate were identical. An increase in the gradient rate due to a change in limit concentration caused an increase in the slope of the relationship between molecular size and flow time.

Mechanism of gradient elution of lysine oligopeptides.—To aid the design of experimental conditions which would resolve oligolysine homologs with greater acuity than that of Fig. 9.4 we investigated the mechanism of elution of these polyvalent molecules. Specifically we asked whether lysine oligopeptides moved throughout the column of Fig. 9.4 with R_f values much less than 1 or whether they moved a significant portion of the column length with R_f values very nearly equal to 1. The latter possibility predicts that increased resolution over that in Fig. 9.4 should result from a decreased gradient rate but not from an increase in column length. The converse is predicted by the possibility that R_f values are much less than 1 throughout the column length.

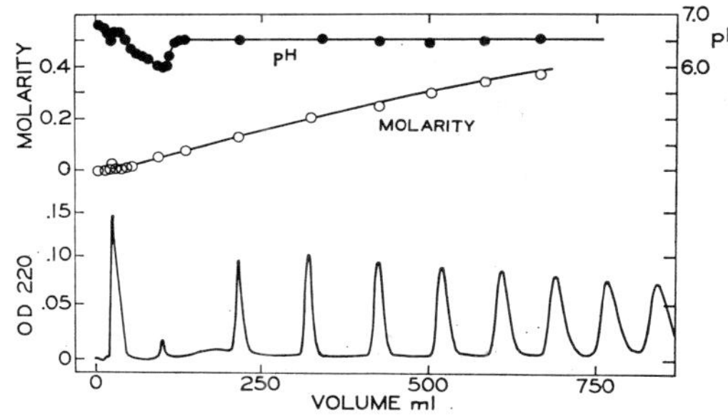

Fig. 9.6. Eluate pH and sodium chloride molarity in a convex gradient chromatogram.

The procedure of Fig. 9.3 was repeated, excepting: the mixing flask contained 0.005 M sodium phosphate buffer, pH 6.9, the limit solution, pH 6.6, was 0.823 M NaCl in 0.005 M sodium phosphate buffer, and the column was washed with several volumes of starting buffer before the sample was applied.

The first possibility, that oligomers moved through a significant portion of column length with R_f values very nearly equal to 1, is not consistent with the data in Fig. 9.3. This possibility predicts that each oligomer should have emerged at a concentration of eluent nearly independent of column length. In the experiment shown in Fig. 9.3, 20-mer appeared in the eluate of columns of lengths 16, 40, and 150 cm at sodium chloride concentrations of 0.74, 0.76, and 0.86 M respectively. It was possible that these calculated values of molarity did not in fact exist in the eluates, since the buffering action of carboxymethyl cellulose tends to modify pH and salt gradients (*17*). We therefore determined the pH and sodium chloride concentration in eluate fractions obtained during chromatography of the acid hydrolysate. A weakly buffered developer was used to facilitate pH measurements; the experiment was otherwise identical to that of Fig. 9.4. The chromatogram, which was superimposable on the chromatogram of Fig. 9.4, and a plot of pH and sodium chloride molarity are shown in Fig. 9.6. The sodium chloride molarity, determined by conductivity measurements, showed an inflection only in the breakthrough peak, and except in the breakthrough fractions was less than the value calculated with equation 2 by a difference of from 0.02 to 0.03 M over the 700 ml range of measurements. The *p*H, measured with a glass electrode, remained constant after the first peak had been eluted. Hence, differences between the calculated values of sodium chloride molarity in the eluate containing 20-mer were significant, and we conclude that 20-mer did not move through a significant portion of the 40 cm column with an R_f value very nearly equal to 1.

The possibility that R_f values were much less than 1 throughout the length of the column was examined. Several oligomers were eluted with sodium chloride solutions of concentration equal to that present in the column during their emergence in the gradient experiment of Fig. 9.4. Values of R_f corresponding to these sodium chloride concentrations were calculated. A 40×0.9 cm carboxymethyl cellulose column loaded with a zone of partial acid hydrolysate of polylysine was developed stepwise with four sodium chloride solutions of 0.14, 0.32, 0.48, and 0.55 molarity; 10 to 15 holdup volumes of each solution were used. The first, small peak was assigned the number 1 and the much larger peaks of nearly equal area which emerged subsequently were assigned integers 2, 3, 4, . . . in order of their emergence. Fig. 9.7 shows the eighth through the thirteenth peaks, which were eluted with the third developing solution. The values of R_f of eluted peaks were calculated; these varied for the 17 peaks from 0.6 to 0.1. A plot of log $(1/R_f - 1)$ against the integral number assigned each peak was linear for peaks eluted by each eluent. That successive peaks eluted by each eluent differed by an increase in one homologous unit is consistent with this linear plot (*18, 19*). That each homolog appeared as one and only one peak is supported by the observations that, after the first, all peaks were of nearly identical area and that no peak emerged at one holdup volume. That the first peak was lysine was indicated by its small size. We conclude that peaks emerged in order of increasing molecular weight and that in fact the integer assigned each peak coincides with the number of lysine residues per molecule contained in the peak. Hence oligomers of lysine as large as 17-mer were eluted under zero gradient conditions with R_f values ranging from 0.5 to 0.6 for the first peak, and from 0.2 to 0.3 for the second, to as low as 0.1 for the last peak eluted in each step. The sodium chloride concentration used to elute the eighth through the thirteenth peaks (identified as 8-mer through 13-mer) was identical to that present in the column during emergence of nonalysine in the gradient experiment of Fig. 9.4. Development with this constant concentration caused the elution of nonalysine at 3.2 holdup volumes or with an R_f value of 0.31. Nonalysine must have moved through the entire column during the gradient experiment with R_f values between 0 and 0.31. Similar reasoning leads to the conclusion that 15-mer must have moved

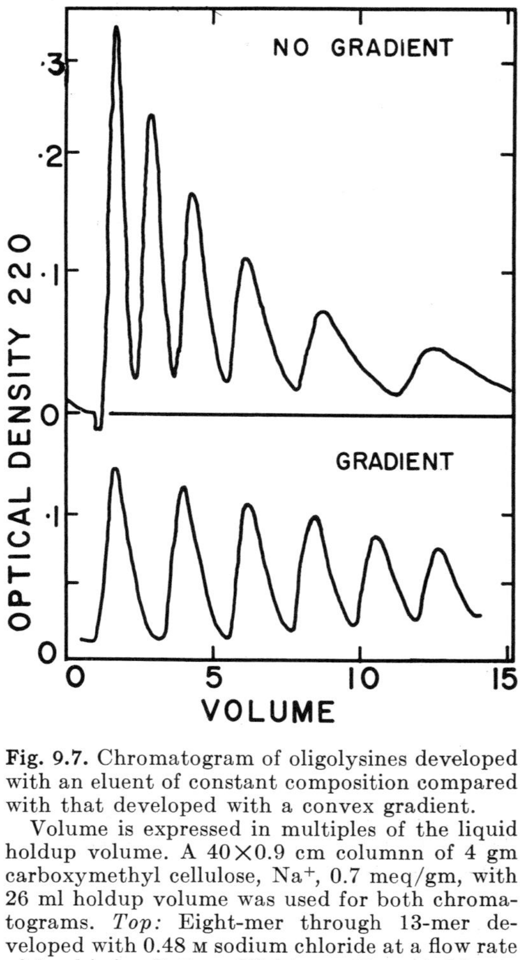

Fig. 9.7. Chromatogram of oligolysines developed with an eluent of constant composition compared with that developed with a convex gradient.

Volume is expressed in multiples of the liquid holdup volume. A 40×0.9 cm columnn of 4 gm carboxymethyl cellulose, Na^+, 0.7 meq/gm, with 26 ml holdup volume was used for both chromatograms. *Top:* Eight-mer through 13-mer developed with 0.48 M sodium chloride at a flow rate of 2 ml/min. *Bottom:* Eight-mer through 13-mer developed with a convex gradient formed by adding 0.823 M sodium chloride at 2 ml/min to a 1000 ml constant volume mixing flask filled initially with water. The concentration of sodium chloride in the column at emergence of the maximum concentration of 9-mer was 0.48 M.

through the entire column in the experiment of Fig. 9.4 with R_f values between 0 and 0.21. We conclude that oligomers of lysine as large as 15-mer can be moved by convex gradients throughout the length of a 40 cm carboxymethyl cellulose column at R_f values considerably less than 1. This conclusion predicts that 8-mer through 13-mer in Fig. 9.7 should be equally well resolved by the gradient and by the no-gradient elution conditions. This prediction is confirmed by the chromatograms shown in Fig. 9.7.

These observations suggest that enhanced resolution may be achieved by increasing column length when columns are developed at gradient rates near that of the experiment shown in Fig. 9.4. This rate is approximately linear, and averages 0.75 mM/min up to a sodium chloride concentration of 0.3 M. Chromatograms 2 and 3 of Fig. 9.8 show that indeed resolution could be improved by a fourfold increase in column length when polylysine was developed with a linear gradient at a rate of 1 mM/min. In Fig. 9.8 the solid areas, presumed to represent identical components, emerged at about the position of dilysine, the dotted areas at about the position of pentalysine. Resolution in both these areas was still further increased by the use of a 150 cm column.

Comparison of the solid and dotted areas of chromatograms 1 and 2 in Fig. 9.8 shows that a fourfold increase in gradient rate from 1 to 4 mM/min caused a marked decrease in resolution. This observation suggests that at a gradient rate of 4 mM/min oligomers from dilysine through pentalysine emerge with R_f values considerably nearer to 1 than they do at 1 mM/min. Hence the gradient rate in the experiment of Fig. 9.4 may be nearly the maximum that will yield improved resolution from an increase in column length.

Analysis of five polylysine preparations

Results of the preceding sections permit two types of analysis of polylysine by gradient elution chromatography on carboxymethyl cellulose. Firstly, moderately long columns and slow convex gradients may be employed for high resolution analysis of polylysines which have molecular weights lower than about 3500. In this range of molecular weights each member of a single homologous series may be detected. Secondly, moderately short columns and fast convex gradients may be used for low resolution analysis of molecular size distribution over a wide range of molecular weights.

Method.—We have analyzed the molecular size distribution in several preparations of polylysine to test the utility of this latter method. Samples of poly-L-lysine were selected that were analogous to the poly-γ-benzyl-L-glutamates of described polydispersity (*1–4*). Poly-ε-carbobenzoxy-L-lysine was prepared by initiation of ε-carbobenzoxy-L-lysine *N*-carboxy anhydride at 4 per cent concentration in dioxane at room temperature with ammonia at *A/I* of 6, 12, and 50, with sodium methoxide at *A/I* of 320, and with lithium methoxide (and/or lithium hydroxide) at *A/I* of 6. Carbobenzoxy groups were removed by reaction of the blocked polymers with anhydrous hydrogen chloride in the dioxane at 90° C.

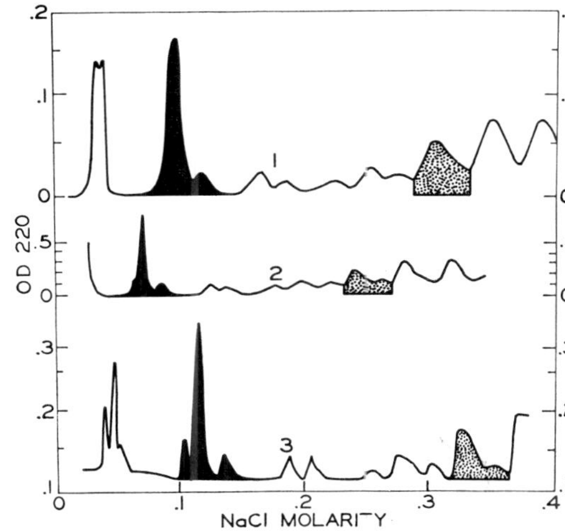

Fig. 9.8. The effects of variations in gradient slope and column length on resolution of polylysine.

Preparation of the polylysine sample, initiated with $LiOCH_3$ in dioxane as *A/I* of 6, is described in the next section.

1. Ten mg sample on 0.9×16 cm column of 1 gm carboxymethyl cellulose, developed with gradient increase of 4 mM NaCl per min at 2.1 ml per min flow rate.
2. Ten mg sample on 0.9×15 cm column of 1 gm carboxymethyl cellulose, developed with gradient increase of 1 mM NaCl per min at 2.0 ml per min flow rate.
3. Thirty mg sample on 0.9×58 cm column of 4 gm carboxymethyl cellulose, developed with gradient increase of 1 mM NaCl per min at 1.0 ml per min flow rate.

Chromatographic conditions were selected to permit low resolution, rapid analysis of the distribution of molecular size over the range contained in the preparations. Ten mg of each polylysine sample in 1 per cent aqueous solution, *p*H 6.8, was applied to the top of a 0.9×15 cm column of 1 gm of carboxymethyl cellulose, 0.7 meq of carboxyl groups per gm, and washed in with water. The preparations were developed with a continuous sodium chloride gradient composed of three distinct parts. These were formed by pumping the following at 2 ml per min into a 370 ml mixing flask initially filled with water: first 370 ml of 0.823 M sodium chloride, then 370 ml of 1.372 M sodium chloride, and finally 370 ml of 1.731 M sodium chloride. These three gradients of continuously increasing sodium chloride concentration are referred to as gradients I, II, and III, respectively. The chromatograms that resulted are shown in Fig. 9.9. Gradients I, II, and III are separated by vertical lines.

Molecular size calibration.—The approximate range in molecular weight of material eluted by each gradient was calculated. In this calculation we assumed that the exponential rates, which varied from about 4 to about 1.5 mM /min, led to a correlation between molecular weight and sodium chloride concentration nearly identical to the correlation produced by a linear gradient of 4 mM/min rate. Under conditions otherwise identical to those of Fig. 9.9, the polylysine preparation obtained by $LiOCH_3$ initiation at *A*/*I* of 6 was eluted with a linear gradient of sodium chloride at a rate of 4 mM/min. The chromatogram is shown in Fig. 9.8 as curve 1.

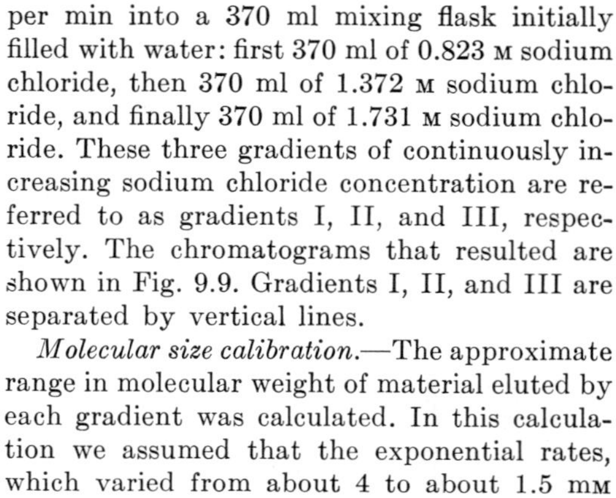

Fig. 9.9. Chromatograms of five poly-L-lysine preparations eluted with three continuous convex gradients.

Ten mg of polylysine were applied to a 0.9×15 cm column of 1 gm carboxymethyl cellulose with an 8 ml holdup volume. A 370 ml mixing flask which was filled initially with water was fed at 2.0 ml per min with 370 ml of 0.823 M NaCl (*left*), then with 370 ml of 1.372 M NaCl (*center*), and finally with 370 ml of 1.731 M NaCl (*right*).

If this preparation contained a homologous series of lysine peptides, then by analogy to the homologous series in the partial acid hydrolysate these homologs would be eluted with a linear gradient at regular intervals. A search for a series of peaks that were eluted similarly to the series in the hydrolysate was made in the following way. Comparative chromatography showed that the first peak of curve 1 overlapped that due to lysine; the second peak (the solid area) overlapped that due to dilysine; the dotted area overlapped that due to pentalysine, and the succeeding two peaks overlapped those due to hexalysine and heptalysine respectively. Integers corresponding to the number of lysyl residues per molecule in the overlapping known peptides were assigned to the peaks overlapping penta-, hexa-, and heptalysine. The logarithm of the peak numbers of these peaks was plotted against the sodium chloride concentration required for their elution. The points corresponding to peaks number five, six, and seven were connected with a straight line. Extrapolation of this line to lower concentrations permitted selection of peaks which most nearly fell into this linear relationship. Peaks number 2, 3, and 4 were thus selected; peak number 2 then corresponded to the major peak in the solid area, an area which had been shown to overlap the position of dilysine. The equation of the straight line established by this procedure was:

$$\text{Log}\,(DP + 1) = 1.9C + 0.1 \qquad (3)$$

where *DP* is the number of lysine residues per molecule minus 1 and *C* is the concentration of sodium chloride calculated to be in the column.

The concentration limits of each part of the continuous gradient, gradients I, II, and III, were computed by means of the gradient equation 2. By combining equations 2 and 3 the data in Table 9.1 showing the approximate size of polylysine eluted by the three sodium chloride gradients were obtained. These calculations of the range in degree of polymerization eluted by each gradient were tested experimentally.

The range in degree of polymerization of peptides eluted by gradient I was evaluated by

Table 9.1—The approximate size of polylysine eluted by three sodium chloride gradients

Number	Molarity of NaCl		Degree of polymerization	
	minimum	maximum	minimum	maximum
I	0.00	0.52	0	11
II	0.52	1.06	11	130
III	1.06	1.48	130	830

comparative chromatography of the partial acid hydrolysate of polylysine with the sample designated in Fig. 9.8 as NH_3 *A/I* 6. The chromatogram of the NH_3 *A/I* 6 sample contains a region absorbing at 220 mμ which is terminated by the vertical line marking the end of gradient I. This region was resolved into several peaks. A single peak at about 40 ml eluate volume was followed by a complex peak, which was composed of two partially resolved areas, at about 80 ml, and this in turn was followed by a complex peak at 100 to 120 ml. Nine additional peaks were eluted at intervals of about 30 ml before the optical density remained constant along the baseline. Components eluted in this first region were compared with the partial acid hydrolysate of polylysine by column chromatography. Lysine overlapped the first peak, dilysine the second, trilysine the third, tetralysine the fourth, pentalysine the fifth, hexalysine the sixth, and heptalysine the seventh. We conclude from this pattern that the eleventh peak, which was eluted at the end of gradient I, chromatographed like undecalysine and hence the limiting degree of polymerization eluted by gradient I was ten. This conclusion supports the prediction in Table 9.1.

The range in degree of polymerization of polypeptides eluted by gradient III was estimated by determining the molecular weight of components under the high molecular weight peak of the sample initiated with ammonia at *A/I* of 12. Following the method suggested by Berger and Yaron in this symposium (*20*), we reacted the isolated peak components with excess carbobenzoxy chloride in cold aqueous sodium hydroxide, *p*H 12, for one hour to obtain poly α,ϵ-carbobenzoxy lysine. The intrinsic viscosity of this product in dimethyl formamide was 0.44 dl/gm at 24.5° C., a value which corresponds to a degree of polymerization of 250 (*21*) or of 200 (*22*). This value of degree of polymerization lies well within the limits calculated for gradient III.

Ammonia initiated polylysine.—The chromatogram of a polylysine initiated with ammonia at *A/I* of 6 is shown at the top of Fig. 9.9. Ninety-seven per cent of the applied nitrogen determined by micro-Kjeldahl analysis was recovered in two widely separated regions of the chromatogram. One partially resolved region was eluted by gradient I; the other, which consisted mainly of a broad peak, was eluted by gradient III. The chromatogram of polylysine initiated with ammonia at *A/I* of 12 also consisted principally of two regions eluted by gradients I and III. Comparison of the two chromatograms revealed that the center of the first region was shifted towards higher molecular weight in the latter preparation, whereas that of the second region remained about the same. The chromatogram of the sample initiated with ammonia at *A/I* of 50 contained no detectable components eluted by gradient I; however, it consisted also of two regions. One was eluted with gradient II and the other with gradient III. It seems reasonable to conclude from this that the average molecular weight of the low molecular weight components shifts towards higher values as the *A/I* is increased whereas the average molecular weight of the high molecular weight components does not change enough to allow detection.

Comparison of these chromatograms revealed that the area in the high molecular weight region increased, whereas the area in the low molecular weight region decreased in the transition from *A/I* of 6 to *A/I* of 12. We conclude that increased *A/I* values lead to relatively greater proportions of higher molecular weight polymers.

The peaks in the low molecular weight region of the ammonia initiated preparations at *A/I* of 6 and 12 were asymmetrical and complex; in contrast, peaks from the partial acid hydrolysate of high molecular weight polylysine were symmetrical and simple. Chromatography of the *A/I* 12 preparation under higher resolution conditions, using a 58 cm column of carboxymethyl cellulose, a liter mixing flask, a limit solution concentration of 0.823 M sodium chloride, and a flow rate of 2 ml/min, permitted separation of two symmetrical peaks near the elution position of each of the first six hydrolysate peaks. We conclude from this that the low molecular weight products have a structural complexity greater than that of a single homologous series. Infra-red spectra of NaCl-KBr pellets of the first nine peaks eluted from the A/I 12 sample under the higher resolution chromatographic conditions showed strong amide carbonyl absorption bands but no detectable absorption bands in the 6.3 and 7.1 μ regions of carboxylate ion absorption. Infrared spectra of similar pellets of each of several peaks isolated from the partial acid hydrolysate of polylysine contained strong absorption bands at 6.3 and 7.1 μ. We conclude that peptide bond cleavage during removal of carbobenzoxy groups did not occur to a detectable extent and therefore cannot explain the structural complexity of the low molecular weight products. Nor can the complexity of structure be explained readily by chain-terminating reactions: components in the low molecular weight region were not detected in the A/I 50 preparation and we would expect low molecular weight compounds formed by chain-terminating reactions in an A/I of 12 preparation also to be present in one of A/I of 50.

Strong base initiated polylysine.—The solution used for initiation of the preparation labeled $LiOCH_3$ A/I 6 in Fig. 9.8 was composed of: CH_3OH, 0.34 ml; H_2O, 0.01 ml (0.5 meq); and $LiOCH_3$, 0.38 meq. The chromatogram of this preparation contained a continuous distribution of material that absorbed at 220 mμ. The region eluted by gradient I was composed of several partially separated peaks, that eluted by gradient II of a very broad peak, and that by gradient III of a nearly horizontal line above the baseline. Peaks eluted by gradient I were complex, as is shown by the higher resolution chromatogram of this preparation in curve 3 of Fig. 9.6. Three peaks were eluted near the elution position of lysine, three (solid area) near that of dilysine, two near that of trilysine, three near that of tetralysine, and two (dotted area) near that of pentalysine. In Fig. 9.8 the area of the second peak of the $LiOCH_3$ preparation is much larger than that of the small second peak in chromatograms of the amine initiated polymers.

The chromatogram of the preparation initiated with sodium methoxide at A/I of 320 consists essentially of a single broad peak eluted by gradient III. A shoulder on the leading edge of this peak begins at about 310 ml of gradient II and drops off slightly during the beginning of gradient III. These results are consistent with the suggestions that such preparations consist very largely of high molecular weight polymers (*1, 23*). It suggests some polydispersity but the resolution is not sufficient to allow a precise evaluation of the inhomogeneity. It should be noted that there is very little material of low or intermediate molecular weight eluted by gradients I or II from the sodium methoxide initiated preparation.

Acknowledgment

This work was supported by research grants (No. E-101) from the National Microbiological Institute of the National Institutes of Health, U.S. Public Health Service.

References

1 E. R. Blout and A. Asadourian, *J. Am. Chem. Soc.* **78,** 955 (1956).
2 M. Idelson and E. R. Blout, *ibid.* **79,** 3948 (1957).
3 J. C. Mitchell, A. E. Woodward, and P. Doty, *ibid.* **79,** 3955 (1957).
4 R. D. Lundberg and P. Doty, *ibid.* **79,** 3961 (1957).
5 E. Katchalski, Y. Shalitin, and M. Gehatia, *ibid.* **77,** 1925 (1955).
6 J. D. Coombes and E. Katchalski, *ibid.* **82,** 5280 (1960).
7 S. G. Waley and J. Watson, *Biochem. J.* **55,** 328 (1953).
8 E. A. Peterson and H. A. Sober, *J. Am. Chem. Soc.* **78,** 751 (1956).
9 G. M. Tener, H. G. Khorana, R. Markham, and E. H. Pol, *ibid.* **80,** 6223 (1958).
10 M. Staehelin, E. A. Peterson, and H. A. Sober, *Arch. Biochem. Biophys.* **85,** 289 (1959).
11 M. Sela and E. Katchalski, *Advances in Protein Chem.* **14,** 391 (1956).
12 C. L. A. Schmidt, P. L. Kirk, and W. K. Appleman, *J. Biol. Chem.* **88,** 285 (1930)
13 S. G. Waley and J. Watson, *J. Chem. Soc.* **1953,** 475.
14 W. G. Miller, *J. Am. Chem. Soc.* **83,** 259 (1961).
15 H. G. Khorana and J. P. Vizsolyi, *ibid.* **83,** 675 (1961).
16 B. Drake, *Arkiv Kemi* **8,** 1 (1956).
17 H. A. Sober, F. J. Gutter, M. M. Wyckoff, and E. A. Peterson, *J. Am. Chem. Soc.* **78,** 756 (1956).
18 A. J. P. Martin, *Biochem. Soc. Symposia* **3,** 4 (1950).
19 E. C. Bate-Smith and R. G. Westall, *Biochim. et Biophys. Acta* **4,** 427 (1950).
20 A. Berger and A. Yaron, Paper 2, this book.
21 J. B. Applequist, Ph.D. thesis, Harvard University (1959).
22 E. Daniel and E. Katchalski, Paper 16, this book.
23 M. Idelson and E. R. Blout, *J. Am. Chem. Soc.* **80,** 2387 (1958).

Discussion of Paper 9: Chromatographic Analysis of Polylysine

HERBERT A. SOBER

As part of a continuing study of the cellulose ion-exchangers which seeks to understand the mechanism of adsorption and the factors determining the resolution of these exchangers, poly-α-amino acids have been used as model substances (E. A. Peterson and H. A. Sober, *J. Am. Chem. Soc.* **78,** 751 [1956]; H. A. Sober and E. A. Peterson in J. T. Edsall (ed.), "J. P. Greenstein Memorial Symposium," Academic Press, New York, 1960, pp. 61–81).

The poly-α-amino acids supplied by Dr. Ephraim Katchalski were prepared from *N*-carboxy-α-amino acid anhydrides in the conventional manner with the polymerization reaction carried out in dimethylformamide using butylamine as initiator. Blocking groups were removed by anhydrous hydrogen bromide and the polymer was precipitated with ether, filtered, and dried. Its average degree of polymerization was estimated from Van Slyke amino nitrogen analyses.

The chromatographic resolution obtained with a polylysine preparation of a number average chain length of 4 on the cation exchanger, CM-cellulose, is shown in Figure 9.10. The analysis was carried out in the Spinco Amino

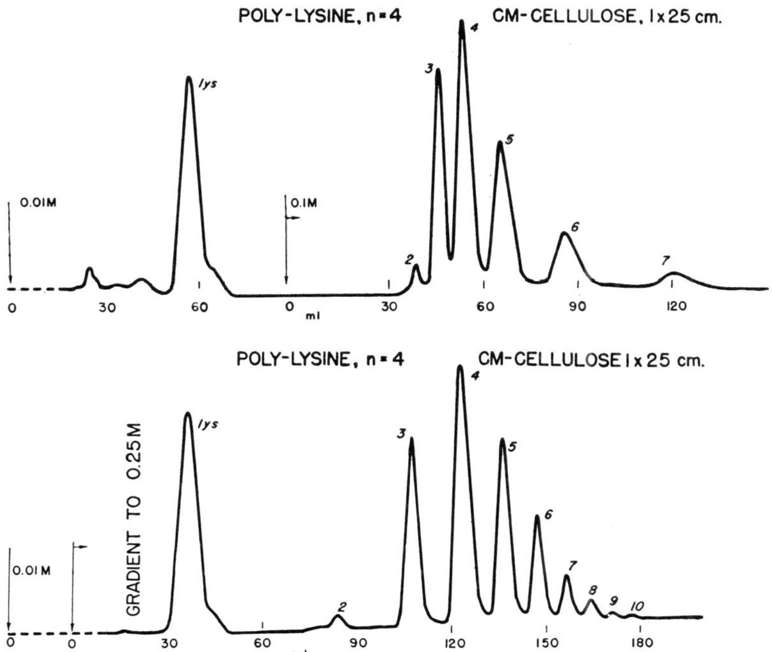

Fig. 9.10. Single buffer and gradient elution chromatography. One mg polylysine·HBr ($n = 4$) — 1 μmole lysine·HCl applied to CM-cellulose (1×25 cm, 0.78 meq/gm, 100–200 mesh) equilibrated with 0.01 M Na citrate *p*H 5. Flow rate, 30 ml/hr. Temp., 60° C. Ordinate, ninhydrin color. Development started at arrows. *Upper,* 0.1 M; *lower,* linear gradient of 198 ml with 0.25 M buffer as limit.

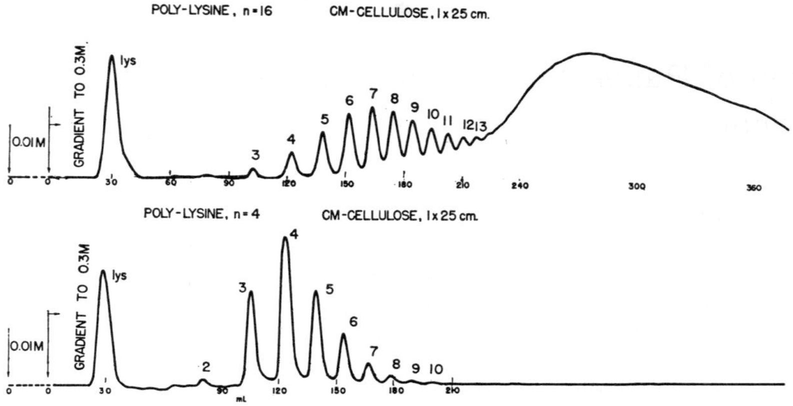

Fig. 9.11. Gradient elution of $n = 16$ and $n = 4$ polylysines. *Upper,* 10 mg polylysine·HBr — 1 μmole lysine·HCl; *lower,* 1 mg polylysine·HBr — 1 μmole lysine·HCl. Developed with exponential gradient (constant volume mixer of 240 ml of 0.01 M buffer, 250 ml of 0.3 M buffer as limit). Remainder as in Figure 9.10.

Acid Analyzer where ninhydrin color was recorded as a function of effluent volume. The monomer, lysine, was added to each sample before application to the column. In the upper part of the figure is shown the result of development with a single buffer, 0.1 M sodium citrate at *p*H 5. It can be seen that oligomers up to 7 residues were resolved. No higher polymers were detected. When a gradient to 0.25 M buffer was used as in the lower part of the figure, resolution was somewhat better. Perhaps more important, however, is the fact that additional components, up to the 10-mer can be seen. In the single buffer elution the peaks corresponding to the higher members of the series tend to broaden with increased retention so that those higher oligomers in low concentration could not be distinguished from the baseline. With the gradient, on the other hand, the increase in eluting power tends to sharpen the peaks toward the end of the chromatogram so that octamer, nonamer, and decamer were clearly visible.

These chromatograms were run at 60°. Lowering the temperature to 30° produced some very interesting results. The effluent position of the penta- and hexamer remained essentially unchanged. At the lower temperature the smaller members of the series emerged earlier, whereas the higher oligomers emerged later than at 60°.

It should be noted that a roughly Gaussian distribution was obtained in this polymer preparation. Calculation from this and other chromatograms of the weight average and the number average degree of polymerization resulted in values of 4.4 and 4.1, respectively, which was in good agreement with the end-group analysis of $n = 4$. The ratio of weight average to number average degree of polymerization was 1.08, indicating a rather narrow distribution of size species (E. Katchalski and M. Sela, *Advances in Protein Chem.* **13,** 243 [1958]).

The chromatographic analysis of a polymer with an average chain length of 16 is shown in Figure 9.11 with the $n = 4$ polymer for comparison. The heterogeneity of the higher molecular weight polymer is obvious. Calculation of the weight average and number average of this preparation indicates that the large unresolved peak at the right must have an average chain length of 19 in order to satisfy the average chain length value of 16 as determined by end-group analysis of the unresolved polymer. It is noteworthy that two modes of sizes appear: one with a maximum at the heptamer and the second with a mean value of a 19-mer. This may well reflect the different rates of polymerization which have been suggested, from indirect evidence, to result from a change of configuration from a random coil to an α-helix (R. D. Lundberg and P. Doty, *J. Am. Chem. Soc.* **79,** 3961 [1957]). It is now possible to test this directly by determining the conformation

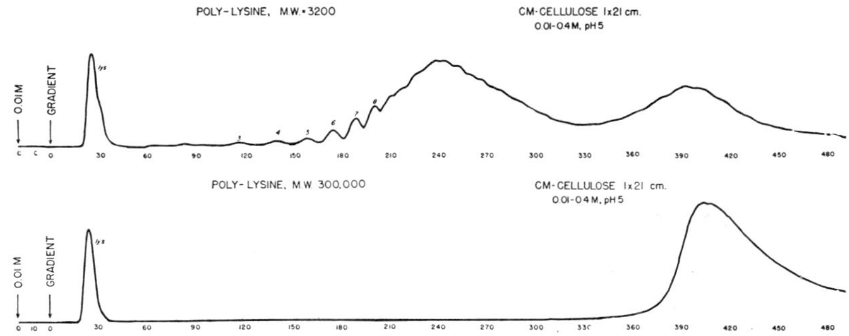

Fig. 9.12. Size distribution in two commercial polylysine preparations. *Upper,* 7.5 mg polylysine·HBr (3200 mol. wt.) 1 μmole lysine·HCl; *lower,* 6 mg polylysine·HBr (300,000 mol. wt.) 1 μmole lysine·HCl. Developed with linear gradient of 496 ml with 0.4 M buffer as limit.

in different solvents of homogeneous polymers of different sizes and comparing the results with polymerization rates. It should be pointed out that even in the case of the $n=16$ polymer the ratio of weight average to number average degree of polymerization was 1.11, indicating an apparent narrow size distribution.

The polymer distribution in two commercial preparations is shown in Figure 9.12. While the chromatographic system is not capable of resolving the components in the larger polymer, it is clear that the smaller polymer, with a molecular weight of 3200, is markedly heterogeneous, showing two distinct and widely separate modes of distribution of polymer sizes.

The same kind of analysis can be performed with the polyglutamic and polyaspartic acids. In this case the anion exchanger, DEAE-cellulose, was used and the chromatogram was developed with a gradient from 0.01–0.5 M sodium citrate at pH 8. As shown in Figure 9.13, in contrast to the polylysine preparation, a Gaussian distribution has not been achieved in the case of the polyglutamic acids but, rather, an excess of the smaller members of the series seems to predominate. The components

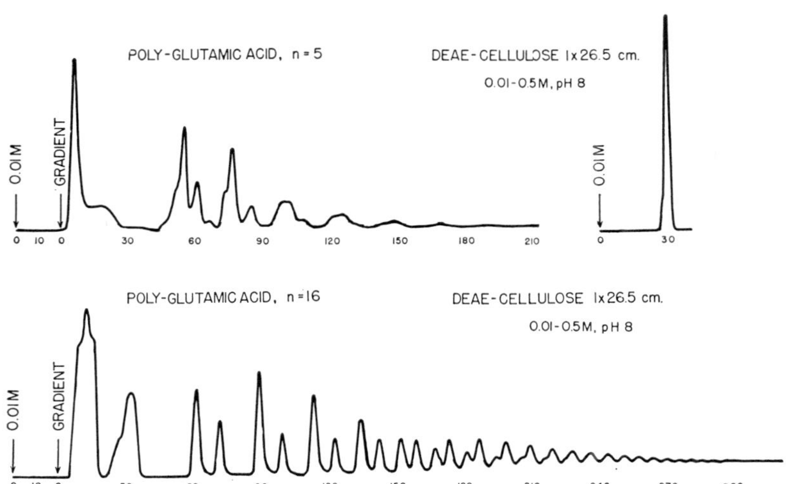

Fig. 9.13. Gradient elution chromatography of polyglutamic acids. Five mg of polyglutamic acid. *Upper,* $n = 5$; *lower,* $n = 16$; *upper right,* 1 μmole glutamic acid. DEAE-cellulose (1×26.5 cm, 0.88 meq N/gm, 100–200 mesh) equilibrated with 0.01 M Na citrate, pH 8. Developed with linear gradient of 396 ml with 0.5 M buffer as limit.

observed have not yet been identified but they are not simple multiples of the monomer. It is exceedingly interesting to note the doublets in the $n = 16$ polyglutamic acid preparation. The ratio of peaks within the doublet is roughly one at 150 cc but varies in opposite directions on either side of this point. The explanation for this doublet formation is obscure and worthy of investigation. It may be related to different initiators, to the butylamine initiating the polymerization according to two different mechanisms, to the formation of carbamates, or to the presence of hydrolyzed monomer as the initiator. In any case, the heterogeneity of these low molecular weight preparations is clear and their use for the determination of the relationship between size of polymer and secondary and tertiary conformation is of questionable validity.

This preliminary work emphasizes that methods are now available for the preparation of homogeneous species within a polymer series. Many of the problems which have had to be approached circuitously can now be studied more directly with pure substances. It should now be possible to examine the effect of chain length on polymerization rates, on the chemical reaction with ninhydrin, on secondary and tertiary conformation as reflected by melting curves, infrared spectroscopy, optical rotation, etc.

To return to the discussion of Paper 9, in our experience with amino acid and with nucleotide polymers, adsorption increases with increasing molecular weight in a homologous system. This seems to be a rather regular function in a given chromatographic system, i.e., with the same load, the same amount of adsorbent, and the same eluting system. However, it is not a simple function of the salt concentration required for elution. We have found that a change in gradient or a change in load will result in the elution of a given polymer in a different salt milieu. Stewart and Stahmann have observed the same effect (see Figure 9.8 of their paper). The relationship between adsorption and polymer size seems to be a complex function of the salt concentration, the gradient slope, interaction between buffer and adsorbent, as well as the adsorbent-load ratio. However, in a single chromatographic system in which the gradient, the load, and the amount of adsorbent is kept constant, the effluent position of a component is relatively invariant and can be used for analytical as well as preparative purposes.

PART III

Properties of Polyamino Acids in Solution

PAPER 10

Some Recent Observations on Polypeptides in Solution

PAUL DOTY and WALTER B. GRATZER

In the eight years that have passed since it was shown that polypeptides could exist in the helical as well as the randomly coiled form in solution (*1*), numerous and diversified investigations on polypeptides in solution have created a new chapter in the physical chemistry of macromolecules and have provided new insight into the basis of protein structure. While it would be useful to review these developments and attempt to evaluate them critically, the magnitude of such an undertaking makes it inappropriate for this kind of presentation. Moreover, having just found that such a review of the single area dealing with optical rotation of polypeptides and proteins (*2*) required more than two hundred pages, our reluctance to proceed further at this point may be understood as well as welcomed. Instead we propose to discuss informally some of our current work which may hopefully indicate the way in which polypeptides continue to contribute to both the physical chemistry of macromolecules and the analysis and understanding of protein structure.

It may first be of value, however, to draw attention again to the uniqueness of the polypeptide molecules. In the coiled form they undergo an endless Brownian dance, the individual segments constantly changing their relations with their neighbors, and no two molecules having precisely the same conformation. But by altering the solvent conditions or the temperature it is possible to cause these molecules to take up rigid and identical conformations, that is, to undergo intramolecular crystallization. As a result of this transition the side chains bear a constant spatial relation to each other; stereochemical regularity is then imposed upon what was previously an undulating, randomly coiled chain in which interactions and inter-group distances had only statistical significance. This property of intramolecular crystallization is shared by only one other class of polymers, the polynucleotides and nucleic acids. Here too the ordered form of the molecules is helical and the structure is stabilized by hydrogen bonds. It would seem extraordinary that no other polymer structures exist in which internal hydrogen bonding can give rise to a periodically ordered conformation, but no others have indeed been found thus far. We are therefore forced to recognize the uniqueness of this capacity in polypeptide chains, one which enables them to meet the exacting and sophisticated demands of protein structure and function.

The examples of work which follow deal with several aspects of this helical form: the establishment of its precise conformation, its contribution to protein structure, its effect on the far ultraviolet absorption spectrum, its effect on the ultraviolet spectra of aromatic groups, and its stability in aqueous solution.

The identification of the helical form in solution as the α-helix

The early work with poly-γ-benzyl-L-glutamate (*2*), in which a series of samples of different molecular weights dissolved in chloroform-formamide displayed intrinsic viscosities in conformity with Simha's relation for rodlike molecules (*3*), clearly established the rodlike nature of these polypeptides in this solvent. The dependence of the sedimentation constant on molecular weight has also been examined (*4*) and found to conform in a similar manner. However, because the prediction for intrinsic viscosity is so specific, requiring a limiting dependence on molecular weight to the power of 1.7, this relation was emphasized. The rodlike nature of the molecules in this system was also demonstrated by flow birefringence methods and this work is to be published shortly (*5*). Numerous other studies have confirmed this result and extended it to other polypeptide systems, although in some the rodlike structures are not completely straight (*6, 7*).

The several proposed helical structures that might be the basis of this rodlike shape can be distinguished by their residue displacement, that is, the length per residue in the helix direction. The determination of this quantity requires measurement of both the molecular weight and length. Using light scattering for both of these, the results on four samples were found to average 1.45 ± 0.09 Å (*2*). Since the residue displacement for the α-helix is 1.50 Å and no other helices gave values near this, the conclusion was drawn that the rodlike form was the α-helix. Independent evaluations of length by means of intrinsic viscosity measurements interpreted by means of Simha's relation gave the same result. Moreover, electron microscope studies on the helices deposited on mica by spraying the solution led to the same conclusion (*8*).

Recently the determination of residue displacement has been undertaken by a different approach, developed by Luzzati and his associates (*9*), who have used low-angle X-ray scattering to determine the mass to length ratio. In concentrated (liquid crystalline) pyridine solutions of poly-γ-benzyl-L-glutamate the 1.50 Å value was observed, but in more dilute (isotropic) solutions in this and two other solvents values around 1.95 Å were obtained. On this basis a more extended helical model (3_{10} helix) is proposed as the form generally present in solution. The low-angle scattering technique is a new one, experience of which is not yet widespread: it is therefore important still to inquire into the errors that could arise, in particular from the departure from uniform densities and over-all cylindrical shape of the helix and from inaccuracies in specific volume values.

It would seem desirable then to balance this new evidence against a re-examination of previous data derived from techniques which may now be regarded as classical. Luzzati has presented a plot of the residue displacement length against molecular weight that includes all published measurements. Such a plot shows a trend toward increased residue displacements with decreasing molecular weight and in the limit a value of about 2.0 Å seems indicated. The conclusion is drawn that such helical molecules are somewhat flexible and only in the limit of vanishing length is the true displacement length exhibited. However, this conclusion does not find support in a careful examination of the data, save for the eventual falling off of the measured displacement length at quite high molecular weights where it has always been recognized that flexibility becomes significant. In particular the values for the lowest molecular weights, which provide most of the justification for extrapolating to a value of about 2 Å, are seen to refer to samples of such low molecular weight that light scattering cannot be considered quantitatively reliable. These samples have dissymmetries of less than 1.1 at 135 and 45° and at these low values the effects of dust, which are always to increase the measured length, cannot be reliably eliminated; neither can precision at the level required be insured. Hence these points should not be included in such an argument. The two other high points are from other sets of data which conform to the lower value (*ca.* 1.5 Å) and hence stand apart from the bulk of the evidence within each investigation.

The recent examination of poly-γ-benzyl-L-glutamate samples of very low polydispersity (*5*), both by light scattering and flow birefringence, again lead unambiguously to the 1.5 Å value of the residue displacement. These results disincline us to modify the conclusion that the α-helix is indeed the form predominantly obtaining in the systems which we have studied.

The role of the α-helix in proteins with special reference to myoglobin

Soon after the demonstration that polypeptides could exist in the helical form in solution it was found that the rotatory dispersion was complex (*10*) and could be fitted by a two-term equation (*11*):

$$[m']_\lambda = \frac{a_o\lambda_o^2}{\lambda^2 - \lambda_o^2} + \frac{b_o\lambda_o^4}{(\lambda^2 - \lambda_o^2)^2}$$

where $[m']$ is the reduced mean residue rotation given by

$$[m'] = \frac{3}{n^2 + 2} \frac{M_o}{100} [\alpha]$$

Here $[\alpha]$ is the specific rotation, n the refractive index, and M_o the mean residue weight. In the randomly coiled, or disordered, form the rotatory dispersion is fitted by only the first term although a quite different value for the constant a_o is required. When this behavior was established it was possible to show (*12*) that the rotatory dispersion of many proteins could be represented by the foregoing two-term equation if a_o and b_o were adjusted to correspond to fractions of the values observed for completely helical polypeptides. That is, the dispersion of proteins could be accounted for if they were considered to be made up of helical and non-helical regions, assuming the latter to resemble the disordered polypeptide chain in rotatory properties.

It has since been found that the β-conformation can occur in proteins and is associated with rotatory characteristics of its own (*13, 14*), which can be incorporated in an optical rotatory analysis of the structure (*14*). It has since been shown that two other conformations can occur, each characterized by different sets of values of the rotatory constants. These appear to occur rather rarely in proteins and in these cases a more complicated and ambiguous analysis is necessary. In general, however, a review of the accumulated data in this area (*2*) shows the serviceability of the original hypothesis. With the determination of the structure of myoglobin, which revealed that 77 per cent of its residues were in the form of right-handed helices (*15*), a direct test of the hypothesis became available, thereby offering an important improvement over most previous work which had been confined largely to tests of self-consistency.

The rotatory dispersion of myoglobin in the visible is dominated by Cotton effects associated with the heme absorption bands (*16, 17*); hence recourse to the ultraviolet was necessary. Preliminary studies in this region, roughly 250–330 mμ, using polypeptides in the helical form, showed that the previous assignment of a value of 212 mμ to λ_o had to be changed to 216 mμ if the Moffitt relation (above) was to be obeyed. Under these circumstances the value of b_o associated with a complete helix changes from -630 to -535. When the dispersion of myoglobin was interpreted in this context a value of 73 per cent helix resulted (*18*). Since the helix in myoglobin is known to be right-handed, this result established the same sense for most helical conformations (i.e., all those characterized by negative values of the constant b_o). The good numerical agreement between this result and the definitive crystallographic values indicates quite clearly that the method of analysis employed is valid and substantially increases the confidence with which the results of similar dispersion studies on other proteins can be viewed.

The far ultraviolet absorption spectra of polypeptides and proteins

Despite a fair degree of success in the rotatory dispersion assessment of helical content, an independent method is needed for the following reason. Where cystine bridges or like constraints occur in proteins, it may be quite possible for some segments lying between such bridges to exist in the opposite, left-handed, helical conformation. Such sections would approximately compensate an equal length of standard, right-handed helix in their optical activity and lead in consequence to an underestimate of the helical content.

In search of an alternative physical method sensitive to the helical conformation an investigation was undertaken of the pronounced hypochromic effect which Imahori and Tanaka (*19*) had observed for the helical conformation of poly-L-glutamic acid in the far ultraviolet absorption band associated with the peptide chromophore. It was possible to show that the helical forms of both poly-L-glutamic acid and poly-L-lysine were not only largely hypochromic with respect to the random, disordered form at 190 mμ, but also showed the development of splitting in the peptide band (*20*). The effect can be seen in Figure 10.1 where the results for poly-L-lysine are shown. This apparent splitting is consistent with the predictions of exciton theory for resonating chromophores in a molecular crystal (*21, 22*). Poly-L-lysine has the unusual property of undergoing a transition from the helix to a β- or pleated-sheet structure when its solution is heated. The upper curve in Figure 10.1 apparently represents this form. It is seen to differ in minor degree from that of the disordered form.

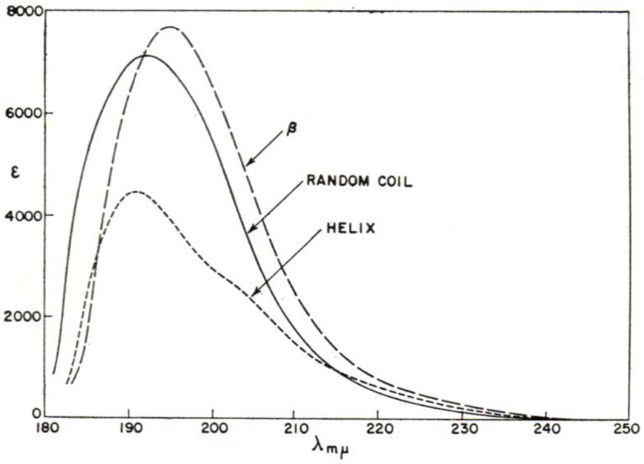

Fig. 10.1. Ultraviolet absorption spectra of poly-L-lysine hydrochloride in aqueous solution: random coil, *p*H 6.0, 25°; α-helix, *p*H 10.8, 25°; β-form, *p*H 10.8, 52°.

This contrasts with infra-red spectra of aqueous solutions where the coil and α-helix are not in general distinguishable, but the β-form shows a distinctive frequency of its own. Thus by combining both kinds of spectra it is in principle possible to analyze the structure of the polypeptide chain in terms of the three primary conformations.

Returning to the relation of conformation to absorbance, the extinction coefficient per residue at 190 mμ has been determined as 6,900 in the random coil, and 4,100 in the α-helix. At 197 mμ, the corresponding values are 6,350 and 3,200.

To exploit this difference in the examination of protein spectra in this region the absorption due to the side chains must be determined and subtracted and, in the first instance, the assumption is made that the absorption of these groups is not affected by their environment in the native protein. On this basis it is possible to predict the absorption of the protein corresponding to 100 per cent and 0 per cent helical content. The measured absorption curve for the protein should then fall between these limits and on the scale which they define the helical content may be estimated.

Such estimates have been made for a series of proteins, and the results to date encourage further efforts in this direction. An unsatisfactory feature is the considerable discrepancy between estimates of helical content based on measurements at different wave lengths in the region of peptide absorption. Thus whereas the estimates for paramyosin based on 190, 197, and 205 mμ all lead to values near 100 per cent, in good agreement with optical rotatory dispersion data, measurements at the same three wave lengths for myoglobin lead to apparent helical contents of 99, 67, and 81 per cent.

It seems likely that these discrepancies are in large measure due to the inadequacy of the above-mentioned assumption regarding the effect of the protein environment on the absorption of side-chain chromophores. It is well known that the absorption bands of the aromatic amino acids in the near ultraviolet are somewhat displaced towards longer wave lengths, compared with their free state in aqueous solution, when they are incorporated in a protein. This is a consequence of the greater polarizability of the hydrophobic environment, and may be reproduced by adding solvents of high refractive index, or for instance detergents, to the aqueous amino-acid solutions.

Preliminary work has shown that similar shifts can be produced in this manner in the 190 mμ region. It will be noted that the aromatic amino acids show allowed transitions with extinction coefficients approaching 50,000 at these wave lengths, so that absorbances at points coresponding to steeply rising curves may be substantially modified by such shifts. By adding to amino-acid solutions concentrations of solvents which match in their effect the shift produced by a protein environment on tyrosine and tryptophan in the near ultraviolet, modified spectra, representing the chromophores within the protein matrix, can be obtained for all amino acids. A set of values of extinction coefficients at selected wave lengths may thus be compiled, and will, it is hoped, lead to the evaluation of the side-chain contribution in more definitive terms.

Difference spectra arising from the incorporation of tyrosine into helical regions

In this and the final section we turn to two applications of specially synthesized polypeptides to current problems in protein chemistry. In the first instance we are concerned with the origin of the familiar blue shift in the vicinity of 280 mμ which proteins undergo upon denaturation (*23*, *24*). This shift involves both the tyrosine and the tryptophan residues of the protein, and is best studied in the form of a difference spectrum with the denatured protein in the reference and the native protein in the sample cell. It was originally suggested that in tyrosine the shift accompanied the breakage of a hydrogen bond between the phenolic hydroxyl and some donor, such as a carboxylate group (*23, 25*). However, subsequent investigations have shown that such shifts in model compounds can arise from a number of causes, notably ionization of vicinal groups and changes in the refractive index of the environment of the chromophore (*26–28*).

The implication of tyrosine in the difference spectra has been particularly thoroughly studied (*29*), and since its association with carboxylate groups has been specifically postulated, it appeared that a copolymer of L-tyrosine and L-glutamic acid offered a useful model for study inasmuch as it could be converted into either the helical or disordered form by changes in *p*H. This transition may be regarded as a primitive model for the denaturation of a protein.

With this in mind a copolymer consisting of 5 mol per cent tyrosine and 95 mol per cent glutamic acid was prepared. Optical rotation measurements confirmed that this polymer was helical below *p*H 4 and randomly coiled above *p*H 6. A careful examination revealed that there was no Cotton effect associated with the 270–290 mμ region in either conformation.

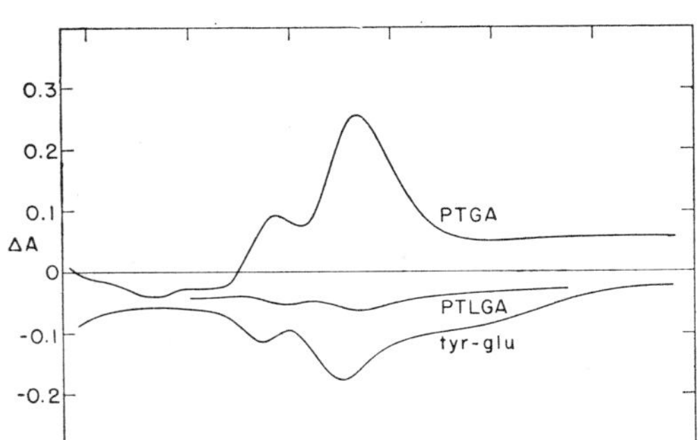

Fig. 10.2. Difference spectra. Copoly-L-tyrosine-L-glutamic acid in 0.1 M phosphate buffer (A_{276} =2.55); reference, *p*H 6.80; sample, *p*H 3.88. Copoly-L-tyrosine-L-lysine-L-glutamic acid in water (A_{276}=2.48); reference, *p*H 6.51; sample, *p*H 3.26. L-Tyrosyl-L-glutamic acid in water (A_{276} =2.10); reference, *p*H 7.97; sample, *p*H 1.67.

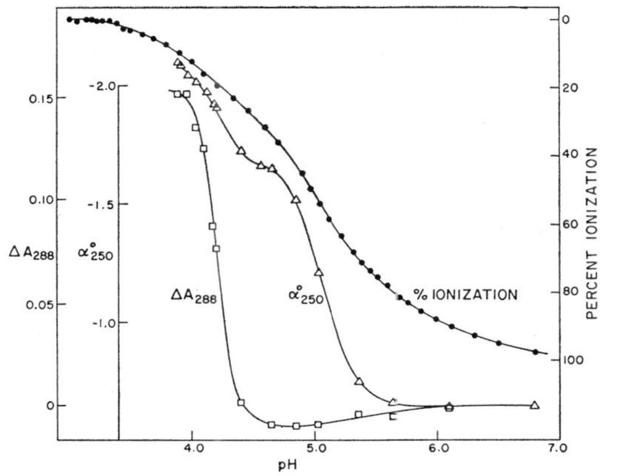

Fig. 10.3. Spectrophotometric and optical rotatory titrations of copoly-L-tyrosine-L-glutamic acid, 0.42 per cent in 0.1 M phosphate buffer; *p*H titration in 0.2 M NaCl.

This excluded the possiblity of any specific asymmetrical interaction between the benzenoid group and the helical backbone. Such an interaction has been reported in the case of some dye molecules (*30*). Likewise no significant hypochromic effect could be discovered upon passing from one form to the other. However, there was a blue shift upon passing from the helical to the coiled form, which expressed itself also in a characteristic difference spectrum, as shown in Figure 10.2. An analogy to protein denaturation is thus indicated.

To examine this effect further, the absorption at the larger peak (287 mμ) was followed as a function of *p*H through the helix-coil transition. The result is shown in Figure 10.3, the degree of ionization and the specific rotation being given for comparison. It is seen that the absorbance change is abrupt and becomes complete before many of the γ-carboxyl groups have ionized. Moreover the breakup of the helix in this region is not continuous with the remainder of the helix-coil transition but is displaced towards lower ionization, suggesting that the 30 per cent change involves inherently weaker helical regions. Since the remainder of the transition coincides with the normal poly-L-glutamic acid transition aqueous salt solutions (ionic strength *ca.* 0.1) it seems justifiable to conclude that the polypeptide helix has decreased stability in the vicinity of the tyrosine groups and breaks up preferentially in the early stages of titration. Thus the blue shift in the tyrosine spectra can be identified with the transfer of the tyrosine residues from a helical environment to that of a disordered state, which is predominantly that of the solvent.

It remains to exclude the vicinal charge effect on the chromophore on ionization of the γ-carboxyl groups as a source of the difference spectra. When the dipeptide L-tyrosyl-L-glutamic acid at *p*H 1.7 (sample) is measured against a solution of the same concentration at *p*H 8, where the carboxyl groups are ionized, a similar but inverted difference spectrum is produced (Figure 10.2). Vicinal ionizations therefore will tend to bring about a red and not a blue shift on raising the *p*H of the polymer.

This scheme outlined above would predict that no blue shift would be observed if a polypeptide were employed that could not develop a completely helical conformation at acid *p*H, since the tyrosine groups would then remain in nonhelical regions. Since it had been previously shown that a copolymer of L-glutamic acid and L-lysine had a helix content of about 15 per cent at neutral *p*H, which increased to a limit of 50 per cent at acid *p*H, a copolymer of this type was prepared with 4 per cent L-tyrosine included. This polymer showed no blue shift, but rather a very small red shift when examined at the two *p*H values (Figure 10.2). The slight red shift was probably the result of a small vicinal ionization effect. This result supports our interpretation of the origin of the blue shift, in terms of a change in the environment of the tyrosine as a direct consequence of the helix-coil transition.

If it is permissible to regard the chromophores in the helical environment as "caged" or shielded by the side chains, these should be less affected by "perturbing" solvents added to the solution to increase its refractive index than the same groups in the random coil. The procedure of Herskovits and Laskowski (*31*) was followed, and the addition of 20 per cent by volume of ethylene glycol did indeed appear to produce a substantially greater difference spectrum in the polypeptide at *p*H 7 than at *p*H 4.

It is to be anticipated in the light of our results that larger blue shifts per aromatic residue may be expected in proteins than those observed here in the helix-coil transition. In the interior of a protein a greater concentration of polarizable groups around the chromophore can be expected than in the polypeptide which we have studied, because of the packing or clustering tendency of side chains. Consequently, the environmental change from the standpoint of polarizability could be greater upon denaturing a protein than upon the helix-coil transition in a synthetic polypeptide. On examining the magnitudes of the blue shift in proteins (*28*) it is seen that this expectation is borne out. The change in molar extinction coefficient at 287 mμ for the polypeptide investigated was about −140. Similar values have been found for pepsin and lysozyme, whereas more than double this value has been reported for ribonuclease, α-chymotrypsin, and bovine serum albumin. This would seem to confirm the general interpretation (*28*) that the blue shift in the spectra of the aromatic groups of proteins upon denaturation resides mainly in a decrease in the polarizability of their immediate environment, an effect that can result from the breakup of a helical conformation containing the chromophore or from a more general unfolding of the molecule with the exposure of these groups to the aqueous environment.

The stability of poly-L-alanine helices in aqueous solution

Since water forms the natural environment of proteins, it is obvious that polypeptides will serve most faithfully as models and analogues if they too are studied in aqueous solution. Yet the polypeptides of most naturally occurring amino acids are not soluble in water. Hence our view of the behavior of the simplest forms of protein analogues is very limited under the circumstances of greatest interest. We have attempted to overcome this situation by preparing block polymers such that a block of one kind of amino-acid residue, insoluble in water, is united at each end to blocks of residues that are polar and thus soluble. When the proper lengths are chosen for the three blocks the resulting polypeptide is soluble in water and the water-insoluble chain forming the middle block has indeed been brought into aqueous solution.

The conformation of the central block is the point of interest and this can be examined most directly by preparing the two flanking blocks from a racemic mixture. They are then optically inactive and the rotatory dispersion can be used to study the central block since it is the only site of optical activity.

To this end we have synthesized a block copolypeptide in which the central block consists of about 130 residues of L-alanine and the two flanking blocks of about 200 residues total of D,L-glutamic acid.

This block copolymer is readily soluble in water and saline solutions. The rotatory dispersion was measured in 0.1 M Na_2HPO_4 and upon analysis by the Moffitt equation using $\lambda_o = 212$ mμ it was found to have a b_o value of −580, assuming that only the L-alanine residues contributed to the rotation. This value reflects a helical content of 92 per cent by the usual standards and thus indicates the helical nature of the poly-L-alanine in aqueous solution. No change could be observed when the temperature was raised to 95°. Thus the helix is extraordinarily stable relative to most proteins.

Similar rotatory dispersion measurements were carried out on solutions to which various denaturing agents had been added. The helical content deduced from such measurements was 78 per cent for 5 M guanidinium hydrochloride, 70 per cent for 8 M urea, and 100 per cent for 0.1 M sodium dodecyl sulfate. Thus the helix is seen to be essentially unaffected by common denaturing agents.

Further dispersion measurements were carried out in two nonaqueous solvents. In hydrazine the helical content was found to be 81 per cent and in trifluoroacetic acid 25 per cent. By comparison poly-γ-benzyl-L-glutamate had been found to be nonhelical in both of these solvents. Thus we are forced to conclude that the poly-L-alanine helix is stable in the presence of denaturing agents and even in organic solvents it survives in varying degrees where many polypeptides become nonhelical.

Although the complete understanding of the basis of the unusual stability of poly-L-alanine helices in aqueous solution must await other studies in which the chain length and composition of the central block is varied, the result itself has some general significance, since it indicates that the α-helix conformation has greater stability when the side chains are nonpolar than when they are polar, as in poly-L-glutamic acid or poly-L-lysine. It is important to note, however, that the stability of the poly-L-alanine helix is not due to side-chain interactions. The methyl group has a van der Waal's radius of 2.0 Å; yet the separation of the centers of nearest neighbors on the surface of the helix is 5.7 Å. The consequences of introducing other nonpolar groups into the central block so that side-chain interactions can indeed occur will clearly be of interest and should permit a useful delineation of the factors involved in the stability of helices composed of nonpolar residues in aqueous environments.

In bringing this discussion to a close it may be useful to return to the polypeptide that has been longest studied in aqueous solution: poly-D,L-alanine. This polymer does have limited solubility in water and early studies by infrared spectroscopy and deuterium exchange were interpreted as indicating that it possessed considerable helical content (*32*). More recent studies however have led to the opposite conclusion (*33*, *34*). In view of the polymerization kinetics of other racemic mixtures of N-carboxyanhydrides it would appear likely that there is a pronounced tendency toward the formation of short sequences of like residues (*35*). Since enantiomorphic residues three removed from each other will tend to interfere sterically when juxtaposed in an α-helix, it seems unlikely that a high helical content will be favored in this instance.

In this context we have examined the far ultraviolet spectra of poly-D,L-alanine and find a substantial hypochromic effect in the vicinity of 190 mμ. Normally interpreted this would indicate about 30 per cent helix although the shape of this absorption band deviates from that encountered in L-polypeptides, so that too much quantitative reliance may not be placed on this finding. A similar result was also noted for poly-D,L-glutamic acid at low *p*H where the L-form is normally completely helical. The agreement in these two cases, together with the above evidence that the L-alanine residues do form stable helices in aqueous solution, strongly suggests that poly-D,L-alanine does indeed possess an appreciable degree of the α-helical conformation. This presumably represents the maximum fraction of the residues that can fit into short helical regions consisting principally of pure sequences of either enantiomorph.

Acknowledgment

The work reported here was supported by grants from the National Science Foundation (G 17556) and the office of Naval Research (Nonr 1866 [25]).

References

1 P. Doty, J. H. Bradbury, A. M. Holtzer, and E. R. Blout, *J. Am. Chem. Soc.* **76**, 947 (1954).

2 P. J. Urnes and P. Doty, *Advances in Protein Chem.* **16** (1961).

3 R. Simha, *J. Phys. Chem.* **44**, 25 (1940).

4 M. J. Schick and P. Doty, unpublished observations.

5 J. T. Yang and P. Doty, *J. Mol. Biol.*, in press.

6 A. Wada and P. Doty, unpublished observations.

7 J. Applequist and P. Doty, this book, Paper 14.

8 C. E. Hall and P. Doty, *J. Am. Chem. Soc.* **80**, 1269 (1958).

9 V. Luzzati, M. Cesari, G. Spach, F. Masson, and J. M. Vincent, *J. Mol. Biol.* **3**, 566 (1961); also this book, Paper 11.

10 P. Doty and J. T. Yang, *J. Am. Chem. Soc.* **78**, 498 (1956).

11 W. Moffitt and J. T. Yang, *Proc. Nat. Acad. Sci.* **42**, 596 (1956).

12 J. T. Yang and P. Doty, *J. Am. Chem. Soc.* **79**, 761 (1957).

13 K. Imahori, *Biochim. et Biophys. Acta* **37**, 336 (1960).

14 A. Wada, M. Tsuboi, and E. Konishi, *J. Phys. Chem.* **65**, 1119 (1961).

15 J. C. Kendrew, R. E. Dickerson, B. E. Strandberg, D. C. Phillips, and V. C. Shore, *Nature* **185**, 422 (1960); also this book, Paper 21.

16 D. Savitz and P. Doty, unpublished observations.

17 G. L. Eichhorn, *Tetrahedron* **13**, 208 (1961).

18 P. J. Urnes, K. Imahori, and P. Doty, *Proc. Nat. Acad. Sci.* **47**, 1635 (1961).

19 K. Imahori and J. Tanaka, *J. Mol. Biol.* **1**, 359 (1959).

20 K. Rosenheck and P. Doty, *Proc. Nat. Acad. Sci.* **47**, 1775 (1961).

21 W. Moffitt, *ibid.* **42**, 736 (1956).

22 W. T. Simpson and D. L. Peterson, *J. Chem. Phys.* **26**, 588 (1957).

23 H. A. Scheraga, *Biochim. et Biophys. Acta* **23**, 196 (1957).

24 E. J. Williams and J. F. Foster, *J. Am. Chem. Soc.* **81**, 865 (1959).

25 M. Laskowski, S. J. Leach, and H. A. Scheraga, *ibid.* **82**, 571 (1960).

26 D. B. Wetlaufer, J. T. Edsall, and B. R. Hollingworth, *J. Biol. Chem.* **233**, 1421 (1958).

27 C. C. Bigelow and I. I. Geschwind, *Compt. rend. trav. lab. Carlsberg* **31**, 283 (1960).

28 S. Yanari and F. Bovey, *J. Biol. Chem.* **235**, 2818 (1960).

29 B. S. Harrap, W. B. Gratzer, and P. Doty, *Ann. Rev. Biochem.* **30**, 269 (1961).

30 L. Stryer and E. R. Blout, *J. Am. Chem. Soc.* **83**, 1411 (1961).

31 T. T. Herskovits and M. Laskowski, *J. Biol. Chem.* **235**, PC56 (1960).

32 A. Berger and K. U. Linderstrøm-Lang, *Arch. Biochem.* **69**, 106 (1957).

33 W. P. Bryan and S. O. Nielsen, *Biochem. et Biophys. Acta* **42**, 552 (1960).

34 See A. Elliott and C. deLozé, discussions, this book.

35 A. Wada, *J. Mol. Biol.* **3**, 507 (1961); also this book, Paper 12.

Discussion of Paper 10: Polypeptides in Solution

A. ELLIOTT

There has been some discussion between sessions about the form of poly-D,L-alanine in aqueous solution, to which I referred in an early paper as an α, folded form (*Nature* **170,** 1066 [1952]). More recently, we have published evidence based on the extrapolation of the specific rotation-composition curve of a series of L and D alanine copolymers which has led us to conclude that the *meso* polymer is in a solvated, random conformation in water as well as in aqueous lithium thiocyanate and in dichloracetic, trifluoracetic, and phosphoric acids (Downie, Elliott, Hanby and Malcolm, *Proc. Roy. Soc.* **A242,** 325 [1957]). A slow rate of deuterium exchange of the NH groups of poly-D,L-alanine) in D_2O has been interpreted as evidence of an α-helix form (Berger and Linderstrøm-Lang, *Arch. Biochem. Biophys.* **69,** 106 [1957]), but later work from the Carlsberg Laboratory (Bryan and Nielsen, *Biochim. et Biophys. Acta,* **42,** 552 [1960]) provides cogent arguments against the case for a well-developed α-helix in aqueous solution. While, therefore, the conformation may not be entirely random, it is reasonable to conclude that it is not helical.

HAROLD A. SCHERAGA

The model which Nemethy and I used for computing the thermodynamics of formation of a hydrophobic bond is indeed based on the contact between nonpolar groups in an aqueous environment (G. Nemethy and H. A. Sheraga, Abstracts of the 138th ACS meeting, p. 4C, New York, N.Y., Sept. 1960). However, these nonpolar groups need not be in the side chain; they can also be in the backbone. In order to be able to set up a partition function to calculate thermodynamic parameters, it seems to me that one has to start with some model for the nonpolar groups in the aqueous environment. We have some confidence in the model we used since it leads to good agreement with experimental thermodynamic parameters for water and for aqueous solutions of hydrocarbons.

DON WETLAUFER

Apropos of a model for hydrophobic bonding, I should like to point out that it is not necessary that two nonpolar side chains have contact in order to produce what is thought of as a hydrophobic bond. A certain amount of stabilization energy (re: the free-swimming hydrocarbon side chain) would probably result simply from placing the side chain folded against a helical segment, for example, even in the absence of an adjacent nonpolar side chain. But rather than consider discrete models for such interactions, it is probably best simply to emphasize that in its most general terms, a hydrophobic bond need only represent a favorable free-energy difference between two states, summing the contributions both from the nonpolar group and its environment in both states.

PAPER 11

The Structure of Poly-L-γ-Benzyl Glutamate in Solution: Helical Conformations Differing from the α-Helix, and Transitions between Helical Forms

V. LUZZATI, M. CESARI, G. SPACH, F. MASSON, and J. M. VINCENT
(translated from the French by C. COHEN)

Since the classical work of Astbury (see Low, ref. *13*), it has been known that in an important class of fibrous proteins the polypeptide chain can adopt two conformations: one called α, where the molecules form a coil stabilized by a set of intramolecular hydrogen bonds; and the other called β, where, in contrast, the polypeptide chains are extended.

Pauling, Corey, and Branson (*25*) generalized the concept of the α-conformation by proposing a series of helical structures (see also ref. *2*) having specified coördinates which could be verified experimentally. Shortly thereafter, Perutz (*26*), followed by others (*2*), recognized certain helical characteristics, in the X-ray diffraction diagrams of poly-benzyl-L-glutamate and a few proteins. From the beginning, one particular structure, called the α-helix by Pauling and Corey, assumed a predominant position.

Poly-benzyl-L-glutamate (hereafter abbreviated as PBLG) has played a special role in the success of the α-helix. Perutz used fibers of PBLG in the first crystallographic proof of the model, and subsequently these were examined in more detail, confirming the structure to be in excellent agreement with that of Pauling and Corey's α-helix. Similarly, Ambrose and Elliott (*1*) made use of oriented samples of this polypeptide in demonstrating the infra-red dichroism, and showed that the hydrogen bonds between the C=O and N—H are effectively parallel to the fiber axis; this orientation is characteristic of helical structures (*2*).

The structure of PBLG has also been studied in dilute solution, by measurements of infrared absorption (*1*), optical rotation (*32, 9*) viscosity (*7, 33, 30*), light scattering (*7, 30*), flow birefringence (*34*), dielectric dispersion (*35*), translational diffusion (*30, 11*), and rotatory diffusion (*31, 24, 22*, and Benoît and Boeckel, in preparation).

Thus it was demonstrated that PBLG can adopt either of two conformations, depending on the nature of the solvent: one, of the α-type, exists in certain moderately polar solvents (dimethylformamide, pyridine, metacresol); in contrast the other, unfolded and irregular, of the "random coil" type, exists in certain very polar solvents (dichloroacetic acid, hydrazine). The properties of the conformation of the α-type clearly indicate that the molecules have a rigid rod shape, and that the polypeptide chain is folded in a regular manner, apparently helical. The temptation was great to identify this structure with the Pauling and Corey α-helix. Reasonable agreement between results of hydrodynamic and light-scattering measurements and the characteristic dimensions of the α-helix seemed, in fact, to justify the conclusion which was thereafter unanimously accepted: that the structure of PBLG, in solutions of these solvents, is that of the α-helix (*8*).

We note here that in undertaking our study of dilute isotropic solutions of PBLG, we wished particularly to calibrate the technique of absolute intensity, low-angle X-ray scattering only recently developed (*15*). On the basis of the preceding work, we were certain to be dealing with the α-helix. Since the beginning of our work, however, the value for the linear mass of the rods, measured by our technique, was found consistently lower than that of the α-helix. This result, of which we were originally sceptical, finally had to be fully accepted, when work with other materials (*18, 20, 21*) proved the discrepancy to be much larger than the experimental error.

Meanwhile, having become interested in the remarkable optical phenomena observed by Robinson, Ward, and Beevers (*28, 29*) in the liquid-crystalline phases of certain PBLG-solvent systems, we began to examine concentrated solutions of PBLG. Our experimental observations in these solutions also brought to light several phenomena far more complex than those we expected to find.

Guided by the experience acquired in the study of association colloids (*17, 12*), of nucleoproteins (*14*), and of desoxyribonucleic acid (*18*), we decided to undertake a systematic study, as a function of concentration, of the structure of the phases formed by PBLG and solvents. Here we describe the results obtained with dimethylformamide, pyridine, and metacresol. A more detailed account of these results has recently been published (*16*).

Experimental technique

The polypeptide.—The various samples of PBLG were prepared according to the procedure described by Blout and Karlson (*4*). Polymerizations were carried out in dioxane solutions using sodium methoxide as initiator. The molecular weight of different samples was determined by light scattering and, in a few cases, by viscosity measurements previously calibrated by light scattering (*30*). We shall characterize the helical structures by the distance h between two successive residues projected onto the helix axis; for the α-helix $h = 1.5$ Å. The solvents used were carefully dehydrated, then purified by several distillations.

Studies of dilute isotropic solutions.—The solutions were prepared by mixing known quantities of polypeptides and solvent. Occasionally it was difficult to obtain homogeneous mixtures; with very viscous solutions heating and mechanical stirring was used. The specimen holder on the X-ray set was filled by a device which eliminated solvent loss due to evaporation. The solutions were studied by means of absolute intensity measurements of low-angle X-ray scattering, employing the technique described elsewhere (*15, 18*).

We recall that the result of this kind of experiment may be represented, after normalization, by a function $j_n(s)$, which is independent of experimental conditions, and depends on the structure of the sample. When this consists of a dilute solution of long rigid rods, $j_n(s)$ is expressed as:

$$j_n(s) = A\frac{1}{2}\exp(-\pi^2R_c{}^2s^2)K_o(\pi^2R_c{}^2s^2) \qquad (1)$$

$$A = c_e\mu(1-\rho_o\psi)^2 \qquad (2)$$

where s measures the diffraction angle ($s = 2 \sin\theta/\lambda$), c_e is the electron concentration (number of solute electrons divided by number of solution electrons), ρ_o is the electron density of the solvent (el. Å^{-3}), ψ is the electron specific volume of the solute in the solvent (el.$^{-1}$ Å^3), μ is the linear mass of the rods (number of electrons per unit length el. Å^{-1}), R_c is the radius of gyration, around the rod axis, of the difference between the electron density of the rods and that of the solvent.

Equation 1 is rigorously valid only at sufficiently small concentrations where correlation effects of position and orientation between the rods are negligible. These effects may be eliminated by performing several experiments at different concentrations and extrapolating the results to $c_e = 0$; we have discussed this problem recently (*18*).

Analysis of the experimental curves is facilitated by the use of a double logarithmic plot. Each experimental curve log $j_n(s)$, plotted as a function of log s, is translated, without rotation, with respect to the theoretical function (see equation 1) log $[\frac{1}{2}\exp(-\pi^2s^2)K(\pi^2s^2)]$, also plotted as a function of log s, until the two curves coincide; the translations of the abscissa and ordinate measure respectively R_c and A. Knowing c_e, ψ, and ρ_o, the value of μ (equation 2) can be determined.

All the experiments were performed at a temperature of approximately 25° C.

Study of concentrated solutions: mesomorphic and crystalline phases.—These solutions were prepared by mixing known quantities of polypeptide and solvent; they were examined after equilibrium was reached. The concentrations were redetermined by dry weight after the X-ray exposures.

The X-ray diagrams were obtained with evacuated focusing cameras, monochromatic radiation, and photographic recording (*14*).

The patterns show several sharp lines in the low-angle region, and occasionally some rather diffuse bands. When the lattice is two-dimensional hexagonal, the spacing (Å^{-1}) of the first line s_1 varies with concentration c (grams of solute per gram of solution) according to the equation (*18*):

$$s_1^{-2} \times 10^{-16} = \frac{\sqrt{3}}{2} \frac{m}{l} \frac{1}{\delta} [c^{-1} - 1 + \delta\bar{v}] \quad (3)$$

where $\bar{v}$ is the partial specific volume of the solute (cm^3 gm^{-1}), δ is the density of the solvent (gm cm^{-3}), and m/l is the linear mass of the rods (gm cm^{-1}). Knowing $\bar{v}$ and δ, one can determine m/l by analysis of the variation of s_1 with c.

We have also examined these samples with the polarizing microscope either mounted in the X-ray sample holder or sealed between a slide and cover slip.

All the experiments were performed at room temperature (about 20° C.), except in a few cases where the temperature is explicitly stated.

Results

Specific volume.—The value of the linear mass derived from absolute intensity, low-angle X-ray scattering experiments depends critically on the accuracy of the specific volume (equation 2).

We have measured the specific volume of PBLG in various solvents using a Pyrex pycnometer, made of a bulb joined to a graduated capillary tube; the volume of the pycnometer is about 10 cm^3.

The results of these measurements are shown in Table 11.1. The mean value was calculated by weighing the experimental values with the concentrations. The average scatter of the points is less than 0.003 cm^3 gm^{-1}. Furthermore, Mitchell, Woodward and Doty (*23*) obtained in dimethyl formamide, $\bar{v} = 0.791 \pm 0.006$ at 25° C., a value which is in good agreement with our measurement.

Table 11.1—Specific volume of PBLG in the various solvents

Solvent	t (°C)	c (%)	$\bar{v}$ (cm^3 gm^{-1})	$\overline{(\bar{v})}$* (cm^3 gm^{-1})
dimethylformamide	20	5.163	0 784	
		2.258	0 791	
		1.091	0 788	
		0.949	0 785	0.786
pyridine	25.1	3.874	0 788	
		2.857	0 787	
		1.554	0 786	0.787
metacresol	25.1	3.452	0 786	
		1.956	0 787	
		1.084	0 793	0.788

* Average weighted by the concentration.

Isotropic solutions—This phase exists in the three solvents at the lowest concentrations. In the polarizing microscope the solution is isotropic at rest.

The X-ray diffraction diagrams contain a diffuse scattering in the low-angle region, which we have examined by the technique described above. The experimental results are given in Figure 11.1, in the double logarithmic plot. The shape of the curve is typically that of a solution of rods; we have drawn in Figure 11.1 the theoretical curve (1) which best fits each experimental curve, keeping in mind effects of interactions. The agreement between the theoretical and experimental curves is excellent; the values of A/c_e are independent of concentration. In each family of curves, the gap between $j_n(s)$ and the theoretical function increases with concentration, and is greatest near the origin, as predicted by the theory (*18*). The effects of interactions depend in addition on the nature of the solvent and the molecular weight of the polypeptide.

Knowing A/c_e, one can compute μ and h; we have performed this calculation for each series of experiments using the mean value of A/c_e (Table 11.2). In order to simplify the discussion of errors, we have plotted h as a function of $\bar{v}$, in Figure 11.2; the most probable value of h is also shown, corresponding to the last column in Table 11.1.

It is appropriate to analyze the precision of these results. We have discussed elsewhere (*18*) the errors inherent in the experimental technique; they are certainly less than 5 per cent. We have also discussed the uncertainty in the choice of R_c, and its consequences on

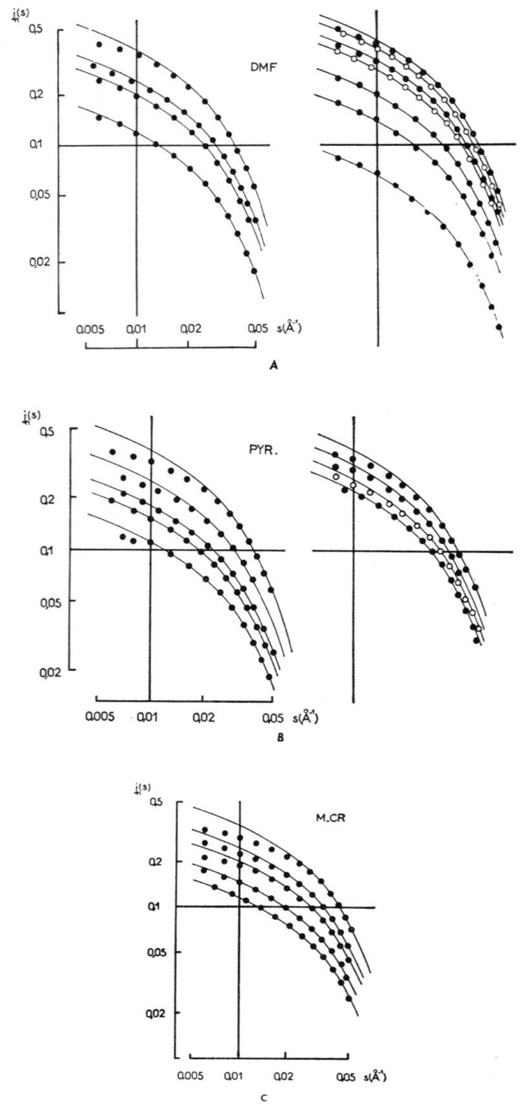

Fig 11.1. Results of absolute intensity, small-angle scattering experiments. Concentrations are given in Table 11.2. A, dimethylformamide; B, pyridine; C, metacresol.

A/c_e: the error of R_c here is not greater than 0.15 Å, which leads to a related error of A/c_e about 5 per cent, on the basis of the most unfavorable hypothesis.

The error due to $\bar{v}$ is not large. Even in allowing the error of $\bar{v}$ to be 0.003 cm³ gm⁻¹ (see Table 11.1), the consequent error of h is not greater than 0.05 Å in the case of dimethylformamide and pyridine, and 0.07 Å in the case of metacresol.

We must still explain the slight dependence of A_e/c^2 on the molecular weight of the polypeptide (Table 11.2), and choose between the different values. We have discussed this problem elsewhere (*16*) and have shown that the proper values correspond to those at low molecular weights. In any case, these are minor differences which may be neglected to a first approximation.

We conclude that *the value of h in the three solvents is about 1.95 Å*; the difference with the value corresponding to the α-helix ($h = 1.5$ Å) greatly exceeds the experimental errror.

Cholesteric phase.—This phase appears turbid and shows remarkable interference effects even to the naked eye. The textures described by Robinson, Ward, and Beevers (*28, 29*) are seen in the polarizing microscope.

The X-ray diagrams show a diffuse band in the region $s = 4 \times 10^{-2}$ Å^{-1}, which shifts toward wider angles, and becomes sharper as the concentration increases. We have shown in Figure 11.3 the position of the center of this band, as a function of concentration, and have indicated its half-width.

With the assumption that the rigid molecules

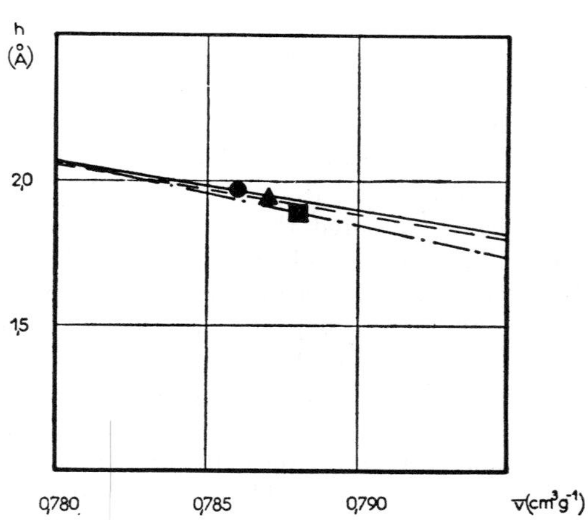

Fig. 11.2. Plot of h as a function of $\bar{v}$ (see Table 11.2). The most probable value of h is given, corresponding to $\overline{(\bar{v})}$ of Table 11.2. Dimethylformamide, ——●——; pyridine, — — ▲ — —; metacresol, — · — ■ · — ·.

are clustered, at least locally, in hexagonal arrays (*29*), the band can be interpreted as the first reflection of this lattice: the spacing of the band can then be expressed (equation 3) as a function of the concentration and the linear mass of the rods. We have plotted in Figure 11.3 s_1 as a function of c for two values of the linear mass: that of the α-helix ($h = 1.5$ Å) and that corresponding to the dilute solution measurements ($h = 2$ Å, see above).

The results depend on the solvent: in pyridine the position of the band is in good agreement with the linear mass of the α-helix; in contrast, in metacresol the agreement is much better with the linear mass obtained from dilute solutions.

In dimethylformamide the cholesteric phase exists only at elevated temperatures (higher than about 40° C.); it is absent at room temperatures.

We should note that the interpretation of the position of a diffuse band in terms of a crystalline lattice cannot be justified in an entirely rigorous manner. In the cases discussed here, however, the progressive shift of the band with concentration and the difference observed in the two solvents seem to justify this simplified treatment.

Paracrystalline phase.—This phase exists in the system PBLG-metacresol at the highest concentrations. The preparations are opaque, very viscous (waxy), and strongly birefringent; no characteristic texture can be observed in the polarizing microscope.

Several sharp lines are found in the small-angle region of the X-ray diagram; the spacings bear the relation $1:\sqrt{3}:\sqrt{4}:\sqrt{7}:\sqrt{9}$, typical of a two-dimensional hexagonal lattice. The intensities of the reflections decrease rapidly with increasing s.

We have shown in Figure 11.3 the spacing of the first line s_1 as a function of concentration, as well as the theoretical curves (3) corresponding to a hexagonal array of rods, where $h = 1.5$ Å and $h = 2.0$ Å. The experimental points are in rather good agreement with $h = 1.5$ Å, corresponding to the α-helix.

Thus this phase is formed by a set of rigid rods (apparently α-helices) packed in a hexagonal lattice, all the solvent filling the gaps between the molecules. We call this

Table 11.2—Experimental results obtained with dilute isotropic solutions

Solvent	ρ_o el. Å^{-3}	M 10^3	c %	c_e %	R_c Å	A	A/c_e	$\overline{(A/c_e)}$
dimethylformamide	0.3108	35	1.86	1.80	4.6	0.060	3.33	
			3.37	3.25		0.103	3.17	
			4.10	3.96		0.123	3.11	
			6.05	5.84		0.192	3.29	3.225
		330	1.00	0.96	4.6	0.033	3.44	
			2.05	1.98		0.070	3.53	
			3.00	2.90		0.101	3.48	
			4.00	3.86		0.143	3.70	
			5.00	4.82		0.163	3.38	
			6.00	5.79		0.192	3.32	
			7.00	6.76		0.217	3.21	3.437
pyridine	0.3116	35	1.76	1.75	4.4	0.056	3.20	
			2.28	2.27		0.076	3.35	
			2.79	2.77		0.089	3.21	
			4.00	3.98		0.125	3.14	
			6.22	6.19		0.189	3.05	3.190
		250	3.30	3.28	4.4	0.111	3.38	
			3.95	3.93		0.130	3.31	
			4.88	4.85		0.156	3.22	
			5.88	5.85		0.192	3.28	3.297
metacresol	0.3317	35	2.5	2.46	3.5	0.051	2.07	
			3.5	3.45		0.065	1.88	
			4.4	4.34		0.089	2.05	
			5.9	5.82		0.111	1.91	
			7.9	7.79		0.156	2.00	1.982

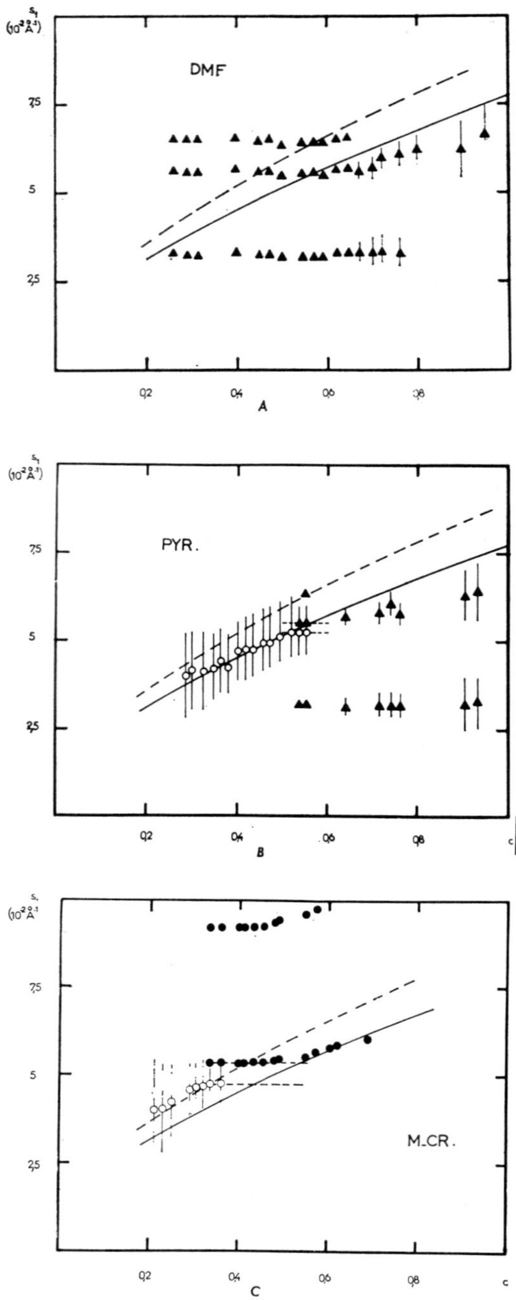

Fig. 11.3. Spacings of the diffraction lines of the concentrated phases. The half-width of the broad bands is given by the length of the vertical bars. Two theoretical curves (equation 3) are drawn, for $h = 1.5$ Å (full line), and $h = 2.0$ Å (dotted line). A, dimethylformamide; B, pyridine; C, metacresol. Cholesteric ○, paracrystalline ●, and complex ▲ phases.

phase "paracrystalline," which is deliberately vague since at the highest concentrations, in addition to the equatorial reflections of the hexagonal lattice, several weak lines and the bands can be observed, indicating a beginning of three-dimensional crystalline organization.

Complex phase.—This phase has been found with two solvents: dimethylformamide and pyridine. The appearance in the polarizing microscope is similar to that of the paracrystalline phase.

The X-ray diffraction diagrams contain a series of sharp lines, whose spacings are typical of the equatorial reflections from a hexagonal lattice. The dimensions of the hexagonal cell, which are much larger than those of the paracrystalline phase (Fig. 11.3), and the intensity of the reflections, which does not decrease regularly as s increases, as in the paracrystalline phase (see *16*), are typical of the complex phase. Furthermore, the dimensions of the lattice are invariant with concentration; the complex phase thus has a fixed composition. This phase is in many respects analogous to the phase that we have called "complex hexagonal" in the soap-water systems (*17, 12*).

The most striking example is given by dimethylformamide. In dilute solutions, only a diffuse low-angle scattering is observed. At higher concentrations (the threshold is about $c = 0.15$) the system of sharp lines of the complex phase appears superimposed over the diffuse band. The relative intensity of the reflections, as well as their spacings, are independent of the concentration; in contrast, the intensity of the set of lines increases relative to that of the diffuse band with increasing concentration, and the diffuse scattering fades out at about $c = 0.6$ (Fig. 11.3). Approaching $c = 0.7$ the lines begin to broaden, and a background of continuous scattering is observed, indicating disorder. At the highest concentrations, the first reflection weakens, and the X-ray diagram becomes similar to that of the paracrystalline phase of other systems. It is difficult to know the polypeptide and dimethylformamide composition of the complex phase since the concentration at which the diffuse low-angle scattering has completely disappeared cannot be determined with adequate precision.

This problem can be solved by study of solutions in pyridine. In this case the complex phase showing all the features described above is preceded, in the sequence of increasing concentrations, by the cholesteric phase, from which it can be seen to separate (Fig. 11.3). By carefully studying the transition zone the concentration can be determined with rather good precision ($c=0.55$).

When the concentration and the dimensions of the hexagonal cell are known (see *16*), the linear mass of the polypeptide within a unit cell can then be determined (equation 3): $\mu=225$ el. $Å^{-1}$, a value several times larger than that for the two types of rods found in the other phases ($h=1.5$ Å, $\mu=77.5$ el. $Å^{-1}$; $h=2.0$ Å, $\mu=58$ el. $Å^{-1}$). The large value of μ as well as the intensity of the first reflection, which is weaker than the following ones (see *16*), indicate that the structure is more complex than that of the other phases. Since the complex phase in pyridine is found to exist between two structures formed with α-helices (cholesteric phase, and dry polypeptide), we assume that it contains several helices of this kind. The number is, in fact, three, since the value of μ in the complex phase is very close to three times that of the α-helix.

Among the different ways three helices may be associated, the most satisfactory seems to be a helical coiling around one common axis, similar to the structure of a three-stranded rope (Fig. 11.4): Crick (*5*) has suggested this kind of structure for polypeptides and has named it "coiled-coil." A proof of this model is provided by the agreement between the observed and calculated structure factors (see *16*).

An additional proof is stereochemical. The structure is compact; the distances which separate neighboring molecules belonging to a secondary helix or to two contiguous helices are approximately the same, without any contacts being too close (Fig. 11.4 and ref. *5*). The different secondary helices are likely to fit one with the others as described by Franklin and Klug (*10*) in the case of tobacco mosaic virus, forming a three-dimensional ordered structure. This fact may explain the fixed composition of polypeptide and solvent, and gives a basis for the presence in the X-ray pictures of several sharp reflections (very weak) whose spacings do not correspond to the equatorial nodes of the hexagonal lattice.

It may be relevant to note that, although this model is quite satisfactory, it is not the only one compatible with the experimental observations.

Conclusions and discussion

Let us briefly summarize the experimental observations. The phase diagrams of the three systems are described schematically in Table 11.3. In all cases the transition between phases occurs with a zone of separation (Fig. 11.3) where two phases are in equilibrium, in agreement with the phase rule. At the highest concentrations ($c>0.7$) the structures become disordered. Solid solutions of the paracrystalline or complex phase with the pure polypeptide

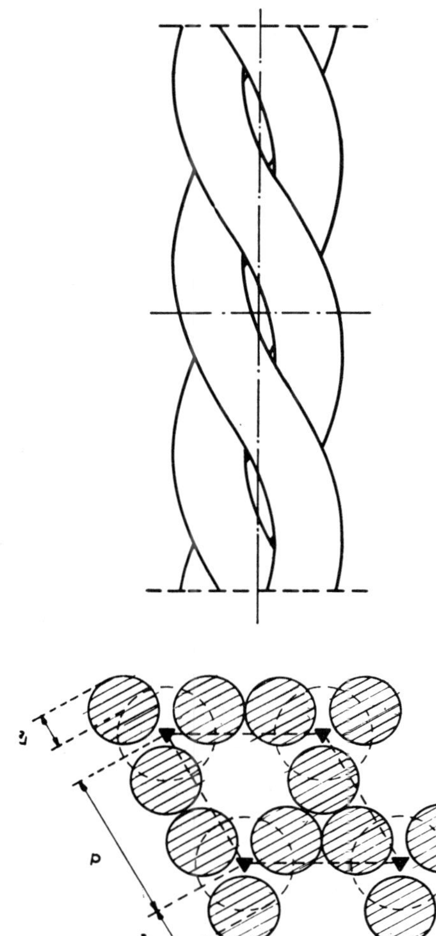

Fig 11.4. Structure of the complex phase.

Table 11.3—Schematic representation of the phase diagrams. The values of h are given in Å

	Iso-tropic	Choles-teric	Paracrys-talline	Com-plex
dimethylformamide	1.96	—	—	$\sim 3\times 1.5$
pyridine	1.93	~ 1.5	—	$\sim 3\times 1.5$
metacresol	1.88	~ 2	~ 1.5	—

Table 11.4—Stability of the helical configuration

Type of structure	h (Å)	Relative instablity per residue (k cal)
α_{II} ribbon	2.75	0.5
3_{10} helix	2.00	1.0
α helix	1.50	0.0
π helix	1.15	0.5
4, 3_{13} helix	1.20	2.4
γ helix	0.98	2.0

are likely to exist in this region.

In all cases the polypeptide behaves as a rigid rod whose linear mass may be determined. We have characterized this parameter by the distance, h, between two neighboring amino-acid residues, projected onto the helical axis. The values of h are shown in Table 11.3. They cluster around 1.5 Å and 2 Å. In isotropic solutions $h \simeq 1.93$ Å in all the cases. The shift to the structure having $h = 1.5$ Å coincides with the transition from the isotropic solution to an organized phase in the case of dimethylformamide and pyridine, and with the transition from the cholesteric to the paracrystalline phase in the case of metacresol.

The value of h, and eventually of R_c, is not sufficient to determine the structure of the rods. But the evidence favoring helical structures is very strong, and there exist only a small number of stable helical conformations of polypeptides. Donohue (*6*) attempted to estimate the relative stability of these structures in analyzing the strains of the peptide backbone, due to the deformation of the chemical bonds.

It is noteworthy that the two values of h which we have found correspond to two of the most stable of these conformations: the α-helix and the 3_{10} helix. It seems unlikely that this coincidence is fortuitous: PBLG apparently adopts these two structures in the conditions of our experiments.

Since the energy of the polypeptide backbone is almost the same in these two structures, the equilibrium is shifted towards one of the two conformations by the polypeptide-solvent interactions. It might be assumed, for example, that the flexibility of the side chains depends on the nature of the solvent. As the concentration increases, the side chains of different molecules come into contact more easily when they are extended. As soon as the side chains are in contact, the peptide backbone is disturbed, and conditions are set favorable to a structure transformation. Similarly the interactions between the polypeptide and the solvent may provide an explanation of the differences among the phase diagrams of the different systems. In order to specify this mechanism more precisely, it would be useful to know the role of the polarity of the solvent on the structure of the helices.

The complex phase is an example of a coiled-coil structure, which has been suggested by Crick (*5*) on a theoretical basis. Furthermore, it is interesting to note that the cholesteric phase may be formed by different helices, according to the nature of the solvent.

It may perhaps be premature to look for biological implications in these observations. Nevertheless, it is remarkable that a change only in concentration can affect the structure of a polypeptide in modifying the helical configuration and in causing a variety of organized structures to appear. It may well be asked whether the description of the structure of proteins in terms of α-helices is not somewhat oversimplified (*8*), and whether transitions between helical structures might not accompany certain morphological modifications of proteins.

Finally, we shall try to explain the obvious contradiction between our results and those of other studies according to which PBLG should be α-helical under the conditions of our experiments.* Thus we shall examine these studies and analyze the interpretations. The experiments may be grouped into three categories.

Light scattering.—The molecular weight, M, and the radius of gyration, R_o, can be determined by this technique.† If the molecules

* The concentrations are in fact higher in the X-ray experiments than in the other studies. We have verified, by light scattering, that no anomaly exists at intermediate concentrations.

† In the case of polydisperse rods, where $f(M)$

have the shape of rigid rods, R_o is proportional to M. The value of h can thus be determined:

$$h = \frac{219 \cdot \sqrt{12} \cdot R_o}{M} \tag{4}$$

h (equation 4) is independent of M. Several authors have studied PBLG by light scattering (*7, 35, 30*) and have taken the experimental results as a proof for the existence of the α-helix. In analyzing these results (Fig. 11.5), however, it can be noticed that, while the mean value of

$$\frac{219 \cdot \sqrt{12} \cdot R_o}{M}$$

is close to 1.5 Å, the experimental points, in fact, vary systematically as a function of M, decreasing as M increases. This variation (noticed moreover by the authors) suggests that the rods are flexible. In such a case the relationship (equation 4) should be extrapolated to $M = 0$ in order to determine h correctly. This extrapolation is not very accurate, but, clearly, leads (Fig. 11.5) to a value of h near 2.0 Å, or even larger. It should be noticed that the effect of polydispersity is to increase the value of equation 4. However, recent measurements (ref. *30*, and work in progress at the Centre de Recherches sur les Macromolécules) show that the correction is smaller than 15 per cent, thus insufficient to account for the gap between the two values of h (1.5 to 2.0 Å).

The correlation of the experimental points with M (Fig. 11.5) could be interpreted by a more elaborate model than the rigid rod—for example, the "worm-like chain" of Porod (*27*), which is suitable to describe the conformation of a slightly flexible rod. In this model the rigidity of the molecule is characterized by a parameter q, called the "persistence length." Equation 4 becomes in this case (*3*):

$$\frac{219 \cdot \sqrt{12} \cdot R_o}{M}$$

$$= h\sqrt{12}\left[\frac{1}{3x} - \frac{1}{x^2} + \frac{2}{x^3} - \frac{2}{x^4}(1 - e^{-x})\right]^{1/2} \tag{5}$$

where $x = Mh/219q$. We have plotted in Fig. 11.5 the function (equation 5) for $q = 505$ Å, and $h = 2.12$ Å, h being 10 per cent higher than 1.93 (as suggested by X-ray measurements) in order to take into account effects of polydispersity. The agreement of the experimental points with the theoretical curve is fairly good—much better in any case than that of a straight horizontal line $h = 1.65$ Å, which would correspond to the α-helix (with the same correction for polydispersity).

Hydrodynamic techniques.—Whether viscosity (*7, 33, 30*), flow birefringence (*34*), translational diffusion (*30, 11*), or rotational diffusion (*31, 24, 22*, and Benoît and Boeckel, in preparation) is considered, the interpretation of these experiments in terms of molecular dimensions involves a certain number of hypotheses: (*a*) the particle is perfectly rigid; (*b*) the structure is compact and impenetrable to the sol-

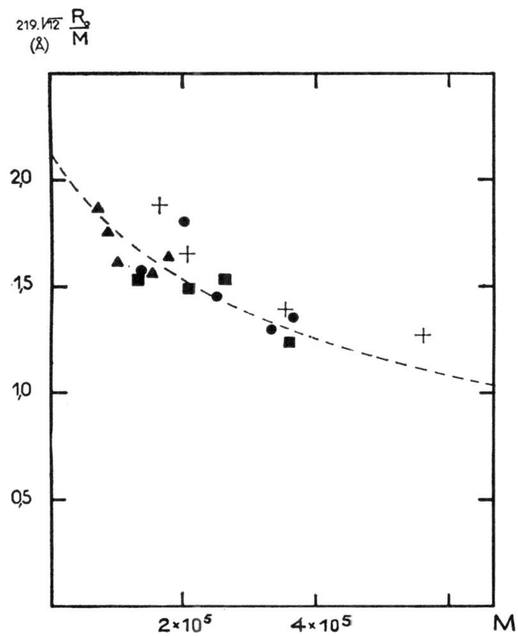

Fig. 11.5. Plot of the light-scattering experimental results.

+ PBLG (raw) in dimethylformamide (ref. **30**).
● PBLG (fractions) in dimethylformamide (**30**).
■ PBLG in chloroform-formamide (**7**)
▲ PBLG in dichloroethane (**35**).

---

is the weight distribution, the following average values are in fact determined:

$$M_w = \int Mf(M)dM \text{ and}$$

$$\langle R_o \rangle = \left[\int MR_o^2(M)f(M)dM \Big/ \int Mf(M)dM\right]^{1/2}$$

vent; (*c*) the hydrodynamic laws of continuous media are still valid at the molecular level; (*d*) the shape of the particle may be described by an ellipsoid or a cylinder of revolution.

Without entering into a detailed analysis of these assumptions (the first, at any rate, does not seem to agree with the light-scattering experiments), it is evident that their agreement with the real behavior of PBLG in solution can only be approximate. It seems doubtful to us that any interpretation of this kind of experiment may lead to a clear-cut choice between $h = 2$ Å and $h = 1.5$ Å.

Optical techniques.—We include in these techniques infra-red absorption (*2*), optical rotatory power (*32, 9*), and dielectric moments (*35*), which in the case of polypeptide provide sensitive tests for α-conformations. Insofar as they have been used up to now, these techniques do not seem to be able to distinguish between two helical conformations (α and 3_{10} helix) whose structures are very similar.

Acknowledgment

The work of M. Cesari in this research was supported by a grant from the Olin Mathieson Chemical Corporation.

References

1 E. J. Ambrose and A. Elliott, *Proc. Roy. Soc.* **A205,** 47 (1951).
2 C. H. Bamford, A. Elliott, and W. E. Hanby, "Synthetic Polypeptides," Academic Press, New York, 1956.
3 H. Benoît and P. Doty, *J. Phys. Chem.* **57,** 958 (1953).
4 E. R. Blout and R. H. Karlson, *J. Am. Chem. Soc.* **78,** 941 (1956).
5 F. H. C. Crick, *Acta Cryst.* **6,** 689 (1953).
6 J. Donohue, *Proc. Nat. Acad. Sci.* **39,** 470 (1953).
7 P. Doty, J. H. Bradbury, and A. M. Holtzer, *J. Am. Chem. Soc.* **78,** 947 (1956).
8 P. Doty, *Rev. Mod. Phys.* **31,** 107 (1959).
9 A. R. Downie, A. Elliott, W. E. Hanby, and B. R. Malcolm, *Proc. Roy. Soc.* **A242,** 325 (1957).
10 R. E. Franklin and A. Klug, *Biochim. et Biophys. Acta* **19,** 403 (1956).
11 L. Freund, Thesis, University of Strasbourg (1961).
12 F. Husson, H. Mustacchi, and V. Luzzati, *Acta Cryst.* **13,** 668 (1960).
13 B. W. Low, in "The Proteins," Vol. IA, Neurath and Bailey (eds.), Academic Press, New York, 1953.
14 V. Luzzati and A. Nicolaïeff, *J. Mol. Biol.* **1,** 127 (1959).
15 V. Luzzati, *Acta Cryst.* **13,** 939 (1960).
16 V. Luzzati, M. Cesari, G. Spach, F. Masson, and J. M. Vincent, *J. Mol. Biol.* **3,** 566 (1961).
17 V. Luzzati, H. Mustacchi, A. Skoulios, and F. Husson, *Acta Cryst.* **13,** 660 (1960).
18 V. Luzzati, A. Nicolaïeff, and F. Masson, *J. Mol. Biol.* **3,** 185 (1961).
19 V. Luzzati and H. Benoît, *Acta Cryst.* **14,** 297 (1961).
20 V. Luzzati, J. Witz, and A. Nicolaïeff, *J. Mol. Biol.* **3,** 367 (1961).
21 V. Luzzati, J. Witz, and A. Nicolaïeff, *ibid.* **3,** 379 (1961).
22 E. Marchal and J. Marchal, *Arch. sci. (Geneva)* **12,** 24 (1959).
23 J. C. Mitchell, A. E. Woodward, and P. Doty, *J. Am. Chem. Soc.* **79,** 3955 (1957).
24 C. T. O'Konski, K. Yoshioka, and W. H. Orttung, *J. Phys. Chem.* **63,** 1558 (1959).
25 L. Pauling, R. B. Corey, and H. R. Branson, *Proc. Nat. Acad. Sci.* **37,** 205 (1951).
26 M. F. Perutz, *Nature* **167,** 1053 (1951).
27 G. Porod, *Monatsh. Chem.* **80,** 251 (1949).
28 C. Robinson, *Trans. Faraday Soc.* **52,** 571 (1956).
29 C. Robinson, J. C. Ward, and R. B. Beevers, *Discussions Faraday Soc.* **25,** 29 (1958).
30 G. Spach, Thesis, University of Strasbourg (1960).
31 I. Tinoco, *J. Am. Chem Soc.* **79,** 4336 (1957).
32 J. T. Yang and P. Doty, *ibid.* **79,** 761 (1957).
33 J. T. Yang, *ibid.* **80,** 1783 (1958).
34 J. T. Yang, *ibid.* **80,** 5139 (1958).
35 A. Wada, *Bull. Chem. Soc. Japan* **33,** 822 (1960).

PAPER 12

Chain Regularity and Dielectric Properties of Poly-α-Amino Acids in Solution

AKIYOSHI WADA

Regularity in the backbone chain of poly-α-amino acid has aroused interest in the biological, physical, and chemical fields not only because the substance is a simple model-material of protein but especially because in the eyes of the physicist it can be seen as a one-dimensional crystal in which the structural units have an ordered orientation.

Numerous experimental techniques—for instance, optical, thermodynamical, hydrodynamical, and electrical—have already been applied to study the molecular configuration of this polymer molecule in solution. One significant result thus obtained is that several regular configurations, such as the α-helix or β-form described by Pauling and Corey (*1*), have been found to exist as stable forms even in the solution state (*2*). This suggests that the one-dimensional crystal-like structure can be studied in an individual molecule, apart from the solute-solute interaction in dilute solutions.

Here, we have recourse to the dielectric method to study such a configuration of the polymer chain. Although this method has been successfully and widely used to study the charge distribution and orientation of the polar groups in simple molecules, it seems to be less useful in the study of polymer molecules for the following reasons:

The electric dipole moment of the molecule is a vector sum of the moments of constituent polar groups. In the usual synthetic polymer molecules, however, the polar groups have a random orientation because of the intramolecular rotational freedom allowed in and between the structural units in the chain, so that the electric dipole moment of the molecule has nothing to distinguish it from those of the freely moving small polar molecules in the solvent. Several theoretical studies on the dielectric properties of the random coil with polar groups have been carried out in regard to the statistical thermodynamic problems of the flexible molecular chain (*3*). The electric dipole moment of such a molecule is, in principle, proportional to the square of the degree of polymerization and lies within an order of several ten Debye units.

The situation, however, is completely different in a synthetic poly-α-amino acid which has a regular chain configuration such as the α-helix or β-form in which the intra- or intermolecular hydrogen bonds freeze up the internal rotaion and in which the polar groups will then orient uniformly in a polymer chain.

In this paper, several problems relating to the dielectric properties of the poly-α-amino acid in solution will be outlined, focusing mainly on the chain regularity of the α-helix.

Notation

The symbols employed in this article are defined below:

a	$L/2$.
a'	Major axis of ellipsoid of revolution.
B	Second dielectric virial coefficient.
b	Radius of cylinder.
b	Radius of the α-helix.
b'	Minor axis of ellipsoid of revolution.
D	Debye unit.
d	Density
f_D	Fraction of D-residue.
f_L	Fraction of L-residue.
H_c	Total fraction of the helix, $(H_l—H_r)$.
H_l	Fraction of the left-handed helix.
H_r	Fraction of the right-handed helix.
k	Boltzmann constant.
L	Length of the α-helix.
L_0	Axial length of the α-helix per residue.
M	Electric dipole moment of the α-helix.

M^* — M in solution.
M_c — Residual weight.
M_w — Weight average molecular weight.
N — Avogadro number.
n — Degree of polymerization.
P — Specific polarization.
R — A function relating to the reaction field.
r — Bond distance.
r — A function relating to the reaction field.
$\vec{u}$ — Unit vector in the direction of the helical axis.
w — Weight fraction.
α_{11}, α_{22}, α_{33} — Principal polarizability.
ϵ^o — Static dielectric constant.
ϵ' — Dielectric constant.
ϵ'' — Dielectric loss factor.
η — Viscosity of solvent.
$\vec{\mu}$ — Moment vector of electric dipole.
$\vec{\mu}^*$ — $\vec{\mu}$ in solution.
$\vec{\mu}^o$ — $\vec{\mu}$ in vacuo.
$\vec{\mu}_o$ — Electric dipole moment of amino acid residue in vacuo.
$\vec{\mu}_m$ — Backbone moment per amino acid residue of the α-helix.
$\vec{\mu}_s$ — Side chain moment.
$\mu_o(\|)$ — Axial component of $\vec{\mu}_c$.
$\mu_m(\|)$ — Axial component of $\vec{\mu}_m$.
$\mu_s(\|)$ — Axial component of $\vec{\mu}_s$.
ν — Frequency.
ν_o — Critical frequency.
τ — Relaxation time.

Chain regularity and orientation of polar groups in poly-α-amino acids

A distinctive feature of the molecular chain of poly-α-amino acids is that, in the structural unit which contains C═O and N—H groups, the former has a proton-accepting character and the latter has a proton-donating character. The inter- and intramolecular hydrogen bonds which can be formed between these groups play a role in stabilizing several preferred structures of the molecular chain. The α-helix is one of such stable configurations having a spiral structure, specified by the translation distance of 1.5 Å and the angular rotation of 100°, fixed by the intramolecular hydrogen bond as shown in Fig. 12.1. In this model it can be seen that all polar groups, C═O and N—H, have the same orientation, almost parallel with the helical axis, so that the electric moments of these groups coöperate with each other.

On the other hand, the β-form, which is another kind of regular chain having intermolecular hydrogen bonds, takes a configuration in which the electric moment of each polar group cancels out the other.

Accordingly, from the dielectric point of view, the picture of the helix in Fig. 12.1, for instance, can be rewritten as a dielectric model as shown in Fig. 12.2. In this model, the elec-

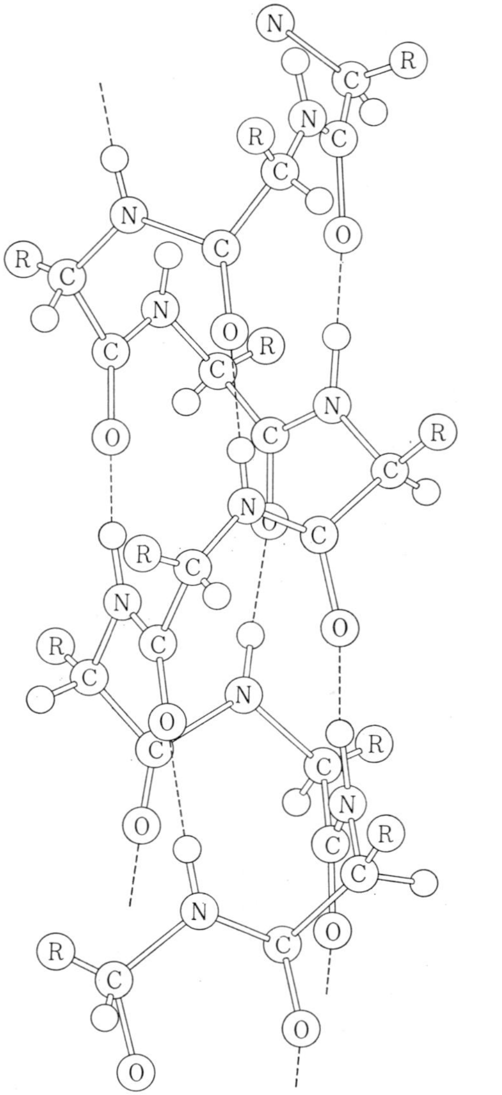

Fig. 12.1. The α-helix (right handed, L-residue). L. Pauling and R. B. Corey, *Proc. Nat. Acad. Sci.* **37,** 235 (1951).

tric dipole moment of the structural unit originates in the

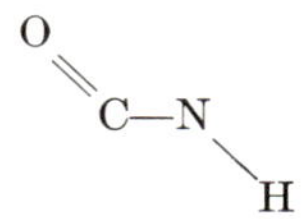

group which is known to be transconfiguration because of the double-bond character of the N—C bond. The moment of this group may be given simply by a vector sum of the bond moments of C=C, N—C and N—H groups which are known to be 1.31, −0.22, and 2.3 D, respectively (*4*). The moment for the trans

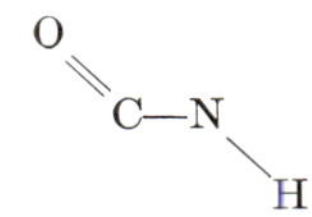

group thus estimated is 3.4 D as illustrated in Fig. 12.3. This value from the simple vector sum of the bond moments, however, differs considerably from the moment 4.39 D which has been experimentally obtained for *N*-methyl acetoamide in carbon tetrachloride (*5*).

A resonance effect

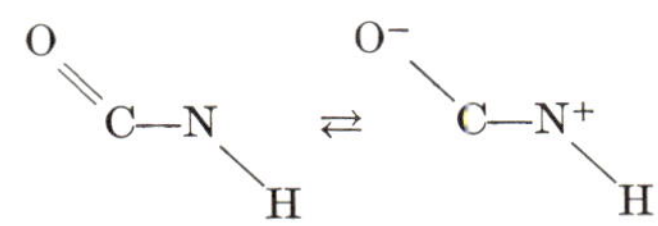

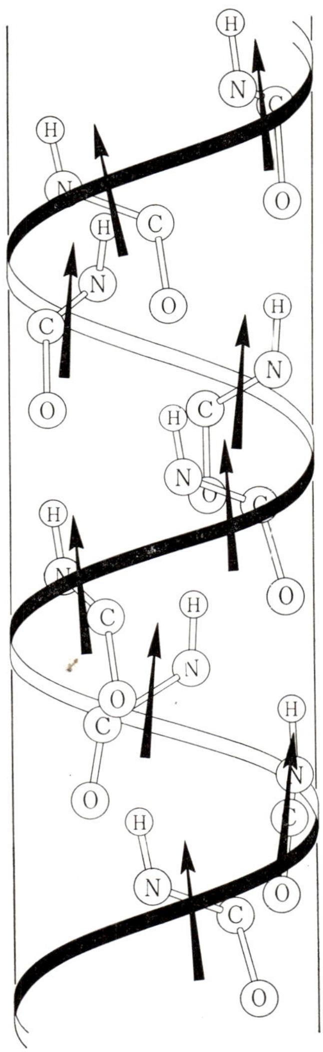

Fig. 12.2. Orientation of polar groups in the α-helix. Residual electric dipole moment which is localized in H—N——C=O group is represented by the black arrow.

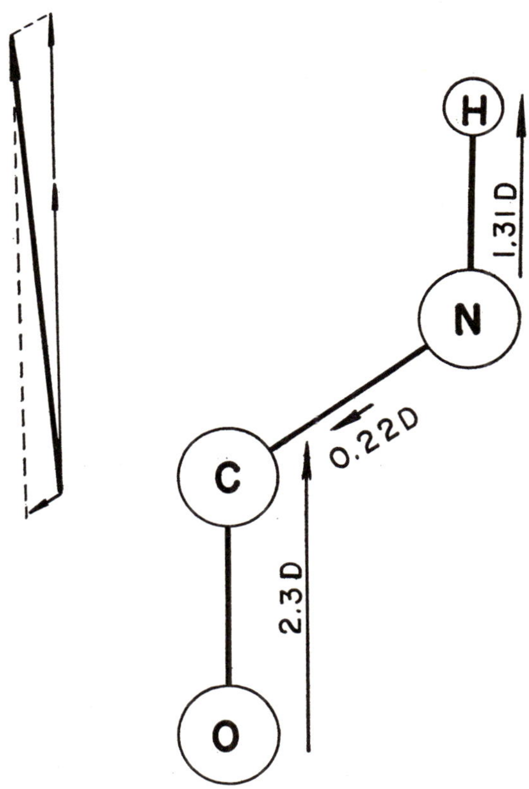

Fig. 12.3. Vectoral relation between the bond moments in the peptide group.

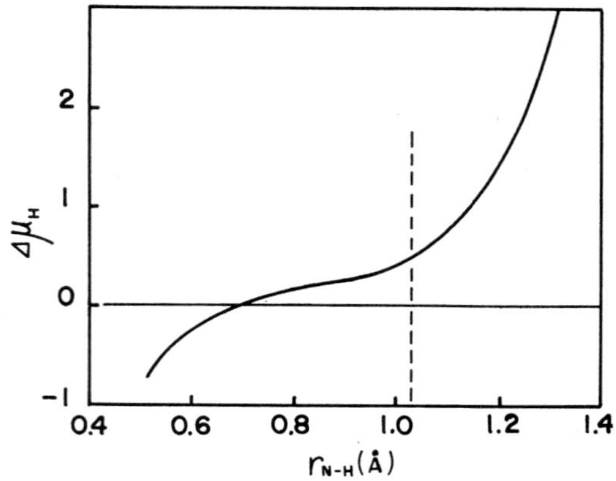

Fig. 12.4. Effect of the hydrogen bonding upon the electric dipole moment of amino acid residue in relation to the N—H bond length, r_{N-H}.

must be partially responsible for this discrepancy, which will add about 1 D to the original dipole moment (*5*).

The effect of the hydrogen bonding should also be considered in the case of the α-helix or β-form. The electronic state of the system; N—H····O can be expressed as a resonance hybrid of the structures

N —H····O
N⁻ H⁺····O
H⁻ H —O⁺.

Putting ψ_1, ψ_2, and ψ_3 the corresponding wave functions for these states, the electronic wave function, ψ, for the system can then be expressed by a linear combination of these wave functions

$$\psi = C_1\psi_1 + C_2\psi_2 + C_3\psi_3 \tag{1}$$

Defining μ_1, μ_2, and μ_3 as the electric dipole moments of the configurations mentioned above, the additional moment $\Delta\mu_H$ due to the hydrogen bonding to that of the nonbonding state is given by

$$\Delta\mu_H = (C_1^2 - C_1^{o^2})\mu_1 + (C_2^2 - C_2^{o^2})\mu_2 + (C_3^2 - C_3^{o^2})\mu_3 \tag{2}$$

where C_1^o, C_2^o, and C_3^o are coefficients for the standard state of no hydrogen bond. For these moments, the following values can be taken: $\mu_1 = 0$, $\mu_2 = er_{N-H}$ and $\mu_3 = er_{N\cdots O}$, where r_{N-H} is N—H bond length, $r_{N\cdots O}$ is N····O distance in N—H····O system, and e is the protonic charge. By the use of these values and according to Coulson and Danielson (*6*), $\Delta\mu_H$ can be estimated as a function of r_{N-H} keeping the $r_{N\cdots O}$ as 2.9 Å, as shown in Fig. 12.4. Although, in a strict sense, the resonance effect should be taken into account in calculating the wave function ψ, here the effect is considered to be independent and additive for simplicity. Thus we can safely conclude that the electric dipole moment of the structural unit in the backbone of the α-helix would be in the range from 4 D to 5 D in vacuo.

Now, we have to give a picture of the electric dipole moment of the α-helix, having n degree of polymerization, taking the effect of the side chain into consideration. When the dipole moment of ith structural unit in the backbone chain is put to be μ_{mi} and that in the side chain μ_{si}, the mean square electric dipole moment of the α-helix is

$$\langle M^2 \rangle = \left\langle \sum_{i,j} (\vec{\mu}_{mi} + \vec{\mu}_{si}) \cdot (\vec{\mu}_{mj} + \vec{\mu}_{sj}) \right\rangle_{\text{Av.}} \tag{3}$$

where the orientation of μ_m is fixed by the hydrogen bonds but μ_s may have several freedoms for internal rotation. So, the following relation will be given in terms of the unit vector of the helical axis $\vec{\mu}$, and the component of the moment $\vec{\mu}_m$ to the direction of the helical axis $\mu_m(\|)$:

$$\sum_{i,j} \langle \vec{\mu}_{mi} \cdot \vec{\mu}_{mj} \rangle_{\text{Av.}} = n^2 \cdot \mu_m{}^2(\|) \tag{4}$$

$$\sum_{i,j} \langle \vec{\mu}_{mi} \cdot \vec{\mu}_{si} \rangle_{\text{Av.}} = n\mu_m(\|) \sum_i \langle \vec{\mu}_{si} \cdot \vec{\mu} \rangle_{\text{Av.}} \tag{5}$$

Postponing the detailed discussion on the effect of the side-chain moment till later, the present discussion is necessarily limited to these two extremes by the lack of structural information about the side chain.

1. With perfect flexibility in the side chain and no interaction between them, that is, a random orientation of the moment in the side chain:

$$\langle M^2 \rangle_{\text{Av.}} = n^2\mu_m{}^2(\|) + 2n\mu_m(\|)\langle \vec{\mu}_s \cdot \vec{\mu} \rangle_{\text{Av.}} \tag{6}$$

2. Without any freedom of internal rotation in the side chain:

$$\langle M^2 \rangle_{\text{Av.}} = n[\mu_m(\|) + \mu_s(\|)] \tag{7}$$

It must be noted here that when free rotation is allowed in the side chain, the influence of the polar groups in it should become negligible in comparison to that in the backbone chain, since the second term in Eq. 6 is proportional n but not n^2 as in the first term.

In addition to the permanent electric dipole moment which has been discussed above, the polarization will more or less affect the dielectric properties of the molecule. When perfect flexibility is assumed in the side chain, a contribution from the orientation polarization of the side chain will be added to the electronic polarization and the total polarizability is

$$\begin{Bmatrix} \alpha_{11} \\ \alpha_{22} \\ \alpha_{33} \end{Bmatrix} = n \begin{Bmatrix} \alpha_{11}^{o} \\ \alpha_{22}^{o} \\ \alpha_{33}^{o} \end{Bmatrix} + \frac{n\mu_s^2}{3kT} \tag{8}$$

where α is a component of the diagonal tensor in the electronic polarizability of the α-helix. The residual electronic polarizability α^o can be calculated by the summation of the atomic refractions of the constituent atoms.

Dieletric dispersion of the α-helix in solutions

To measure the dielectric constant of solutions, the heterodyne beat method and the resonance method which have been successfully and widely used in the high-frequency range is not adequate for the purpose of the present investigation. Because the solute, that is the α-helix, is a rigid and highly asymmetric molecule and the relaxation time of over-all molecular rotation in the solvent should exceed the order of millisecond, the apparatus must be designed to cover the low-frequency range—as low as several ten cycles, where sufficient time can be allowed for very slow orientation polarization. The bridge method which has been developed, for instance, by Cole and Gross (*7*) and by von Hippel and his co-workers (*8*), seems to be appropriate for the present purpose (*9*).

Poly-γ-benzyl-L-glutamate (PBLG) was selected for this work because of its solubility in a variety of solvents and existence of experimental data by many experimental techniques.

The dielectric constant of this PBLG in ethylene dichloride (EDC) or chloroform (CF), in which the PBLG is known to exist in the stable α-helical configuration (*2*), depends on the measurement frequency and it has produced a typical dielectric dispersion curve as shown in Fig. 12.5a. The dielectric absorption is also plotted in Fig. 12.5b (*10*). In general, such dielectric dispersion and absorption curves can be expressed by Debye equations (*11*), the calculated curve for the case of the single relaxation time being plotted in broken lines in the same figures. The disagreement between the experimental data and the results of simple theoretical calculation suggests heterogeneity in the relaxation time of the orientation polarization which will be discussed later.

Electric dipole moment of the α-helix

The α-helix itself can be considered, as mentioned above, as the large rigid electric dipole, the moment of which, M, will be given by the following equation

$$M = \mu_o(||) \cdot n \tag{9}$$

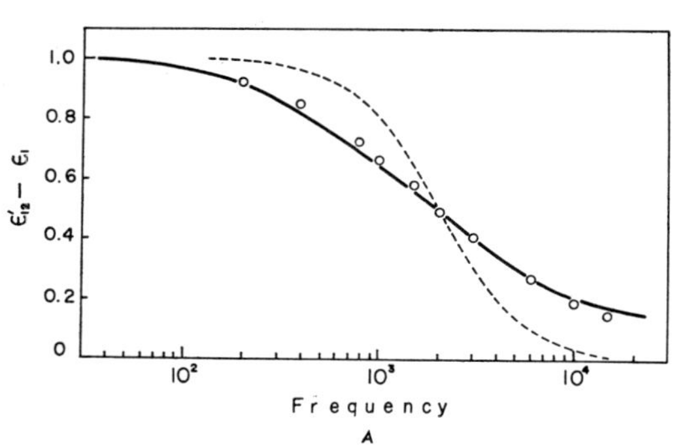

Fig. 12.5a. Representative dielectric dispersion curve of the α-helix of PBLG in ethylene dichloride at 25° C. (○). Broken line is plotted for the Debye theoretical equation of single relaxation time.

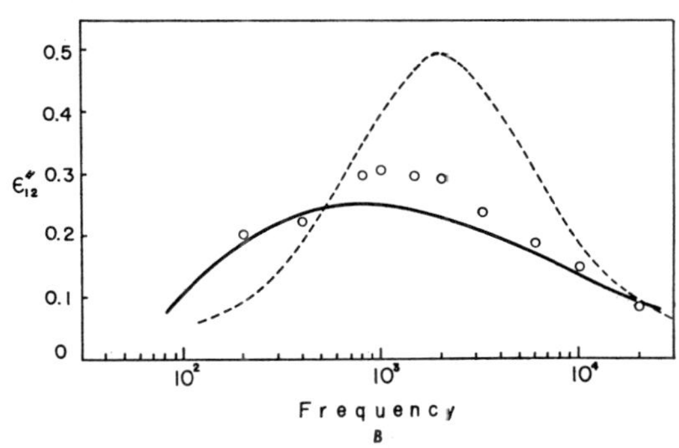

Fig. 12.5b. Representative dielectric absorption curve of the α-helix of PBLG in ethylene dichloride at 25° C. (○). Broken line is plotted for the Debye theoretical equation of single relaxation time.

where $\mu_o(\|)$ is the component of the electric dipole moment of the residue parallel with the helical axis. Namely, $\mu_o(\|)$ lies between $|\mu_m(\|)| + |\mu_s(\|)|$ and $|\mu_m(\|)| - |\mu_s(\|)|$ depending on the flexibility and orientation of the side chain. The component perpendicular to the axis will be cancelled out (except for the negligible contribution from the incomplete turn at the chain end) by the symmetrical requirement.

When solute-solvent interaction is absent, the specific polarization of the solution P_2^{oo}, that has been extrapolated to zero frequency and to zero concentration, has the following relation to the dipole moment of the solute M:

$$P_2^{oo} = (4\pi N/9M_o)[\{\alpha_{11} + \xi(\epsilon_1)(\alpha_{22} + \alpha_{33})\} + M^2/kT] \quad (10)$$

where N is the Avogadro number, M_o is the molecular weight of a structural unit, α_{11} is the diagonal element of the polarizability in the direction of the cylindrical axis, α_{22} and α_{33} are those perpendicular to the axis originating in the manner mentioned in the last section, and $\xi(\epsilon)$ is a correction factor for the internal field in the direction perpendicular to the cylindrical axis. From Eqs. 8 and 9, the first term in Eq. 10 is proportional to the first power of the degree of polymerization and to the second power in the second term. This means that the first term becomes negligible as the degree of polymerization becomes higher. Consequently a relation

$$P_2^{oo} = (4\pi NM^2/9nM_okT) \quad (11)$$

can be used instead of Eq. 10 fairly accurately for medium and high molecular weight α-helix, since the first term becomes less than 10 per cent of the second over ten degrees of polymerization. The static dielectric constant of the solution ϵ_{12}^o bears the following relation to P_2^{oo} when the cavity field is assumed to be the same as the applied field, which is quite plausible for such a long and thin dipole:

$$\begin{aligned} P_2^{oo} &= \lim_{w_2\to 0} P_2^o = \lim_{w_2\to 0} [(\epsilon_{12}'^o - 1)/3d_{12} - P_1]/w_2 + P_1 \\ &= \lim_{w_2\to 0} [(\epsilon_{12}'^o - \epsilon_1)/3d_{12}\cdot w_2] \qquad (12) \\ \epsilon_{12}'^o &= \lim_{\nu\to 0} \epsilon_{12}' \end{aligned}$$

where P is specific polarization, d is density, w is weight fraction, and the subscripts 1, 2, and 12 denote the solvent, solute, and solution, respectively.

The electric dipole moment M can then be given, in theory, from the static dielectric constant and Eqs. 11 and 12. But in practice, to get ϵ_{12}^o or P_2^o, the dielectric constants of solution measured at several frequencies must be extrapolated to zero frequency. In this process the following problems must be considered:

1) The relation between ϵ_{12} and ν is not linear.

2) The dispersion curve does not agree with Debye's theoretical equation in which a single relaxation time is assumed.

3) In the low-frequency range, the polarization effect on the electrode surface will not be negligible.

Several devices, which unfortunately are not always sufficiently consistent, are proposed for this extrapolation and the P_2^o thus obtained is plotted against weight fraction w_2 to extrapolate to zero concentration. In Fig. 12.6, the P_2^o vs. w_2 plots for ethylene dichloride and chloroform solution are shown. In these solvents, especially in chloroform, the α-helix is

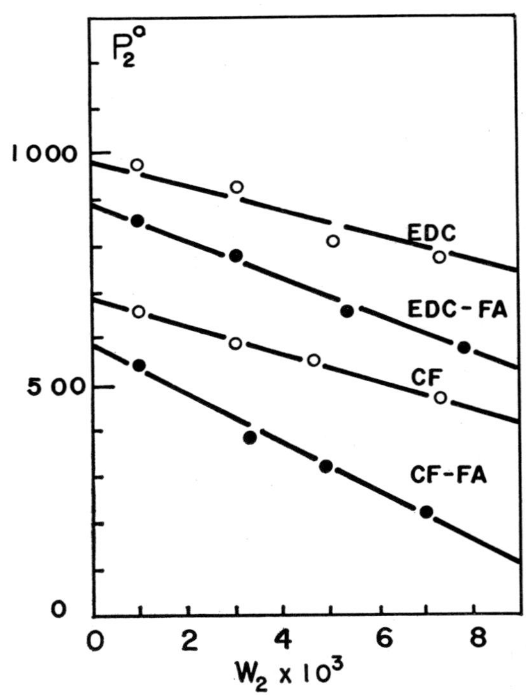

Fig. 12.6. Concentration dependence of P_2^o of PBLG in EDC, EDC-Fa, CF, and CF-FA.

reported to exist in an associated state, and those saturated by formamide are known to dissociate them (2). Comparing both, that is pure and formamide-saturated solvents, it can be seen that they give almost the same intercept but different slopes, that is, a negative slope in the latter case where the α-helix is believed to be nonassociated state. This slope is related to the coefficient of the second term of the series in which the specific polarization is written in powers of inverse molar volume V_m:

$$P = A + B/V_m + C/V_m^2 \qquad (13)$$

The coefficient B, that is, the so-called "second dielectric virial coefficient," has the following relation to the dipole-dipole interaction (12).

$$B = \frac{4\pi N^2}{9kT\Omega}\int (M_1 \cdot M_2)\cdot \exp\left[-u_2/kT\right] d\tau_2 \qquad (14)$$

where u_{12} is the potential energy of interaction between the molecules of 1 and 2, and $\Omega = \int d\tau_2/V_m$. In this equation the effect of dipole-induced dipole interaction is ignored for simplicity.

A linear relation between the degree of polymerization (DP) and P_2^{oo} or M was obtained (9, 13) as would be expected from the dielectric model of the α-helix mentioned above, as shown in Fig. 12.7a and 12.7b. When the weight average molecular weight such as that measured by light scattering is used, the dipole moment in Eq. 2 is also given by the same average.

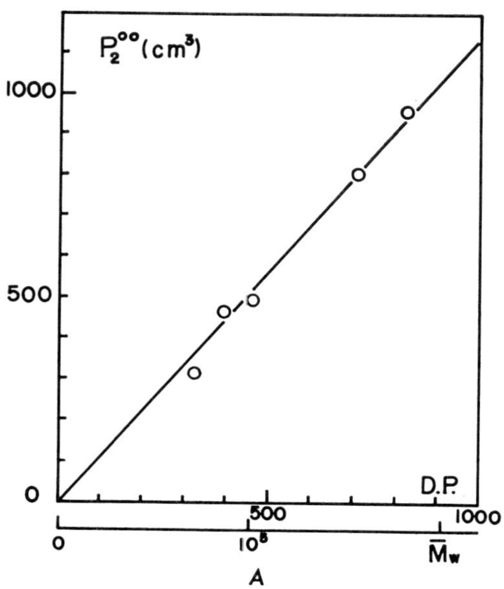

Fig. 12.7a. Dependence of the specific polarization P_2^{oo} of the α-helix upon the degree of polymerization (EDC, 25° C.).

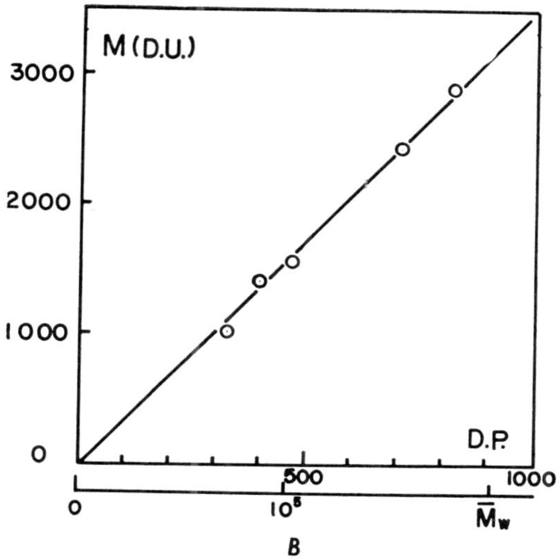

Fig. 12.7b. Dependence of the electric dipole moment of the α-helix upon the degree of polymerization (EDC, 25° C.).

In these plots the slopes will provide $4\pi N\mu_o^2(\|)/9M_okT$ in Fig. 12.7a and $\mu_o(\|)$ in Fig. 12.7b, respectively. This leads to a value of 3.6 D as a moment $\mu_o(\|)$ of the α-helix in EDC.

Before comparing this experimental result with the value expected from the structure of the α-helix, the influence of the environmental dielectrics to the dipole moment of the rodlike molecule must be considered. When we take into account, in approaching the real molecule, the polarizability in the structural units besides the rigid dipole in the α-helix, the moment in the dielectric medium should not be the same as that in vacuo. Let us estimate next the induced moment in the α-helix taking a simplified model as follows.

A simplified dielectric model for the α-helix in which each structural element, the amino-acid residue, is represented by a point dipole μ_o and a polarizability α distributed on every L_oA of the axis of the cylindrical cavity with radius b, is shown in Fig. 12.8. The reaction field on a residue may be a sum of the contribution of its own dipole moment and that of

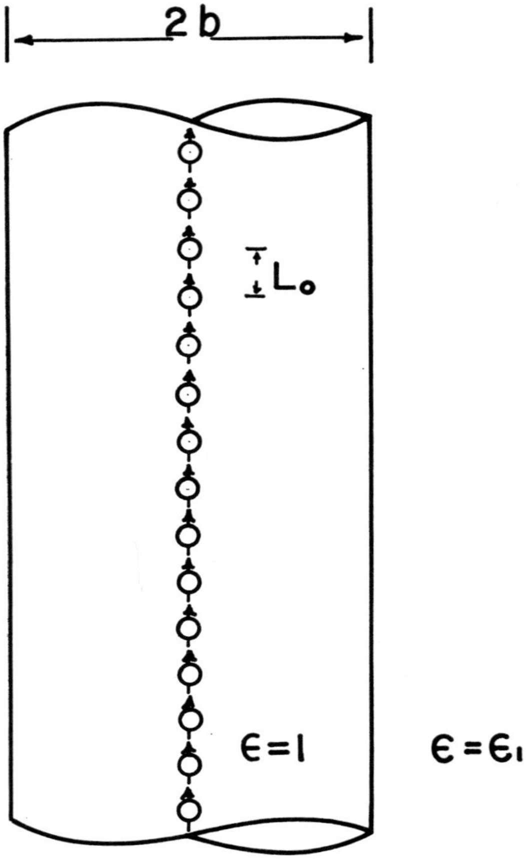

Fig. 12.8. Dielectric model of the α-helix in which each amino-acid residue is represented by a point dipole μ_o (arrow) and polarizability α (open circle).

neighboring residues. Namely, the reaction field at the kth residue $\mu_o{}^*R_k$ is

$$\mu_o{}^*R_k = \mu_o{}^*\left[r_k + 2\sum_{i=1}^{n} r_{k+i}\right] \tag{15}$$

where $\mu_o r_{k+i}$ is an axial component of the reaction field at the position of $(k+i)$th residue induced by the electric dipole moment μ_o put in ith position and parallel with the cylindrical axis. When the cylindrical cavity is assumed to be of infinite length to exclude troublesome end effect, R_k is independent with k, that is, $R_k = R_j = \cdots = R(\epsilon)$. Then the dipole moment in the dielectric medium M^* can be written in a simple form

$$M^* = n\mu_o{}^* = \frac{n\mu_o}{1 - \alpha_{11}{}^oR(\epsilon_1)} \tag{16}$$

After an electrostatic consideration of the cylindrical cavity, the function $R(\epsilon_1)$ is estimated (*10*). A ratio of the dipole moment in a medium of dielectric constant ϵ_1 to that in vacuo ($\epsilon_1 = 1$) is given as a measure of the solvent effect for several radii of the cylindrical cavity as shown in Fig. 12.9.

In practice, $\mu_o{}^*(\|)$ has been obtained as 3.6 D in ethylene chloride ($\epsilon_1 \fallingdotseq 10$) and 2.9 D in chloroform ($\epsilon_1 \fallingdotseq 5$). Considering these values and assuming $2b = 14.8$ Å, the residual moment $\mu_o(\|)$ in vacuo can be estimated experimentally as 2.3–2.5 D. The moment originated in the backbone, $\mu_m(\|)$ in vacuo has been assumed to be in the range 4 D and 5 D as mentioned above, so the difference between this and the experimental value will be due to the moment in the side chain $\mu_s(\|)$ which must orient in a direction which will cancel that of the backbone chain. According to the experimental results of the polarized infrared absorption spectrum by Tsuboi (*14*), the carbonyl group, which is only a main electric dipole in the side chain ester group, has been observed to be at an angle of 53° with the helical axis. If this direction is one that cancels the moment in the backbone chain, the resulting residual moment in vacuo may possibly lie in between 2.6 and 3.6 D, by which the experimental result will be plausible.

Another possibility, which may affect the

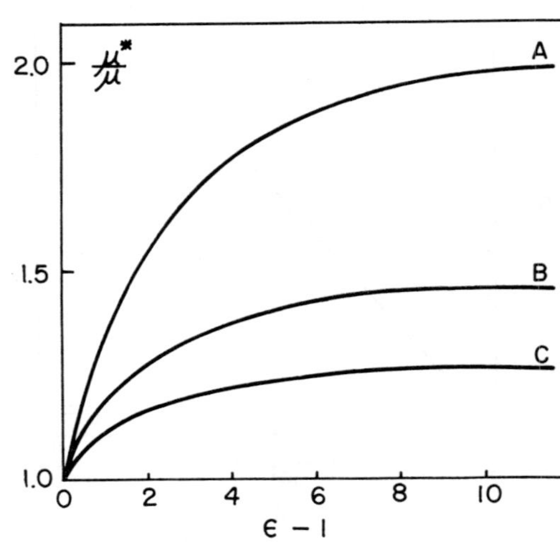

Fig. 12.9. Dependence of μ^*/μ upon the dielectric constant of medium. (A) $b = 4.5$ Å, (B) $b = 6.0$ Å, (C) $b = 7.5$ Å.

dipole moment in solution, is a short-range interaction between solute and solvent, but because of its complex nature we will leave this to another report.

Relaxation time of orientation polarization

Among the sets of information on the dielectric properties of poly-α-amino acid in solutions, the critical frequency, that is a frequency where the dielectric constant, $\epsilon_{12}'-\epsilon_1$, is half that in a static field, $\epsilon_{12}'^0-\epsilon_1$, is related to the relaxation time of the orientation polarization of the solute. A very low critical frequency obtained in the present case means that the α-helix is hydrodynamically resistant to rotation in solution, so that the molecular shape is fairly asymmetric and not flexible.

The hydrodynamically equivalent ellipsoid of revolution may be a suitable model by which to consider the critical frequency in regard to the molecular shape of the α-helix. This model has successfully been used by Doty and his co-workers (*2*) to elucidate the intrinsic viscosity of this molecule in relation to Simha's equation.

The relaxation time, τ, of the rotation of an elongated ellipsoid of revolution about the minor axis and then the critical frequency ν_o can be theoretically given by Perrin's equation as follows (*15*):

$$\nu_o = (2\pi\tau)^{-1} \qquad (17a)$$

$$\tau = 2\tau_o/3[2\rho^2/n(2\rho) - \rho^2]$$
$$\tau_o = 4\pi\eta a'b'^2/kT \qquad (17b)$$

where $\rho = a'/b'$ is axial ratio, a' and b' are the radii of the major axis and minor axis of the ellipsoid, respectively, and η is viscosity of the solvent. To combine this formula with the critical frequency of a cylindrical molecule of length L_o and radius b, the volume and the length of the hydrodynamically equivalent ellipsoid of revolution are equated to those of the cylinder, where the radius of the minor axis is given by $b' = (3/2)^{1/2}b$.

The critical frequencies obtained for the α-helix of PBLG in EDC are plotted against degree of polymerization in Fig. 12.10 (*16*). In the same figure, the theoretical relation given by Eq. 17, in which $2b = 15$ Å is used because it best fits the data, is plotted in a solid line. The negative slope observed in the plot of the critical frequency vs. concentration may be due to

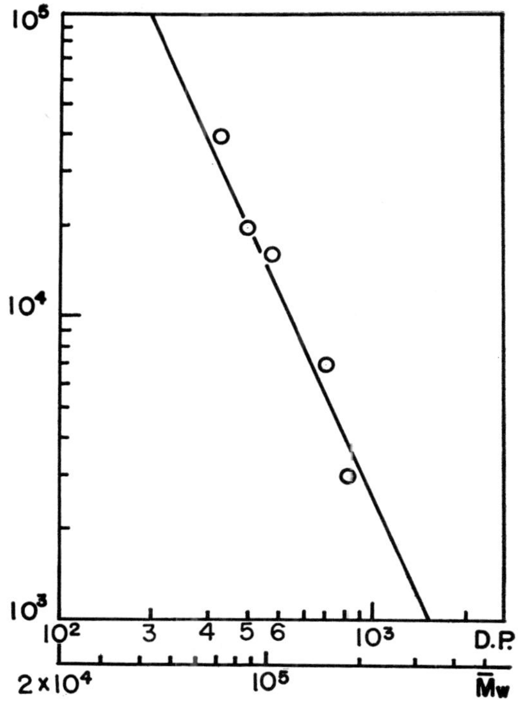

Fig. 12.10. Double logarithmic plot of the critical frequency of PBLG in EDC against degree of polymerization.

the increase in the viscosity of the solution or dipole-dipole interaction.

All of the above results are consistent with and tend to support the belief that the α-helix is fairly rigid and that the molecular model which has been used is correct.

Dispersion curve and polydispersity of molecular weight

The dispersion and absorption curves obtained for the synthetic polypeptide solutions are found to differ from the theoretical curve calculated by Debye's equation in which a single relaxation time is assumed as mentioned in the last section. The discrepancy may be responsible for (1) the heterogeneity of the relaxation time due to the molecular weight distribution which is inevitable in such a synthetic polymer, and/or for (2) dipole-dipole interaction. By the fact that the shape of the dispersion and absorption curves does not depend particularly on the concentration, and also from the similar situation of the curve even at infinite dilution, the molecular weight distribution will be the principal factor affecting the heterogeneity in the relaxation time. To solve this problem, let us consider a solution in which the weight fraction of the n-mer molecule is $w_2(n)$ and the electric dipole moment is $M_n^*(=n\mu_o^*)$. When the Debye equation is assumed to be applicable to each individual molecule of the relaxation time $\tau(n)$ and the solute-solute interaction is absent as in dilute solutions, the dielectric constant $\epsilon_{12}'-\epsilon_1$ and loss factor ϵ_{12}'' of the solution can be written as a superpose of those of the individual molecules by a function of the frequency of the applied field.

$$\epsilon_{12}' - \epsilon_1 = d_{12} \sum_{n=1}^{\infty} w_2(n) P_2(n) = (4\pi N d_{12}/3kT) \sum_{n=1}^{\infty} (M_n^{*2} w_2(n)/n \cdot M_o) \left(\frac{1}{1 + \{2\pi\nu \cdot \tau(n)\}^2} \right) \tag{18}$$

$$\epsilon_{12}'' = (4\pi N d_{12}/3kT) \sum_{n=1}^{\infty} (M_n^{*2} w_2(n)/n \cdot M_o) \left(\frac{2\pi\nu \cdot \tau(n)}{1 + \{2\pi\nu \cdot \tau(n)\}^2} \right) \tag{19}$$

By the use of Perrin's equation of Eq. 17, $\tau(n)$ can be written as a function of n,

$$\tau(n) = (8\pi\eta/3kT) L_o^3 \lfloor \eta^3 / \{2/n(nL_o/b') - 1\} \tag{17b'}$$

Defining number distribution function of the molecule as $f(n)$, so that $f(n)dn$ is the fraction of the α-helix of n-degree of polymerization in the range dn in solution, Eqs. 18 and 19 can be written in an integral form as follows:

$$\lim_{w_2 \to 0} \left(\frac{\epsilon_{12}' - \epsilon_1}{w_2} \right) = \frac{4\pi N d_1 \mu_o^{*2}}{3M_o kT} \int_o^{\infty} \frac{n^2 f(n)}{1 + (2\pi\nu H)^2 n^6} \Big/ \int_o^{\infty} nf(n)dn \tag{20}$$

$$\lim_{w_2 \to 0} \frac{\epsilon_{12}''}{w_2} = \frac{4\pi N d_1 \mu_o^{*2}}{3M_o kT} \int_o^{\infty} \frac{2\pi\nu H n^5 f(n)}{1 + (2\pi\nu H)^2 n^6} \Big/ \int_o^{\infty} nf(n)dn \tag{21}$$

where

$$H = (8\pi\eta L_o^3/3kT)[2 \cdot \ln(nL_o/b') - 1]^{-1}$$

Using Eqs. 20 and 21 in place of the Debye equation and by an appropriate distribution function, such as the most probable distribution by Flory for the present case, both of the measured curves can be understood to be reasonable results. The theoretical curves thus obtained for the polydispersed α-helix of the most probable distribution with the following constants: $2b = 14.8$ Å, $n_n = 1/1np = 500$, $L_o = 1.5$ Å and $\mu_o^* = 3.4$ D, are plotted as a solid line in Figs. 12.5a and 12.5b (*10*).

The static dielectric constant of such a polydispersed α-helix is related to the weight average degree of polymerization, n_w, instead of Eq. 11,

$$\lim_{w_2 \to o} \frac{\epsilon_{12}' - \epsilon_1}{w_2} = \frac{4\pi N d_1 \mu_o^{2*}}{3kTM_o} \cdot \frac{\int_o^{\infty} n^2 f(n)dn}{\int_o^{\infty} nf(n)dn} = \frac{4\pi N d\mu_o^{*2} \bar{n}_w}{3kTM_o} \tag{22}$$

Consequently, the observed specific polarization is a weight-averaged one, and when the weight-average molecular weight is used for the calculation of the dipole moment in Eq. 22, the result must also be given in the same average.

Relation between P_2^{oo} and ν_o of the perfect α-helix

Among the two sets of information which are given by one dielectric dispersion curve without any subsidiary measurement, specific polarization relates to the orientation of polar groups in the molecule, on the one hand, and the critical frequency is a function of the hydrodynamic shape of the molecule, on the other hand. Theoretically, P_2^{oo} and ν_o are given from Eqs. 11 and 17 in terms of the degree of polymerization as follows:

$$P_2^o = 4\pi N n \mu_o^2 / 9_k T M_o \tag{23a}$$

$$\nu_o = (3kT/16\pi^2\eta L_o^3)(n)^{-3}(2/n(L_o/b) \cdot n - 1) \tag{23b}$$

The relation between them in the case of the perfect α-helix can then be obtained by eliminating the degree of polymerization from these equations and is plotted in Fig. 12.11. Hereafter this line plotted for the perfect α-helix is termed "the theoretical $P—\nu$ curve." Actually, the specific polarization and the critical frequency measured for several PBLG samples of different molecular weight are situated quite consistently along the theoretical $P—\nu$ curve as shown by closed circles in Fig. 12.11. This curve may be especially useful in determining whether or not the orientation of the polar groups and the hydrodynamic shape are related linearly as such in the α-helix, although the possibility of coincidence may not be excluded. The study of the structure of the associated α-helix in dioxane–dimethyl formamide mixed solvent should be mentioned as a suitable example of the application of this device.

When the α-helices associate head-to-tail with a linear and rigid junction, the $P—\nu$ relation of the associated molecule should still be on the theoretical $P—\nu$ curve for the perfect α-helix. In other words, a point corresponding to P_2^{oo} and log ν of the α-helix associated in such a manner must be located on the theoretical $P—\nu$ curve and will move along it with the changing degree of association.

On the other hand, any deviation from this theoretical $P—\nu$ curve means that the association is of a different type from the one mentioned above. The plots experimentally obtained are shown in Fig. 12.12, where a double plot is made against solute concentration and the fraction of dimethyl formamide by which the degree of association has been known to be changed. As a qualitative conclusion produced from this experimental result, the type assigned for this association process is one in which a head-to-tail type of association is superposed on a side-by-side one (*17*).

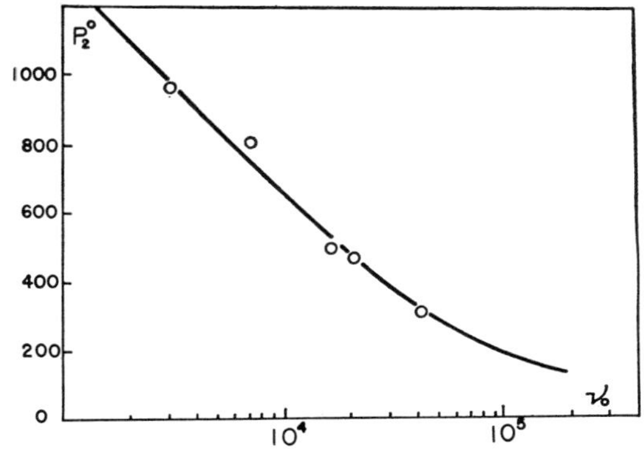

Fig. 12.11. Relation between specific polarization and logarithm of critical frequency of a perfect α-helix in EDC.

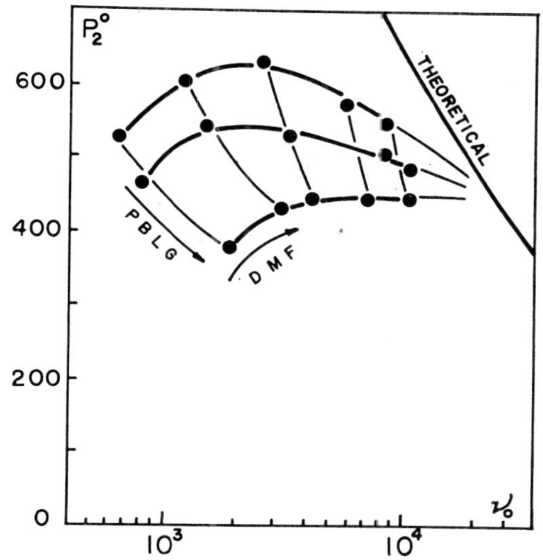

Fig. 12.12. Plot of P_2^o vs. log ν_o for the associated α-helix. Arrows show direction of dissociation.

D,L-*Copolymer and defect of chain regularity*

Since the dielectric information is a function of the orientation of polar groups in the molecule, and since it does not depend upon the helical sense of the backbone chain, it is useful to study the D,L-copolymer in respect to the regularity in its backbone or side chain. From kinetic studies and by measuring the optical rotation, Doty and Lundberg have found that the screw sense of the preformed α-helix has been maintained by the succeeding residues at this point in the polymerization process even when these are antipodes (*18*). And they found also that the reversal of the screw sense occurs after about four antipodes have been added in sequence (*18*). From the same point of view, Downie *et al.* have examined a number of copolymers of different D:L composition by measuring the optical rotation and have suggested, in conclusion, the possible existence of a significant amount of random-coil structure in the range near the equipmolar D:L composition (*19*).

In analyzing the optical rotatory dispersion data, the difference between the fraction of

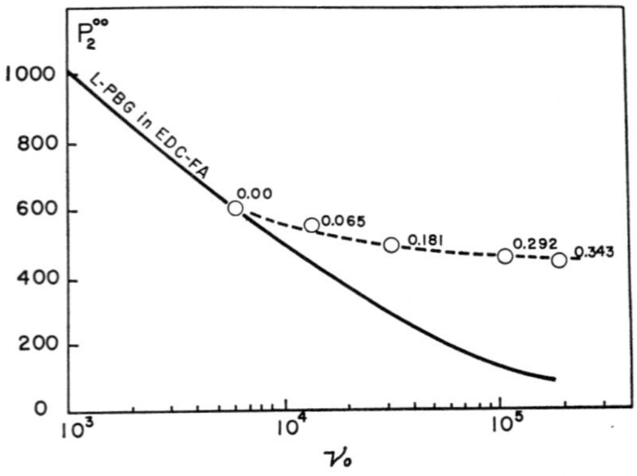

Fig. 12.13. Deviation of P_2^{oo} vs. log ν_o plot of D,L-copolymer (open circles) from that of the perfect α-helix (rigid line). Numbers show the fraction of D-residue.

right-handed and left-handed helices, H_r—H_l can be given from both the intercept and the slope of the Moffitt modified plot, respectively. In Fig. 12.13, H_r—H_l are plotted against f_L where a nice agreement between the results calculated from I and S are obtained.

It can be concluded from the above-mentioned result that the D-residue copolymerized with an excess amount of L-residue is forced to form a helical structure, the screw sense of which is the original one of the α-helix of L-residue. This situation seems to hold until 25 per cent D-fraction is added. Above this fraction a change in the helical structure, which may be due to the formation of the left-handed helix and/or random coil, seems to occur as suggested in the drop of the curve in Fig. 12.4.

Before discussing the dielectric behavior of the D,L-copolymer, let us consider its polymerization kinetics in a simplified model. According to the kinetic studies (*18, 20*) of Doty and Lundberg, and Blout, the reaction rate between the same kind of optically isomeric residues such as

$$\sim\sim\sim\sim\sim\sim L + L \rightarrow \sim\sim\sim\sim\sim\sim L$$
$$\sim\sim\sim\sim\sim\sim D + D \rightarrow \sim\sim\sim\sim\sim\sim D$$

is found to be much higher (by five times) than the rate of cross reaction between antipodes such as

$$\sim\sim\sim\sim\sim\sim L + D \rightarrow \sim\sim\sim\sim\sim\sim D$$
$$\sim\sim\sim\sim\sim\sim D + L \rightarrow \sim\sim\sim\sim\sim\sim L$$

Furthermore, the rate of addition of a residue to the favorable helical end (that is, right-handed for L-residue and left-handed for D-residue, respectively, for PBG) is also known to be about five times higher than that to the uncoiled chain end.

A reversal of the helical sense will not occur too often, for when the polymer chain has once formed a helical structure of any sense, the probability of having a sequence of the necessary residues for inversion is very small. In conformity with such kinetic information, a model of the product of D,L-copolymerization will be derived as follows.

At the beginning of the polymerization reaction, each optically isomeric residue will tend to react with the same kind of residue. Although the selection will not be very positive in this stage, the growing chain will finally get a chance to form a helical structure of either screw sense. Once the growing chain has formed a helical configuration of either screw sense, it will be more particular about the residue to be added, and two kind of polymers, one which has a right-handed screw sense and is composed mainly by L-residues and another left-handed

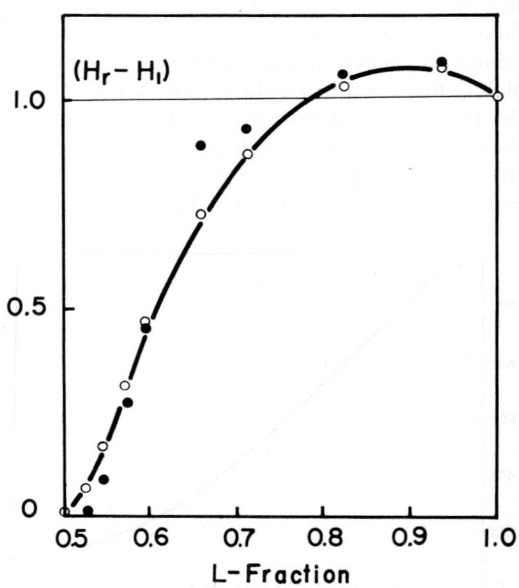

Fig. 12.14. Differences between the fraction of right- and left-handed helices obtained from the intercept (○) and slope (●) of the Moffitt modified plot.

mainly composed of D-residues, will therefore result.

Considering the selective reactivity between the same kind of optically isomeric residue, and as a result of simple statistical calculation, the fraction of the right- and left-handed helices are given as follows:

$$H_r = [f_L^{m-1}/(f_L^{m-1} + f_D^{m-1})] \times H_o \quad (24a)$$
$$H_l = [f_D^{m-1}/(f_L^{m-1} + f_D^{m-1})] \times H_o \quad (24b)$$

where f_L and f_D are the fractions of L- and D-residues copolymerized, respectively, m is a critical number of the same kind of residue in sequence which is necessary to stabilize a helical structure, and H_o is the total fraction of the helical structure, that is $H_o = H_r + H_l$.

From the kinetic studies on the L-polymer and D,L-copolymer, this critical number is believed to be around four. Accordingly, the theoretical curve computed for $m=4$ has been found to be in close agreement with the experimental results as shown in Fig. 12.14. The deviation seen as it approaches a mesopolymer may be due to a defect in the chain regularity that may correspond to the random-coil formation as Downie *et al.* have suggested (*19*).

Let us try next to examine the D,L-copolymer of benzyl glutamate from the dielectric point of view, by which the orientation of polar groups in the molecule should become clear.

At first, a noteworthy feature in the P—ν plot of the D,L-copolymer in contrast to that of the α-helix of pure L-polymer is the small change in the specific polarization in spite of an increase in the critical frequency as shown in Fig. 12.15. In comparing the electric dipole moment and the critical frequency individually with those of the α-helix of pure L-polymer as in Fig. 12.7 and Fig. 12.10, it is belived that the deviation originates in the specific polarization, which corresponds to the large electric dipole moment, but not in the critical frequency up to $f_D = 0.4$.

The residual electric dipole moments of each sample are plotted against its L-fraction in Fig. 12.16, where a steady increase in μ_o with D-fraction is shown in the D-fraction range $f_D = 0$–0.35.

This appearance in μ_o cannot be attributed to the configuration of the backbone chain since the polar groups in it have already made a maximum contribution in the pure L-polymer as its moment vector is almost parallel with

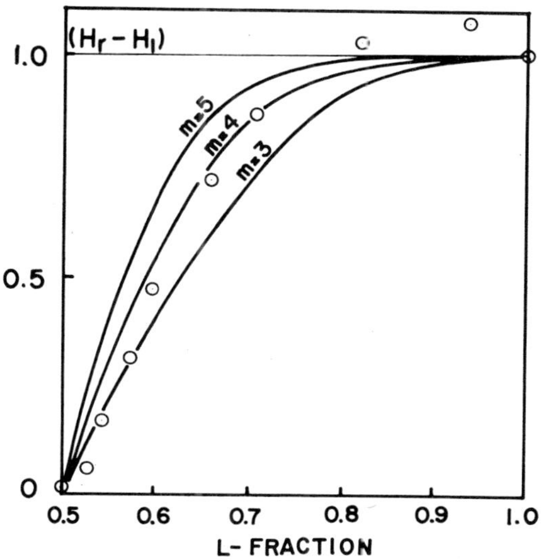

Fig. 12.15. Theoretical plot of $(H_q - H_l)$ vs. L-fraction for several values of m. Open circles obtained from the intercept of the modified plot.

the helical axis.

The only explanation for the dielectric behavior can be found in the orientation of the polar groups in the side chain. As mentioned in the former section, the electric dipole moment of the polar groups in the side chain of L-PBG (the C=O group) is given to be at an angle of 53° with the helical axis, and in the case of the right- (or left-) handed helix made by the L-residue (or D-residue) will be in a direction to offset the moment in the backbone chain. So, the total moment of the residue becomes $|\mu_o| = |\mu_m| - |\mu_s|$. When the antipode is incorporated into it (that is, the case of D-residues in the right-handed helix or of the L-residues in the left-handed helix), the side chain of these will be in a direction which will reinforce the backbone moment, in contrast to the former case, and it becomes $|\mu_o| = |\mu_m| + |\mu_s'|$.

The above-mentioned situation most clearly explains the plot of μ_o against f_L and the intercept of the straight extrapolation to $f_D = 0.5$ will give the residual moment of the pure backbone chain of the α-helix. Unfortunately, as $|\mu_s'|$ will not necessarily be equal to $|\mu_s|$ because the two bonds thrusting out from the β-carbon of the peptide chain are not symmetrical with the plane perpendicular to the helical axis, the effect of the side chain will not be cancelled completely even in the ideal single-

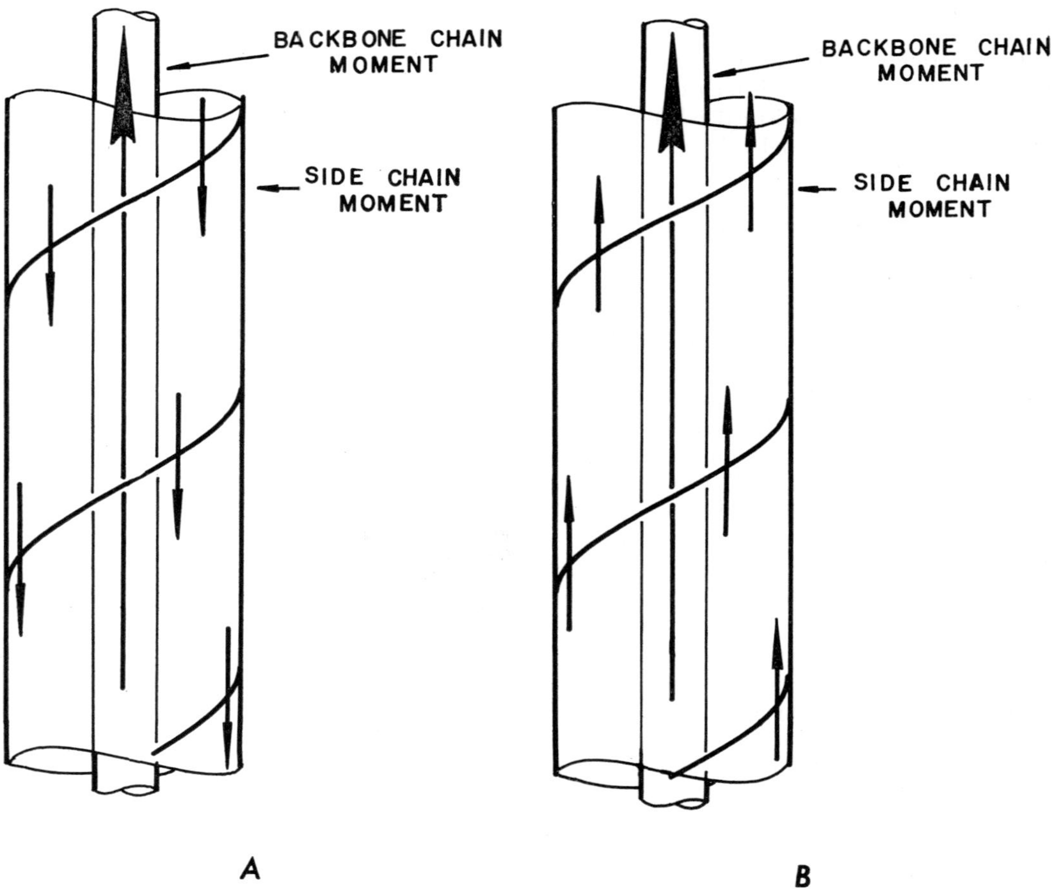

Fig. 12.16. An illustrative drawing of the orientation of polar groups in the backbone and side chain in the right-handed α-helix. a) Helix of L-isomer. b) Helix of D-isomer.

screw sense helix of meso-polymer. Also the other influences such as those arising from the flexibility of the side chain or those from the junction of L- and D-blocks may not be entirely negligible.

However, the extrapolated value 6.5 D is still believed to give an approximate value of the residual moment of the backbone of the α-helix because this is a theoretically supported value as mentioned in the former section.

The fact that the drop in $(H_r—H_l)$ vs. f_L curve starts before that in μ_o vs. f_L plot seems to indicate that the regularity in the backbone chain is maintained after a change in the helical structure has occurred. Although the drop in μ_o as it approaches $f_D = 0.5$ is clearly due to some defect in chain regularity, its relatively large value suggests that the random-coil formation must occur at the end of the chain but not in the middle of the chain. For, when the molecule has uncoiling and flexible parts in the middle of the molecular chain, not only is the moment of these parts negligible but also the molecular dipole moment, given by a vector sum of the moments of each ordered part, will be small and considerable change in μ_o would result. Therefore the disorder must exist at the chain end; otherwise it must be very small in quantity or have a rigid structure.

The disorder at the chain end seems quite

reasonable as the result of the competition between the L- and D-residues until such a sequence as LLLL or DDDD occurs to form the helical structure.

Though the possibility of a disorder in the middle of the molecule may not be entirely excluded, it must be rigid or else the quantity must be very small in order to explain the present experimental result.

Assuming the 6.5 D is the backbone moment of the α-helix, only the maximum fraction of the disordered part can be estimated from the value μ_o as about 35 per cent at $f_D = 0.5$. When the disorder is in the middle of the molecule and if it behaves as a flexible joint between the rigid dipole, the amount of the random coil is calculated to be much smaller, but further discussion may be useless until we have more information about the disordered configuration.

To summarize, the following conclusions concerning the structure of the D,L-copoly-α-benzyl glutamate have resulted:

1. Only the one-screw sense helix exists in the range of $f_D = 0$–0.25.

2. Regularity in the orientation of polar groups in the backbone chain is maintained up to $f_D = 0.35$.

3. Some disorder appears when more than 35 per cent of antipodes is introduced into the α-helix. This seems to originate in the unfolding part at the end of the molecule or in a few disordered points in the molecule.

4. But even in the copolymer polymerized from the equimolar mixture of L- and D-isomeric monomers, the major part of the substance seems to be composed of the polymers having an ordered configuration which may be a mixture of right- and left-handed helices of L-residue-rich and D-residue-rich polymers, respectively.

The present study on the D,L-copolymer has also produced a model of the ordered configuration in the backbone and side chain of the α-helix in conformity with the discussion in another section, as follows:

5. The side chain of the α-helix has a regular orientation as does its backbone chain.

6. In the right-handed helix, the electric dipole moment of the side chain is in an opposite direction to that in the backbone chain in the L-residue and in the same direction in the D-residue, as illustrated in Fig. 12.17.

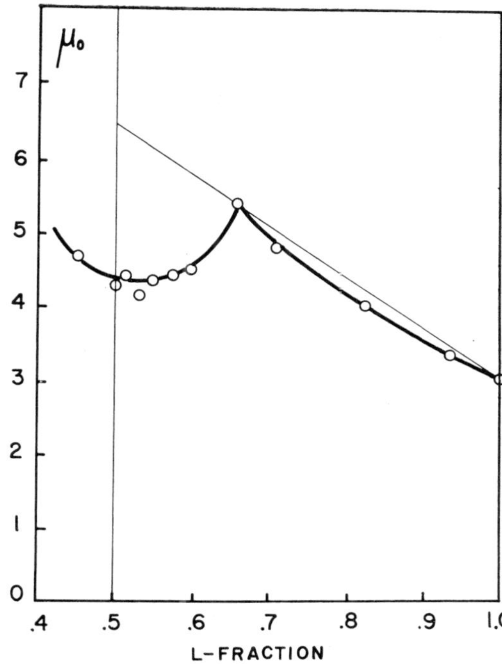

Fig. 12.17. Plot of μ_o, the electric dipole moment per residue, against the L-fraction. Narrow line of straight extrapolation gives 6.5 D at $f_D = 0.5$.

7. The electric dipole moment of the backbone chain of the α-helix is about 6.5 D per residue in EDC.

In summary, conclusive evidence is now shown that chain regularity exists not only in the backbone chain of the poly-α-amino acid but also in its side chain. This information is believed to provide a basis for the study of the specificity of the biological molecules, such as the protein and nucleic acid, in relation to their side-chain structure, and it also suggests the possibility of developing a synthetic polymer having specific individuality.

References

1 L. Pauling, R. B. Corey, and H. R. Branson, *Proc. Nat. Acad. Sci. U. S.* **37**, 205 (1951).
2 P. Doty, J. H. Bradbury, and A. M. Holtzer, *J. Am. Chem. Soc.* **78**, 947 (1956); P. Doty, A. M. Holtzer, J. H. Bradbury, and E. R. Blout, *ibid.* **76**, 4493 (1954).
3 For instance, P. Debye and F. Buche, *J. Chem. Phys.* **19**, 589 (1951).
4 C. P. Smyth: "Dielectric Behavior and

Structure," McGraw-Hill Book Co., Inc., N. Y., 1955.

5 S. Mizushima, T. Shimanouchi, S. Nagakura, K. Kuratani, M. Tsuboi, H. Baba, and O. Fujioka, *J. Am. Chem. Soc.* **72,** 3490 (1950).

6 C. A. Coulson and U. Danielson, *Arkiv Fysik* **8,** 239 (1954).

7 R. H. Cole and P. M. Gross, Jr., *Rev. Sci. Instr.* **20,** 252 (1949).

8 A. von Hippel, Progress Rept. No. 1, Lab. Insulation Res. MIT, p. 7 (1947).

9 A. Wada, *Bull. Chem. Soc. Japan* **33,** 822 (1960).

10 A. Wada, *J. Chem. Phys.* **31,** 495 (1959).

11 P. Debye, "Polar Molecules," Chemical Catalog Co., N. Y., 1929.

12 A. D. Buckingham, *J. Chem. Phys.* **23,** 2370 (1955).

13 A. Wada, *ibid.* **30,** 328 (1959).

14 M. Tsuboi, private communication.

15 F. Perrin, *J. phys. radium* **5,** 497 (1934).

16 A. Wada, *J. Chem. Phys.* **30,** 329 (1959).

17 A. Wada, *J. Polymer Sci.* **45,** 145 (1960).

18 P. Doty and R. D. Lundberg, *Proc. Nat. Acad. Sci.* **43,** 213 (1957); R. D. Lundberg and P. Doty, *J. Am. Chem. Soc.* **79,** 3961 (1957).

19 A. R. Downie, A. Elliott, W. E. Hanby, and B. R. Malcolm, *Proc. Roy. Soc.* **A242,** 325 (1957).

20 M. Idelson and E. R. Blout, *J. Am. Chem. Soc.* **80,** 2387 (1958).

PAPER 13

The Relation between Conformation and Light Absorption in Polypeptides and Proteins

IGNACIO TINOCO, Jr., ANITA HALPERN, and WILLIAM T. SIMPSON

The study of optical properties of proteins and other polypeptides has provided a significant fraction of the information about their structure. We would like to discuss the quantitative relations between visible and ultraviolet light-absorption and polypeptide structure. In particular we will relate the frequency of absorption and the intensity of absorption to conformation. The theoretical results will then be compared with experiments on the absorption of poly-L-glutamic acid and poly-L-lysine in the helix and random-coil forms.

Theory

Absorption frequency

Unique group.—The change in maximum absorption frequency when a molecule or part of a molecule undergoes a change in environment can characterize both the environment and the molecule. We will discuss the general effects which in principle always lead to a frequency shift; specific reactions with the environment which sometimes cause additional effects will not be discussed.

The influence of environment on the energy states of a molecule has been the subject of many papers. For examples see Bayliss and McRae (*3*, *4*), Ooshika (*22*), McRae (*18*), Longuet-Higgins and Pople (*16*), Longuet-Higgins (*17*), Coulson (*6*), and Robinson (*26*). The change in energy when a molecule is taken from a vacuum to a solvent, or from one solvent to another (as discussed in the papers above) is exactly analogous to the energy change which accompanies changes in conformation of the molecule (*25*, *29*). To obtain the frequency shifts associated with changes in conformation or solvent we must find the shift in the upper and lower energy levels for the transition. The lower energy level is the total energy of the system (where system means the molecule plus the environment) with the chromophoric molecule in its ground state. The upper energy level is the total energy of the system with the molecule in its excited state. It is clear that the medium around the molecule and the conformation of the molecule will determine the energy levels and therefore the absorption frequency. As we are excluding inter- and intra-molecular electron transfer, simple electrostatic forces and London forces will determine the energy.

Let us consider a single molecule and find its absorption frequency in a particular medium (solvent) as compared to that in a vacuum or dilute gas.

The electrostatic interaction with the solvent can be characterized by interactions between net charges, permanent dipole moments, and polarizabilities. When a molecule undergoes a transition to an excited state and absorbs light, both its permanent dipole moment and polarizability may change (Fig. 13.1a). If this molecule is in a solvent (Fig. 13.1b), it will interact with the charges, dipoles, and polarizabilities of the solvent particles. The electrostatic frequency shift will be directly proportional to the difference in permanent dipole moment and polarizability of the chromophoric molecule in its ground and excited states.

The change in permanent dipole moment may be zero, positive, or negative. For rigid, nonpolar molecules it is zero. For many simple polar molecules it is small, because the total number of electrons in the molecule is large compared to that involved in the transition.

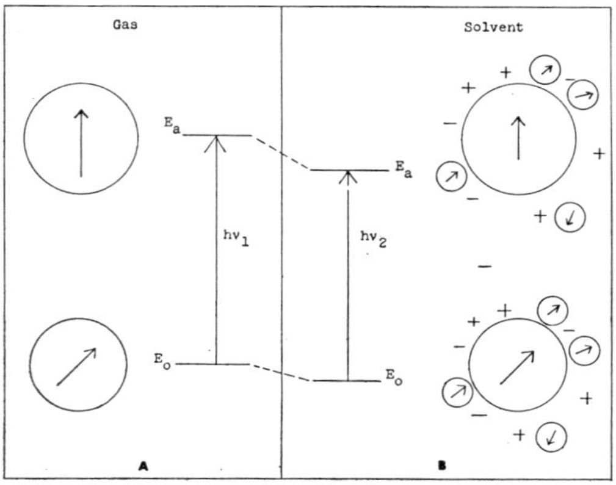

Fig. 13.1. A schematic representation of the electrostatic energy shifts which occur when a molecule is placed in a solvent. A molecule in the ground state will have a certain permanent dipole moment (single-headed arrow) and polarizability (circle). The same molecule in the excited state will have a different permanent dipole moment and polarizability. The difference in interaction of the ground and excited states of the chromophore with the molecules and ions of the solvent leads to a frequency shift.

For long, conjugated polar molecules, however, a large separation of charge may occur in the excited state and thus lead to a frequency shift (*5*). The difference in polarizability will probably always be small. For proteins and polypeptides the electrostatic frequency shift is difficult to estimate.

The London interaction (induced dipole–induced dipole) with the solvent will generally be the dominant one. It will always lead to a red shift relative to the gas phase absorption frequency (*16*). This red shift is directly proportional to the polarizability of the solvent.

Although we have been discussing a molecule in a solvent, the same reasoning applies to a particular group in a protein. We therefore expect a blue shift (a shift to shorter wave lengths) when a group surrounded by polarizable material in the native protein is thrust into the less polarizable solvent by a denaturation process. This is in complete agreement with the conclusions of Yanari and Bovey (*32*) based on extensive model experiments.

Identical groups.—Identical groups have a special kind of interaction (resonance interaction) among themselves in addition to the interactions described above. This interaction is different in that it can lead to splitting of absorption bands into two or more new bands. The requirement for significant interaction among the identical groups are (1) identical energy levels and (2) fixed geometry. The first requirement is a crucial one because it means that any perturbation that is not the same for all identical groups will destroy the resonance interaction. Vibrations within the groups and collision of each group with the solvent will blur the identity of the electronic levels and will thus suppress the interaction. A quantitative discussion is given by Simpson and Peterson (*27*), Moffitt and Witkowski (*21*), and McRae (*19*). The requirement of fixed geometry among the groups limits large effects, which can be positive or negative, to crystals and rigid polymers.

Summary.—A unique group will have an absorption frequency which is directly proportional to the polarizability of the medium around it: the larger the polarizability, the smaller the frequency. The absorption frequency due to interaction between identical groups in a rigid array may, owing to resonance, be substantially different from that of the unique group.

Absorption intensity

Unique group or identical groups.—The change in the intensity of absorption with conformational changes has not been studied as extensively as the frequency shifts (*25, 28, 29, 31*). However, it is easier to predict intensity changes, because a distinction between unique and identical groups need not be made. Resonance interaction between identical groups does not cause a change in absorption intensity. The changes in absorption always involve lending intensity to, or borrowing intensity from other absorption bands in the system. Borrowing from oneself leads to no net change; similarly, interaction among identical energy levels leads to no net change in intensity.*

The intensity of absorption measured as the area of the absorption band is characterized by an oscillator strength. The oscillator strength is proportional to the frequency of the transition and to the square of the transition dipole mo-

* This statement does not agree with the intensity equations given in references *28* and *29*; corrected equations are given in references *25* and *31*.

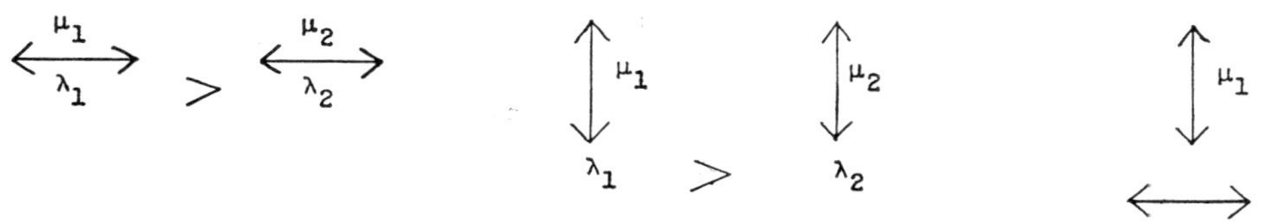

A: Transition 1 borrows intensity from transition 2. B: Transition 2 borrows intensity from transition 1. C: No borrowing occurs.

Fig. 13.2. The change of intensity of an absorption band caused by the interaction of transition dipoles (double-headed arrows). For colinear transitions (2A) the higher wave-length band shows hyperchromism. For parallel, coplanar transitions (2B) the higher wave length band shows hypochromism. For perpendicular transitions (2C) or random transitions there is no intensity change.

ment. The transition dipole moment is a measure of the electronic mobility associated with the transition and characterizes the polarizability of the transition.

Any change in frequency (except that due to resonance interaction) will also cause a change in intensity. The main change in absorption, however, will come from the interaction of the specific chromophoric transition dipoles with all the other transition dipoles in the system. For two colinear transition dipoles, the lower frequency (longer wave length) transition will borrow intensity from the upper (Fig. 13.2a). For two parallel coplanar transition dipoles the upper frequency transition will borrow intensity from the lower (Fig. 13.2b). For randomly oriented or perpendicular transition dipoles there is no interaction (Fig. 13.2c). In a polymer every group near enough to the chromophore to cause appreciable effect must be considered. As each group has many transitions which can interact with the chromophore, these too must be summed. The transitions in the vacuum ultraviolet which are inaccessible to our experimental methods can be approximated by a polarizability.

We can now see from Fig. 13.2 that any change in conformation such as a helix-to-coil transition or the denaturation of a protein may lead to an increase or decrease in absorbancy. This change is measured by the area of the absorption band (the oscillator strength). For the special case in which the shapes of the absorption curves of the two conformations do not change, the oscillator strengths are proportional to extinction coefficients.

In distinction to the solvent-sensitive frequency shifts, large changes in intensity with solvent only occur for specific solvent-solute complexes; molecules which have random relative orientation to each other do not contribute to the intensity changes.

Theoretical details

The equations characterizing the absorption of a system composed of interacting parts have been derived previously, so we will mainly state the results.

The electrostatic energy and frequency shift are found by writing down the energies of interaction of charges, dipoles, and polarizabilities. The results for the transition frequency (ν_2) of a unique group i in a rigid system relative to the frequency (ν_1) of the isolated (dilute gas phase) group is:

Electrostatic frequency shift for a rigid system

$$h(\nu_2 - \nu_1) = (\boldsymbol{\mu}_{iaa} - \boldsymbol{\mu}_{ioo})\cdot \sum_{j\neq i}\left[\frac{\rho_j \boldsymbol{R}_{ij}}{R_{ij}^3} + \boldsymbol{T}_{ij}\cdot\boldsymbol{\mu}_j - (1/2)\,\boldsymbol{T}_{ij}\cdot\boldsymbol{\alpha}_j\cdot\boldsymbol{T}_{ji}\cdot\boldsymbol{\mu}_{ioo}\right] \quad (1)$$
$$- (1/2)(\boldsymbol{\alpha}_{iaa} - \boldsymbol{\alpha}_{ioo}) : \sum_{j\neq i}\left[\frac{\rho_j^2 \boldsymbol{R}_{ij}\boldsymbol{R}_{ji}}{R_{ij}^6} + \boldsymbol{T}_{ij}\cdot\boldsymbol{\mu}_j\boldsymbol{\mu}_j\cdot\boldsymbol{T}_{ji}\right]$$

where,

$\boldsymbol{\mu}_{iaa}$ = the permanent dipole moment of group i in the excited state c.

$\boldsymbol{\mu}_{ioo}$ = the permanent dipole moment of group i in the ground state.

ρ_j, $\boldsymbol{\mu}_j$, $\boldsymbol{\alpha}_j$ = the charge, permanent dipole, and polarizability of group j.

$\boldsymbol{R}_{ij}$ = the distance vector between group i and j.

$\boldsymbol{T}_{ij} = [1 - 3\boldsymbol{R}_{ij}\boldsymbol{R}_{ij}/R_{ij}^2]R_{ij}^{-3}$

$\boldsymbol{\alpha}_{ioo}$, $\boldsymbol{\alpha}_{iaa}$ = the polarizability of group i in the ground and excited states.

For a flexible system we must average overall configurations; we weight the contribution of each configuration according to its probability.

$$\langle h(\Delta\nu)\rangle = \frac{h\int \Delta\nu e^{-V/kT}}{\int e^{-V/kT}} \quad (2)$$

The value of V is the potential energy of the configuration. One extreme case of a flexible system is a completely random system in which there is no correlation between i and j. For this system (which may approximate a molecule in a solvent) we obtain:

Electrostatic frequency shift for a random system

$$h(\nu_2 - \nu_1) = -\frac{n_s}{\langle R^6\rangle}\left[(\mu_{aa} - \mu_{oo})\mu_{oo}\alpha_s + (\alpha_{aa} - \alpha_{oo})\mu_s^2\right] - \frac{n_i}{6\langle R^4\rangle}(\alpha_{aa} - \alpha_{oo})\rho^2 \quad (3)$$

where,

n_s = the average number of solvent molecules at a distance $\langle R^6\rangle$ with permanent dipole moment μ_s and average polarizability α_s around the chromophore.

n_i = the average number of ions at a distance $\langle R^4\rangle$ with charge ρ around the chromophore.

The London frequency shift (*16*) comes from the interaction of the transition dipoles in the chromophore with all other transition dipoles in the system.

London frequency shift for a rigid system

$$\begin{aligned} h(\nu_2 - \nu_1) &= \sum_{j\neq i}\sum_{a\neq o}\sum_{b\neq o}(\boldsymbol{\mu}_{ioa}\cdot \boldsymbol{T}_{ij}\cdot \boldsymbol{\mu}_{job})^2/h(\nu_a + \nu_b) \\ &- \sum_{j\neq i}\sum_{b\neq o,a}(\boldsymbol{\mu}_{ioa}\cdot \boldsymbol{T}_{ij}\cdot \boldsymbol{\mu}_{job})^2/h(\nu_b - \nu_a) \\ &- \sum_{j\neq i}\sum_{b\neq o,a}\sum_{c\neq o}(\boldsymbol{\mu}_{iab}\cdot \boldsymbol{T}_{ij}\cdot \boldsymbol{\mu}_{joc})^2/h(\nu_b + \nu_c - \nu_a). \end{aligned} \quad (4)$$

One can replace the sums of transition moments by polarizabilities.

$$\boldsymbol{\alpha}_{\text{static}} = 2\sum_{b\neq o}\frac{\boldsymbol{\mu}_{ob}\boldsymbol{\mu}_{ob}}{h\nu_b} \quad (5)$$

With certain approximations one obtains

$$h(\nu_2 - \nu_1) = -\sum_{j\neq i}\sum_{l=1}^{3}(\alpha_{jl}/2)\left[(h\nu_a/8)\sum_{m=1}^{3}\alpha_{im}G_{imjl}^2 + \mu_{ioa}^2 G_{ijl}^2\right] \quad (6)$$

where,

α_{jl}, α_{im} = the l and m components of the polarizabilities of groups j and i.

$$G_{imjl} = \left[\boldsymbol{e}_{im}\cdot \boldsymbol{e}_{jl} - \frac{3(\boldsymbol{e}_{im}\cdot \boldsymbol{R}_{ij})(\boldsymbol{e}_{jl}\cdot \boldsymbol{R}_{ij})}{R_{ij}^2}\right]R_{ij}^{-3}$$

$\boldsymbol{e}_{im}$, $\boldsymbol{e}_{jl}$ = unit vectors along directions m in group i and l in group j.

$\boldsymbol{R}_{ij}$ = distance between groups i and j.

If we average Eq. 6 over all equally probable orientations of groups i and j we obtain: London frequency shift for a random system

$$h(\nu_2 - \nu_1) = -\frac{n_s\alpha_s}{\langle R^6\rangle}\left[(3h\nu_a/8)\alpha + \mu_{oa}^2\right] \quad (7)$$

where,

n_s = the average number of solvent molecules at a distance $\langle R^6\rangle$ with average polarizability α_s around the chromophore.

α, ν_a, μ_{oa} = the polarizability, frequency of transition, and transition dipole moment of the chromophore.

The resonance energy shifts depend on the interaction among the transition dipoles of the identical groups. The number of absorption bands that occur depend on the geometry of the system. We will only discuss a long helix. A helix will have two absorption bands for every absorption band in the monomer; one is polarized parallel to the helix axis, the other perpendicular (*20*). Helix resonance bands

$$\begin{aligned} \nu_\perp &= \nu_a + (1/hN)\sum_{i=1}^{N}\sum_{j\neq i}^{N}\mu_{oa}^2 G_{ij}\cos\alpha\pi(j\text{-}i)P \\ \nu_\parallel &= \nu_a + (1/hN)\sum_{i=1}^{N}\sum_{j\neq i}^{N}\mu_{oa}^2 G_{ij} \end{aligned} \quad (8)$$

where,

ν_a = the frequency of the monomer.

μ_{oa} = the transition moment of the monomer.

$$G_{ij} = \left[\boldsymbol{e}_i\cdot \boldsymbol{e}_j - \frac{3(\boldsymbol{R}_{ij}\cdot \boldsymbol{e}_j)(\boldsymbol{R}_{ij}\cdot \boldsymbol{e}_i)}{R_{ij}^2}\right]R_{ij}^{-3}$$

$\boldsymbol{e}_i$, $\boldsymbol{e}_j$ = unit vectors in the directions of the transition moments.

P = number of residues/turn of helix.

The intensity ratio for these bands is $\nu_\perp\mu_{oa}^2{}_\perp/\nu_\parallel\mu_{oa}^2{}_\parallel$. Eq. (8) applies to a system in which the vibrations and solvent perturbation effects are small; for large solvent and vibrational perturbations no splitting occurs (*27*). A better approximation to G_{ij} can be made by using monopole interactions instead of point dipole interactions. The definition of G_{ij} then becomes

$$G_{ij} = \sum_s \sum_t \rho_{is}\rho_{jt}/R_{isjt}$$

where ρ_{is}, ρ_{jt} are transition monopoles for groups i and j.

The change in intensity caused by a change in conformation depends on both electrostatic and London effects. The latter, which may be the most important, will be discussed here.

Intensity change for a rigid system.

$$F_{oa}/f_{oa} = 1 - (4K\lambda_a^2/N) \sum_{i=1}^{N} \sum_{j\neq i} \sum_{b\neq a} \frac{f_{ob}\lambda_b^2 G_{ij} \boldsymbol{e}_i \cdot \boldsymbol{e}_j}{(\lambda_a^2 - \lambda_b^2)} \quad (9)$$

where,

F_{oa} = the oscillator strength for group i (which can be unique or one of an identical set of N groups) in a rigid conformation.

f_{oa}, f_{ob} = the oscillator strength for the isolated groups.

$$f_{oa} = (8\pi^2 m/3he^2)\nu_a\mu_{oa}^2 = 4.32 \times 10^{-9} \int \epsilon_{oa} d\omega$$

ν_a, ν_b = absorption frequencies for transitions a and b. $K = 3e^2/8\pi^2 mc^2$.

The transitions in the far ultraviolet can be approximated by polarizabilities. The polarizability at wave length λ_a is

$$3\alpha(\lambda_a) = 2K\lambda_a^2 \sum_{b\neq o,a} \frac{f_{ob}\lambda_b^2}{(\lambda_a^2 - \lambda_b^2)} \quad (10)$$

therefore,

$$F_{oa}/f_{oa} = 1 - (4K\lambda_a^2/N) \sum_{i=1}^{N} \sum_{j\neq i} \sum_{b\neq a} \frac{f_{ob}\lambda_b^2 G_{ij} \boldsymbol{e}_i \cdot \boldsymbol{e}_j}{\lambda_a^2 - \lambda_b^2} - 2\sum_{j\neq i} \sum_{l=1}^{3} \alpha_{jl}(\lambda_a) G_{ijl} \boldsymbol{e}_i \cdot \boldsymbol{e}_{jl} \quad (11)$$

where, α_{jl} = the component in the l direction of the polarizability of group j. It is understood that the sum over b in the second term includes the transitions closest to a; the remaining transitions are approximated by the last term.

Experiment

Experiments relating light absorption and protein conformation have been mainly concerned with frequency shifts determined by difference spectra of the aromatic amino acids (*14*, *32*). As these papers point out, an explanation of these frequency shifts can be made either in terms of the general London (polarizability) forces or in terms of specific hydrogen bonds.

Experiments on conformation-dependent intensity changes in polyglutamic acid have been reported by Imahori and Tanaka (*12*) and Doty (*7*); Glazer and Smith (*9*) and Foss (*8*) have reported analogous effects in proteins. We have studied the helix-coil transition of poly-L-lysine by absorption measurements, and we have measured the absorption spectra of poly-L-glutamic acid in a vacuum spectrograph.

Experimental details

Polylysine.—Poly-L-lysine · HBr (Pilot Chemical Co. Lot No. L-5, molecular weight = 85,000) was dialyzed exhaustively versus (1) 0.01 M NaOH and (2) distilled water to remove the bromide ions and leave the free base. The resultant solution had a *p*H between 10.5 and 11.0. The concentration of this stock solution was determined by lyophilyzing to constant weight. Analysis of the product gave *N*, 20.9 per cent; *C*, 56.7 per cent (calculated: *N*, 21.8 per cent; *C*, 56.2 per cent).

Polyglutamic acid.—Poly-L-glutamic acid (Pilot Chemical Co. Lot No. G-10, molecular weight = 84,000) was used without further treatment.

Small molecules.—Solutions of small molecules were prepared by weight. The molecules were: *N*-methyl acetamide (Eastman Organic Chemicals), alanine amide and *N*-acetyl alanine amide (synthesized in this laboratory), and ethylenediamine dihydrochloride (Eastman Organic Chemicals).

Absorption measurements.—The absorption to 1850 Å of polylysine and the small molecules was measured with a Carey Model 14 Spectrophotometer; quartz cells with 1 or 5 mm pathlengths were used. The absorption of polyglutamic acid was measured to 1770 Å with a vacuum spectrograph (*11*), having a dispersion of 8.5 Å/mm at the photographic plate. A fluorite cell with a 0.001 inch path-length was used. Both polymers were studied in water with either HCl or NaOH added to adjust the *p*H. Hydroxide-ion absorption is significant below 2000 Å and requires careful control of the *p*H of the blank for basic solutions. This absorption also prevented measurements above *p*H 11.

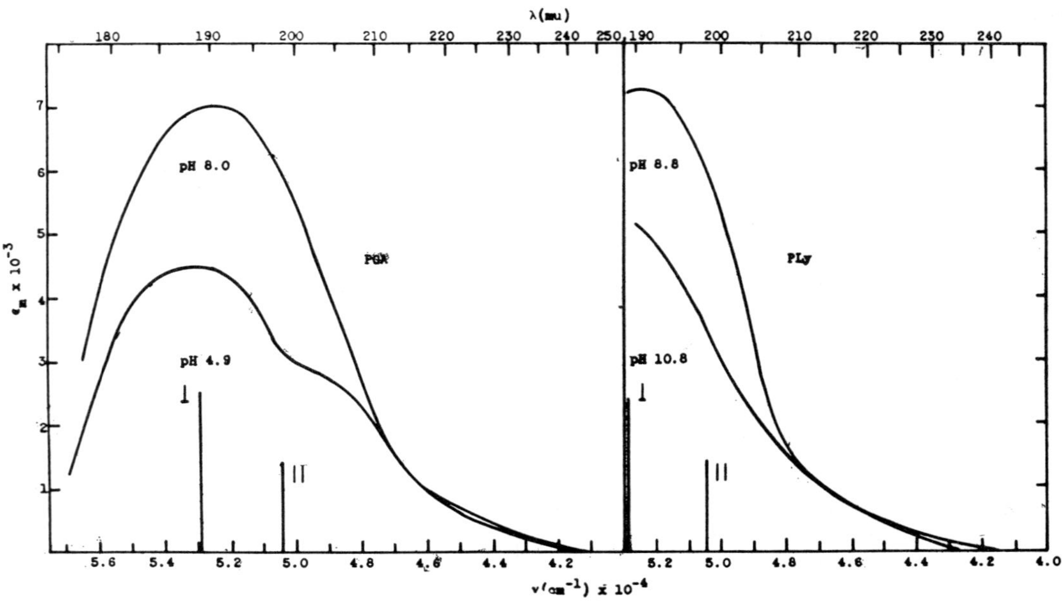

Fig. 13.3. The absorption curves for poly-L-glutamic acid and poly-L-lysine in the helix (*p*H 4.9 for PGA, *p*H 10.8 for PLy) and random coil (*p*H 8.0 for PGA, *p*H 8.8 for PLy) forms. The vertical lines show the calculated positions of the helix bands relative to the coil band. The relative lengths of the lines show the calculated relative intensity of the helix bands; the band at 200 mμ is predicted to be polarized perpendicular to the helix axis.

Results

The molar extinction coefficients for poly-L-glutamic acid and poly-L-lysine in the helix and random-coil forms are plotted versus frequency in Fig. 13.3. The absolute polylysine absorption was determined, but only the relative absorption was determined for polyglutamic acid. For polyglutamic acid the maximum extinction coefficients of Imahori and Tanaka were used.

Both polymers absorb significantly less light as helices, i.e., show over-all hypochromism, but there is a slight increase in absorption coefficient for the helix above 215 mμ. There is also a slight blue shift in the maximum helix absorption. The helix absorption curve is broader than the coil absorption and in the polyglutamic acid spectrum there is an indication of a new band. Essentially identical spectra have been obtained by Doty (*7*) and co-workers for polyglutamic acid. They find that the amide absorption of the polylysine helix also seems to show two bands. Detection of the helix band splitting seems to depend on the equipment used. Doty (*7*) and our work with the photographic method find the helix splitting for polyglutamic acid, although Imahori and Tanaka (*12*) did not. Doty also finds the splitting for polylysine, while our work with the Carey Model 14 does not. Although the shoulder on the main amide peak in the helix spectrum is small, and it is not seen with all spectrophotometers, it definitely indicates a new feature in the helix spectrum not present in the coil. Measurements at low temperatures may be needed to completely resolve this new feature.

The main features of these spectra can be

quantitatively understood in terms of interactions between the transition dipoles. The resonance interaction among the 1900 Å absorptions of each amide group could lead to the helix splitting, while the interaction of this transition with the lower wave-length amide and side-chain absorption bands leads to the over-all decrease in helix absorption. The increase in helix absorption coefficient above 215 mμ can be caused by the $n \rightarrow \pi^*$ transition (220 mμ) in the carbonyl of the amide group. This transition can interact with other bands as shown in Fig. 13.2, or the difference in hydrogen bonding of the carbonyl in the helix and coil can cause a shift in the $n\pi$ band which makes it visible in the helix spectrum. The slight blue shift may be due to many factors which will not be analyzed here.

The *p*H dependence of the maximum absorption of polylysine is shown in Fig. 13.4. On the same graph the *p*H dependence obtained by Imahori and Tanaka (*12*) for polyglutamic acid is shown. The two polymers show analogous behavior; in the charged, random-coil form they have similar absorption which decreases markedly in the uncharged, helix form. The polylysine transition is quite sharp; it has a mid-point at *p*H 9.1 and it seems to be essentially complete within 0.7 *p*H units. These data can be compared with the measure of the polylysine helix-coil transition obtained by Applequist (*1*) using optical rotation. He finds a similar sharp transition for polylysine in water, but the mid-point is near *p*H 10. The main difference in the materials used seem to be that we removed the bromide counterions, while he did not. Either bromide counterions stabilize the helix more than chloride ions do, or optical density and optical rotation are not measures of the same structural parameters.

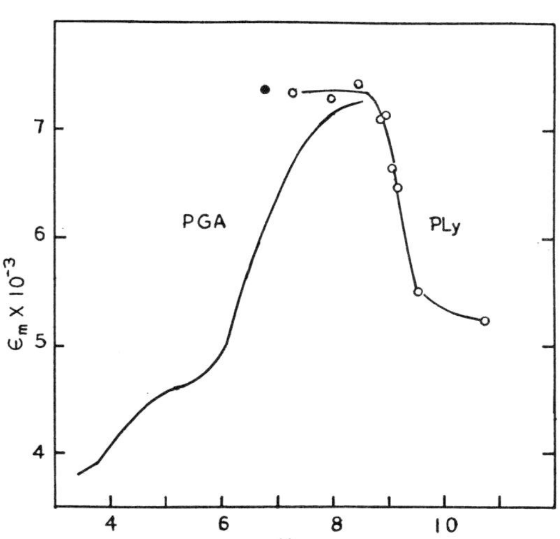

Fig. 13.4. The *p*H dependence of the absorption coefficient at 1900 Å for poly-L-glutamic acid (*12*) and poly-L-lysine.

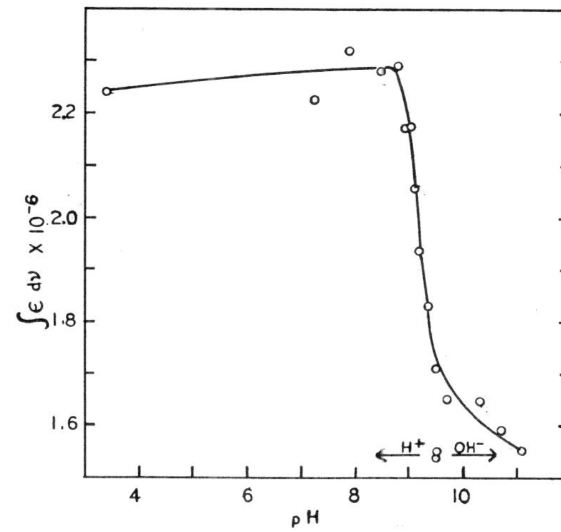

Fig. 13.5. The *p*H dependence of the integrated absorption to 1925 Å for poly-L-lysine.

A more complete picture of the *p*H dependence of the helix-to-coil transition for polylysine is given in Fig. 13.5. The area under the absorption curve measured to 1925 Å is plotted versus *p*H. More experimental points were available at this higher wave length than for ϵ_{max}.

Comparison of experiment with theory

We want to make a quantitative comparison between our experimental results for the helix absorption and calculations based on the interaction between transition moments. The experimental results are summarized in Fig. 13.3 and Table 13.1. The oscillator strengths were determined by measurement of the area under the ϵ_m versus frequency curves with a planimeter. For polyglutamic acid the absorption curve was completed by extrapolation; for polylysine the area to the maximum was measured and the resultant value was multiplied by 2. We shall ascribe the total oscillator

Table 13.1—The optical properties of several amides

Amide	λ_{max} (Å)	$\epsilon_m \times 10^{-3}$	f	F_{helix}/f_{coil}
Poly-L-lysine (coil)	1900	7.34	.240	.70
Poly-L-glutamic acid (coil)	1900	7.2*	.265	.64
N-methyl acetamide	1900	7.4	—	—
L-alanine amide†	<1900	4.9	—	—
N-acetyl-L-alanineamide†	1900	11.2	—	—
"Dipeptide" amide	—	6.3	—	—

* Imahori and Tanaka (*12*).
† These samples were not dried before measurement; therefore their values of ϵ_m and f may be low.

strengths that were measured to the amide absorption. The NH_2 and NH_3^+ groups in polylysine have an extinction coefficient of less than 150 at 1900 Å. The COOH absorption is negligible, but the tail of the COO^- absorption at 1730 Å (*2*) may be as high as 1000 at 1900 Å (see also *12*). Although the COO^- contributes 14 per cent to ϵ_{1900} for polyglutamate, it contributes much less to the oscillator strength. Scattering corrections (which are small) were not made because anomalous scattering in the absorption region (*13*) is difficult to appraise quantitatively. The values for the simple amides are included to show that the peptide group in the random-coil form has the same absorption as an analogous single amide group. It was also found that the absorption of the simple amides was essentially independent of pH. The *N*-methyl acetamide

$$(CH_3CONHCH_3)$$

can be compared directly with the polymers. To find the comparable absorption of *N*-acetyl alanine amide $[CH_3CONHC^*H(CH_3)CONH_2]$ we subtract the absorption of alanine amide $[NH_2C^*H(CH_3)CONH_2]$. This is done because an unsubsituted amide ($CONH_2$) absorbs less than a substituted amide (CONHR). The result is listed as "dipeptide amide" in Table 13.1.

To calculate the absorption of the helix relative to that of a single amide group (which is the same as the amide absorption of the random coil) we need to know the spectrum of the amide (*10, 11, 2*) and the geometry of the helix (*23*). These data are shown in Fig. 13.6 and Table 13.2 for the polyalanine α-helix. We choose polyalanine as our model because the position of each nucleus is fixed. The position of each group or transition is specified in a cylindrical coördinate system by a, γ, z. The direction of each transition is specified in polar coördinates by θ, ϕ. Further details of these specifications are given by Tinoco and Woody (*30*). Currently available estimates of the parameters specifying three amide transitions are shown in Fig. 13.6. The remaining amide transitions are approximated by a polarizability centered at the amide N. It is assumed that the two components of the polarizability in the plane of the amide ring are equal, but that they are 25 per cent larger than the component perpendicular to the ring (*15*). The CH_3 and H polarizabilities are assumed to be symmetrical. The values of polarizability (α) given in Table 13.2 are the average polarizabilities at 1900 Å which were estimated from sodium *D* line polarizabilities.

TABLE *13.2*—The parameters specifying the geometry and polarizability of the groups in polyalanine

Group	a(Å)	γ(rad)	z(Å)	θ(rad)	ϕ(rad)	α(Å³)
$NV_1(\pi\pi^*)$	1.49	1.03	−.007	0.900	4.55	—
$n'\pi^*$	1.76	1.46	−.81	1.57	2.56	—
$n\sigma$	1.76	1.46	−.81	1.82	5.13	—
amide	1.57	.508	.86	—	—	4.25
CH_3(R.H.L.)	3.20	−.337	.83	—	—	3.05
H(R.H.L.)	3.08	.222	−.45	—	—	.585
CH_3(L.H.L.)	3.44	.281	−.64	—	—	3.05
H(L.H.L.)	2.90	.262	−.59	—	—	.585

ABSORPTION FREQUENCY

The splitting of the random-coil absorption into two helix bands is predicted if the resonance interaction of the amides is stronger than the perturbations of the vibrations and solvent collisions (*24*). Moffitt (*20*) calculated the splitting for a very long α-helix to be 2800 cm^{-1}. One helix band is polarized parallel to the helix axis and the other perpendicular. The ratio of the intensities of the bands is proportional to the square of the ratio of the parallel and perpendicular components of the transition moment.

The resonance interaction energy can be calculated in two ways. One can treat transition moments approximately as point dipoles, or more exactly one can use equivalent point monopoles distributed over the π system. For the 1900 Å, NV_1 (π—π^*) amide transition in polypeptides we have used both methods. In one calculation the point dipole (shown in the top panel of Fig. 13.6) interacts with 14 identical dipoles on each side of it. In the other calculation, point monopoles to give an equivalent dipole are placed on the *N*, *C*, and *O* atoms; these point charges then interact with 14 groups of charges on each side. The results are given in Table 13.3. The results of the more accurate point monopole calculation are also shown as the vertical lines in Fig. 13.3. The length of these lines is proportional to the calculated oscillator strengths of the parallel and perpendicular bands, $\nu_{\|}\mu_{\|}^2$ and $\nu_{\perp}\mu_{\perp}^2$. The agreement between calculation and experiment is satisfactory. A convincing proof, however, that the calculation is correct would need to come from measurements of the polarization of the absorption bands.

Table 13.3—The calculated absorption wave lengths for a polypeptide α-helix. Two different approximations were used

	Point dipole	Point monopole
splitting (cm^{-1})	1981	2590
$\lambda_{\|}$(Å)	1960	1982
$\lambda_{\perp}$(Å)	1887	1886
$\mu_{\|}(D)$	1.88	1.88
$\mu_{\perp}(D)$	2.40	2.40

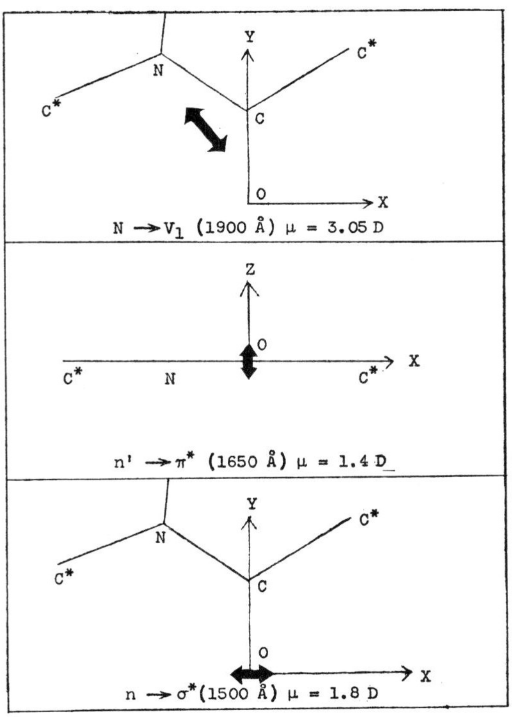

Fig. 13.6. The directions and magnitudes of the transition moments in an amide group.

ABSORPTION INTENSITY

The decrease of total absorption intensity of the helix relative to the random coil is caused by the interaction of the 1900 Å absorption with all the other absorption bands in the polymer. For each group the effect of all the bands in the far ultraviolet can be approximated by a polarizability. Point dipoles and point polarizabilities were used; each group interacts with 28 other groups. The results for the contribution of each transition or group to the hypochromism ($1-F_{helix}/f_{coil}$) are shown in Table 13.4. The CH_3 and H contributions were calculated both for right-handed and left-handed helices of L-amino acids to show that they are not the same (the two helix conformations are diastereoisomers) and because both conformations exist in nature. The agreement between calculation and experiment is again satisfactory. Although the calculation is by no means good enough to distinguish between right- and left-handed helices, it does indicate that if a more polarizable group replaced the methyl group a distinction might be made.

The dependence of absorption intensity on helix length can also be obtained from our calculations. Experimental results are not available, but the information is useful in trying to relate absorption intensity to helix content. If all helix regions in a molecule are long, then the amount of helix is a linear function of the absorption.

$$\% \text{ helix} = 100 \frac{f_{\text{coil}} - F}{f_{\text{coil}} - F_{\text{helix}}} \cong 100 \frac{\epsilon_{\text{coil}} - \epsilon}{\epsilon_{\text{coil}} - \epsilon_{\text{helix}}}$$

If all helix regions are not long, then the length distribution of the helices and the absorption as a function of length must be known. The calculated results for the relative absorption as a function of helix length are shown in Fig. 13.7. It is encouraging to notice that one turn of the helix has 80 per cent of the hypochromism of a very long helix. This means that the absorption will be directly proportional to the amount of helix even for short helices.

Table 13.4—The contribution of each group and transition to the decrease in absorption (hypochromism) of polyalanine α-helix

Group	$1\text{-}F_{\text{helix}}/f_{\text{coil}}$
$n'\pi^*$ (1650 Å)	0.097
$n\sigma$ (1500 Å)	0.070
amide	0.224
CH_3 (R.H.L.)	−0.032
H(R.H.L.)	0.007
CH_3(L.H.L.)	0.038
H(L.H.L.)	−0.005
R.H.L.-Polyalanine	0.366
L.H.L.-Polyalanine	0.424
Polylysine (expt.)	0.30
Polyglutamic acid (expt.)	0.36

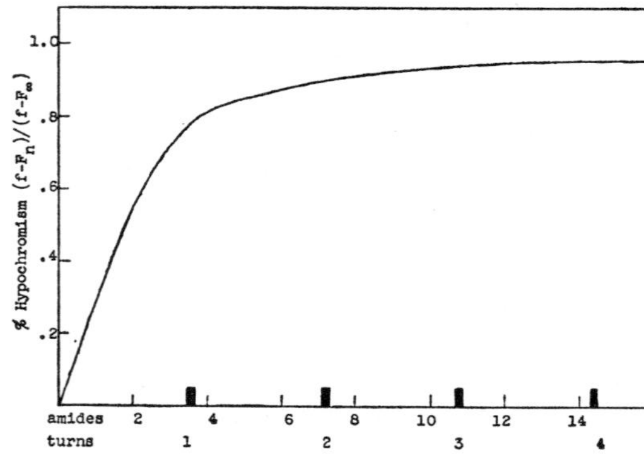

Fig. 13.7. The calculated decrease in intensity (hypochromism) for polyalanine α-helix as a function of length.

Summary

Good agreement between theory and experiment is found for the α-helix absorption. Both the helix intensity and the helix band splitting can be interpreted quantitatively in terms of interactions among the amide and side-chain transition dipoles and polarizabilities. Predictions about the polarization of the helix bands and the helix-length dependence of the absorption intensity have been made.

Acknowledgment

Supported in part by research grant A-2220 from the National Institute of Arthritis and Metabolic Diseases, U.S. Public Health Service, by an unrestricted grant from Research Corporation, and in part by contract with the U.S. Air Force through the Air Force Office of Scientific Research of the Air Research and Development Command.

References

1 J. B. Applequist, Ph.D. thesis, Harvard (1959).
2 E. E. Barnes and W. T. Simpson, to be published, 1961.
3 N. S. Bayliss and E. G. McRae, *J. Am. Chem. Soc.* **74,** 5803 (1952).
4 N. S. Bayliss and E. G. McRae, *J. Phys. Chem.* **58,** 1002, (1954).
5 L. G. S. Brooker *et al.*, see the series of articles in *J. Am. Chem. Soc.* beginning with **73,** 5326 (1951)
6 C. A. Coulson, *Proc. Roy. Soc.* **255 A,** 69 (1960).
7 P. Doty, *J. Polymer Sci.* **49,** 129 (1961).
8 J. G. Foss, *Biochim. Biophys. Acta* **47,** 569 (1961).
9 A. N. Glazer and E. L. Smith, *J. Biol. Chem.* **235,** PC43 (1960).
10 J. S. Ham and J. R. Platt, *J. Chem. Phys.* **20,** 335 (1952).
11 H. D. Hunt and W. T. Simpson, *J. Am. Chem. Soc.* **75,** 4540 (1953).
12 K. Imahori and J. Tanaka, *J. Mol. Biol.* **1,** 359 (1959).
13 P. Latimer and E. Rabinowitch, *Arch. Biochem. Biophys.* **84,** 428 (1959).
14 S. J. Leach, and H. A. Scheraga, *J. Biol. Chem.* **235,** 2827 (1960).
15 C. G. LeFevre and R. J. W. LeFevre, *Revs. Pure and Appl. Chem.* (*Australia*) **5,** 261 (1955).
16 H. C. Longuet-Higgins and J. A. Pople, *J. Chem. Phys.* **27,** 192 (1957).

17 H. C. Longuet-Higgins. *Proc. Roy. Soc.* **255A,** 63(1960).
18 E. G. McRae, *J. Phys. Chem.* **61,** 562 (1957).
19 E. G. McRae, *Australian J. Chem.* **14,** 329 (1961).
20 W. Moffitt, *J. Chem. Phys.* **25,** 467 (1956).
21 W. Moffitt and A. Witkowski, *ibid.* **33,** 872 (1960).
22 Y. Ooshika, *J. Phys. Soc. Japan* **9,** 594 (1954).
23 L. Pauling *et al., Acta Cryst.* **8,** 853 (1955).
24 D. L. Peterson and W. T. Simpson, *J. Am. Chem. Soc.* **79,** 2375 (1957).
25 W. Rhodes, *ibid.* **83,** 3609 (1961).
26 G. W. Robinson, *J. Mol. Spectroscopy* **6,** 58 (1961).
27 W. T. Simpson and D. L. Peterson, *J. Chem. Phys.* **26,** 588 (1957).
28 I. Tinoco, Jr., *J. Am. Chem. Soc.* **82,** 4785 (1960).
29 I. Tinoco, Jr., *J. Chem. Phys.* **33,** 1332 (1960).
30 I. Tinoco, Jr., and R. W. Woody, *ibid.* **32,** 461 (1960).
31 I. Tinoco, Jr., *ibid.* **34,** 1067 (1961).
32 S. Yanari and F. A. Bovey, *J. Biol. Chem.* **235,** 2818 (1960).

Discussion of Paper 13: Conformation and Light Absorption

AKIYOSHI WADA

In regard to the *p*H dependence of the ultraviolet extinction coefficient of poly-L-glutamic acid (PGA) in aqueous solution given originally by Drs. Imahori and Tanaka (*J. Mol. Biol.* **1,** 359 [1959]), and here by Dr. Tinoco, it seems to be not completely established that the whole change in the extinction coefficient is produced by the helix-coil transition. In Fig. 13.8 the change of optical rotation, $[\alpha]_{546}$, of PGA with *p*H which were given by Drs. Blout and Idelson (*J. Am. Chem. Soc.* **80,** 4909 [1958]) and by myself (*Mol. Phys.* **3,** 409 [1960]) independently, is shown superimposed upon the extinction coefficient and titration curves given by Drs. Imahori and Tanaka.

According to the Moffitt-Yang-Doty-Imahori equation the optical rotation at any fixed wave length of $\lambda(\gg\lambda_o)$ can be given in terms of the helix content x_H as follows:

$$[\alpha]_\lambda = A + Bx_H$$

where A and B are constants which may be determined experimentally. So, at least as far as the Moffitt equation is concerned, the optical rotation should be a good measure of the helix-coil transition. Assuming this to be the case, then it appears from the comparison with the extinction coefficient curve that only the plateau in the ϵ_{1900} versus *p*H plot between *p*H 4.5 and 6.0 corresponds to the helix-coil transition and not the whole change from *p*H 3.0 to 8.0.

Since the extinction coefficient curve below *p*H 4.5 and above *p*H 6.0 have good parallel relation with the titration curve, the change of ϵ_{1900} in these regions may reflect a charge effect, or something like that, localized in the residues but not a change in the whole molec-

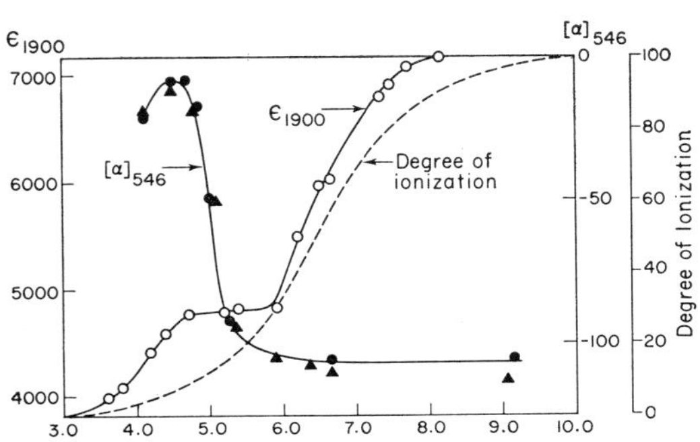

Fig. 13.8. The *p*H dependence of the extinction coefficient, optical rotation, and degree of ionization of PGA in aqueous solution.

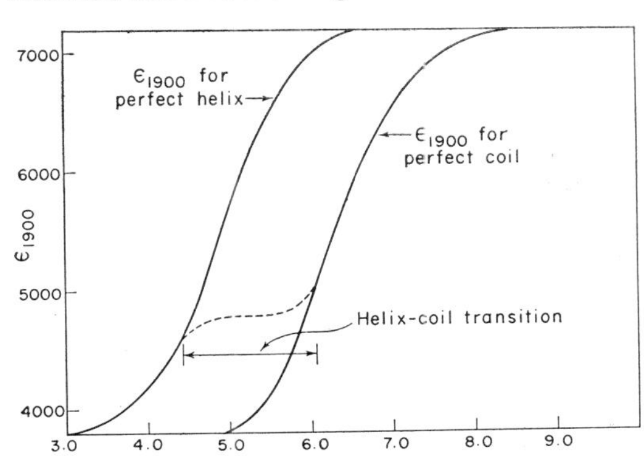

Fig. 13.9. Hypothetical plots of the extinction coefficients (ϵ_{1900}) for the perfect helix and perfect coil, to illustrate the plateau produced by the transition between them.

ular conformation. A similar plateau in the extinction coefficient curve can also be observed just in the *p*H region of the helix-coil transition of poly-L-lysine measured by Dr. Tinoco, though it is not so clear as in the case of PGA.

In conclusion, it seems reasonable to consider that the extinction coefficient of PGA changes with the degree of ionization or *p*H as shown in Fig. 13.9 where the change for the perfect helix and the perfect coil are shown by rigid lines. The helix-coil transition will then give a plateau as shown by the dotted line between them. In this case, the relation: $\epsilon_{helix} > \epsilon_{coil}$ is given at the same *p*H.

JON APPLEQUIST

I would like to expand on Dr. Wada's remarks as they apply to poly-L-lysine. As Dr. Tinoco has pointed out, there is a disparity in the appearance of the transition as observed by extinction coefficient and optical rotation measurements, as may be seen by comparison of Fig. 13.5 in Dr. Tinoco's paper with Fig. 14.8 of our paper (Applequist and Doty, this book, Paper 14). The shift in the midpoint of the transition for poly-L-lysine to a *p*H one unit lower than that observed in the optical rotation experiments seems like a rather large effect to attribute to the difference in counterions in the two experiments. This explanation would most likely require that the Cl^- ions be bound to polylysine more strongly than Br^- ions, but this is the reverse of the order of the binding strengths of these ions to cationic resins (R. Kunin, "Ion Exchange Resins," 2nd ed., Wiley, New York, 1958, p. 60). Our sedimentation experiments (Paper 14) in 1 M NaCl and 1 M NaBr were consistent with the existence of normal binding order for cationic polylysine.

This leads one to suspect that the absorption intensity of the band at 190 mμ and the optical rotation do not provide measures of the same quantity, particularly in view of Dr. Wada's remarks. The discrepancy for poly-L-lysine is very similar to that for poly-L-glutamic acid, as may be seen from the fact that the change in integrated absorption intensity in Fig. 13.5 follows rather closely the titration curve of poly-L-lysine in Paper 14, Fig. 14.8, with a bare suggestion of a plateau in the former, as pointed out by Dr. Wada. This coincidence may be fortuitous to some extent, since the absorption experiments were carried out at much higher dilution than the titration measurements, and some concentration dependence is expected for both effects in the absence of added salt. However, it does suggest that the discrepancy between the optical rotation and ultraviolet absorption results may be due to an effect of the charged groups on the ultraviolet absorption which is independent of the change in conformation of the molecule.

I. TINOCO, JR.

The comments of Drs. Wada and Applequist emphasize the need for a valid measure of helix content. Dr. Wada assumes that the optical rotation measures the amount of helix and therefore that the optical density changes markedly with the state of ionization. At present it seems more likely to me that the optical density measures the conformation and that the optical rotation does not. Dr. Wada postulates that ϵ_{1900} for an amide in polyglutamic acid (as helix or coil) changes by nearly a factor of 2 as the carboxyl ionizes (Fig. 13.9). A similar effect is postulated for polylysine. However, this effect is not seen in analogous simple molecules. Both $NH_2C^*H(CH_3)CONH_2$ and $CH_3CONHC^*H(CH_3)COOH$ show only a small change in amide absorption on ionization of the carboxyl or amino group. Of course there may be a very large effect, in the absorption polyelectrolytes, but neither experimental nor theoretical evidence exists at present for it. The semi-quantitative explanation given for hypochromism in polypeptides also fits the data for polynucleotides. In nucleic acids there is a change in conformation and light absorption with no change in ionization of the groups in the polymer.

A possible explanation for the discrepancy between optical density and optical rotation measures of helix conformation is their different dependence on length. If the optical rotation is due to interaction among the amide groups, it will be much more sensitive to the number of amides in an ordered array than will optical density. That is, rupture of 10 per cent of the hydrogen bonds in the helix may lead to a large change in optical rotation. Thus a sharp transition in the optical rotation would not necessarily correspond to a similar transition in the polymer.

It is clear that more experiments are needed to determine the actual transitions in the polypeptides and to learn what the optical properties measure.

K. IMAHORI and J. TANAKA

Ths discussion of Dr. Wada could be focused on two points: (1) The helix-coil transition observed by the optical rotation is not in agreement with that obtained by ultraviolet absorption. (2) It could be expected that the extinction coefficient of helical conformation is larger than that of random coil.

For the first point, we should say our rotation data are not in contradiction to those of ultraviolet absorption. One should not neglect our ORD results at pH 5.3 (Fig. 4 of our paper, *J. Mol. Biol.* **1,** 359 [1959]). As was cited in the text the helix content of PGA at this pH is calculated to be as high as 71 per cent from both slope and intercept [$(\alpha)_{546} \approx -22°$]. We can say that our data on the helix-coil transition obtained by two ways are in fairly good agreement. But, our result is in contradiction to that of Dr. Wada. According to Fig. 8.1 of Dr. Wada, $(\alpha)_{546}$ at pH 5.3 should exist around $-94°$ and the helix is almost completely melted at this pH. We cannot explain the contradiction of these results. But, if we check carefully the transition curve of Dr. Blout to which Dr. Wada referred, we can see that Dr. Blout's curve is somewhat shifted to the alkaline side of Dr. Wada's. The probable explanation may be that the transition curve is quite sensitive to the molecular weight, polydispersity, degree of debenzylation, etc., and the condition of the measurement (concentration, ionic strength, temperature, etc.).

For the second point, Dr. Wada attributed the large change of the extinction coefficient to the degree of ionization of carboxyl group. But, as was explained in our text, we have measured the ionization effect and corrected the data taking the ionization effect into consideration. According to Dr. Wada's Fig. 8.2 he expects the extinction coefficient of the carboxyl group to be 3800 for an unionized form and 7200 for an ionized form. This expectation is quite extraordinary since, according to our data, acetic acid has an extinction coefficient of 40 and 1000 for its unionized form and ionized forms, respectively. These values are in agreement with the data of Ley and Arends (*Z. Phys. Chem.* **B17,** 177 [1932]) and Rusoff, Platt, Klevens, and Burr (*J. Am. Chem. Soc.* **67,** 673 [1945]). We have also measured the extinction coefficient of polyacrylic acid (DP 10,000) both in unionized form and ionized form. The extinction coefficient of the former form was about 75 (2150 Å) and that of the latter was 560 (1880 Å). Thus, the possibility of hyperchromism of carboxyl group by a polymer formation is excluded.

In conclusion, we should say that the reexamination of the rotational work, especially that of rotatory dispersion, at the transition region is badly needed using the same sample and under the same condition with those for ultraviolet absorption. But, we do not agree with Dr. Wada's suggestion that the large change of the extinction coefficient could be attributed to the ionization of carboxyl group.

I. TINOCO, JR.

I think that Dr. Wada implied an indirect effect of the carboxyl ionization on the amide absorption, instead of a direct effect of the carboxyl absorption.

TATSUO MIYAZAWA

In connection with the pH dependence of the extinction coefficient of polyglutamic acid, I would like to present the unpublished results (T. Miyazawa and E. R. Blout) of the optical rotatory dispersion measurements. The pH of the solution was adjusted by adding hydrochloric acid to the aqueous solution of sodium

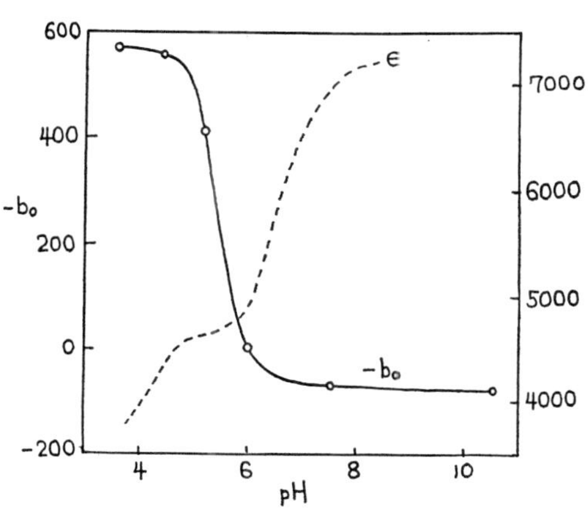

Fig. 13.10

poly-L-glutamate (R-4273-112) and the solute concentration was 0.21–0.23 per cent by weight. The Moffitt plot was made using the value of $\lambda_o = 212$ mμ. The values of b_o thus obtained are plotted versus *p*H in Fig. 13.10. It may be seen that, in view of the values of b_o, the helix-coil transition takes place in a narrow region of *p*H = 5–6. In the same figure is reproduced the *p*H dependence of the extinction coefficients measured by Imahori and Tanaka. The helix-coil transition appears to be responsible for the plateau observed between *p*H = 5–6.

W. B. GRATZER

I should like to mention some very preliminary results which we have obtained on the polarization spectra of α-helical polypeptides (W. B. Gratzer *et al., Proc. Nat. Acad. Sci.* **47,** 1785 [1961]). Oriented films of poly-L-alanine and poly-L-glutamic acid were prepared on quartz discs and polarization spectra were measured on a Beckman DK-2 recording spectrophotometer with a polarizing prism in the sample beam. All the films which we have prepared so far have a rather high scattering background, for which an approximate correction can be made by a logarithmic extrapolation into the region of absorption. With the polarization vector parallel to the helix axis, an absorption profile with a prominent feature at 205 mμ is observed, corresponding to the shoulder previously noted by Rosenheck and Doty (to be published; quoted by P. Doty in *J. Polymer Sci.* **49,** 148 [1961]) in poly-L-glutamic acid, and also described by Professor Tinoco. When the light is polarized at right angles to the helix axis, this feature is absent and a single strong peak appears just above 190 mμ.

The directions of polarization of these two bands agree with the predictions of Moffitt (*Proc. Nat. Acad. Sci.* **42,** 736 [1956]). Moreover, the separation of the maxima in the polarization spectra is about 3300 cm^{-1}, which accords closely with Moffitt's predicted value of 2800 cm^{-1}.

The effective absorbances, when correction has been made for background effects, are low in these experiments, and the orientation may be very imperfect. It is not therefore at this stage possible to make meaningful calculations of extinction coefficients and oscillator strengths. We are at the moment experimenting with better methods of orientation.

PAPER 14

α-Helix Formation in Poly-ε-Carbobenzoxy-L-Lysine and Poly-L-Lysine

JON APPLEQUIST and PAUL DOTY

The purpose of the work reported here is to contribute to present understanding of the conditions affecting the stability of the α-helical conformation of synthetic polypeptides in solution. An interest in poly-L-lysine exists because of its water solubility, and hence because of the bearing studies of this type may have upon the problems of protein secondary structures in aqueous solution. Attention here is focused on effects of *p*H, ionic strength, and temperature in bringing about a helix-coil transition in poly-L-lysine, and on the mechanism of this transition.

Poly-ε-carbobenzoxy-L-lysine (PCBL) enters this study as the precursor of poly-L-lysine in the synthetic sequence, due to the fact that the ε-amino groups must be blocked during polymerization. PCBL is insoluble in water, showing solubility characteristics rather similar to those of the more thoroughly characterized material, poly-γ-benzyl-L-glutamate (PBLG) (*1*). For this reason it was considered desirable to obtain information as to the conformation of PCBL in a solvent known to favor the helical form for PBLG, in order to discover any effects due to the change in side-chain structure. The solvent chosen was *N,N*-dimethylformamide (DMF).

Experimental

Materials.—The *N*-carboxyanhydride of ε-carbobenzoxy-L-lysine was obtained by reaction of phosgene with ε-carbobenzoxy-L-lysine, which had been prepared by the method of Noguchi and Katchalski (*2*). PCBL was obtained on polymerization of this material with sodium methoxide initiation in dioxane solution, following Blout and Karlson (*3*). In addition, a number of samples of PCBL, whose preparation codes (Table 14.1) are prefixed with MI-, were donated by Dr. Martin Idelson. Cleavage of the carbobenzoxy groups was accomplished by the HBr-glacial acetic acid treatment of Ben-Ishai and Berger (*4*). The resulting poly-L-lysine hydrobromide showed no absorption peak in the 250–260 mμ range, indicating complete removal of the carbobenzoxy groups.

Optical rotation.—Rotation measurements were made with a Rudolph Photoelectric Spectropolarimeter, Model 200 SV/80/650-610, using a Sylvania concentrated zirconium arc lamp as light source. The wave-length range covered in dispersion studies was about 325–650 mμ. PCBL solutions were at concentrations of 0.5–1 per cent, and poly-L-lysine solutions at 0.1–0.5 per cent. The precision in most of the specific rotation values reported is about ±1 per cent.

Light scattering.—Light-scattering measurements were made with a Brice-Phoenix Model 1000 photometer at 436 mμ at angles from 30° to 135° with respect to the incident beam. An Erlenmeyer type of scattering cell was used, and was calibrated as described by Phillips (*5*). Solvents and DMF solutions were clarified by filtration through ultrafine sintered glass. Aqueous solutions were clarified by centrifugation at 100,000 *g* for 90 minutes. Refractive index increments were measured with a Brice-Phoenix visual differential refractometer at 436 mμ. For PCBL in DMF the result is $dn/dc = 0.123 \pm 0.002$ cc/gm. For poly-L-lysine hydrochloride in 1.0 M NaCl, $dn/dc = 0.245 \pm 0.002$ cc/gm, where the concentration is expressed in terms of the hydrochloride.

Viscosity.—The viscosities of solutions were

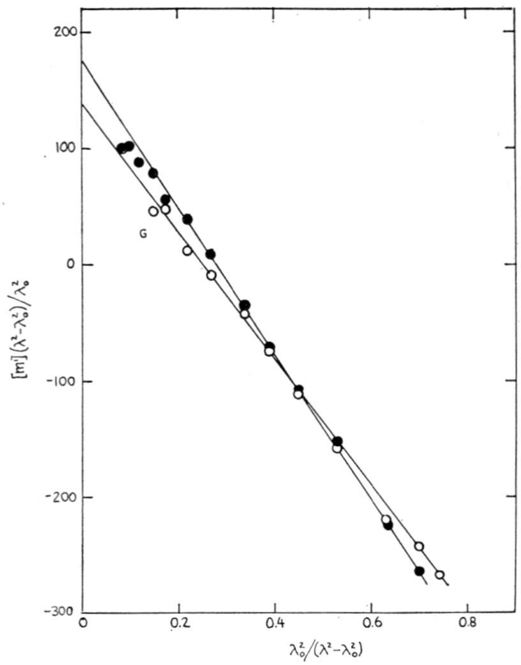

Fig. 14.1. Optical rotatory dispersion of two samples of PCBL in DMF. For sample J, $b_o = -625$. For sample G, $b_o = -545$.

measured as a function of concentration in Ubbelohde viscometers having solvent flow times of at least 90 sec. The temperature of the surrounding bath was 24.4°, regulated to ±0.01°.

Flow birefringence.—The extinction angle of flow birefringence was measured in the apparatus manufactured by the Rao Instrument Company, Brooklyn, New York. Measurements were made over the gradient range of about 2000 to 8000 sec^{-1}. Measurements were found to be essentially independent of concentration below 0.2 gm/dl, and results reported here were obtained in that range.

Sedimentation.—Sedimentation velocities were measured in a Spinco Model E Analytical Ultracentrifuge operating at 59,780 rpm, using schlieren optics. The partial specific volume $\bar{v}$ of poly-L-lysine hydrochloride in 1.0 M NaCl was found to be 0.722 ± 0.005 cc/gm by pycnometric density measurements. For the polylysine cation, one obtains $\bar{v} = 0.781$ cc/gm, taking into account the contribution of 0.507 cc/gm to the total partial specific volume by the Cl^- ions (*6*).

Concentration determination.—Concentrations of PCBL solutions were determined by dry-weight measurements. Concentrations of poly-L-lysine solutions are based on nitrogen content as determined by the micro-Kjeldahl method (*7*) or by the biuret spectrophotometric method (*8*). Such concentrations are expressed in terms of free polylysine, unless specified as including the associated hydrogen halide.

Potentiometric titrations.—Polylysine solutions at a concentration of 0.015 N were titrated under a nitrogen atmosphere with 1 M KOH delivered from a micrometer ultramicroburet. The *p*H was followed with a Beckman Model G *p*H Meter using a general purpose glass electrode. Reverse titration with 1 M HBr was found to produce titration curves identical within experimental error.

*Poly-ε-carbobenzoxy-*L*-lysine*

Rotatory dispersion.—Moffitt and Yang (*9*) showed that the optical rotatory dispersion of helical polypeptides follows the equation

$$[m'] = \frac{a_o\lambda_o^2}{\lambda^2 - \lambda_o^2} + \frac{b_o\lambda_o^4}{(\lambda^2 - \lambda_o^2)^2} \quad (1)$$

where

$$[m'] = \frac{3}{n^2 + 2}\,\frac{M_o}{100}\,[\alpha] \quad (2)$$

$[m']$ is the effective residue rotation, $[\alpha]$ is the specific rotation in cc deg gm^{-1} dm^{-1}, λ is the wave length at which the rotation is measured, and a_o, b_o, and λ_o are constants. The value $\lambda_o = 212$ mμ has been found to fit the dispersion for all polypeptides tested. M_o is the residue weight and n the index of refraction of the solution. For PCBL in DMF we have from equation (2) $[m'] = 1.945\,[\alpha]$. The dispersion data for two samples of PCBL in DMF are plotted in the usual way in Figure 14.1, giving straight lines which, according to equation (1), have slopes equal to b_o and vertical intercepts equal to a_o. Again, λ_o was set equal to 212 mμ. For sample J we find $a_o = 170$ and $b_o = -625$, in good agreement with the values of these parameters for other polypeptides believed to have the α-helical conformation (*9*). The paramter b_o has been found to be particularly characteristic of the helix backbone, and the value found here would indicate that the ma-

Table 14.1—Light-scattering and viscosity data on poly-ε-carbobenzoxy-L-lysine in dimethylformamide

Sample	Preparation code	M_w	Degree of polymerization	Virial coeff. $B\times10^4$ (cc/gm²)	$[\eta]$ (dl/gm)	Scattering radius (Å)
A	MI-II-439-50	57,000	216	2.6	0.35	170
B	MI-II-126-50	147,000	560	—	1.02	330
G	MI-II-406-25	329,000	1260	2.1	2.3	431
J	MI-II-358-200	775,000	2950	1.9	4.1	635
K	JA-III-19	950,000	3610	2.6	4.25	875

terial is 100 per cent helical. The dispersion data for sample G give the values $a_o=135$ and $b_o=-545$. Since for randomly coiled polypeptides $b_o=0$, it might appear that some randomly coiled material is present in this sample. However, in view of the scatter in the data for this case, this conclusion is only tentative.

Light Scattering.—Light-scattering measurements as a function of scattering angle θ and solute concentration are used to calculate the weight-average molecular weight M_w and molecular dimensions of the solute. For details of the method of calculation, one is referred to the recent review by Stacey (*10*). Data obtained from the customary Zimm plots for five samples of PCBL in DMF are shown in Table 14.1.

The two samples of highest molecular weight showed great enough angular dissymmetry of scattering to yield information about the molecular shape. Doty and Steiner (*11*) have tabulated theoretical values of the particle scattering factor $P(\theta)$ for various simple molecular models. Curves of $1/P(\theta)$ vs. $\sin^2(\theta/2)$ taken from these tables are shown in Figure 14.2. Curves 1 and 3 of the figure are the theoretical curves for a random coil with $R=2140$ Å (root-mean-square end-to-end distance) and for a rigid rod of length 3030 Å, respectively. Both curves have the same limiting slope as the experimental curve for sample K; therefore, both models have the correct scattering radius* for this sample. However, the large deviations of the experimental data at high angles rule out either of these models as the correct one. Downward deviations of the experimental data from a theoretical curve can be the result of polydispersity (*12*), but a polydisperse random coil is inconsistent with the rotatory dispersion behavior and with flow birefringence evidence

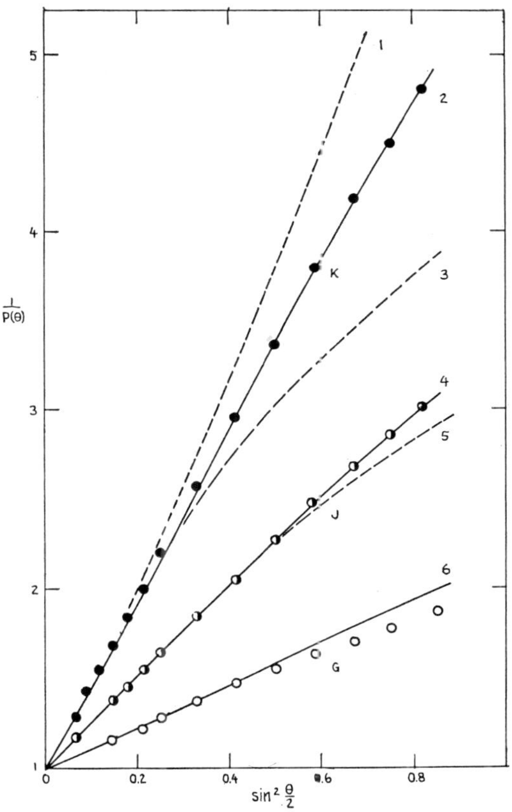

Fig. 14.2. Theoretical and experimental particle-scattering factors for PCBL samples G, J, and K. Numbered curves correspond to the following: (1) Random coil with $R=2140$ Å. (2) Worm-like model with $a=262$ Å, $x=35$. (3) Rigid rod of length 3030 Å. (4) Worm-like model with $a=262$ Å, $x=21$. (5) Rigid rod of length 2200 Å. (6) Rigid rod of length 1490 Å.

* The term "scattering radius" used here refers to the same quantity designated erroneously as the "radius of gyration" in most current literature on light scattering.

presented below. The actual molecular shape is therefore probably a rod with some random-coil character, due to flexibility of the rod or to occasional bends. To test this hypothesis, we may resort to the "worm-like" model of Kratky and Porod (*13*). They represent a polymer molecule as a continuously curved filament whose stiffness is characterized by the persistence length a, which turns out to be just half the length of the statistical unit used by Kuhn to characterize stiff polymer chains (*14*). Application of the Kratky-Porod model to the present data is made possible by Peterlin's calculation of its particle scattering factor (*15*) in terms of the parameters a and x, where $x = L/a$, L being the length of the fully extended chain. Peterlin's equation for $P(\theta)$ is

$$P(\theta) = e^p \sum_{j=0}^{\infty} \frac{(-p)^j}{j!} F[(p + j)x] \qquad (3)$$

where

$$F(x) = \frac{2}{x^2}(x - 1 + e^{-x}) \qquad (4)$$

and

$$p = \frac{16\pi^2 a^2}{3\lambda^2} \sin^2(\theta/2)$$

Here λ is the wave length of the incident light in the scattering medium. Peterlin has shown that the shape of the $1/P(\theta)$ vs. $\sin^2(\theta/2)$ curve is sensitive to the value of x when $1/P(\theta)$ becomes larger than about three. The initial slope of the curve is a function of a and x. Using equation (3), then, neglecting all terms in the sum higher than the third, curve 2 in Figure 14.2 is calculated for a chain with $x = 35$, $a = 262$ Å. The agreement with the experimental data is seen to be rather precise. The best value of x is determined to within about 10 per cent, although an unknown effect of polydispersity could make this value too low. The length of the chain is $L = ax = 9160$ Å, which when combined with the molecular weight gives a linear density of 104 mol. wt. units per Å. Since the theoretical value for the α-helix is 175, it does not appear that a model of an α-helix with some flexibility introduced is adequate to explain the behavior completely.* This point will be discussed further after some data from flow birefringence studies have been presented.

It is seen in Figure 14.2 that the scattering factor for sample J is barely in the range where a distinction can be made among different models. Thus curve 4 is calculated from equation (3) for a chain with $x = 21$ and $a = 262$ Å. Curve 5 corresponds to a rigid rod of length 2200 Å, and is clearly a less satisfactory representation of the experimental data than curve 4. The latter was found to be rather insensitive to the value of x chosen unless a was assumed to have the same value as for sample K. If this assumption is valid, then x should be proportional to the molecular weight, and should be about 28, from the data on sample K. The value found is a bit low, possibly due to polydispersity, but it is clear that a model with essentially the same properties of flexibility fits both of these samples reasonably well.

Curve 6 in Figure 14.2 is calculated for a rod of length 1490 Å. The agreement with the experimental data for sample G is reasonably good. The downward deviations of the latter may be due to polydispersity, but cannot be accounted for by flexibility. In this range, however, a chain of the flexibility indicated above would not be distinguishable from a rigid rod by this criterion. These data are, then, at least not inconsistent with the above model.

The Kratky and Porod treatment permits calculation of the mean square end-to-end distance R of the flexible molecules by means of the relation

$$R^2 = 2a^2(x - 1 + e^{-x}) \qquad (5)$$

Values of R for the samples studied are given in Table 14.2. For the three samples of lowest

Table 14.2—Data for the "wormlike" model

Sample	End-to-end distance R (Å)	Contour length L (Å)	Linear density M_w/L
A	590	590	93
B	1160	1160	127
G	1490	1490	221
J	1650	5500	141
K	2160	9160	104

* The zigzag chain model investigated by Hermans and Hermans (*J. Phys. Chem.*, **62,** 1543 [1958]) might also be used to fit the data. This possibility was not pursued extensively because of some features of the model which seem no less arbitrary than those of the Kratky-Porod model.

molecular weight R should be roughly the same as the length of a rod having the same scattering radius, if the stiffness observed at high molecular weight is also characteristic of these samples. The values of R and L in Table 14.2 for these samples are therefore both calculated as for a rigid rod.

Table 14.2 also includes values of the linear density (M_w/L) for all samples, to be compared with the value of 175 expected for the α-helix. The wide variation in values is beyond the range of probable experimental error, and this may be an indication that the applicability of the worm-like model at high molecular weight does not extend through the whole range of molecular weight. In this case, the values of L are in doubt for all samples except K and not too much meaning can be attached to the calculated linear densities.

Viscosity.—It is now possible to correlate the intrinsic viscosity of PCBL in DMF with the molecular weight. The relation between log $[\eta]$ and log M_w is seen in Figure 14.3 to be linear over part of the range, but shows pronounced downward curvature at high molecular weights. In the low region the slope is near unity, as is characteristic of free-draining coils (*16*). Thus it appears that a genuine rodlike shape is not present anywhere in this molecular weight region. The lower slope at higher molecular weight must be taken as indication of a solvent-entrapping nature of the polymer chains (*16*). Such behavior is surprising, considering the high helix content indicated by the rotatory dispersion, but is conceivable if the helix contains numerous bends.

The open circles in Figure 14.3 represent values of η_{sp}/c at $c = 0.2$ gm/dl in dichloroacetic acid (DCA). (These data were supplied with the samples by Dr. Idelson.) Evidence from other polypeptides has shown that the helical conformation is destroyed by this solvent (*1*). As in the case of PBLG, it is known that the sensitivity of the viscosity to molecular weight is decreased, due to the collapse of the helical structure. The effect in PCBL is less pronounced than in PBLG, but at least there is evidence that the helix is a significant part of the structure over the whole molecular weight range in DMF. The curvature of the plot in DCA may well be real, as such curvature was also detected in PBLG in the same solvent (*17*).

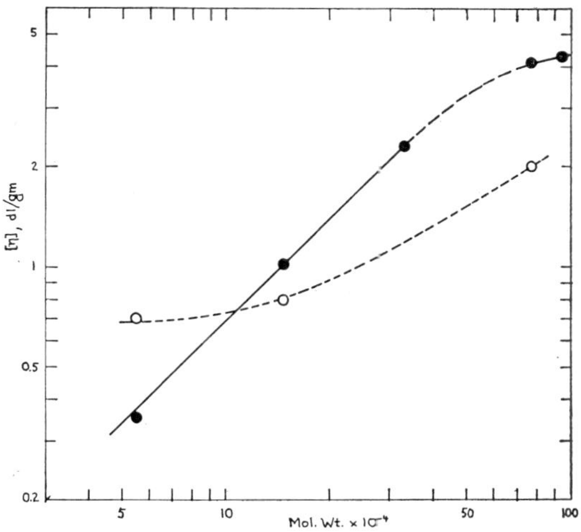

Fig. 14.3. Log $[\eta]$ versus log M_w for PCBL in dimethylformamide (●), and log η_{sp}/c for 0.2 per cent solution in dichloroacetic acid (○).

Flow birefringence.—The angle between the direction of flow and the optic axis in a streaming, birefringent liquid is called the extinction angle χ and has been measured on a number of solutions of PCBL in DMF. χ has been tabulated as a function of G/Θ, the ratio of the flow gradient to the rotational diffusion constant, for ellipsoidal solute molecules by Scheraga, Edsall, and Gadd (*18*). Assuming the ellipsoidal model to be at least approximately valid, one may use the value of Θ obtained from the tables to calculate the molecular length R by means of the Perrin equation (*19*):

$$\frac{R^3}{8} = a^3 = \frac{3kT}{16\pi\eta\Theta}\left[2 \ln (2a/b) - 1\right] \qquad (6)$$

Here a and b are the semi-major and semi-minor axes, respectively, η is the solvent viscosity, k is the Boltzmann constant, and T is the absolute temperature. R is used as the symbol for the ellipsoid length because, as this model is applied to flexible rodlike molecules, the length should correspond roughly to the mean separation of the ends. The axial ratio a/b need be known only approximately for use in equation (6), and has been estimated for this purpose by means of Simha's equation relating the intrinsic viscosity to the axial ratio (*20*). Values of a/b ranging from 42 to 93 were obtained for the samples studied.

The gradient dependence of molecular length of a number of samples of PCBL in DMF is

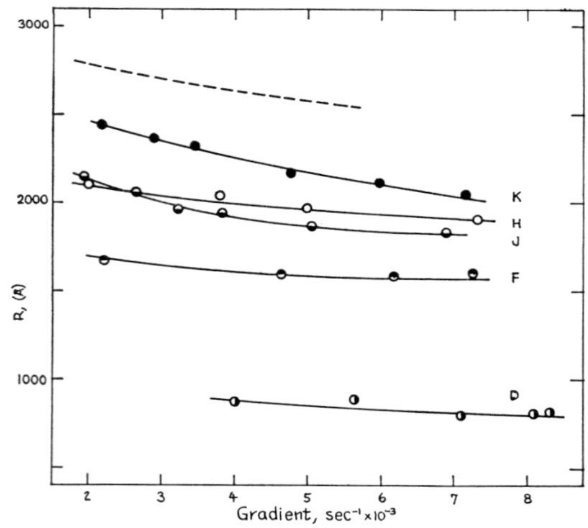

Fig. 14.4. End-to-end distance estimated from flow birefringence versus flow gradient for five samples of PCBL in DMF. Dashed curve shows data for native collagen by Boedtker and Doty (*21*).

illustrated in Figure 14.4. It is worth noting that the variation in length observed is roughly comparable to that found by Boedtker and Doty (*21*) for native collagen (dashed curve in Figure 14.4) which was shown by other methods to have a very narrow molecular weight distribution. This indication that the polydispersity in PCBL is low is consistent with expectations from the nature of the polymerization reaction (*3, 22*).

An over-all picture of the molecular weight dependence of the molecular size is seen in Figure 14.5, where values of R (taken at the high gradient limit in the case of flow birefringence) are plotted versus DP. It is seen that the flow birefringence and light scattering values are in reasonable agreement. (The agreement at high DP is much poorer if a rigid-rod model is assumed in calculating the light scattering R.) The solid curve was calculated for the Kratky-Porod model, using equation (5), with values of a and x obtained for sample K, and assuming x to be proportional to the molecular weight. The curve parallels the experimental data favorably. However, the rather consistent deviations in the low molecular weight range support suspicion that the model which applied well at high molecular weight is not as valid over the whole range. Nevertheless, the support for an extended but flexible molecule over the whole range is clear.

Discussion.—The evidence from rotatory dispersion has shown that the backbone structure of PCBL in DMF is basically the same as that deduced for PBLG in the same solvent. From the light scattering, viscosity, and flow birefringence behavior it is concluded that the molecules are indeed considerably stiffer than random coils, but appear to be more flexible than the helices of PBLG. The angular dependence of the particle-scattering factor and the molecular weight dependence of the mean molecular extension both support the Kratky-Porod "worm-like" model for the molecule. Such a model is best reconciled with the helical structure by the assumption that there are imperfections in the helix in the form of sequences of broken hydrogen bonds. Such imperfections need not affect significantly the number of residues in the helical form, but would profoundly affect the shape of the molecule, giving rise to occasional bends in the axis. Such a bent helix would not correspond exactly to the wormlike model to which PCBL has been compared here, and this may help account for the failure to obtain complete agreement with the wormlike model. The agreement is good enough to demonstrate the existence of imperfections that were not apparent in the structure

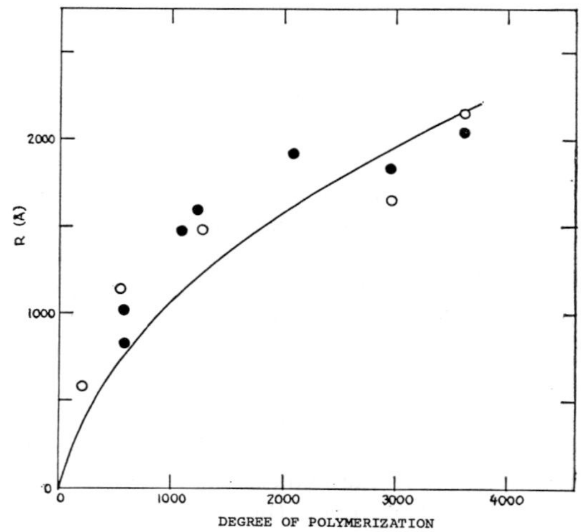

Fig. 14.5. End-to-end distance of PCBL in DMF determined by light scattering (○), and by flow birefringence (●). Solid curve calculated for wormlike model with persistence length 262 Å.

of PBLG in the same solvent (*1*).

The extent of flexibility of the PCBL structure is perhaps better understood on comparison of its persistence length of 262 Å with values found for other "stiff" chain molecules. For deoxyribonucleic acid a persistence length of 500 Å has been found by Peterlin (*15*), and for cellulose trinitrate a value of 117 Å is given by Doty (*23*). The light scattering and hydrodynamic behavior of both of these materials indicates that the molecular shape is that of a free-draining coil.

One might speculate as to the source of the flexibility of the PCBL helix relative to the PBLG helix in the same solvent. This may be the result of a packing problem among the amino-acid side chains. The side chains in PCBL are somewhat bulkier than in PBLG, and their freedom of motion in the helix is somewhat more restricted due to their proximity to each other. Such an effect might significantly influence the stability of the intramolecular hydrogen bonds, leading to more frequent breaks in the PCBL helix than in the PBLG helix.

Support for the conclusion that the PCBL helix is less stable than the PBLG helix comes also from recent work of Fasman, Idelson, and Blout (*24*) on the helix-coil transition induced by addition of dichloroacetic acid to solutions of the polypeptides in chloroform. It was found that the helical structure of PCBL disappears when the solvent contains 36 per cent (by volume) of dichloroacetic acid, while 68 per cent dichloroacetic acid is required to bring about the transition in PBLG (*25*). It is interesting that the phenomenon studied by these workers using optical rotation as the main tool was very sensitive to structural features which were apparent in the present study only when macromolecular measurements were extended to materials of the highest attainable molecular weights.

Poly-L-lysine

Description of the sample.—The results reported in this section were obtained on a single preparation of poly-L-lysine, designated III-23. This was obtained as the hydrobromide by decarbobenzoxylation of PCBL sample K, having a DP of 3610, as determined by light scattering. A special effort was made to produce a polylysine sample of high molecular weight in order that the hydrodynamic properties would be sensitive to the helix-coil transition.

Poly-L-lysine hydrobromide is normally soluble in water over the whole *p*H range. Solutions of *p*H above 11 appear to be metastable, as precipitation occurs gradually, although not too rapidly for a number of measurements to be made. After the sample had been stored as a lyophilized solid for three months, however, it was found to dissolve only very slowly in water, and even then to show an anomalously high viscosity which decreased slowly with time. Such solutions gave copious precipitation above *p*H 11. Only after such solutions had stood at neutral *p*H for several weeks were they found to reproduce the viscosity and solubility behavior of the freshly prepared poly-L-lysine. Evidently the solid is converted on standing to a form less easily disaggregated by water. The phenomenon is very likely similar to the α-β transformation which occurs on aging of poly-L-lysine films at low humidity, reported by Blout and Lenormant (*26*). The behavior is reflected also in the infrared spectra. Solutions in D_2O at *p*H 11 normally show the amide I absorption band at 1630 cm^{-1}, characteristic of the deuterated α-form (*27*). However, when the poly-L-lysine is aggregated, as evidenced by copious precipitation at this *p*H, the suspension in D_2O shows the amide I band at 1610 cm^{-1}, indicating predominance of the β-form (*28, 29, 30*). At neutral *p*H the amide I band always appears at 1640 cm^{-1}.

Occasionally poly-L-lysine was found to undergo degradation under mild conditions, as evidence by anomalously low viscosities and sedimentation velocities. In some cases this was known to be due to contamination by a mold. In any case it was easily detected by the broadening of the sedimentation boundary which accompanies the randomization of the molecular weight distribution.

Molecular weight.—For light-scattering, sedimentation, and viscosity measurements to determine the molecular weight, a single stock solution of poly-L-lysine hydrochloride in 1.0 M NaCl at *p*H 4.5 was used. The absence of aggregation or degradation was established by the normal dependence of viscosity on concentration and the appearance of a sharp sedimentation boundary moving with the velocity found for other comparable solutions. From a

Zimm plot of the light-scattering results one obtains $M_w = 245{,}000$ for poly-L-lysine hydrochloride, assuming that all of the counterions are included in the particle whose weight is so determined. Subtracting the weight of the chloride ions, one obtains 192,000 for the molecular weight of the poly-L-lysine cation.

Sedimentation and viscosity data provide another estimate of the molecular weight using the Flory-Mandelkern relationship for randomly-coiled polymer molecules (*31*):

$$M^{2/3} = \frac{s^o N \eta_o [\eta]^{1/3}}{2.5 \times 10^6 (1 - \bar{v}\rho)} \qquad (7)$$

Here s^o and $[\eta]$ are the limiting sedimentation constant and intrinsic viscosity, respectively, in the same solvent, N is Avogadro's number, η_o is the solvent viscosity, $\bar{v}$ is the partial specific volume of the solute, and ρ is the solvent density. In order to obtain reasonable agreement with the light-scattering result, it is necessary to make the somewhat tenuous assumption that the polyion sediments independently of its counterions, in which case the values of $[\eta]$ and $\bar{v}$ applicable are those for the polyion, 2.6 dl/gm and 0.781 cc/gm, respectively. With $s^o = 1.92$ S, one then obtains for the polyion $M = 194{,}000$, in agreement with the light-scattering result. The agreement is probably fortuitous, since it is likely that some of the counterions are bound to the polyion, having a significant effect on the appropriate values of both $[\eta]$ and $\bar{v}$; e.g., if total binding of counterions is assumed, the calculated molecular weight is reduced by 58 per cent.

In the absence of more complete knowledge of the nature of the macromolecular component in either the light-scattering or sedimentation experiments, we accept tentatively the molecular weights calculated, corresponding to a DP of 1500. Thus it appears that about one peptide bond per molecule was broken during the decarbobenzoxylation.

Optical rotatory dispersion.—The changes in optical rotation and rotatory dispersion observed in poly-L-glutamic acid solutions with varying pH (*27*) led to the expectation of similar changes in solutions of poly-L-lysine in the neighborhood of pH 10, which is approximately the pK_a of the ε-amine groups. The expectation was borne out, as shown by the curves of specific rotation versus pH in Figure 14.6, for sample III-23. It is seen that $[\alpha]_D$ above pH 11, where the side chains are neutral, is about 100° more positive than at neutral pH, where the amine groups are positively charged through acquisition of a proton. The magnitude and direction of this change are consistent with the transition from the randomly coiled form in the charged state to the helical form in the uncharged state. Further evidence for this conformational change is obtained from the rotatory dispersion, as displayed in the form of the Moffitt plot in Figure 14.7, for poly-L-lysine at various pH in water. In using the Moffitt equation, the proportionality constant between $[m']$ and $[\alpha]$ differs from unity by less than 1 per cent for aqueous solutions of poly-L-lysine; hence the two are regarded as identical for present purposes. The zero slope of the Moffitt plot at neutral pH corresponds to one-term Drude dispersion. The increasingly negative values of b_o as the pH is raised support the conclusion that the helix is formed as the side chains lose their charge. All of the dispersions in Figure 14.7 except the uppermost one were measured before any precipitate formed at high pH. For these experiments the b_o value becomes constant at $-480°$ at high pH. The

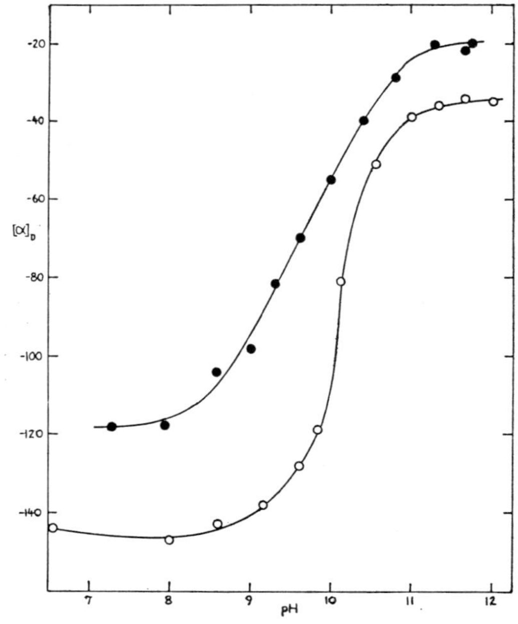

Fig. 14.6. Specific rotation versus pH for poly-L-lysine in water (○), and in 1 M NaBr plus 0.2 M $NaHCO_3$ (●).

uppermost curve gives a value of $b_o = -650°$, and was taken after a precipitate had been removed on standing at *p*H 11.9 (incidentally in the presence of 0.2 M NaBr). The precipitate amounted to about 25 per cent of the total poly-L-lysine, and apparently constituted a fraction which was unable to form the helix, perhaps due to its state of aggregation or its molecular weight. It is evident that the fraction remaining in solution has the dispersion characteristic of a completely helical polypeptide (*9*). The other parameters relevant to this material are $a_o = -30°$ and $[\alpha]_D = -19°$.

Estimation of helix content from dispersion data.—Various methods have been proposed for interpreting rotatory dispersion data in terms of helix content (*32, 33, 34*). The differences between the methods depend on the particular dispersion parameters chosen; in each

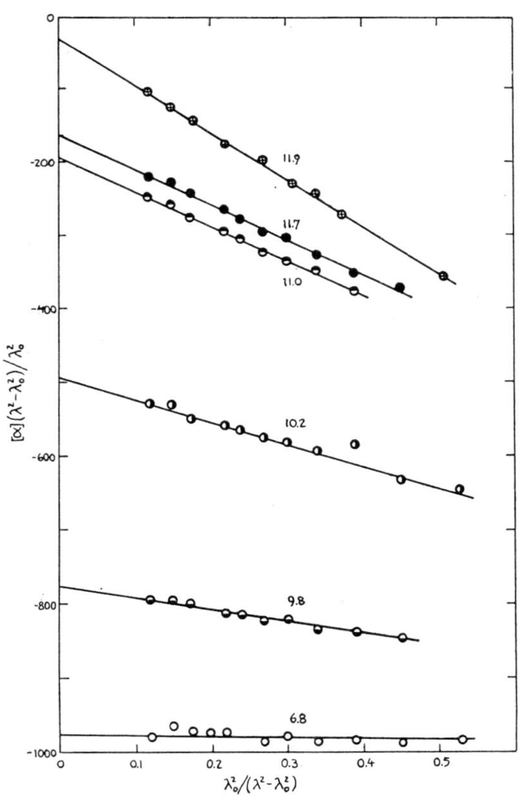

Fig. 14.7. Optical rotatory dispersion of poly-L-lysine in water at *p*H's indicated. Uppermost curve determined after removal of precipitating fraction, showing $b_o = -650°$.

case the value of a dispersion parameter is calibrated against a scale of helix content. The present study has the advantage of a "built-in" calibration scale for the parameters, since the dispersion is known at both extremes of the helix-coil transition. Using the subscripts *h* and *c* to indicate the values of the parameters at the helix and coil extremes of the transition, respectively, we have

$$[m']_c = \frac{a_{oc}\lambda_o^2}{\lambda^2 - \lambda_o^2}$$

$$[m']_h = \frac{a_{oh}\lambda_o^2}{\lambda^2 - \lambda_o^2} + \frac{b_{oh}\lambda_o^4}{(\lambda^2 - \lambda_o^2)^2}$$

If f_h is the fractional helix content, then the rotation at any stage of the transition is given by

$$\begin{aligned}[m'] &= [m']_c + f_h([m']_h - [m']_c) \\ &= [a_{oc} + f_h(a_{oh} - a_{oc})]\frac{\lambda_o^2}{\lambda^2 - \lambda_o^2} \\ &\quad + f_h b_{oh}\frac{\lambda_o^4}{(\lambda^2 - \lambda_o^2)^2}\end{aligned}$$

Equating the coefficients in the last equation with those in equation (1), one obtains for f_h

$$f_h = \frac{[m'] - [m']_c}{[m']_h - [m']_c} = \frac{a_o - a_{oc}}{a_{oh} - a_{oc}} = \frac{b_o}{b_{oh}} \quad (8)$$

Thus equation (8) provides three different (but not independent) methods of calculating f_h. It is seen that all three methods must give exactly the same result if the quantities involved are determined accurately. The significance of this conclusion is that the rotation at a single wave length can serve as well to calculate f_h as the complete dispersion. This has not been the case in all polypeptide and protein studies, because the end points of the rotation scale are not always known. In such cases the form of equation (8) involving b_o has been useful (*32*), since b_{oh} has been found to be constant for a number of polypeptide-solvent systems.

Equation (8) is admittedly an oversimplification, because it implies that a partially helical polypeptide consists of an ideal mixture of helices and random coils. It will be shown below that such is not the case, in that the size of the helical regions varies during the course of the transition. Although a size dependence of the helix rotation has not been found, it is to be expected that there would be such an effect in short helices due to the increased significance of the end residues. For want of any

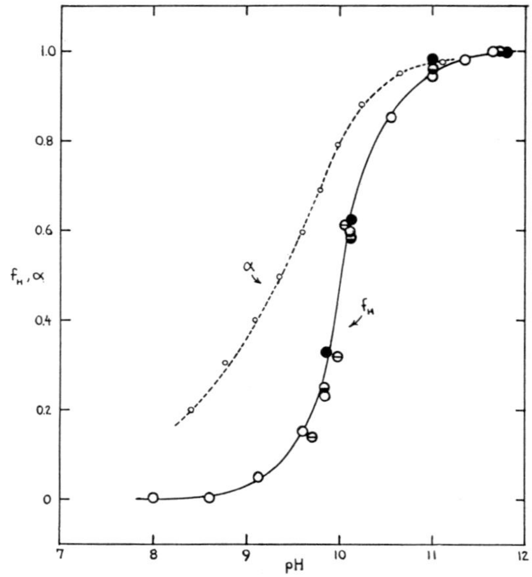

Fig. 14.8. Helix content of poly-L-lysine in water as determined by: (○) $[\alpha]_D$; (◒) a_o; (● and ⊖) b_o, in two experiments. Dashed curve shows degree of dissociation α determined in water.

quantitative information, the effect of helix length is ignored in calculating f_h.

In Figures 14.8 and 14.9 are shown the values of f_h taken from the rotation data in Figure 14.6 and the dispersion data. The three methods of equation (8) give results which agree to within about 3 per cent. The titration curves are shown for comparison to emphasize the marked effect of added salt in stabilizing the helix at a given degree of ionization. The effect was already apparent in Figure 14.6, but was partially masked by the shift of the effective pK_a to a higher value in the presence of added salt. The explanation for this result is simply that the shielding of the electrostatic repulsions between charged side chains by the added salt favors formation of the helix. It is worth noting that the salt stabilization was not at all apparent in poly-L-glutamic acid, according to the work of Idelson and Blout (*22*) and Wada (*35*). They found that the addition of salt at constant pH *decreased* the helix content; thus the stabilization of the helix was completely masked by the effect of added salt to increase the net charge on the molecule at a given pH.

Discussion of titration curves.—Peller (*36*) has shown on theoretical grounds that as the transition from helix to coil occurs, the charge on the polypeptide molecule must increase as a result of the decreased electrostatic interactions between residues in the coil.

Thus the degree of dissociation α is different for helical and coiled residues at a given pH, and the measured value of α is

$$\alpha = f_h \alpha_h + f_c \alpha_c \quad (9)$$

Here $f_c = 1 - f_h$, and α_h is always greater than α_c. Now from Figure 14.8 it is seen that 40 per cent of the titration in water occurs over a range where the helix content is less than 5 per cent. This portion of the curve therefore corresponds to titration of the random coil, and should behave according to the equation for polyelectrolytes (*37*):

$$p\text{H} = p\text{K} + \log \frac{\alpha_c}{1 - \alpha_c} - 0.868(1 - \alpha_c)w \quad (10)$$

In order to determine the constant w, the pK is taken as 10.44, as determined by Katchalsky, Shavit, and Eisenberg (*38*). Thus we find from the first 40 per cent of the titration curve $w = 2.15 \pm 0.07$. The titration described by equation (10) is shown as the dashed curve in Figure 14.10, and, as expected, agrees with the experimental points only below $\alpha = 0.4$. The

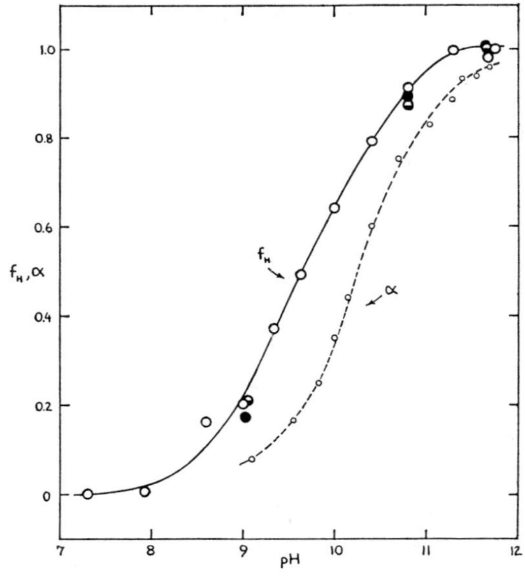

Fig. 14.9. Helix content of poly-L-lysine in 1 M NaBr plus 0.2 M $NaHCO_3$ as determined by: (○) $[\alpha]_D$; (◒) a_o; (●) b_o. Dashed curve shows degree of dissociation α determined in 1 M KBr.

solid curve in Figure 14.10 is calculated by means of equation (9), assuming $\alpha_h = 1$ throughout the entire titration. This assumption implies that helical regions are completely uncharged, wherever they occur in the titration, and that all the charge is in the randomly coiled regions. The agreement with the experimental data on the basis of this rather extreme assumption is fairly good, and it is concluded that the degree of charge is indeed very small in the helix as compared to the coil under these conditions (no added salt). The effect predicted by Peller thus appears to be quite significent.

The thermally induced transition.—The study of the helix-coil equilibrium with respect to temperature is of interest in connection with the interpretation of the phenomenon as a phase transition (*39*). In Figure 14.11 $[\alpha]_D$ for poly-L-lysine in water is shown as a function of temperature at three different *p*H values (determined at 22° C.). Measurements were not carried above 40° C. due to the onset of increased turbidity at this temperature. The dispersion parameters were determined at each temperature (Table 14.3), and it is clear that a transition from helix to coil occurs as the temperature is raised. The transition is, however, much too gradual to be termed a first-order phase transition in the sense of the term originally specified by Ehrenfest.* The behavior is to be contrasted with that observed for poly-γ-benzyl-L-glutamate in mixed solvents (*25*), where the transition was much sharper and occurred in the reverse direction. The significance of this will be discussed shortly. First, in order to evaluate f_h as a function of temperature, the behavior of $[\alpha]_D$ at *p*H 9.72 deserves

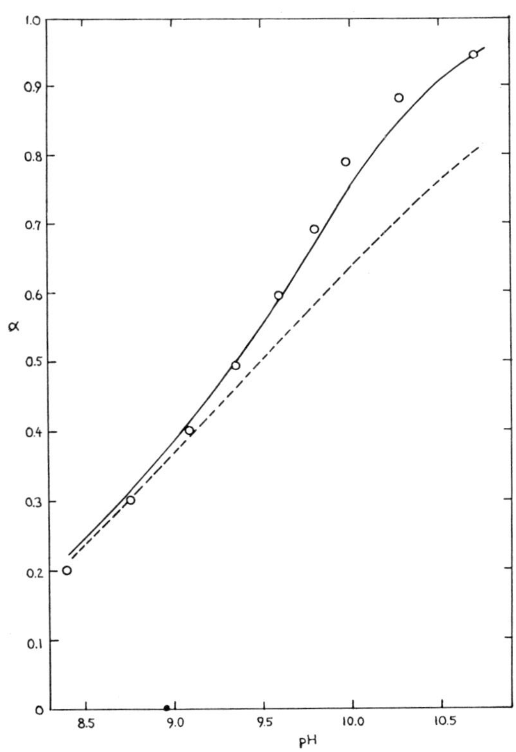

Fig. 14.10. Titration curve for poly-L-lysine in water, $c = 0.19$ per cent. Circles show experimental points. Dashed curve is calculated for a random coil using equation (10). Solid curve is correction to dashed curve from equation (9), assuming $\alpha_h = 1$.

Table 14.3—Optical rotation of poly-L-lysine in water at various temperatures

*p*H	t°C.	$[\alpha]_D$	b_o	f_h	K
10.10	3.0	−35	−480	0.73	2.74
	5.6	−35	(−470)	0.72	2.56
	9.8	−39	(−450)	0.69	2.21
	11.6	−42	(−430)	0.67	1.99
	21.4	−51	−400	0.61	1.56
	31.4	−57	(−340)	0.52	1.10
	36.2	−63	(−310	0.48	0.92
	42.0	−66	−260)	0.40	0.68
	3.1	−68	−280	0.42	0.74
9.98	23.2	−77	−210	0.32	0.47
	33.0	−84	(−178)	0.27	0.38
	41.6	−94	0	0	0
9.72	3.0	−119	0	0	0
	25.6	−108	−90	0.14	0.17
	38.8	−103	−60	0.09	0.09

Values of b_o shown in parentheses were determined indirectly from a correlation between $[\alpha]_D$ and b_o found to be linear and independent of temperature for $b_o < -200$.

* Ehrenfest (*40*) defined a first-order phase transition as a discontinuity in the entropy and volume of a system at some temperature.

One might expect the gradualness to be due in part to a decrease in the degree of charge with increasing temperature, but it is a characteristic of weak acids in their buffer regions that the degree of dissociation α varies very little with temperature. A calculation based on a heat of dissociation of 11.5 kcal/mole for the ϵ-amino groups (*41*) shows that the change in α over the 40° temperature interval involved here is only about 10^{-9}, ignoring the effect of the conformation change. The latter effect is of the same order of magnitude. Hence no significant change in α occurs.

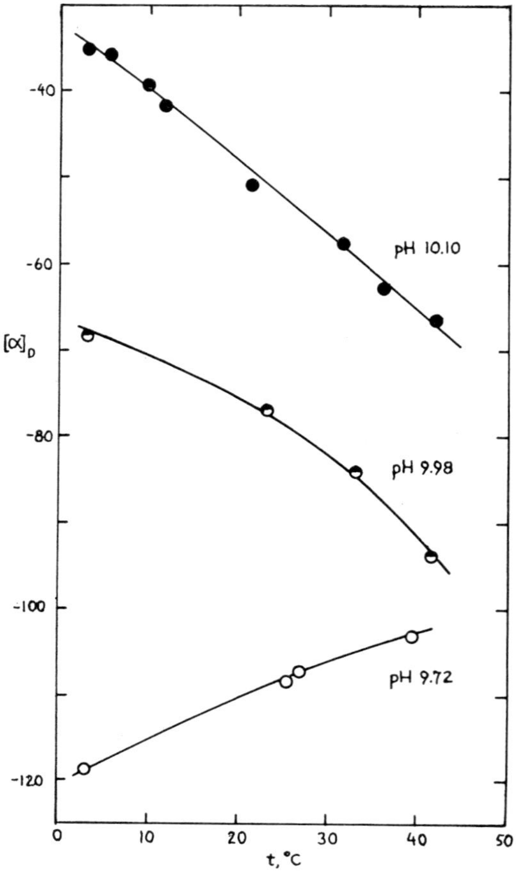

Fig. 14.11. Specific rotation versus temperature for poly-L-lysine in water. The pH values are those determined at 22° C.

comment. Since b_o did not exceed $-90°$ (in magnitude) in this range, the changes observed must be due to the positive temperature coefficient of rotation characteristic of random coils (*33*). Because of this effect, the end points of the rotation scale for determining f_h are not fixed. However, since the value of b_{oh} appears to be relatively independent of the solvent conditions, the use of b_o to estimate f_h is very likely valid at any temperature. We therefore take $f_h = -b_o/650$. (Three data calculated this way from the b_o values at room temperature are included in Figure 14.8.)

The behavior of the helix-coil transition has been explored theoretically by a number of authors, who have shown that the transition should lack sharpness due to its nature as a one-dimensional phase change (*42*). In the treatment by Zimm and Bragg (*42*), the sharpness of the transition depends upon two parameters: ΔH, the enthalpy change per mole of residues on going from coil to helix, and σ, which may be defined as the equilibrium constant for the formation of interruptions in a helical region by a process that maintains a constant number of hydrogen bonds. To show how the sharpness depends on these parameters, we define a function $K = f_h/(1-f_h)$, which has the form of an equilibrium constant. It is essential to note that K does not represent the equilibrium ratio of distinct molecular species (unless, of course, the transition occurs in an all-or-none step) and therefore need not show the same temperature dependence as a true equilibrium constant. From the equations of Zimm and Bragg relating f_h to the temperature for very long chains, we deduce the following relationship, valid at the midpoint of the transition:*

$$d \ln K/d(1/T) = -\Delta H/R\sigma^{1/2} \qquad (11)$$

Thus the sharpness of the transition is characterized by the quantity $\Delta H/\sigma^{1/2}$, which replaces the ΔH of the usual van't Hoff equation. From the data in Table 14.3 it is found that a plot of $\log K$ versus $1/T$ is linear over the range of the measurements at the two highest pH values. The slopes of the lines give the values of $\Delta H/\sigma^{1/2}$ given in Table 14.4. Although it is not possible to determine both ΔH and σ from these measurements, it is useful to discuss the consequences of assuming that σ has the same

Table 14.4—Thermodynamic quantities for the coil-helix transition of poly-L-lysine in water

pH	$\Delta H/\sigma^{1/2}$, kcal/mole
9.98	−3.8
10.10	−5.6

* Equation (11) follows from equations (22), (40), and (43) of reference (*42*). A similar result obtained by Peller (*43*) is in the form

$$d \ln K/d(1/T) = -x\Delta H/Ry_{hc}$$

where x is the number of residues in a statistical unit, taken as unity in the Zimm-Bragg theory, and y_{hc}^2 is the equilibrium constant for the formation of interruptions in a helical region by a process that maintains a constant number of *hydrogen-bonded* (*helical*) *residues*. The significance of this distinction between the two relationships is not yet clear.

value as found for poly-γ-benzyl-L-glutamate in mixed solvents, since it has been pointed out by Zimm and Bragg (*42*) that σ should be relatively independent of the particular solvent. The value is $\sigma = 2 \times 10^{-4}$, according to Zimm, Doty, and Iso (*44*). For ΔH one then finds -54 cal/mole and -80 cal/mole at pH 9.98 and 10.10 respectively. These values are much smaller than the -1500 cal/mole estimated by Schellman (*39*) as the enthalpy of the hydrogen bond, and, if they are real, may indicate that some other important interaction is present in the poly-L-lysine helix. The hydrophobic bonding between side chains discussed by Kauzmann (*45*) appears to be a likely possibility, since these are expected to have *positive* enthalpies of formation in the range of one to two kcal/mole. An additional positive contribution to ΔH is the electrostatic interaction, since about 20 per cent of the ε-amino groups bear a charge in these experiments.

The average number of residues in a helical region at the midpoint of the transition is given as $\sigma^{-1/2}$ by Zimm and Bragg (*42*). Thus another consequence of the value assumed for σ above is that this number is about 70 residues. Since the DP of the poly-L-lysine is about 1500, we can draw the qualitative conclusion that the transition occurs by way of a range of states in which short helical and coiled regions are intermingled. This indication of the mechanism of the transition invites comparison with the hydrodynamic properties of poly-L-lysine in the transition region. If the intermediate state is indeed an interrupted helix, then its sedimentation, viscosity, and flow birefringence behavior should be distinguishable from that of an all-or-none mixture. The following section will show the results of such studies.

Flow birefringence.—The molecular weight of the polylysine used in this study is great enough to allow confirmation by flow birefringence of the helical structure indicated above. This was done on a solution of 0.25 per cent polylysine in 1 M NaBr at pH 11.9. The average molecular length determined over the flow gradient range of 2300–4000 sec^{-1} is 2320 Å, based on an ellipsoidal model with axial ratio 150. The variation in calculated length over this range was about 10 per cent, indicating a rather narrow molecular weight distribution. The length required by the α-helical model for a DP of 1500 is 2250 Å, in good agreement

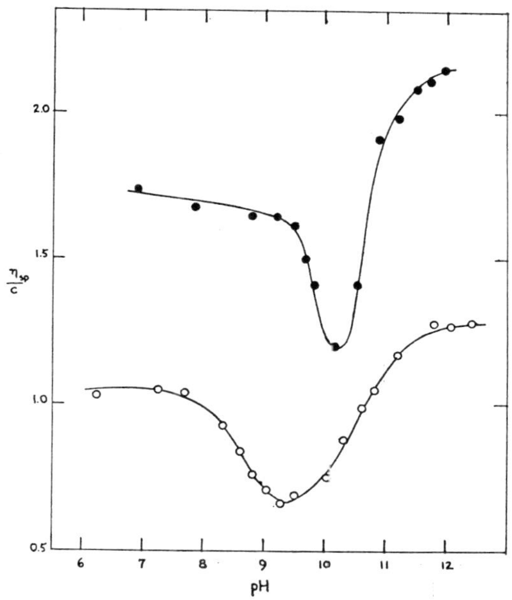

Fig. 14.12. Reduced specific viscosity versus pH for two samples of poly-L-lysine obtained by partial degradation of sample III-23. Upper curve is for 0.093 per cent solution in 1 M NaCl. Lower curve is for 0.51 per cent solution in 1 M NaBr plus 0.2 M $NaHCO_3$.

with the experimental value. This agreement is gratifying, in that it lends credence to both the model and the molecular weight measurement.

It is significant that the flow birefringence in the above solution disappeared below a pH of approximately 10.7 as acid was added. According to Figure 14.9, the polypeptide is about 90 per cent helical at this pH, yet the rodlike structure is absent. This is consistent with a picture of the transition occurring by way of "interrupted" helices.

Viscosity.—The large change in hydrodynamic properties demonstrated by flow birefringence is also reflected in the viscosity, as seen in Figure 14.12. Two samples were studied here, both resulting from partial degradation of sample III-23. To suppress expansion of the random coil and thus enhance the difference in viscosity between the helix and coil, 1 M salt was added. As expected from studies of poly-γ-benzyl-L-glutamate and poly-L-glutamic acid, the viscosity of the helix is greater than that of the coil in the molecular weight region of these samples (*1, 27*). The

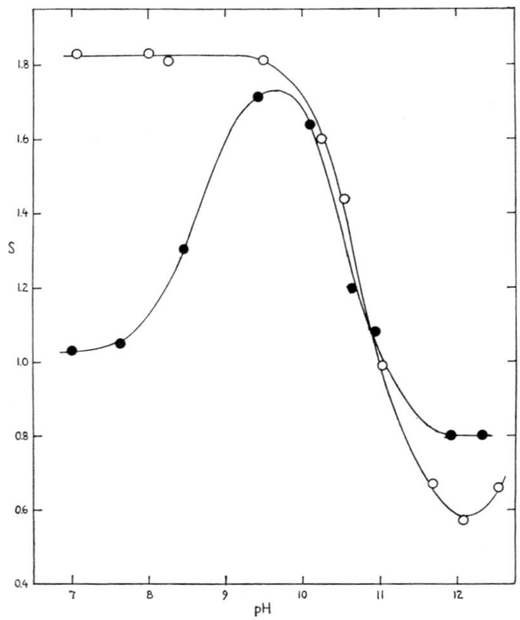

Fig. 14.13. Sedimentation constant of poly-L-lysine at 0.32 per cent concentration in (○) 1 M NaBr plus 0.2 M $NaHCO_3$; (●) 1 M NaCl plus 0.2 M $NaHCO_3$.

pronounced minima in the viscosities are consistent with the proposal of partially helical states in the transition region, for such states are expected to be more compact (in the sense of having a shorter end-to-end distance) than either the random coil or the intact helix.

Sedimentation.—Early efforts in this work to distinguish between the possible mechanisms of the transition were based on the likelihood that in an all-or-none mechanism the two extreme forms could be separated by virtue of their differing sedimentation rates, provided the rate of restoration of equilibrium was sufficiently slow. Since in fact one observes only a single sharp sedimentation boundary at all pH, this possibility was eliminated. The evidence presented thus far supports the existence of interrupted helical states in the transition, and it will be seen that the sedimentation study is also consistent with this conclusion.

Sedimentation constants were determined as a function of pH in solutions containing 0.32 per cent polylysine III-23, 0.2 M $NaHCO_3$, and either 1 M NaCl or 1 M NaBr. The results are shown in Figure 14.13. The behavior is similar in the two solvents at high pH, but differs widely below pH 10. The maximum occurring in s in the NaCl solution is at pH 9.6, which is precisely the midpoint of the transition as judged by the optical rotation (Figure 14.9). It would seem that the maximum is ascribable at least in part to the same minimum in the frictional constant that was reflected in the viscosity measurements, and is therefore another manifestation of the partially helical states.

That the behavior of the sedimentation constant is governed by other effects, in addition to the variation in frictional constant, is evident from the disparity in behavior between the two solvents. It would appear that the binding of Br^- to the highly charged polyion masks the effect of the minimum in frictional constant when NaBr is the added salt. Binding of Cl^- appears to be less significant, though the extent of such binding cannot be evaluated. The changes which affect s as the pH is varied may be summarized as follows: first, change in frictional constant due to the conformational change; second, change in particle weight due to decreasing anion binding with decreasing positive charge; and third, change in partial specific volume, due both to the varying state of charge and to the varying contribution of the bound anions. Since none of these effects is completely understood, it is not possible to say to what extent the observed behavior is a reflection of the conformational change. Any attempt to rule out a rapidly equilibrating all-or-none mechanism on the basis of these data would, of course, be difficult; however, the evidence against such a mechanism from the experiments discussed above seems adequate.

Summary

Poly-ε-carbobenzoxy-L-lysine in dimethylformamide solution was found to have a helical conformation ($b_o = -625°$), the axis of the helix having occasional bends. The shape of the molecule was judged from the light-scattering envelopes of high molecular weight samples and from viscosity and flow birefringence measurements on samples with DP's ranging from 570 to 3610.

It has been shown that high molecular weight poly-L-lysine (DP 1500) in aqueous solution undergoes a transition from a helical to a randomly coiled conformation under the influence of decreasing pH or increasing temperature. Strong evidence has been presented in favor of an interrupted helical conformation as the intermediate state of the molecule. These conclusions have resulted from the following experimental results:

1) $[\alpha]_D$ for poly-L-lysine in water increases from $-144°$ at pH 7 to $-35°$ at pH 12, the change occurring over a range of about one pH unit. Addition of 1 M salt plus buffer broadens the transition and shifts it to lower pH. The magnitude and direction of the change are as expected for the formation of completely helical material at high pH. Dispersion studies at high pH showed $b_o = -650$, in agreement with the value of this quantity for other completely helical polypeptides.

2) The change in rotation with increasing temperature in the transition pH region indicated a very gradual disappearance of the helical structure. At pH 10.10 in water the helix content was found to vary only from 70 per cent to 40 per cent over a 40° temperature range. In terms of the Zimm-Bragg theory, this was found to correspond to a value of $\Delta H/\sigma^{1/2} = -5.6$ kcal/mole at this pH.

3) Flow birefringence evidence at pH 11.9 indicated a rodlike shape for the molecule, with a length of 2320 Å. This compares favorably with a length of 2250 Å expected for the α-helix in a molecule of DP 1500.

4) Measurements of viscosity as a function of pH in the presence of 1 M salt showed pronounced minima in the transition region. This was taken as indication of an interrupted helical conformation in the transition region, in agreement with the expectation from the results of the thermal transition.

5) Sedimentation velocity measurements also reflected the minimum in frictional constant detected in the viscosity measurements.

Acknowledgment

One of the authors (J. A.) is indebted to the National Science Foundation for predoctoral fellowship support during the course of this study (1955–58). This work was reported in part at the 133rd American Chemical Society Meeting, April, 1958, at San Francisco, and in full in the Ph.D. thesis of Jon Applequist, Harvard University, 1959.

References

1 P. Doty, J. H. Bradbury, and A. M. Holtzer, *J. Am. Chem. Soc.* **78,** 947 (1956).
2 Quoted in E. Katchalski and M. Sela, *Advances in Protein Chem.* **13,** 394 (1958).
3 E. R. Blout and R. H. Karlson, *J. Am. Chem. Soc.* **78,** 941 (1956).
4 D. Ben-Ishai and A. Berger, *J. Org. Chem.* **17,** 1564 (1952).
5 G. W. Phillips, Ph.D. thesis, Harvard University (1956).
6 K. Fajans and O. Johnson, *J. Am. Chem. Soc.* **64,** 668 (1942).
7 T. S. Ma and G. Zuazaga, *Ind. Eng. Chem., Anal. Ed.* **14,** 280 (1942).
8 G. A. Gornall, C. J. Bardawill, and M. M. David, *J. Biol. Chem.* **177,** 751 (1949).
9 W. Moffitt and J. T. Yang, *Proc. Nat. Acad. Sci.* **42,** 596 (1956).
10 K. A. Stacey, in "A Laboratory Manual of Analytical Methods of Protein Chemistry," P. Alexander and R. J. Block (eds.), Pergamon, New York, 1961, Vol. 3, p. 245.
11 P. Doty and R. F. Steiner, *J Chem. Phys.* **18,** 1211 (1950).
12 H. Benoit, A. M. Holtzer, and P. Doty, *J. Phys. Chem.* **58,** 635 (1954).
13 O. Kratky and G. Porod, *Rec. trav. chim.* **68,** 1106 (1949).
14 W. Kuhn, *Kolloid Z.* **87,** 3 (1939).
15 A. Peterlin, *Makromol. Chem.* **9,** 244 (1953).
16 P. Flory, "Principles of Polymer Chemistry," Cornell, Ithaca, N. Y., 1953, pp. 602–611.
17 J. C. Mitchell, A. E. Woodward, and P. Doty, *J. Am. Chem. Soc.,* **79,** 3955 (1957).
18 H. A. Scheraga, J. T. Edsall, and S. Gadd, *J. Chem. Phys.* **19,** 1101 (1951).
19 F. Perrin, *J. phys. radium* (7) **5,** 497 (1934).
20 R. Simha, *J. Phys. Chem.* **44,** 25 (1940).
21 H. Boedtker and P. Doty, *J. Am. Chem. Soc.* **78,** 4267 (1956).
22 M. Idelson and E. R. Blout, *ibid.* **80,** 2387 (1958).
23 P. Doty, *J. Cellular Comp. Physiol.* **49,** 27 (1957).
24 G. D. Fasman, M. Idelson, and E. R. Blout, *J. Am. Chem. Soc.* **83,** 709 (1961).
25 E. R. Blout, P. Doty, and J. T. Yang, *ibid.,* **79,** 749 (1957).
26 E. R. Blout and H. Lenormant, *Nature* **179,** 960 (1957).
27 P. Doty, A. Wada, J. T. Yang, E. R. Blout, *J. Polymer Sci.* **23,** 851 (1957).
28 E. R. Blout and M. Idelson, *J. Am. Chem. Soc.* **80,** 4909 (1958).
29 P. Doty, K. Imahori, and E. Klemperer, *Proc. Nat. Acad. Sci.* **44,** 424 (1958).
30 H. Lenormant, A. Baudras, and E. R. Blout, *J. Am. Chem. Soc.* **80,** 6191 (1958).
31 L. Mandelkern and P. Flory, *J. Chem. Phys.* **20,** 212 (1952).
32 C. Cohen and A. G. Szent-Gyorgyi, *J. Am. Chem. Soc.* **79,** 248 (1957).

33 J. A. Schellman, *Compt. rend. trav. Lab. Carlsberg, Sér. Chim.*, **30**, Nos. 21–27 (1958).
34 J. T. Yang and P. Doty, *J. Am. Chem. Soc.* **79**, 761 (1957).
35 A. Wada, *Mol. Phys.* **3**, 409 (1960).
36 L. Peller, *J. Phys. Chem.* **63**, 1199 (1959).
37 J. T. Edsall and J. Wyman, "Biophysical Chemistry," Vol. I, Academic Press, New York, 1958. p. 518.
38 A. Katchalsky, N. Shavit, and H. Eisenberg, *J. Polymer Sci.* **13**, 69 (1954).
39 J. A. Schellman, *Compt. rend. trav. lab. Carlsberg, Sér. Chim.* **29**, No. 15, (1955).
40 P. Ehrenfest, *Proc. Acad. Sci. Amsterdam* **36**, 153 (1933). See also W. H. Keesom, "Helium," Elsevier, Amsterdam, 1942, p. 257.
41 E. J. Cohn and J. T. Edsall, "Proteins, Amino Acids, and Peptides," Reinhold, New York, 1943, p. 89.
42 B. H. Zimm and J. K. Bragg, *J. Chem. Phys.* **31**, 526 (1959). See also references cited therein.
43 L. Peller, *J. Phys. Chem.* **63**, 1194 (1959).
44 B. H. Zimm, P. Doty, and K. Iso, *Proc. Nat. Acad. Sci.* **45**, 1601 (1959).
45 W. Kauzmann, *Advances in Protein Chem.* **14**, 37 (1959).

Discussion of Paper 14: α-Helix Formation in Polylysine

JEN TSI YANG

Flow birefringence is an extremely sensitive technique for detecting the polydispersity of rigid particles; the hydrodynamic lengths as calculated from the Perrin equation drop markedly with increasing rate of shear if the system is polydisperse. A wormlike stiff chain, however, tends to deform (stretch) under applied shearing stress; this could make an otherwise polydisperse system appear less so. These uncertainties can be resolved by using the recently proposed intrinsic viscosity method (at zero gradient), provided our chief purpose is to determine an average length (J. T. Yang, *J. Polymer Sci.*, **54**, S14 [1961]). It can be shown that:

$$L \text{ (in cm)} \equiv 2a = 6.82 \times 10^{-8}([\eta]_o M)^{1/3}[p^2/\nu(p)]^{1/3}$$

where L is the length (= major axis) and p the axial ratio of an equivalent hydrodynamic ellipsoid, ν the Simha's viscosity increment, and M the molecular weight of the particles. It has been shown that the term in the bracket is very insensitive to axial ratios provided they are large. Using the above equation we have recalculated the data on poly-ε-carbobenzoxy-L-lysine in dimethylformamide (Table 14.1, Paper 14, this book) shown as in Table 14.5. These calculated lengths lie somewhat above the theoretical curve for a wormlike model in Applequist and Doty's Figure 14.5.

Table 14.5

Sample	Axial ratio (assumed)	Hydrodynamic length (Å)
A	10	360
B	30	830
G	50	1500
J	100	2570
K	150	2820

Similar calculations can be applied to the data of Drs. Daniel and Katchalski (this book, Paper 16) on the same sample. For the nine samples listed in their Table 16.3 we have found the following hydrodynamic lengths: I, 120–140 Å; II, 260–290 Å; III, 350–400 Å; IV, 490–460 Å; V, 600–600 Å; VI, 860–870 Å; VII, 1310–1310 Å; VIII, 1270–1400 Å; IX, 1950–2010 Å. The two figures for each sample refer to the use of M_{SD} and M_{LS} (in their Table 16.2), respectively.

B. H. ZIMM

The gradualness of the temperature change of the properties of salt solutions of polylysine tends to give the misleading impression that the nature of the transition is fundamentally different from that in polybenzyl-glutamate. I think that quite the contrary is the case. The transitions in polylysine, polyglutamic acid, and polybenzylglutamate are probably very similar in that all proceed through a state of partially interrupted helices. The apparent differences in the temperature behavior must arise because the heat of helix formation per residue is nearly zero in the two former cases, while it has a positive value of about one kilocalorie in the latter case. (In contrast, Schellmann's estimate is a *negative* value of about one kilocalorie.) The evidence for this statement comes from the sharpness of the transition as the degree of charge is varied. Quantitative estimation of the electrostatic energy of the charges in polyglutamic acid shows that the sharpness of the transition cannot be explained by a large electrostatic energy; the amount of this energy per residue in 0.1 M salt solution is nearly the same as kT. Therefore the sharpness must result from the highly coöperative nature of helix formation. The quantitative treatment of the polyglutamic acid case by S. A. Rice and myself (*Mol. Physics* **3**, 391 [1960]) showed that the molecular states in the transition region were very similar to those in polybenzylglutamate in spite of the dissimilarity of their temperature profiles. The data on polylysine look very similar to those on polyglutamic acid, so that the transition probably proceeds again through similar molecular states.

PAPER 15

Optical Rotatory Studies of Dye-Polypeptide Complexes

LUBERT STRYER

Optical rotatory dispersion has proven to be an important technique in the elucidation of the structure of synthetic polypeptides and of proteins. Its utility is based on the large changes in the rotatory contribution of the peptide chromophore that accompany transitions of conformation. We have sought to extend the usefulness of this technique in structural studies by investigating the optical rotatory properties of extrinsic chromophores bound to synthetic polypeptides (*1*, *2*, *3*). The striking finding is that certain *symmetric, optically inactive dye molecules acquire optical rotatory power on binding to helical macromolecules.*

Optical rotatory measurements in the visible absorption bands of the bound dyes were made possible by the use of a high-intensity light source and a photoelectric polarimeter. It has been known for many years that the rotatory dispersion in the absorption band of an optically active chromophore undergoes a characteristic change in magnitude and in sign termed the Cotton effect (Figure 15.1 and reference *4*). One of the dyes used in our study was Acridine Orange, an optically inactive, planar, symmetric molecule, that binds strongly to both the helical and the random-coil conformations of poly-L-glutamic acid (molecular weight, about 50,000). Upon binding to the random-coil conformation of poly-L-glutamic acid, there is a shift in the absorption maximum of the Acridine Orange, while the rotatory dispersion of the dye:random polypeptide complex is almost identical with that of the random-coil poly-L-glutamic acid alone (Figure 15.2). In marked contrast, when Acridine Orange is bound to the helical form of poly-L-glutamic acid, there is a large change in the rotatory dispersion of the complex. In fact,

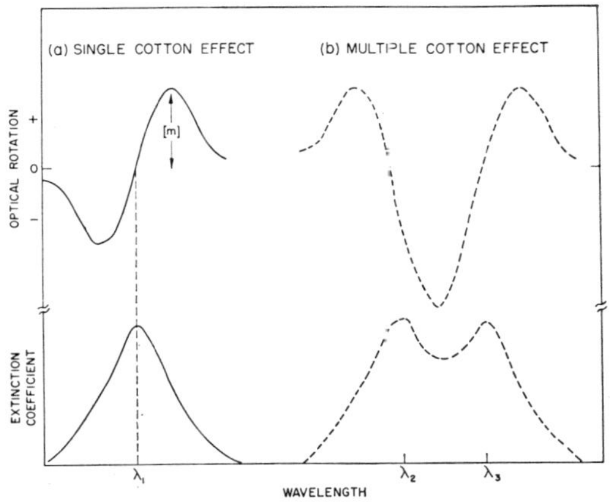

Fig. 15.1. Idealized representation of optical rotatory dispersion in the absorption bands of optically active chromophores. (*a*) A single Cotton effect, characterized by a magnitude [*m*], a positive sign, and an inflection point at λ_1, is shown to correspond with the isolated absorption band whose absorption maximum is at λ_1. (*b*) A multiple Cotton effect is shown for two optically active chromophores at wave lengths near one another; the multiple Cotton effect is composed of a negative Cotton effect at λ_2 and a positive Cotton effect at λ_2.

a negative Cotton effect is observed with an inflection point at 468 mμ, which corresponds to the absorption maximum of the bound Acridine Orange at 465 mμ. Thus, the Acridine Orange chromophore has acquired optical rotatory power upon binding to the helical polypeptide. The change in the magnitude of this Cotton effect with *p*H closely parallels the helix to random-coil transition of polyglutamic acid. When the Acridine Orange is bound to the helical polypeptide of opposite screw-sense, poly-D-glutamic acid, the opposite sign of Cot-

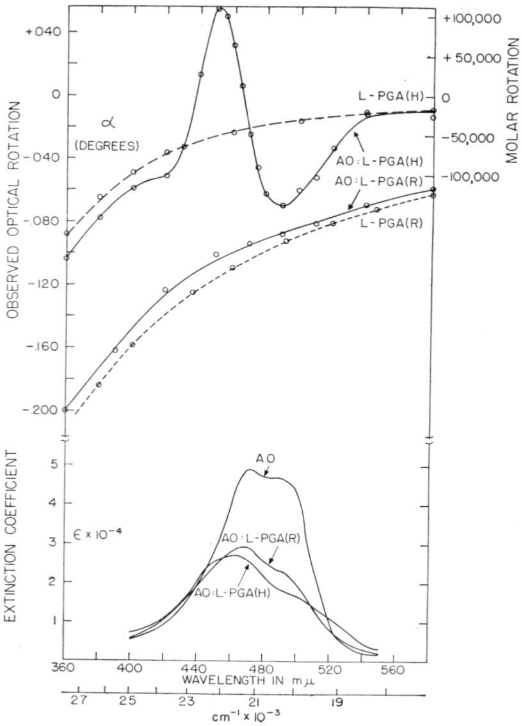

Fig. 15.2. Optical rotatory dispersions and absorption spectra of complexes of Acridine Orange and poly-L-glutamic acid. The absorption spectrum of unbound Acridine Orange (AO) markedly changes upon binding to either the random conformation of poly-L-glutamic acid [AO:L-PGA(*R*)] or the helical conformation [AO:L-PGA(*H*)]. Only the complex of Acridine Orange and the helical conformation of L-PGA [AO:L-PGA(*H*)] exhibits a Cotton effect.

ton effect is observed.

Under certain conditions, Acridine Orange displays a multiple Cotton effect upon binding to helical polyglutamic acid. In addition to the 468 mμ Cotton effect, there is one of opposite sign at 510 mμ. This second Cotton effect is observed only under conditions of low *p*H, high ionic strength, and high polypeptide to dye ratio. Evidently, the 510 mμ Cotton effect arises from a different mode of binding of Acridine Orange to the helical polypeptide.

Similar effects are observed for several other dyes such as Acriflavine, pseudoisocyanine, Rhodamine B, Rhodamine G, and a sub tituted toluidine-thiazole dye. For all of these dyes, induced optical rotatory power is observed when the dye is bound to the helical, but not to the random, conformation of polyglutamic acid. The signs of the Cotton effects vary, indicating that the dyes either are bound in different ways or possess different dichroisms. For a particular absorption band of these dyes, the sign of the induced Cotton effect is opposite for complexes with polyglutamic acid of opposite screw-sense of helix.

We have proposed three schematic models showing how bound dye might acquire optical rotatory power (Figure 15.3). In the first of these models, a single dye chromophore acquires optical rotatory power by interacting with the local asymmetric environment of the α-helix. In the other two models, interaction between two or more dye molecules is necessary for Cotton effects in the dye absorption bands. A dissymmetric dye ensemble could arise in two ways. In Model II, the polypeptide α-helix serves to bind the dyes in a repetitive fashion to form a dye "super-helix" which possesses a single screw-sense. A different type

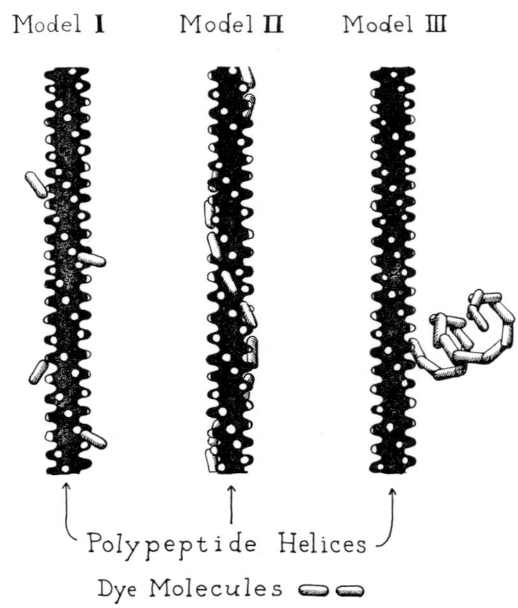

Fig. 15.3. Suggested modes of interaction between dye and polypeptide that could give rise to the induced Cotton effects in the absorption bands of the bound dye. Model I, unaggregated dye interacts near the asymmetric α-carbon atom of the peptide residue. Model II, dye end-to-end aggregate giving rise to dye super-helix. Model III, tangential dye helix.

of dye helix is shown in Model III, in which the ordered structure of the side chains of the α-helical polypeptide acts as a seeding center to favor one of the screw-senses of a dye helix which forms tangential to the polypeptide α-helix.

There is good evidence that the induced Cotton effects exhibited by the micellar pseudoisocyanine:helical polyglutamic acid complexes (Figure 15.4) arise from the tangential dye helix depicted in Model III. Pseudoisocyanine micelles have been shown by X-ray diffraction to have a helical structure (*5*). Such a pseudoisocyanine micelle, by itself, has no optical rotation, since there are equal amounts of left- and right-handed dye helix. One of these screw-senses of dye helix is favored, however, when the pseudoisocyanine micelle binds to helical polyglutamic acid, and the dye then acquires optical rotatory power. This suggested origin of the induced optical rotatory power of the micellar pseudoisocyanine bound to helical polyglutamic acid receives further support from a correlation of the signs of the observed Cotton effects and the dichroisms of the micellar dye. The two shorter wave-length visible absorption bands have a dichroism perpendicular to the dye fiber axis, while for the long wave length, high intensity J band it is parallel (*6*). This agrees with the signs of the Cotton effects, as the two shorter wave-length Cotton effects have a sign opposite that of the one observed in the J band. The micellar pseudoisocyanine: helical polyglutamic acid complex is of additional interest because it provides the first examples of Cotton effects in absorption bands arising from exciton interaction among identical chromophores. The narrow half-width of the Cotton effect at 574 mμ corresponds to the very sharp J band, which has its origin in the interaction among the helically arranged pseudoisocyanine molecules in the micelle. It is interesting to note that the exciton splitting for this dye helix, of the order of 2000 cm^{-1}, is quite similar to what Moffitt had calculated for the peptide chromophore in the α-helix (*7*). The molar rotation of the Cotton effect in the J band is exceptionally large, and may be related to the helical arrangement of the pseudoisocyanine.

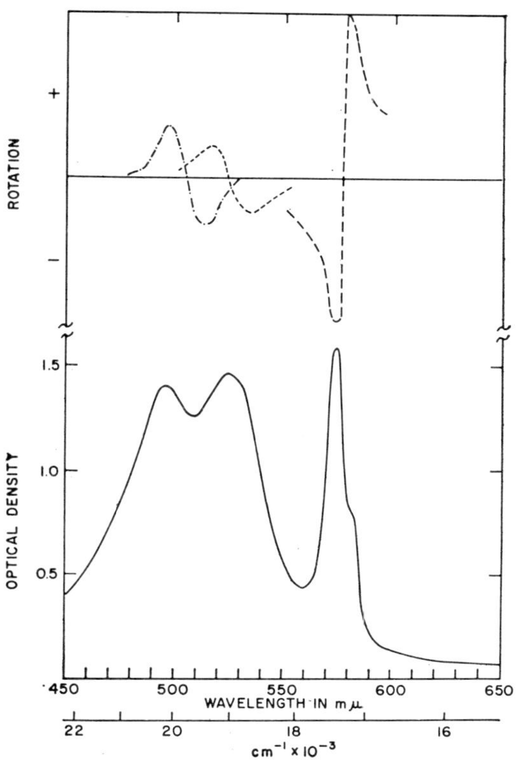

Fig. 15.4. The relationship of the absorption bands (*lower curve*) of the micellar pseudoisocyanine:helical poly-L-glutamic acid complex to the resolved Cotton effects (*upper curves*). The dichroism of the absorption bands corresponds with the signs of the Cotton effects.

The sensitivity of these induced Cotton effects to the conformation of polyglutamic acid suggests that this approach may prove useful in elucidating certain aspects of protein structure. There are a number of difficulties, however, which must be considered. First, it is necessary that the dye bind to the particular region of the protein that is of interest. Second, it is not known whether ordered structures other than α-helix will give rise to thse induced Cotton ef-effects. Our only experience has been with macromolecules that are either helical or in the random coil. Furthermore, we do not know whether polypeptide α-helices of the same screw-sense but composed of opposite amino-acid isomers would give rise to induced Cotton effects of the same sign. If this were so, then it would prove to be a valuable technique in determining the helical screw-sense. Finally, difficulties will arise in the interpretation of these induced Cotton effects because of uncertainty in the mode of binding of the dyes and in the characteristics of their electronic transitions. Taking these present limitations into account, this approach nevertheless appears promising. Recently, we have found that the longer wave-length Cotton effect in Acridine Orange is more sensitive than the shorter wave-length Cotton effect to disruption of α-helix in copolymers of L-serine and L-glutamic acid (*8*). A technique for distinguishing between short and long helical segments obviously would be desirable, since present optical rotatory parameters only reveal the total helix content and not its distribution. A different use of induced Cotton effects has been initiated by Ulmer and Vallee in the study of the active site of dehydrogenase enzymes (*9*). The optical rotatory properties of Acridine Orange:insulin complexes are being studied by Beychok (*10*). It appears likely that optical rotatory dispersion of bound extrinsic chromophores will be of value in the study of the subtler aspects of protein structure, but the soundest use of the technique will require a more complete understanding than we now possess of how dyes interact with macromolecules.

Acknowledgment

This work was supported in part by U.S. Public Health Service Grant A2558.

References

1 E. R. Blout and L. Stryer, *Proc. Nat. Acad. Sci.* **45**, 1591 (1959).
2 L. Stryer and E. R. Blout, *Abstracts of the Biophysical Society*, Fourth Annual Meeting, February, 1960.
3 L. Stryer and E. R. Blout, *J. Am. Chem. Soc.* **83**, 1411 (1961).
4 For example, see W. Kuhn, *Ann. Rev. Phys. Chem.* **9**, 417 (1958).
5 A. Rich and M. Kasha, unpublished results.
6 W. Hoppe, *Kolloid-Z.* **109**, 21 (1944).
7 W. Moffitt, *Proc. Nat. Acad. Sci.* **42**, 596 (1956).
8 L. Stryer and E. R. Blout, unpublished results.
9 D. D. Ulmer and B. L. Vallee, *J. Biol. Chem.* **236**, 730 (1961).
10 S. Beychok and E. R. Blout, to be published.

PAPER 16

The Hydrodynamic Behavior and Molecular Configuration of Poly-ε, N-Carbobenzoxy-L-Lysine in Dimethylformamide Solution

EZRA DANIEL and EPHRAIM KATCHALSKI

The recent literature contains only few articles dealing with the hydrodynamic properties of poly-α-amino acids in solution. Most of these are concerned with the viscosity of high molecular weight polypeptides in different solvents but do not give any information as to other hydrodynamic properties such as diffusion and sedimentation (*1, 2*). The elucidation of the shape and general behavior of macromolecules in solution requires, however, an evaluation of the concentration and molecular weight dependence of various hydrodynamic parameters.

Measurements of viscosity and light scattering carried out on samples of poly-γ-benzyl-L-glutamate of different chain length (*3, 4*), revealed that the molecules of this polypeptide behave in weakly interacting solvents, like dimethylformamide, as rigid, rodlike molecules possessing the dimensions of an α-helix (*5*). On the other hand, in strongly interacting solvents such as dichloroacetic acid and trifluoroacetic acid the polypeptide chain exhibits properties of a random coil. The above conclusions have recently been corroborated by the finding that the sheer dependence of the intrinsic viscosity of poly-γ-benzyl-L-glutamate in *m*-cresol, a solvent favoring α-helix formation, is in complete accord with the viscosity theories developed for asymmetric rigid particles (*6, 7*).

In the following a survey is given of the viscosity, sedimentation, and diffusion behavior in dimethylformamide (DMF) of nine samples of poly-ε,*N*-carbobenzoxy-L-lysine (PCBL) in the molecular weight range of 20,000 to 500,000. An analysis of the molecular weight dependence of the various hydrodynamic parameters obtained was carried out taking into account the various theories dealing with corpuscular particles and flexible threadlike molecules. From this analysis it was concluded that the PCBL molecules exist in DMF solution as helical rods possessing a marked flexibility.

Experimental

Materials.—The samples of poly-ε,*N*-carbobenzoxy-L-lysine (PCBL) used (see Table 16.1) were prepared by the polymerization of ε,*N*-carbobenzoxy-α,*N*-carboxy-L-lysine anhydride (*1, 8, 9*) in dioxane using diethylamine or triethylamine as initiators. The reaction mixtures contained 5 gm of monomer in 60 ml anhydrous dioxane and the polymerization was carried out at room temperature and atmospheric pressure for 48 hours with magnetic stirring. The polymers formed were precipitated by pouring the mixture into 500 ml water acidified with a few

Table 16.1—Preparation of the poly-ε,*N*-carbobenzoxy-L-lysine (PCBL) samples

Sample No.*	Initiator	*I/A*
I	$(C_2H_5)_2NH$	1:200
II	$(C_2H_5)_3N$	1:100
III	$(C_2H_5)_2NH$	1: 50
IV	$(C_2H_5)_3N$	1: 36
V	$(C_2H_5)_3N$	1: 1
VI	$(C_2H_5)_3N$	1: 1
VII	$(C_2H_5)_3N$	1: 20
VIII	$(C_2H_5)_3N$	1: 10
IX	$(C_2H_5)_3N$	1: 24

* The various samples were prepared by the polymerization, at room temperature, of ε,*N*-carbobenzoxy-α,*N*-carboxy-L-lysine anhydride in dioxane using $(C_2H_5)_2NH$ or $(C_2H_5)_3N$ as initiators. The molar ratio of the initiator to the anhydride (*I/A*) is given in the last column of the table.

drops of concentrated hydrochloric acid. The polypeptides which separated out were washed with water and dried in vacuo over phosphorous pentoxide.

The dimethylformamide (DMF) used as solvent in all the measurements recorded below was a B.D.H. product (n_D^{20} 1.430 to 1.431).

Viscosity.—Viscosity measurements were made with a 10 ml volume Ostwald-Fenske viscometer at 20 ± 0.05°. The time of efflux for pure solvent (DMF) at this temperature was 232 seconds; no kinetic energy correction was necessary. For each of the PCBL samples tested a plot was made of the reduced specific viscosity, η_{sp}/c, versus polymer concentration c, given in grams per 100 gm solvent. Extrapolation to $c = 0$ gave the corresponding intrinsic viscosities $[\eta]$, which were converted into dl/gm units, using the density of DMF, $\rho_{20} = 0.950$ gm/ml.

The radius of the capillary of the viscometer was 2.21×10^{-2} cm, as determined by weighing a measured length of mercury introduced into the capillary. The volume of the bulb, between the two etched lines, was 3.0 ml. As it takes 232 seconds for DMF in the bulb to run out, the rate of flow through the capillary is 0.013 ml/sec. From these data a maximal velocity gradient, $G_{max} = 1.53 \times 10^3$ sec^{-1} was calculated for the flow of solvent at the walls of the capillary.*

Sedimentation.—Measurements of sedimentation velocity were carried out in a Spinco Model E ultracentrifuge equipped with schlieren phase-plate optics. The temperature of the rotor was determined and controlled by the RTIC unit of the instrument. Conventional 12 mm cells with a 4° sector were used for solutions with concentrations down to 0.10 gm/100 gm. Solutions with lower concentrations were run in 30 mm cells. The value of the intrinsic sedimentation constant, s_o, was determined by plotting s against the concentration c and extrapolating to $c = 0$.

The approach to equilibrium method of Archibald (*10*) was used to obtain molecular weights of three samples of PCBL. In this case, the ultracentrifuge patterns were analyzed by means of a two-way Hauzer Profil P219, Switzerland, microcomparator. Following the practice of Klainer and Kegeles (*11*), the concentrations of the initial polypeptide solutions were determined in arbitrary units in a synthetic boundary cell (*12*). The molecular weights were calculated according to the procedure given by Schachman (*13*) using the data at the meniscus.

Diffusion.—The diffusion coefficients were determined in a Spinco Model E ultracentrifuge using a synthetic boundary cell according to Ehrenberg (*14*). This procedure has been used in our laboratory for several years and was found to yield diffusion coefficients for bovine serum albumin, bovine pancreatic ribonuclease and potassium chloride in fair agreement with those recorded in the literature (*15, 16*). All measurements with PCBL in DMF were carried out at 9,000 to 12,000 r.p.m. The diffusion coefficients were calculated from the schlieren patterns obtained, by the "maximum ordinate-area method (D_A)," by the "maximum ordinate method (D_i)," as well as by the "statistical method (D_m)" (*17*).

Partial specific volume.—The partial specific volume of PCBL in DMF, ($\overline{V}$), was derived from density measurements of solvent and solutions in a pyknometer of about 10 ml capacity (*18*). A value of 0.847 ml/gm was obtained for PCBL sample IV. The same partial specific volume was assumed for the other PCBL samples investigated. This value is in fair agreement with the reciprocal of the density of solid PCBL (0.82 ml/gm) (*19*).

Light scattering.—Light-scattering measurements were carried out in a Brice-Phoenix Universal 1000 series light-scattering photometer using a mercury arc lamp. The scattering of unpolarized blue light (436 mμ) at 90° from square cells of 14 or 25 ml capacity was recorded. For dissymmetry studies a 45 ml semi-octagonal cell was used.

The constants of the instrument were calibrated with benzene assuming a Rayleigh ratio of $R_{90} = 48.4 \times 10^{-6}$ cm^{-1} (*20*). The solvent (DMF) was distilled in vacuo and further purified by centrifugation for 1 hour at 15,000 r.p.m. in a Servall centrifuge. Centrifugation was repeated till a dissymmetry value of one was obtained. The PCBL samples were dissolved in the clarified solvent and the solutions obtained purified by centrifugation at 15,000

* For the low-velocity gradients used ($G_{max} = 1.53 \times 10^3$ sec.$^{-1}$) it might be assumed that the intrinsic viscosities of the PCBL preparations measured are equal to those at zero velocity gradient (see ref. *50*).

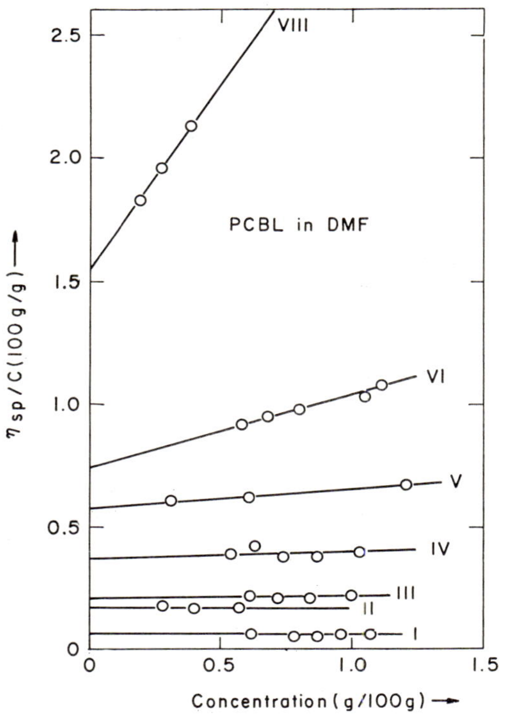

Fig. 16.1. The concentration dependence of the reduced specific viscosity, η_{sp}/c, of PCBL in DMF. Measurements were carried out at 20°.

r.p.m. until constant dissymmetry ratio. Dilutions were made in the cell by adding clarified solvent. The PCBL concentration was occasionally checked by analysis for carbobenzoxy groups (*20a*).

A refractive index increment of 0.123 ml/gm was obtained at 20°, with the aid of a dipping refractometer, Bausch and Lomb Optical Company, N.Y. This value is in agreement with that given by Applequist (*21*).

Results

Viscosity.—The viscosities of the various preparations of PCBL in DMF at 20° are presented in Fig. 16.1, where the reduced specific viscosity, η_{sp}/c, is plotted versus polypeptide concentration, c. The data given indicate that for each of the PCBL samples the variation of η_{sp}/c with c can be expressed by the well-known Huggins' equation (*22*):

$$\eta_{sp}/c = [\eta] + k'[\eta]^2c \tag{1}$$

where the hydrodynamic correction factor k' attains a value of 0.5 to 0.8. Normal values of k' were obtained also by Applequist (*21*) ($k'=0.3$ to 0.5 at 24.4°) for the same solute-solvent system, and by Doty *et al.* (*4*) for poly-γ-benzyl-L-glutamate in DMF. The intrinsic viscosities of the different PCBL samples synthesized are given in the second column of Table 16.2.

Sedimentation.—The concentration dependence of the sedimentation constant, s, of three representative PCBL samples (I, III, and VII) is given in Fig. 16.2. It will be noticed that sample VII possessing the highest intrinsic sedimentation constant shows the most pronounced variation of s with c. The intrinsic sedimentation constant, s_o, of the various polymer samples are presented in Table 16.2. A comparison of the values of the intrinsic sedimentation constant with the corresponding intrinsic viscosities reveals that the increase in $[\eta]$ is accompanied by a parallel increase in s_o.

A correction of the sedimentation constants

Table 16.2—Average molecular weights and hydrodynamic characteristics of the PCBL samples

Sample No.	$[\eta]$ (dl/gm)	S_o (Svedbergs)	$D\times10^7$ ($cm^2\ sec^{-1}$)	M_{SD}*	M_A*	M_{LS}*	DP*
I	0.06	1.90	11.5	20,000	18,000	33,600	76
II	0.18	2.15	6.5	41,000	—	55,100	156
III	0.23	2.65	5.2	63,000	78,000	95,100	240
IV	0.39	2.90	4.0	89,000	—	75,700	340
V	0.60	2.80	3.45	100,000	—	101,000	382
VI	0.78	3.20	2.0	196,000	—	189,000	748
VII	2.00	3.70	1.8	252,000	263,000	244,000	962
VIII	1.63	3.50	1.55	277,000	—	377,000	1,057
IX	2.73	4.13	0.95	534,000	—	572,000	2,038

* M_{SD} denotes the average molecular weights derived from sedimentation and diffusion, M_A denotes those obtained by the Archibald approach-to-equilibrium method, and M_{LS} those derived from light-scattering measurements. DP is the average degree of polymerization calculated from M_{SD}.

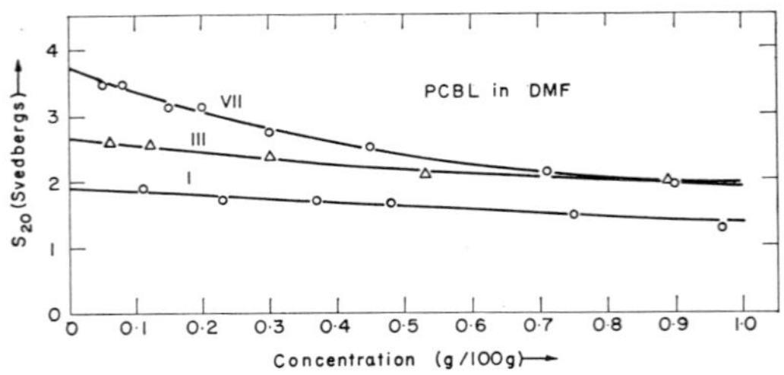

Fig. 16.2. The concentration dependence of the sedimentation coefficient s_{20} of PCBL I, III, and VII in DMF.

of samples III and VII for the viscosity of the solution through multiplication by η/η_o, where η denotes the viscosity of the solution and η_o is that of the solvent (*23, 24*), is given in Fig. 16.3. It is of interest that the curve giving $s\eta/\eta_o$ as a function of c for the low molecular weight sample (III) has a small and negative slope, while the corresponding slope of the high molecular weight sample (VII) is large and positive. These findings indicate according to Burgers (*25*) and Schachman and Kauzmann (*24*) that the molecules of the high molecular weight sample are highly asymmetric.

Diffusion.—The concentration dependence of the diffusion coefficient D_A for three representative PCBL samples (I, III, and VII) is given in Fig. 16.4. The figure shows that D_A for each of these PCBL preparations remains practically constant in the concentration range studied. Similar behavior is shown by the other PCBL samples. Furthermore, the diffusion coefficients D_i and D_m are practically equal to the corresponding values of D_A. The diffusion coefficients D of the various PCBL preparations are listed in Table 16.2.

The diffusion coefficient is given according to Onsager and Fuoss (*26*) (see also ref. *27*) by the expression

$$D = \frac{RT}{f}\left(1 + \frac{dln\gamma}{dlnc}\right) \tag{2}$$

where f is the frictional coefficient and γ is the activity coefficient of the solute. The constancy of D for PCBL in DMF thus indicates that the hydrodynamic factor, f, and the thermodynamic factor

$$\left(1 + \frac{dln\gamma}{dlnc}\right)$$

are either invariant to concentration changes or approximately cancel one another. Similar behavior has been found in other polymer-solvent systems (*27*). An independent evaluation, to be described below, of f from sedimentation studies and of

$$\left(1 + \frac{dln\gamma}{dlnc}\right)$$

from light-scattering measurements confirmed the above assumption as to the approximate proportionality of the two factors.

Molecular weights.—The average molecular weights of the PCBL samples used were determined by the following three methods: (*a*) *Sedimentation-diffusion:* The molecular weights, M_{SD}, obtained by this method, were derived from the well-known Svedberg equation (*18*). The diffusion coefficients and the intrinsic sedimentation constants were taken from Table 16.2. Values of $\overline{V} = 0.847$ ml/gm and $\rho = 0.950$ gm/ml

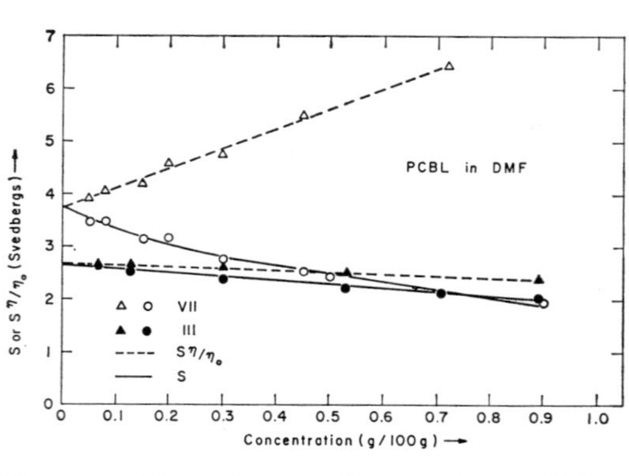

Fig. 16.3. Plot of the sedimentation coefficient s_{20} (full line), and the corrected sedimentation coefficient $s_{20}\ \eta/\eta_o$ (broken line), versus concentration for the PCBL samples III and VII.

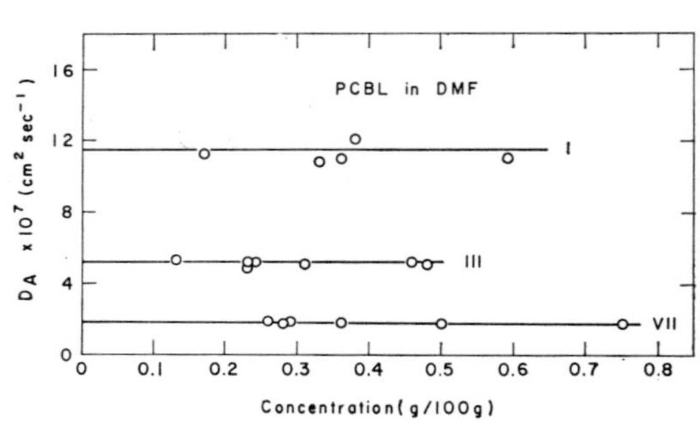

Fig. 16.4. The concentration dependence of the diffusion coefficient D_A of PCBL samples I, III, and VII, measured in DMF at 20°.

were applied. (*b*) *Light scattering:* The dissymmetry method (*20*) was applied. The scattering factor $P(90)$ was derived from the intrinsic dissymmetry factor (I_{45}/I_{135}), assuming a rodlike shape for the poly-ε-carbobenzoxy-L-lysine molecules of preparations I, II, III, IV, and V, and a random coil for samples VI, VII, VIII, and IX (see Discussion). The reduced intensity, R_{90}, was determined as usual and the molecular weights, M_{LS}, were calculated by means of the equation (see ref. *20*)

$$K\frac{c}{R_{90}}P(90) = \frac{1}{M_{LS}} + 2BP(90)c \qquad (3)$$

where K is a characteristic constant and B is the second virial coefficient. A value of $K = 2.79 \times 10^{-7}$ cm² mole gm⁻², similar to that given by Applequist (*21*), was used. (*c*) *Archibald's approach to equilibrium method* (*10*): The method was applied only to three representative samples (I, III, and VII). The values of M_A given in Table 16.2, were obtained by extrapolating the reciprocal of the apparent molecular weight to $c = 0$ (*28*) (see Fig. 16.5).

Table 16.2 shows that the average molecular weights obtained by the three different experimental techniques are on the whole in fair agreement. No satisfactory explanation is available for the very marked discrepancy between the M_{SD} and M_{LS} values of sample VIII. The average degrees of polymerization, DP, given in the last column of Table 16.2 were calculated from the corresponding average values M_{SD}, using a value of 262 for the molecular weight of an ε,N-carbobenzoxy-lysine residue. In the following the molecular weights derived from sedimentation and diffusion will be denoted by M and used in the derivation of the molecular weight dependence of the hydrodynamic parameters.

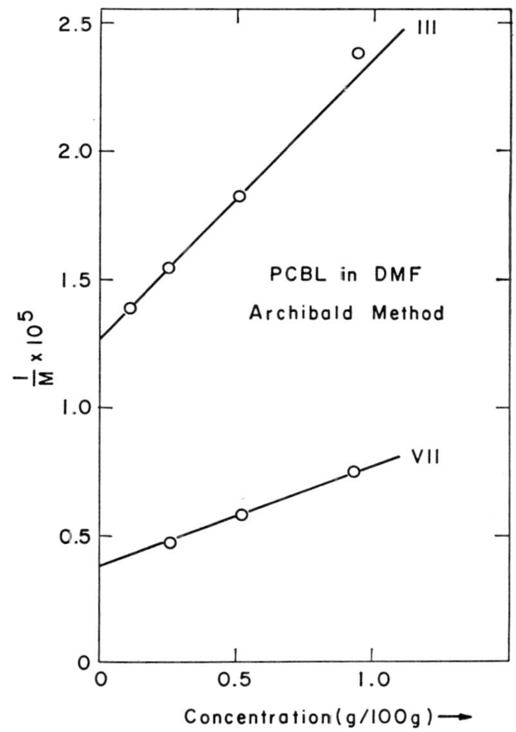

Fig. 16.5. Plot of the reciprocal of the apparent molecular weight derived by the Archibald approach to equilibrium method for PCBL samples III and VII in DMF at 20°.

An estimate of the magnitude of the second virial coefficient, B, could be obtained from the slopes of the curves describing

$$\frac{Kc}{R_{90}}P(90), \ \frac{1}{M_{\text{app}}}$$

(derived from Archibald's approach to equilibrium method, see ref. *28*) or D/S as a function of c (*29, 30*). All three methods yielded values of B in the range of 2.0×10^{-4} to 5.5×10^{-4} mole cm³ gm⁻² for the various PCBL samples tested. The finding that the value of B derived from D/S as a function of c approximately equals that obtained by the other two methods, when taken in conjunction with the previous finding that the diffusion coefficient is practically independent of concentration, is in accord with the suggestion made above that the hydrodynamic and thermodynamic factors approximately cancel one another in the system PCBL-DMF investigated.

No detailed study was made of the polydispersity of the PCBL samples prepared. Some indication concerning the sharpness of the molecular weight distribution could, however, be obtained from the various diffusion schlieren patterns. A comparison of the diffusion curves for PCBL I and VIII, in normal coördinates, with the corresponding normal curves is given in Fig. 16.6. The close fit between the experimental and ideal curves indicates that both preparations possess a rather sharp molecular weight distribution (*27*). A sharp distribution has in fact been predicted for polyamino acids

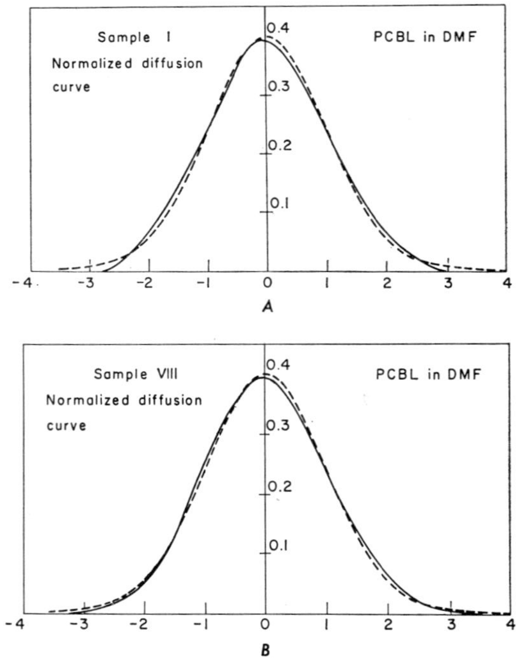

Fig. 16.6. Comparison of the normalized diffusion curve of PCBL (full line) with an ideal curve (broken line). *A:* PCBL sample I. *B:* PCBL sample VIII.

derived from *N*-carboxyamino acid anhydrides as a result of the mechanism and kinetics of polymerization (*1, 31*). The observation that the position of the maximum of the normalized experimental diffusion curve coincides with that of the normal curve is in accord with the previous finding that the diffusion coefficients of PCBL in DMF do not vary significantly with concentration.

Molecular weight dependence of the hydrodynamic parameters.—The variation with the molecular weight of the intrinsic viscosity is presented in Fig. 16.7. The data are those of Table 16.2, and are plotted on a log-log scale. The straight line obtained indicates that the following empirical correlation holds for the range of average molecular weights investigated.

$$[\eta] = 2.24 \times 10^{-7} \times M^{1.26} \quad (4)$$

The molecular weight dependence of the intrinsic sedimentaiton and diffusion coefficients (see Table 16.2) is presented graphically in Fig. 16.8. The straight lines obtained may be presented by the equations:

$$s_o = 1.23 \times 10^{-14} M^{0.27} \quad (5)$$

$$D = 1.51 \times 10^{-3} M^{-0.73} \quad (6)$$

The variation with molecular weight of the dissymmetry ratio I_{45}/I_{135} is given in Fig. 16.9. The parameters $[\eta]$, s_o, D, and I_{45}/I_{135} used in the following theoretical treatment are those taken from the curves of Figures 16.7, 16.8, and 16.9 respectively. This smoothing procedure ensures the detection of the actual trend in the calculated molecular parameters.

Discussion

In the following an attempt will be made to clarify the over-all shape of the poly-ε,*N*-carbobenzoxy-L-lysine (PCBL) molecules in DMF solution from the hydrodynamic data obtained. Optical rotatory dispersion data have indicated that PCBL exists in DMF solution, similarly to other optically active poly-α-amino acids, in a helical conformation (*21*). It should be borne in mind, however, that the characteristic anomalous optical rotatory dispersion is determined by relatively short asymmetric helical segments (*2, 32, 33*), while the hydrodynamic

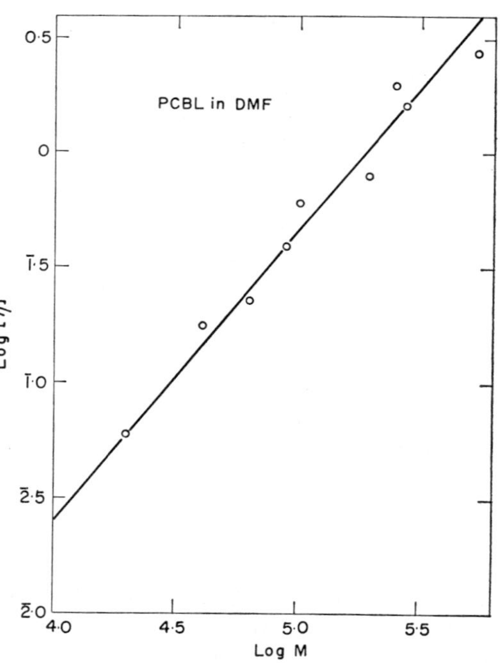

Fig. 16.7. Double logarithmic plot of the intrinsic viscosity, $[\eta]$, versus molecular weight, M, for PCBL in DMF at 20°.

properties are the result of the over-all shape of the macromolecule. A relatively small number of imperfections in a helical rod may thus markedly affect the hydrodynamic properties without altering significantly the optical rotatory behavior. Recent measurements of the viscosity, light scattering, and diffusion of poly-γ-benzyl-L-glutamate in DMF (*34*) revealed that the helical rodlike molecules of this polypeptide (see ref. *4*) possess an appreciable degree of flexibility. Fasman *et al.* (*9*) used the composition of solvent ($CHCl_3$-dichloroacetic acid) at which helix-coil transition occurs as a measure of helix stability. By this criterion the helices of the poly-γ-benzyl-L-glutamate molecules are more stable than those of PCBL. Helices of low stability are expected to possess a higher degree of flexibility than those of helices with great stability.

A number of theories have been developed to account for the hydrodynamic behavior of macromolecules in solution (for review see refs. *35, 36*). These may be divided into theories dealing with corpuscular particles and those describing the behavior of threadlike macromolecules. Since no a priori assumption could be made as to the shape of the PCBL molecules in DMF solution, theories of both types were applied in order to determine the model which best fits the experimental data.

Corpuscular particles are usually substituted by equivalent ellipsoids of revolution which exhibit the same hydrodynamic properties. The equations most commonly used to describe the

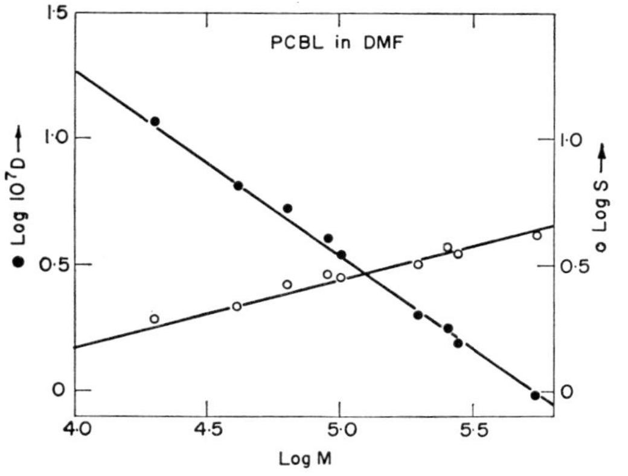

Fig. 16.8. Double logarithmic plots of the intrinsic sedimentation and diffusion coefficients (*s* and *D*) versus molecular weight for PCBL in DMF at 20°.

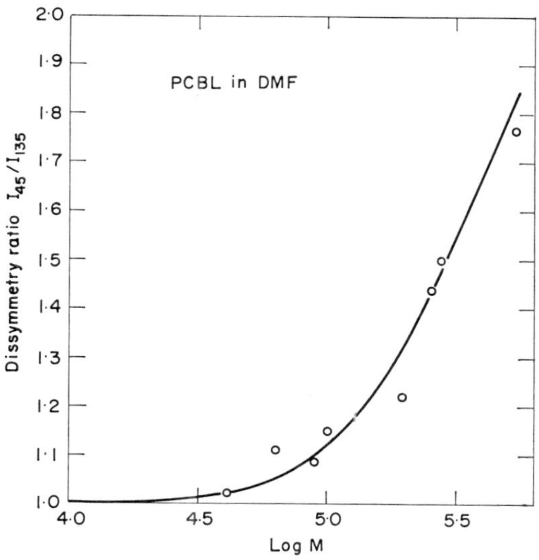

Fig. 16.9. Plot of the dissymetry ratio I_{45}/I_{135} versus the logarithm of molecular weight for PCBL in DMF.

intrinsic viscosities, $[\eta]$, and friction coefficients, f, of such ellipsoids are those of Simha (*37*), Eq. 7, and of Perrin (*38*), Eq. 8, respectively (see also ref. *39*).

$$[\eta] = \frac{N\nu(p)}{100} \cdot \frac{V_e}{M} \tag{7}$$

$$f = (162\pi^2)^{1/3} V_e^{1/3} \eta_0 / F(p) \tag{8}$$

In these equations V_e is the volume of the hydrodynamic ellipsoid, M is the molecular weight of the solute as used in expressing concentration, N is the Avogadro number, η_o is the viscosity of the solvent, and $\nu(p)$ and $F(p)$ are functions of the axial ratio p.

In order to avoid any assumptions as to the volume of the effective hydrodynamic ellipsoid, V_e, Scheraga and Mandelkern (*39*) suggested that the axial ratio p be derived from s_o, $[\eta]$ and M by the elimination of V_e from Equations 7 and 8 and a study of the parameter β (see Eq. 9) obtained in this way.

$$\beta = \frac{Ns_o[\eta]^{1/3}\eta_o}{M^{2/3}(1 - \bar{V}\rho)} = \frac{N^{1/3}}{(16200\pi^2)^{1/3}} F(p)\nu^{1/3}(p) \tag{9}$$

Values of β for prolate ellipsoids as a function of p have been tabulated by the above authors. The various samples of PCBL yielded numerical values of $\beta = 2.44 \times 10^6$ to 2.61×10^6, corresponding to axial ratios of about 11 to 18.

The volumes of the effective hydrodynamic

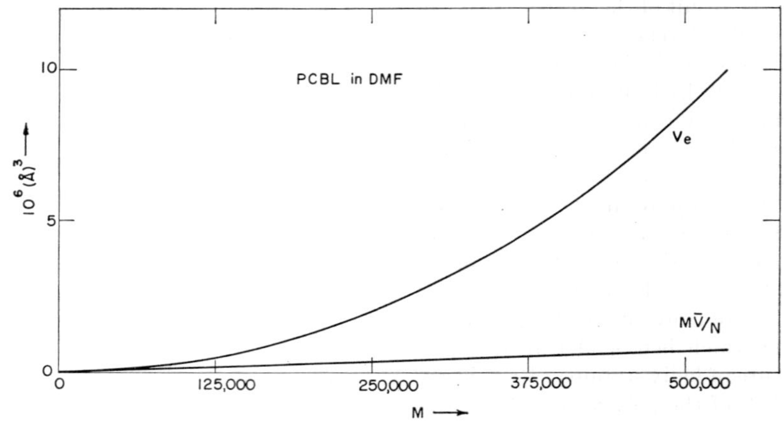

Fig. 16.10. Plot of the effective volume of the equivalent ellipsoids, V_e, and the molecular volume, $M\overline{V}/N$, versus molecular weight.

ellipsoids, V_e, were finally obtained by inserting the numerical values of $\nu(p)$, corresponding to the axial ratios calculated, into Eq. 7. A plot of V_e versus molecular weight is given in Fig. 16.10. For comparison the molecular volumes $M\overline{V}/N$ derived from the partial specific volume, $\overline{V}$, were plotted on the same figure. Fig. 16.10 shows that the effective volumes V_e of the low molecular weight PCBL samples (average molecular weights up to 100,000) are similar to those calculated from the partial specific volume $\overline{V}$. For the PCBL samples with high average molecular weights the values of V_e exceed the corresponding values of $M\overline{V}/N$, the disagreement increasing with molecular weight. These results suggest that the hydrodynamic properties of the low molecular weight samples may be adequately represented by an ellipsoid of revolution with a volume close to that of the unsolvated polymer. In the high molecular weight preparations, on the other hand, the volume of the equivalent ellipsoids is so much greater than that of the unsolvated polymer molecules (in the case of PCBL IX, for example, $V_e \simeq 15 \times M\overline{V}/N$) that the corpuscular model does not seem to hold, and a flexible coiled model, which is known to be associated with high values of V_e, seems to represent more genuinely the configuration of the high molecular weight PCBL molecules.

The dimensions of the equivalent prolate ellipsoids of revolution were derived for the low molecular weight samples according to Simha or Perrin (see Table 16.3) from the molecular volume $M\overline{V}/N$ and the corresponding axial ratios $p = L/d$, where L is the axis of revolution and d is the equatorial axis. The data given in Table 16.3 show that the equatorial axes of the equivalent ellipsoids vary within the limits of about 20 to 22 Å, for the different PCBL samples whose average molecular weights lie in the range of 20,000 to 100,000. The molecular diameters obtained seem reasonable when compared with that reported for poly-γ-benzyl-L-glutamate, 18.2 Å, in its helical conformation (*4*), taking into consideration that the side chains of the latter polyamino acid are considerably shorter than those of PCBL. The major axes, L, of the equivalent ellipsoids, on the other hand, show a marked increase with molecular weight. The values of L derived from the hydrodymamic data are in fair agreement with those calculated from light-scattering data. The latter were obtained from the corresponding dissymmetry ratios, I_{45}/I_{135}, assuming rodlike molecules (*20*).

The data given in Table 16.3 show that the length of the equivalent ellipsoids is proportional to the average molecular weight. The translation along the major axis per amino acid is obtained by dividing the major axis L by the degree of polymerization, DP. The values thus calculated (see last column of Table 16.3) are in close agreement with the value of 1.5 Å predicted for the α-helix by Pauling and Corey (*5*). This finding strongly supports the assumption that the molecules of the PCBL preparations with average molecular weights up to

Table 16.3—Axial ratios and dimensions of the equivalent prolate ellipsoids of revolution

Sample No.	Axial ratios (p) Å Calcd. according to:		Minor axes (d) Å Calcd. according to:		Major axes (L) Å Calcd. according to:			Translation per residue* (Å)
	Simha	Perrin	Simha	Perrin	Simha	Perrin	Light scattering	
I	6	5	20.7	22.0	124	110	—	1.63
II	12	12	20.9	20.9	251	251	250	1.61
III	17	19	21.3	20.7	362	393	385	1.51
IV	22	26	22.1	20.9	486	543	495	1.43
V	25	29	22.1	21.0	551	609	580	1.44

* The last column of the table has been obtained by dividing the major axis calculated according to Simha by the degree of polymerization.

100,000 behave in DMF solution as rigid α-helical rods.

The conclusion drawn above as to the flexibility of the PCBL molecules with molecular weights above 100,000 is further supported by the value of the exponent of M in the viscosity molecular weight relation, $[\eta]=KM^{\alpha}$. The value of $\alpha=1.26$ obtained is considerably lower than that to be expected ($\alpha=1.7$, ref. *40*) for prolate ellipsoids of revolution with lengths proportional to the molecular weight. For coiled threadlike macromolecules the exponent α varies in the range 0.5 to 1.0 (*41, 42*). Free-draining coils give $\alpha=1.0$, while nondraining macromolecules give $\alpha=0.5$.

The hydrodynamic behavior of random coils in solution has been theoretically investigated by various authors. The different methods of approach were reviewed and extended by Peterlin (*36*). The main theories are those of Debye-Bueche (*41*) dealing with the porous sphere model, and of Kirkwood and Riseman (*42*) and Kuhn and Kuhn (*43, 44, 45*) assuming a necklace model. With the aid of Equations 10, 11, and 12 of Debye-Bueche-Peterlin (*46*) and the tabulated values of $F_1(\sigma)$ and $\psi(\sigma)$ as a function of σ (*46, 41*), an evaluation was made of the function $\psi/\phi^{1/3}$, the shielding ratio σ, the function $\psi(\sigma)$, the sphere diameter d_s, and the root-mean-square end-to-end distance R_{BDB} for the various high molecular weight PCBL preparations (see Table 16.4).

$$3\pi\left(\frac{6}{\pi N}\right)^{1/3}\frac{D\eta_o}{kT}([\eta]M)^{1/3} = \phi^{1/3}/\psi = 1/F_1(\sigma) \quad (10)^*$$

$$d_s = \frac{kT}{3\pi\eta_o D\psi(\sigma)} \quad (11)$$

$$d_s = 1.054R_{BDB} \quad (12)$$

According to the theory the shielding ratio, σ, equals zero for macromolecules with complete solvent permeability; it approaches infinity for solvent-immobilizing coils. The low values of σ obtained (1.95 to 1.8) are thus in accord with the previous conclusion drawn from the viscosity-molecular weight relation concerning the high permeability to solvent of the PCBL molecules in DMF solution. As expected the calculated root-mean-square end-to-end distance, R_{BDB}, increases with molecular weight. The numerical values of R_{BDB} are in fair agreement with the corresponding values R_{LS} derived from the dissymmetry ratio, I_{45}/I_{135}, determined by

* In Equations 10 and 14 $[\eta]$ is given in cm^3/gm

Table 16.4—Molecular parameters for PCBL in DMF

Sample No.	$\psi/\phi^{1/3}$	σ	R_{BDB} (Å)	R_{LS} (Å)	R_K (Å)	A_m (Å)	1.5DP/A_m*
V	0.615	1.95	337	410	289	146	3.9
VI	0.609	1.90	580	670	511	232	4.8
VII	0.607	1.90	696	770	626	271	5.3
VIII	0.601	1.80	787	800	689	299	5.3
IX	0.595	1.80	1275	1090	1157	438	7.0

* The last column of the table gives the number of statistical elements per molecule.

light scattering, assuming a random-coil model.

In an attempt to evaluate the experimental data with the aid of the Kirkwood-Riseman theory (*42*) the values of the expression

$$\frac{1}{3\pi}\frac{kT}{D\eta_o}\frac{1}{([\eta]M)^{1/3}}\left(\frac{\pi N}{6}\right)^{1/3}$$

obtained for the different PCBL samples were compared with those to be expected according to theory. Since the experimental values (0.595 to 0.615) are greater than the maximum theoretical value, 0.55 (see ref. *46*), we were unable to use the Kirkwood-Riseman theory for the estimation of the macromolecular hydrodynamic parameters. A similar discrepancy between theory and experiment was recorded previously for polymethyl-methacrylate in acetone and for polyacrylonitrile in DMF (*46*).

The calculation of the length of the statistical segment A_m according to Kuhn and Kuhn, and Peterlin (*45, 46*) was effected with the aid of Equations 13 and 14:

$$x^3 - (1.954C - 1.17)B^{-1}x + 0.954B^{-1} = 0 \quad (13)$$

$$B = \frac{0.43NP^2l_o^3}{M_o[\eta]}; \qquad C = \frac{Nl_o[s]}{0.136M_o}; \qquad (14)^*$$

$$A_m = x^2Pl_o; \qquad [s] = \frac{\eta_o s_o}{(1 - \bar{V}\rho)}$$

where P denotes the degree of polymerization, M_o denotes the molecular weight of the monomer residue, and l_o is the actual length of a monomer residue. In accord with the assumption discussed above that the PCBL molecules attain an α-helix conformation in DMF solution, it was assumed that $l_o = 1.5$ Å. The high values of A_m obtained for the high molecular weight samples investigated ($A_m = 146$ Å to 438 Å, see Table 16.4) demonstrate the relatively high rigidity of the polypeptide molecules in solution. The data given in the table show that the length of the "rigid helical statistical segment," A_m, increases with molecular weight. This finding suggests that the number of imperfections per unit mass decreases on increasing molecular weight. The stability of the helical configuration seems, therefore, to increase with chain length. A similar conclusion was drawn previously on experimental and theoretical grounds (*47, 48*). It should be remarked that the increase in the molecular weight of PCBL is accompanied not only by an increase in the length of the statistical element, A_m, but also by a simultaneous increase in the number of statistical elements per molecule. An increase in the degree of polymerization is thus accompanied by a corresponding increase in the flexibility of the molecule. The root-mean-square end-to-end distances $R_K = xPl_o$ calculated by the Kuhn and Kuhn theory are in fair agreement with those derived according to Debye-Bueche (*41*). It is pertinent to note that the number of statistical elements per peptide chain is rather small even in the PCBL sample (IX) possessing an average molecular weight of 534,000 (see last column of Table 16.4). A quantitative description of the hydrodynamic behavior of the PCBL samples investigated by the now available theories dealing with random coils having a large number of statistical elements should, therefore, hardly be expected.

The evaluation of our experimental data by theories developed for corpuscular particles as well as those for threadlike molecules reveals that the low molecular weight PCBL molecules (with average molecular weights below 100,000) show in DMF solution the characteristic behavior of rigid rods. The high molecular weight PCBL preparations, on the other hand, possess some of the characteristics of permeable, extended non-Gaussian coils. A further increase in the degree of polymerization should be expected to lead to a transition from the free-draining non-Gaussian coils to solvent-immobilizing Gaussian coils.

The optical rotatory behavior of PCBL in DMF (*21*) indicates the presence of a macromolecular asymmetric secondary structure. This finding when taken in conjunction with the conclusions drawn above as to the dimensions of the macromolecules investigated suggests that the polypeptide molecules possess in DMF solution an α-helix conformation. It is the flexibility of the helical polypeptide backbone which determines the over-all shape of the PCBL molecules, i.e., their "tertiary structure," as detected by the various hydrodynamic methods applied.

A comparison of the hydrodynamic properties of PCBL with those of poly-γ-benzyl-L-glutamate (*4, 34*) in DMF solution suggests that the helical conformation of the former polyamino acid is less rigid than that of the latter. A similar conclusion was drawn by Applequist (*21*) who interpreted his light-

* In Equations 10 and 14 $[\eta]$ is given in cm^3/gm.

scattering data by means of the Porod wormlike model (*49*). It is thus plausible to assume, in accord with previous findings (*9*), that the chemical nature of the side chains may markedly affect the stability as well as flexibility of the polypeptide helical structure in solution.

Acknowledgment

The authors are indebted to Mr. M. Weinstein for carrying out the light-scattering measurements. Thanks are due also to Mr. A. Lustig for technical assistance. This investigation was supported by a research grant (A-3083) from the National Institutes of Health, United States Public Health Service.

References

1 E. Katchalski and M. Sela, *Advances in Protein Chem.* **13,** 243 (1958).
2 E. Katchalski and I. Z. Steinberg, *Ann. Rev. Phys. Chem.* **12** (1961).
3 P. Doty, A. M. Holtzer, J. H. Bradbury, and E. R. Blout, *J. Am. Chem. Soc.* **76,** 4493 (1954).
4 P. Doty, J. H. Bradbury, and A. M. Holtzer, *ibid.* **78,** 947 (1956).
5 L. Pauling, R. B. Corey, and H. R. Branson, *Proc. Nat. Acad. Sci.* **37,** 205 (1951).
6 J. T. Yang, *J. Am. Chem. Soc.* **80,** 1783 (1958).
7 J. T. Yang, *ibid.* **81,** 3902 (1959).
8 E. Katchalski, I. Grossfeld, and M. Frankel, *ibid.* **70,** 2094 (1948).
9 G. D. Fasman, M. Idelson, and E. R. Blout, *ibid.* **83,** 709 (1961).
10 W. J. Archibald, *J. Phys. Coll. Chem.* **51,** 1204 (1949).
11 S. M. Klainer and G. Kegeles, *J. Phys. Chem.* **59,** 952 (1955).
12 E. G. Pickels, W. F. Harrington, and H. K. Schachman, *Proc. Nat. Acad. Sci.* **38,** 943 (1952).
13 H. K. Schachman, "Methods in Enzymology," Academic Press, Inc., New York, 1957, Vol. IV, p. 32.
14 A. Ehrenberg, *Acta Chem. Scand.* **11,** 1257 (1957).
15 J. M. Creeth, *J. Phys. Chem.* **62,** 66 (1958).
16 A. R. Gordon, *Ann. N.Y. Acad. Sci.* **46,** 285 (1945).
17 A. L. Geddes, "Physical Methods in Organic Chemistry," Interscience, New York, 1949, Vol. I, Pt. I, p. 551.
18 T. Svedberg and K. O. Pederson, "The Ultracentrifuge," Oxford, Clarendon Press, 1940.
19 H. L. Yakel, *Acta Cryst.* **6,** 724 (1953).
20 P. Doty and R. F. Steiner, *J. Chem. Phys.* **18,** 1211 (1950).
20a A. Patchornik and S. Ehrlich-Rogozinski, *Anal. Chem.* **33,** 803 (1961).
21 J. B. Applequist, Ph.D. thesis, Harvard University, Cambridge, Mass., 1959.
22 M. L. Huggins, *J. Am. Chem. Soc.* **64,** 2716 (1942).
23 M. A. Lauffer, *ibid.* **66,** 1195 (1944).
24 H. K. Schachman and W. J. Kauzmann, *J. Phys. Coll. Chem.* **53,** 150 (1949).
25 J. M. Burgers, *Proc. Acad. Sci. Amsterdam* **44,** 1075, 1177 (1941); **45,** 9, 126 (1942).
26 L. Onsager and R. M. Fuoss, *J. Phys. Chem.* **51,** 158 (1938).
27 C. O. Beckmann and J. L. Rosenberg, *Ann. N.Y. Acad. Sci.* **46,** 329 (1945).
28 G. Kegeles, S. M. Klainer, and W. J. Salem, *J. Phys. Chem.* **61,** 1286 (1957).
29 P. J. Flory, "Principles of Polymer Chemistry," Cornell University Press, Ithaca, N. Y., 1953, p. 306.
30 E. Katchalski and I. Z. Steinberg, in preparation.
31 C. H. Bamford, A. Elliott, and W. E. Hanby, "Synthetic Polypeptides," Academic Press, New York, 1956.
32 I. Z. Steinberg, W. F. Harrington, A. Berger, M. Sela, and E. Katchalski, *J. Am. Chem. Soc.* **82,** 5263 (1960).
33 M. Goodman and E. E. Schmitt, *Ann. N.Y. Acad. Sci.* **88,** 669 (1960).
34 G. Spach, Ph.D. thesis, Strasbourg University (1960).
35 C. Sadron, *Progress in Biophys. and Biophys. Chem.* **3,** 237 (1953).
36 A. Peterlin, *Makromol. Chem.* **34,** 89 (1959).
37 R. Simha, *J. Phys. Chem.* **44,** 25 (1940).
38 F. Perrin, *J. phys. radium VII,* **7,** 1 (1936).
39 H. A. Scheraga and L. Mandelkern, *J. Am. Chem. Soc.* **75,** 179 (1953).
40 R. Simha, *J. Chem. Phys.* **13,** 188 (1945).
41 P. Debye and A. M. Bueche, *ibid.* **16,** 573 (1948).
42 J. G. Kirkwood and J. Riseman, *ibid.* **16,** 565 (1948).
43 H. Kuhn and W. H. Kuhn, *J. Polymer Sci.* **5,** 519 (1950).
44 H. Kuhn and W. H. Kuhn, *ibid.* **9,** 1 (1952).
45 H. Kuhn, W. H. Kuhn, and A. Silberberg, *ibid.* **14,** 193 (1954).
46 A. Peterlin, *J. Colloid. Sci.* **10,** 587 (1955).
47 B. H. Zimm and J. K. Bragg, *J. Chem. Phys.* **31,** 526 (1959).
48 B. H. Zimm, P. Doty, and K. Iso, *Proc. Nat. Acad. Sci.* **45,** 1601 (1959).
49 O. Kratky and G. Porod *Rec. trav. chim.* **68,** 1106 (1949).
50 H. A. Scheraga, *J. Chem. Phys.* **23,** 1526 (1955).

PAPER 17

Conformational Studies on Oligomeric and Polymeric Peptides

M. GOODMAN, E. E. SCHMITT, I. LISTOWSKY, F. BOARDMAN, I. G. ROSEN, and M. A. STAKE

Oligomer studies

We have synthesized oligomeric peptides derived from γ-methyl-L-glutamate (*1, 2, 3*) and β-methyl-L-aspartate (*4*). These peptides were subjected to conformational studies using optical rotations (*1, 5*) and dispersions (*2, 5*). The compounds were synthesized via mixed anhydride and *p*-nitrophenyl ester methods (*3, 5*). Each peptide was carefully purified by extraction and crystallization techniques (*5*).

Since optical activity has been extremely useful in the elucidation of secondary structure in proteins (*6–10*) and polypeptides (*11–16*), it was considered likely that optical rotations could be applied to the examination of the spatial arrangements of oligomeric peptides. We have demonstrated positive optical activity enhancements in so-called "helix-forming" solvents in the L-glutamate oligomer series (*2, 3*). These enhancements we attributed primarily to the onset of intramolecular hydrogen bonding (Fig. 17.1). In a later section of this paper, we discuss refinements of this interpretation based on model compounds. With a "random coil" solvent such as dichloroacetic acid, no positive enhancement of the optical activity is observed (Fig. 17.1). As stated before (*17*), association phenomena complicate our results in dioxane.

Rotatory dispersion data (*2, 3*) support the contention that intramolecular hydrogen bonding is a cause of abnormal effects (*18–21*). In dimethylformamide with our oligomeric peptides, the Moffitt plot b_o becomes increasingly more negative above the pentamer (*2, 3*) (Fig. 17.2). In addition, the λ_c of the Drude plot is constant (approximately 180 mμ) for the di- through tetrapeptide. Starting with the pentamer a shift to longer wave lengths is encountered. Moreover, commencing with the heptamer the modified Drude plot of $[\alpha]\lambda^2$ versus $[\alpha]$ becomes nonlinear (*2, 3, 22*).

Model compound studies

In addition to oligomeric peptide synthesis, we prepared a series of model compounds (I–VI) which can be used to calculate the random-coil rotation in the poly-γ-methyl-L-glutamate series (*23*). Brand (*24*) and Doty (*25*) suggested that a peptide end group contributes to the total optical activity of a peptide differently from an internal residue. We have verified this contention and correlated model compound rotations with the oligomeric peptides derived from γ-methyl-L-glutamate.

OMe
|
Z-glu-gly-OMe (I)

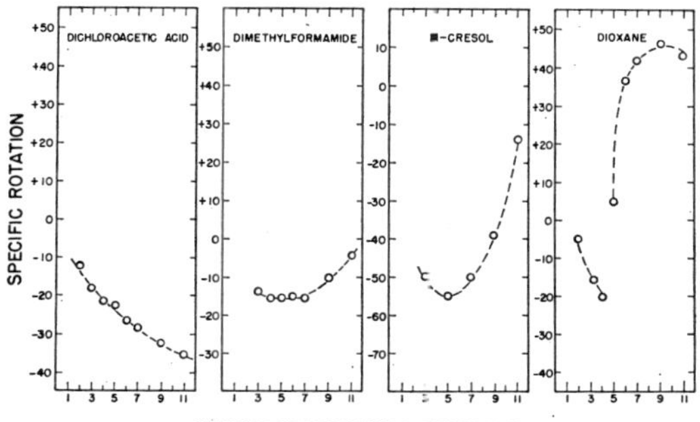

Fig. 17.1. Optical activity (measured at 589 mμ) of the oligomeric peptides derived from γ-methyl-L-glutamate in various solvents at 25°.

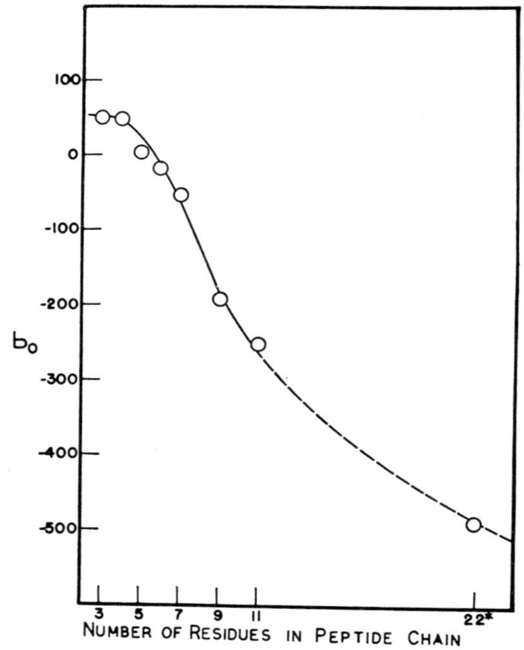

Fig. 17.2. The b_o coefficients of the oligomeric peptides derived from γ-methyl-L-glutamate in dimethylformamide at 25°.

* The b_o coefficient for a lower molecular weight polymer (*A/I* 22) is also included in this plot.

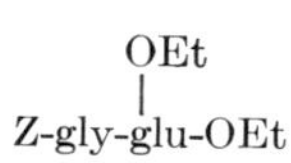 (II)

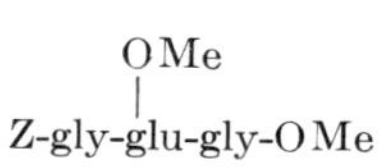 (III)

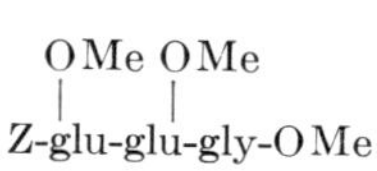 (IV)

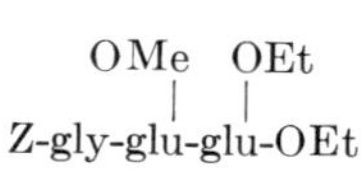 (V)

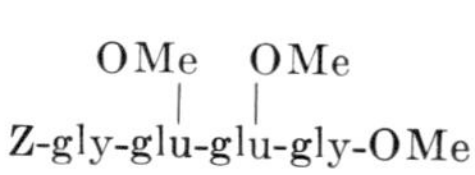 (VI)

where compounds I, II, and III represent models for the N-terminal, C-terminal, and internal residues, respectively, in the poly-γ-methyl-L-glutamate series. Compounds IV and V represent terminal glutamyl residues adjacent to internal glutamates. Compound VI simulates a model substance containing adjacent internal glutamyl residues. With compounds (IV–VI), the effect on optical activity of adjacent optically active residues can be studied. Using molar rotations (φ) defined as:

$$\phi = \frac{[\alpha]_{\mathrm{D}}^{25} \times MW}{10000} \quad \frac{\text{deg}}{\text{cm mole}} \tag{1}$$

We have related these compounds to the oligomeric series of peptides by the following equation:

$$\phi_{\text{total}} = \phi_{N\text{terminal}} + \phi_{C\text{terminal}} + n\phi_{\text{internal}}. \tag{2}$$

Using the additivity principle of equation 2, we constructed graphs of the total molar rotations of peptides plotted against the number of internal residues. The molar rotation of compound III in a given solvent determines the slope of lines obtained using the "simple" model compounds (I–III). Half of the molar rotation of substance VI gives the slope based on the "complex" model compounds (IV–VI).

In dioxane and dimethylformamide (Fig. 17.3), close agreement exists between the calculated molar rotations (compounds I–III; lines *a* and *e*) and the experimental oligomeric peptide values as long as secondary structural effects do not enter. Exact correspondence occurs when the "complex" substances (IV–VI; lines *b* and *f*) are used. Small deviations from the calculated rotations based on compounds (IV–VI) are observed at the pentamer in dimethyl formamide while large deviations commence at the nonamer. In dioxane, large deviations in rotation are noted at the pentamer. These observations are in accordance with the concept that secondary structure appears at these stages (*2, 3, 22*). As noted above, association occurs in dioxane (*17*) obscuring our interpretations somewhat.

Similarly, in *m*-cresol (Fig. 17.3) the large deviation of the oligomeric peptide rotations from the calculated values (line *i*) commences at the nonapeptide. At this point, helical structures begin. The slope of the line made by the dipeptide through tetrapeptide corresponds to the slope calculated from simple model compounds (I–III, line *g*) while the slope of the line made by the pentamer through the heptamer is more shallow and corresponds to the slope based on the complex model compounds (IV–VI; line *h*). Although two slopes are observed through the heptamer, we feel certain that only a random coil exists.

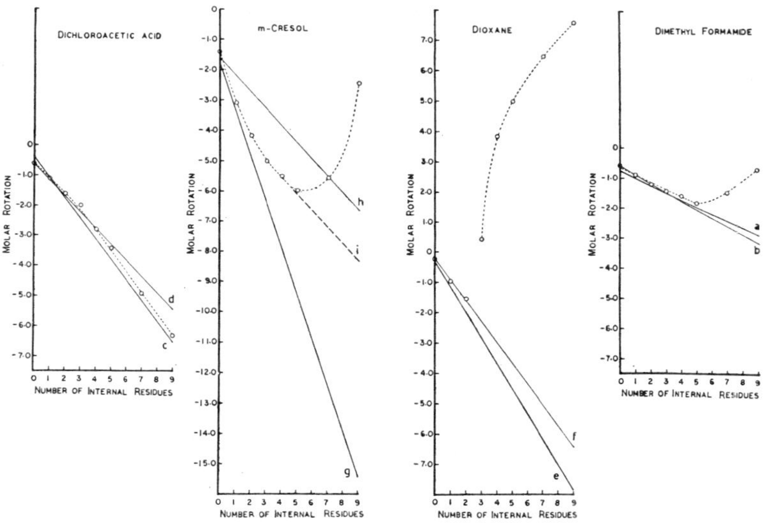

Fig. 17.3. Molar rotations based on the model compounds (I–III; lines *a*, *c*, *e*, *g*) and the model compounds (IV–VI; lines *b*, *d*, *f*, *h*) and oligomeric peptides derived from γ-methyl-L-glutamate, in various solvents.

In dichloroacetic acid (Fig. 17.3), where secondary structural effects do not enter, two distinct slopes are also noted. From the dimer through the tetramer, a shallow slope is obtained corresponding to the slope based on the complex model substances (IV–VI; line *d*). A steeper slope is observed from the pentamer through the undecamer, which is nearly identical with the slope based on the simple model compounds (I–III; line *c*). The existence of two slopes in dichloroacetic acid is best explained as a steric hindrance phenomenon. We believe that the end groups of the peptides solvate completely and, thereby, prevent the approach of additional solvent molecules to the neighboring internal amide bonds. It is not until the pentamer that a fully solvatable internal residue appears (residue 3). This residue is not bonded to an end group. In *m*-cresol, on the other hand, side-chain solvation effects are probably dominant. Two internal adjacent glutamyl side chains create sufficient steric hindrance to prevent complete solvation of internal amide residues.

β-Methyl-L-aspartate studies

Recently we initiated a program of oligomeric and model compound peptide synthesis based on β-methyl-L-aspartate (*4*). Preliminary results indicate some striking differences from the glutamate series as can be seen by comparing Table 17.1 with the glutamate data (*3*).

With the glutamate peptides, abnormal rotatory effects commenced above the pentamer. The situation in the aspartate series shows that abnormal effects exist in all the compounds

Table 17.1—Peptides derived from β-methyl-L-aspartate

	Dioxane			Dichloroacetic acid			Dimethylformamide		
	$[\alpha]_D^{25}$	b_o	λ_c	$[\alpha]_D^{25}$	b_o	λ_c	$[\alpha]_D^{25}$	b_o	λ_c
Dipeptide	− 1.39	−125	340	+17.94	−101	182*	−25.74	−46	232
Tripeptide	−28.58	−213	263	− 1.01	−142	326*	−34.47	−85	239
Tetrapeptide	−50.06	−245	256	−13.1	−152	325	−44.44	−111	237
Pentapeptide	−53.2	−222	254	−19.1	−188	273	−48.1	−150	242

* Nonlinear plot.

made to date. A reasonable explanation for these differences and correlations with model compounds must await synthesis of higher members of the oligomeric series.

Acidic solvent studies

The optical activity of the internal model peptide (compound III) is essentially identical $[\alpha]_D^{25} = \sim -15°$) in the following acidic solvents: (*a*) dichloroacetic acid; (*b*) trifluoroacetic acid; (*c*) formic acid; (*d*) acetic acid. The rotation of a given polymer of γ-methyl-L-glutamate, then, should also be the same barring conformational or specific interaction effects. The observation of the identical rotations for the model compounds may not be general. It is possible that compound VI will yield different slopes for each acid. Certainly other amino acids may give vastly different results. With these reservations in mind, we examined poly-γ-methyl-L-glutamate (DP_n 140) in dichloroacetic acid and trifluoroacetic acid. Table 17.2 contains our results.

Table 17.2—Rotatory properties of a poly-γ-methyl-L-glutamate (DP_n140) in acidic solvents

Solvent	$[\alpha]_{289}^{25}$	b_o	λ_c
Dichloroacetic acid	−31.3	+30	187
Trifluoroacetic acid	−76.1	+68	203
Dichloroacetic acid–benzene (50% *v/v*)	−10.4	−190	309*
Trifluoroacetic acid–benzene (50% *v/v*)	−92	+150	183

* Slope was nonlinear at higher wave lengths.

It is quite obvious from these data that the specific rotations are quite different from the values predicted using model compounds. With both solvents the effect of polydispersity on the optical activity is unknown. However, it is worth noting that the deviations from the predicted rotation are in opposite directions. In the case of dichloroacetic acid the deviation is in the positive direction, while with trifluoroacetic acid it is in the negative direction. The polymer in both solvents is essentially a random coil. An explanation for the deviations may be as follows: with dichloroacetic acid some residual intramolecular hydrogen bonding (helix type) may persist, while in trifluoroacetic acid specific interactions may contribute to the asymmetry in a manner different from the configurational optical activity calculated using the model compounds. Possible indirect support for these suggestions may be seen from the data where the acidic solvents have been combined with benzene. In dichloroacetic acid the rotatory dispersion data indicate appreciable helix formation. In trifluoroacetic acid a random coil still appears to be present, however, with different rotatory properties which may be explained on the basis of side chain–main chain interactions (*25, 26*).

Temperature dependence studies

A gradual decrease of the optical rotation deviation of the oligomeric peptides from those calculated using model compounds was observed in dimethylformamide as the temperature was increased from 5° to 70°. At 70°, a deviation exists only at and above the nonapeptide. The molar rotation difference between experimental and calculated is much smaller for the nonapeptide at 70° than at lower temperatures. These observations are attributed to the fact that secondary structural contributions to the optical rotation in the oligomeric peptides are diminished with increasing temperature.

References

1 M. Goodman and E. E. Schmitt, *J. Am. Chem. Soc.* **81,** 5507 (1959).
2 M. Goodman, E. E. Schmitt, and D. Yphantis, *ibid.* **82,** 3483 (1960).
3 M. Goodman and E. E. Schmitt, *Ann. N.Y. Acad. Sci.* **88,** 669 (1960).
4 M. Goodman and F. Boardman, to be published.
5 M. Goodman, E. E. Schmitt, and D. Yphantis, *J. Am. Chem. Soc.*, to be published.
6 S. J. Leach, *Rev. Pure Appl. Chem.* **9,** 33 (1959).
7 C. Tanford, *J. Am. Chem. Soc.* **83,** 1628 (1961).
8 C. Tanford and V. G. Taggart, *ibid.* 1634 (1961).
9 C. Cohen, *Nature* **175,** 129 (1955).
10 J. G. Foss and J. A. Schellman, *J. Phys. Chem.* **63,** 2007 (1959).
11 P. Doty, A. M. Holtzer, J. H. Bradbury, and E. R. Blout, *J. Am. Chem. Soc.* **76,** 4492 (1954).
12 E. R. Blout, R. H. Karlson, P. Doty, and B. Hargitay, *ibid.* **76,** 4492 (1954).

13 A. R. Downie, A. Elliott, W. E. Hanby, and B. R. Malcolm, *Proc. Roy. Soc.* **A242,** 325 (1957).
14 P. Doty and R. D. Lundberg, *Proc. Nat. Acad. Sci.* **43,** 213 (1957).
15 A. Elliott, W. E. Hanby, and B. R. Malcolm, *Nature* **178,** 1170 (1956).
16 E. R. Blout and R. H. Karlson, *J. Am. Chem. Soc.* **80,** 1259 (1958).
17 M. Goodman, *J. Polymer Sci.* **49,** 150 (1961).
18 W. Moffitt, *J. Chem. Phys.* **25,** 467 (1956).
19 E. R. Blout, in C. Djerassi "Optical Rotatory Dispersion," McGraw-Hill, New York, 1960, Chap. 17.
20 P. Doty, A. Wada, J. T. Yang, and E. R. Blout, *J. Polymer Sci.* **23,** 851 (1957).
21 I. Tinoco and R. J. Woody, *J. Chem. Phys.* **32,** 461 (1960).
22 J. T. Yang and P. Doty, *J. Am. Chem. Soc.* **79,** 761 (1957).
23 M. Goodman, I. Listowsky, and E. E. Schmitt, *ibid.,* to be published.
24 E. Brand, B. F. Erlanger, and H. Sachs, *ibid.* **23,** 3508 (1951).
25 P. Doty and E. P. Geiduschek in "The Proteins," H. Bailey and K. Neurath (eds.), Acad. Press, N. Y., 1953, p. 393.

PAPER 18

Characteristic Amide Bands and Conformations of Polypeptides

TATSUO MIYAZAWA

Polypeptide chains have been known to take various conformations. For example, polyglycine has two crystalline forms (*1*); polyglycine I is in the extended conformation whereas polyglycine II is in a helical conformation with a threefold screw axis (*2*). Poly-L-alanine (*3*) takes either the α-helix (*4*) or the extended conformation in the solid film. In solution poly-γ-benzyl-L-glutamate (*5*) or poly-α,L-glutamic acid (*6*) exists either in the α-helical conformation or in the randomly coiled conformation. The infrared spectra of these polypeptides naturally depend upon their chain conformations and the amide I band (1600–1700 cm^{-1}) and the amide II band (1500–1550 cm^{-1}) have been used for conformation diagnoses since Ambrose and Elliott pointed out certain frequency-conformation correlations (*7*). It has been found (*8*) that optically active polypeptides in the folded conformation exhibit the amide I band (parallel dichroism) at 1652–1657 cm^{-1} and the amide II band (perpendicular dichroism) at 1545–1551 cm^{-1}, whereas those in the extended conformation exhibit the amide I band (perpendicular dichroism) at 1628–1635 cm^{-1} and the amide II band (parallel dichroism) at 1521–1525 cm^{-1}.

In addition to the amide I and II bands the peptide CONH group exhibits a number of other characteristic amide bands. The frequencies and approximate descriptions of the amide bands of liquid *N*-methylacetamide (*9–11*) are summarized as shown below.

amide A	3280 cm^{-1}	Fermi resonance {N—H stretching, 2×amide II}
amide B	3090 cm^{-1}	
amide I	1653 cm^{-1}	C═O stretching
amide II	1567 cm^{-1}	N—H in-plane bending, C—N stretching
amide III	1299 cm^{-1}	C—N stretching, N—H in-plane bending
amide IV	627 cm^{-1}	O═C—N in-plane bending
amide V	725 cm^{-1}	N—H out-of-plane bending
amide VI	600 cm^{-1}	C═O out-of-plane bending
amide VII	206 cm^{-1}	C—N torsion

The amide I, II, III, and IV bands are due to the in-plane vibrations of the CONH group and the nature of these vibrations have been studied by theoretical calculations of *N*-methylacetamide (*12*). On the basis of these calculations the amide I and II vibrations of polypeptides have been treated in terms of the vibrational interactions among peptide groups, and as expected from this treatment the amide I and II bands of the α-helix or the extended conformation have been found to be split into two components (*13*). The amide I and II frequencies of various chain conformations have been systematically explained and also certain new frequency-conformation correlations have been found (*14*). These correlations for the amide I and II bands are useful for conformation diagnoses of polypeptides and some fibrous proteins; however, they may not serve well if a number of various conformations coexist (*15*). The amide IV–VII vibrations are not localized in the CONH group and strong vibrational interactions will take place among peptide groups. Accordingly the infrared bands due to these vibrations are expected to depend upon the conformational changes much more than do the amide I and II bands. In this connection theoretical treatments of the normal vibrations of helical conformations will be informative. In the present paper the nature of all the fundamental amide vibrations and the dependence of these vibrations upon the helical chain conformations will be discussed. The results of the normal coördinate treatments of polyglycine I will be described. Finally the infrared spectra of poly-γ-methyl-L-glutamate and poly-γ-methyl-DL-glutamate will be compared with each other.

Amide A and B bands

Monosubstituted amides and polypeptides show two characteristic bands in the 3000 cm^{-1} region, namely, the strong band at *ca.* 3300 cm^{-1} (the amide A band) and the weaker band at *ca.* 3100 cm^{-1} (the amide B band) (*16*). These bands disappear on N-deuteration and are undoubtedly associated with the N—H bond. The nature of the amide B band has been discussed (*17*). Badger and Pullin have proposed the explanation that the amide A and B bands arise from the Fermi resonance between the fundamental N—H stretching vibration and the first overtone of the amide II vibration (*19*). In connection with this explanation, the amide bands of *N*-methylacetamide or *N*-methylformamide have been measured in the gaseous state, liquid state, and the crystalline state. For the case of *N*-methylacetamide: the amide A bands were observed at 3470 cm^{-1} (gas), 3280 cm^{-1} (liquid), and 3250 cm^{-1} (crystal at −60°); the amide B bands at 3090 cm^{-1} (liquid), and 3100 cm^{-1} (crystal); and the amide II bands at 1487 cm^{-1} (gas), 1567 cm^{-1} (liquid), and 1583 cm^{-1} (crystal). Thus as the hydrogen bonds become stronger the amide A frequency decreases whereas both the amide B and amide II frequencies increase. The frequency shifts of the amide A, B, and II bands are thus qualitatively in accord with the Fermi resonance explanation. A quantitative treatment has been made (*10*), however, and the frequencies of the A and B bands have been derived to be

$$\nu_A = \nu_a^o/2 + \nu_{II}^o + (\nu'^2 + \nu''^2)^{\frac{1}{2}} \quad (1)$$

$$\nu_B = \nu_a^o/2 + \nu_{II}^o - (\nu'^2 + \nu''^2)^{\frac{1}{2}} \quad (2)$$

$$\nu'' = \nu_a^o/2 - \nu_{II}^o$$

where ν_a^o and ν_{II}^o are the unperturbed amide A and II frequencies and ν' is the interaction constant arising from the mechanical anharmonicity. The intensity ratio of the A and B bands has been found to be

$$I_B/I_A = \{(\nu'^2 + \nu''^2)^{\frac{1}{2}} - \nu''\}/\{(\nu'^2 + \nu''^2)^{\frac{1}{2}} + \nu''.\} \quad (3)$$

From the observed frequencies of ν_A, ν_B, and ν_{II}, the unperturbed N—H stretching frequencies and the interaction constants of *N*-methylacetamide have been found to be $\nu_a^o = 3236$ cm^{-1} and $\nu' = 80$ cm^{-1} in the liquid state and $\nu_a^o = 3184$ cm^{-1} and $\nu' = 74$ cm^{-1} in the crystalline state. Making use of these values, the intensity ratios are calculated to be 0.3 for the liquid state and 0.7_5 in the crystalline state, in good agreements with the observed ratios. Satisfactory agreements between the calculated and observed intensity ratios have also been obtained for the case of *N*-methylformamide. These analyses (*10*) render substantial support to the explanation proposed by Badger and Pullin. For the case of cis amide group two characteristic bands (*16*) are observed at *ca.* 3200 cm^{-1} and *ca.* 3100 cm^{-1} and these bands are considered to arise from the Fermi resonance between the fundamental N—H stretching vibration and the combination tone of the C=O stretching vibration (*ca.* 1650 cm^{-1}) and the N—H in-plane bending vibration (*ca.* 1450 cm^{-1}) (*20*).

It may be remarked that the difference between the unperturbed N—H stretching frequency ν_a^o and the observed ν_A frequency is as great as 45 cm^{-1} in the liquid state or 65 cm^{-1} in the crystalline state. Accordingly in discussing the strengths of hydrogen bonds it is important to refer to the ν_a^o frequency rather than to the observed ν_A frequency.

The amide A band of the folded conformation shows parallel dichroism, whereas the amide A band of the extended conformation shows perpendicular dichroism. The amide A frequencies of the two conformations, however, are not much different (for example, the folded form of poly-L-alanine: 3293 cm^{-1}, the extended form: 3283 cm^{-1}) (*3, 7*). The amide A and B frequencies of polypeptides in various conformations are considered to depend upon the strengths of the hydrogen bonds, the Fermi resonance, and the vibrational interactions among peptide groups through intergroup hydrogen bonds.

Nature of the amide I, II, and III vibrations

It has long been recognized that the trans-peptide CONH group exhibits the amide I band at *ca.* 1650 cm^{-1} and the amide II band at *ca.* 1550 cm^{-1}. The amide I band is undoubtedly associated with the C=O bond. The amide II band has once been assigned to the N—H in-plane bending mode (*21*) or to the C—N stretching mode (*22*); however, Lenormant's experimental observations (*23*) have led Fraser and Price to suggest that the amide I band is due to the antisymmetric (out-of-phase) stretching mode of the O=C—N group and that the amide II and III bands arise from the interaction between the symmetric (in-phase) stretching mode of the O=C—N group and the N—H in-plane bending mode (*24*). Miyazawa *et al.* have made a series of infrared studies on

$$(H—CONH—)_2,\ (CH_3—CONH—)_2,\ (—CONH—CH_3)_2,\ H—CONH—CH_3,\ CH_3—CONH—CH_3,\ CH_3—CONH—C_2H_5,$$

and their N-deuterated derivatives and have systematically assigned the normal vibrations of all these molecules (*9, 25*). The observed amide bands are partly shown in Table 18.1. The amide I–IV bands of the CONH group are due to the in-plane vibrations, whereas the amide V and VI bands are due to the out-of-plane vibrations. The corresponding characteristic bands of the COND group are also shown in Table 18.1.

In order to obtain quantitative informations about the nature of the in-plane amide vibrations, the normal vibrations of *N*-methylacetamide and its N-deuterated derivative have been calculated (*12, 26*). The potential constants were transferred from other molecules whose potential constants have been uniquely determined. The in-plane frequencies thus calculated agreed well with the observed ones even though no adjustments in the potential constants have been made in order to obtain better frequency agreements. The vibrational potential energy distributions (useful measures of the nature of normal vibrations) and the vibrational modes of the characteristic amide vibrations have been calculated as shown in Table 18.2 and in Fig. 18.1.

Amide I vibration (*ca. 1650 cm^{-1}*).—For this vibration the C—N stretching motion takes place out-of-phase from the C=O stretching motion; however, the amplitude ratio $r(C—N)/r(C=O)$ is as small as 0.4. As shown in Fig. 18.1 the N—H in-plane bending amplitude is the greatest of all. However, the potential constant associated with this displacement is small. Because of the large potential constant associated with the C=O stretching displacement, 80 per cent of the vibrational potential energy is associated with the C=O stretching mode and this vibration may well be described as the C=O stretching vibration in terms of the potential energy.

Amide II vibration (*ca. 1550 cm^{-1}*).—For this vibration the amplitude ratio of the C=O stretching and C—N stretching displacements is as small as $r(C=O)/r(C—N) = -0.2$, and energetically the contribution of the C=O stretching mode is negligible. As seen in Fig. 18.1 the N—H in-plane bending displacement is quite prominent while the carbonyl oxygen atom and the methyl groups stay at rest. Sixty per cent of the vibrational potential energy is

Table 18.1—Characteristic amide frequencies (in cm^{-1}) in the liquid state (*9*)

	I	II	III	IV	V	VI
$CH_3—CONH—CH_3$	1653	1567	1299	627	725	600
$CH_3—CONH—C_2H_5$	1655	1565	1295	619	710	602
	I′	II′	III′	IV′	V′	VI′
$CH_3—COND—CH_3$	1642	1475	960	627	510	(627)
$CH_3—COND—C_2H_5$	1640	1473	960	619	525	(619)

Table 18.2—Observed and calculated amide frequencies (in cm^{-1}) and calculated potential energy distributions (in percentage) of *N*-methylacetamide and its *N*-deuterated derivative (*12*)

	Frequencies		Potential energy distributions*				
	Obs.	Calc.	$\nu_{C=O}$	ν_{C-N}	β_{N-H}	$\beta_{O=C-N}$	Others
I	1653	1644	0.8	0.1	0.1	0.0	0.0
II	1567	1509	0.0	0.4	0.6	0.0	0.0
III	1299	1296	0.1	0.3	0.3	0.1	0.2
IV	627	602	0.0	0.0	0.0	0.4	0.6
I′	1642	1630	0.8	0.2	0.0	0.0	0.0
II′	1475	1421	0.0	0.7	0.2	0.1	0.0
III′	960	928	0.1	0.1	0.2	0.0	0.6
IV′	627	601	0.0	0.0	0.0	0.4	0.6

* ν, stretching mode; β, bending mode.

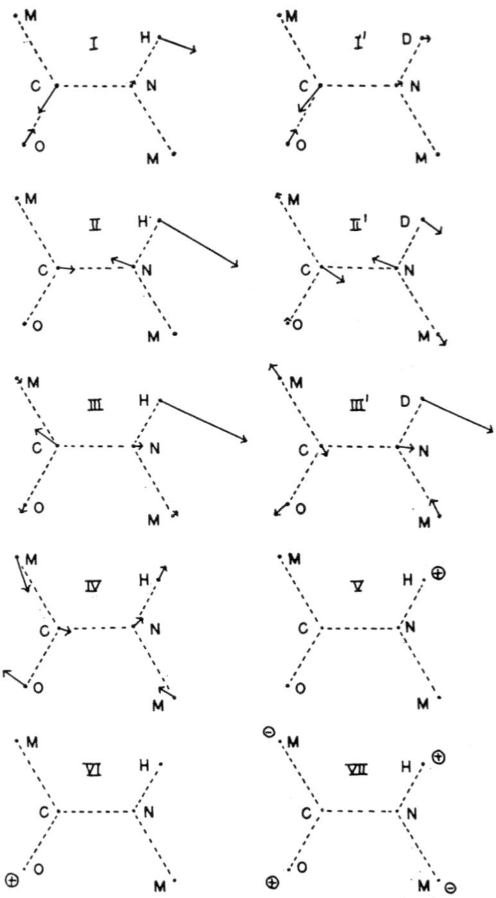

Fig. 18.1. Calculated normal modes of the amide I, II, III, and IV vibrations of *N*-methylacetamide and of the amide I′, II′, and III′ vibrations of N-deuterated *N*-methylacetamide and approximate representations of the amide V, VI, and VII vibrations.

associated with the N—H bending mode and 40 per cent with the C—N stretching mode, and thus these two motions are both important energywise in the amide II vibration (Table 18.2).

Amide III vibration (*ca. 1300* cm^{-1}).—For this vibration the amplitude ratio of the C=O and C—N stretching displacements is 0.5:1.0. The amplitude of the N—H bending displacement is the largest one as in the case of the amide II vibration. Thirty per cent of the vibrational potential energy is associated with each of the N—H bending and C—N stretching modes, and 10 per cent with each of the C=O stretching and O=C—N bending modes. The vibrational interaction with other motions may not be neglected—for example, 20 per cent of the potential energy is associated with the C_α—C stretching mode.

Amide I′ vibration (*ca. 1650* cm^{-1}).—The amide I′ vibration of the COND group corresponds to the amide I vibration of the CONH group, and is essentially the C=O stretching vibration. For the amide I′ vibration the contribution of the N—D bending mode is negligible, and because of this the amide I′ frequency is lower than the amide I frequency by about 10 cm^{-1}.

Amide II′ vibration (*ca. 1450* cm^{-1}).—For the amide II′ vibration of the COND group the stretching amplitude ratio $r(C{=}O)/r(C{-}N)$ is as small as +0.2 and energetically the C=O stretching motion is negligible. Instead the amplitude of the N—D in-plane bending displacement is prominent and 20 per cent of the vibrational potential energy is associated with the N—D bending mode. This vibration, however, is primarily due to the C—N stretching motion in terms of the potential energy distributions.

Amide III′ vibration (ca. 950 cm^{-1}).—Weak amide III′ band is observed at *ca.* 950 cm^{-1} for N-deuterated amides and lactams and this band has been considered to be primarily due to the N—D bending mode. Actually, however, the normal coördinate treatment indicates that the contributions of C_α—C and N—C_α stretching motions are as important as the N—D bending motion in terms of the energy distributions, although the N—D bending amplitude is the largest one.

Amide I and II bands of various conformations of polypeptides

As mentioned previously the amide I and II bands have been used for structure diagnoses of polypeptide chains. In order to study the causes for the frequency-conformation correlations, some theoretical considerations have been made (*13*). The vibrational modes of the amide I and II vibrations have been calculated as shown in Fig. 18.1 and it will be noticed that for these vibrations the displacements of the α-carbon atoms are negligible as compared with the displacements of the C, O, N, and H atoms of the peptide group. In the polypeptide chain, $\cdots$—CONH—C_α—CONH—C_α—CONH—$\cdots$, adjacent CONH groups have one α-carbon atom in between, and since the amide I (or II) vibration does not involve appreciable displacements of α-carbon atoms the interactions among the amide I vibrations of adjacent peptide groups will be small and may be treated by the first-order perturbation theory. Taking into account the vibrational interactions through hydrogen bonds as well as the interactions between peptide groups in the chain, the frequency has been found to be

$$\nu(\delta, \delta') = \nu_o + \sum_s D_s \cos(s\delta) + \sum_{s'} D_s' \cos(s'\delta') \quad (4)$$

where ν_o is the unperturbed frequency, and the second and third terms arise from the intrachain and interchain interactions, respectively. The phase difference between the vibrational motions of adjacent groups in the chain is δ and the constant D_s is determined by the potential and kinetic energy interactions between *s*th neighbors in the chain. D_s' and δ' pertain to the interchain vibrational interactions and have meanings similar to those of δ and D_s respectively. For an infinite helical conformation of polypeptide chains, infrared absorption arises either from the vibrations with the phase difference of $\delta = 0$ (parallel band) or from the degenerate vibrations with the phase difference of $\delta = \theta$ (perpendicular band), where θ is the angle of rotation about the helix axis on passing from a peptide group to the adjacent one in the chain.

Extended conformation

The extended polypeptide chains have been found to exist in buckled conformations as either the parallel-chain pleated sheet or the antiparallel-chain pleated sheet (*27*). In order to interpret the infrared spectra of polypeptides in these extended conformations it has been found to be useful to refer to the spectra of Nylon 66 which exists in the fully extended conformation.

Nylon 66

This polymer assumes the parallel-chain, fully extended conformation and a unit cell contains two peptide groups from one chain. Thus for infrared-active vibrations adjacent peptide groups through interchain hydrogen bonds move in phase. Taking into account only the adjacent group interactions through hydrogen bonds, the frequency of a localized vibration is derived from equation 4 to be

$$\nu = \nu_o + D_1' \quad (5)$$

The *D* terms due to the intrachain vibrational interactions have been neglected since each peptide group of Nylon 66 is separated from others in the chain by four or six methylene groups. The amide I and II bands of highly crystalline Nylon 66 are observed at 1640 cm^{-1} and 1540 cm^{-1}, respectively (*28*).

The value of D_1' in equation 5 may be calculated if the unperturbed frequency ν_o is known. For the amide II vibration the unperturbed frequency has been observed at 1535 cm^{-1} for incompletely N-deuterated poly-γ-benzyl-L-glutamate. In this case vibrational interactions among amide II vibrations do not take place since each CONH group is surrounded by COND groups and is isolated from other

CONH groups. The isolated CONH group is, however, in the same potential field as the CONH groups in the undeuterated molecule and its amide II frequency is equal to ν_o. The value of D_1' for the amide II vibration of Nylon 66 is thus calculated to be 5 cm^{-1}.

For the amide II vibration (see Fig. 18.1) the carbonyl oxygen atom stays at rest while the amide hydrogen atom oscillates with a large amplitude. The vibrational interaction between the amide II vibrations through hydrogen bonds is therefore caused by the displacement of the amide hydrogen atom perpendicular to the NH · · O hydrogen bond, and thus the interaction constant is considered to be proportional to the cosine function of the dihedral angle between the planes of adjacent peptide groups through hydrogen bonds. In transferring the calculated value of $D_1' = 5$ cm^{-1} of Nylon 66 to various conformations of polypeptides, the dihedral angle will be taken into account, and the interaction constant will be taken as $D_1' = 5$ cm^{-1} for the parallel-chain pleated sheet (the angle: 0°), as $D_1' = -5$ cm^{-1} for the antiparallel-chain pleated sheet (the angle: 180°), and as $D_3 = 2$ cm^{-1} (the angle: 60°) for the α-helix.

For the amide I vibration (see Fig. 18.1), on the other hand, the vibrational interaction through hydrogen bonds is caused by the displacement of the carbonyl oxygen atom along the NH · · O hydrogen bond and the interaction constant will not depend upon the dihedral angle between the planes of adjacent groups through hydrogen bonds. For the amide I vibration, therefore, a common interaction constant (through hydrogen bonds) will be used for all the conformations having linear intergroup hydrogen bonds.

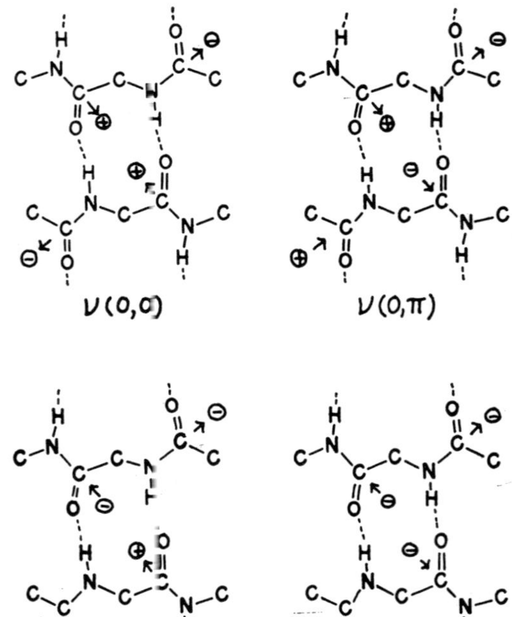

Fig. 18.2. The vibrational modes of the antiparallel-chain pleated sheet. The arrows represent the components of the transition moments of the peptide groups in the plane of the paper. The plus or minus signs represent the components of the transition moments perpendicular to the plane of the paper, the former pointing upward and the latter pointing downward.

Antiparallel-chain pleated sheet

For the antiparallel-chain pleated sheet (*27*) with the fiber period of 7.0 Å, the internal rotation angles are calculated as $t(C_\alpha—C) = -145°$ and $t(N—C_\alpha) = 142°$ (see Appendix). The polypeptide chain is thus twisted from the fully extended conformation and the plane of each peptide group is tilted from the fiber axis by an angle of about 30°.

Infrared absorption bands of the crystalline region arise from in-phase motions of corresponding groups in all the units cells. A unit cell of the antiparallel-chain pleated sheet contains four peptide groups, two from one chain and the other two from an adjacent chain. Accordingly there are four vibrations as shown schematically in Fig. 18.2. The $\nu(0, 0)$ vibration is infrared inactive. The $\nu(0, \pi)$ vibration is symmetric with respect to the twofold screw axis parallel to the z axis (Fig. 18.2) and gives rise to a parallel band. The $\nu(\pi, 0)$ vibration is symmetric with respect to the twofold screw axis parallel to the y axis whereas the $\nu(\pi, \pi)$ vibration is symmetric with respect to the twofold axis parallel to the x axis. Thus both the $\nu(\pi, 0)$ and $\nu(\pi, \pi)$ vibrations give rise to perpendicular bands.

Taking into account the vibrational interactions between adjacent groups in the chain and through interchain hydrogen bonds, the frequencies of the infrared active vibrations are derived from equation 4 as

$$\begin{aligned} \nu(0, \pi) &= \nu_o + D_1 - D_1' \\ \nu(\pi, 0) &= \nu_o - D_1 + D_1' \qquad (6) \\ \nu(\pi, \pi) &= \nu_o - D_1 - D_1' \end{aligned}$$

Amide I bands.—Two amide I bands have been observed for polyglycine I (*29, 30*) at 1685 cm^{-1} and 1632 cm^{-1} and similarly at 1695 cm^{-1} and 1634 cm^{-1} for silk gut (*Bombyx mori*) (*31*). The weak parallel band at *ca.* 1690 cm^{-1} and the strong perpendicular band at *ca.* 1630 cm^{-1} have been found to be due to the $\nu(0, \pi)$ and $\nu(\pi, 0)$ amide I vibrations, respectively. The band due to the $\nu(\pi, \pi)$ amide I vibration has not been observed.

The amide I frequencies of polyglycine I may now be discussed in terms of the vibrational interactions. In applying equation 6 the values of ν_o and D_1' are assumed to be the same as those of Nylon 66. From the observed frequencies of Nylon 66 and the $\nu(0, \pi)$ and $\nu(\pi, 0)$ bands of polyglycine the unknown constants are calculated to be $\nu_o = 1658$ cm^{-1}, $D_1 = 8$ cm^{-1}, and $D_1' = -18$ cm^{-1}. The unobserved $\nu(\pi, \pi)$ amide I frequency is calculated to be 1668 cm^{-1} (Table 18.3). This band is expected to exhibit perpendicular dichroism; however, its theoretical intensity is estimated to be only one-third of the intensity of the weak $\nu(0, \pi)$ amide I band (*32*).

D_1' ($= -18$ cm^{-1}) is the constant for the adjacent group interactions through hydrogen bonds, and its negative sign indicates that the potential energy associated with the in-phase motions ($\delta' = 0$) is lower than the out-of-phase ($\delta' = \pi$) motions. In fact the C═O stretching displacement of a CONH group will favor the polar resonance form NH$^+$═C—O$^-$ of adjacent groups through hydrogen bonds and will lower the potential energy associated with the C═O stretching displacements of adjacent groups.

It has been recognized that the $\nu(\pi, 0)$ amide I band of the extended conformation at *ca.* 1630 cm^{-1} shows very strong perpendicular dichroism. In fact this band is due to the transition moment exactly perpendicular to the fiber axis and is not overlapped by the parallel band at *ca.* 1690 cm^{-1}. The apparent dichroism of the $\nu(\pi, 0)$ amide I band serves as a useful measure of the orientation of the fiber axis in oriented samples.

Amide II bands.—The parallel $\nu(0, \pi)$ amide II band of polyglycine I has been observed at *ca.* 1530 cm^{-1}. Referring to the amide II frequency of Nylon 66, the intrachain interaction constant is readily found to be $D_1 = -10$ cm^{-1}, and the perpendicular amide II frequencies are calculated to be $\nu(\pi, 0) = 1540$ cm^{-1} and $\nu(\pi, \pi) = 1550$ cm^{-1}. The intensity of the $\nu(\pi, \pi)$ amide II band is expected to be about one third of the strong parallel amide II band. In fact many polypeptides in the extended conformations exhibit ill-defined shoulder at *ca.* 1550 cm^{-1}. Accordingly a weak perpendicular band

Table 18.3—Observed and calculated frequencies (cm^{-1}) of the amide I and II bands of polypeptides in various conformations (*14*)

Conformation	Designation	Amide I		Amide II	
		ν_{obs}*	ν_{calc}	ν_{obs}	ν_{calc}†
Random coil		1656^s	1658	1535^s	1535
α-helix	$\nu(0)$	1650^s	(1650)	1516^w	(1516)
	$\nu(2\pi/n)$‡	1652^m	1647	1546^s	1540
Parallel-chain pleated sheet	$\nu(0, 0)$	1645^w	1648	1530^s	1530
	$\nu(\pi, 0)$	1630^s	1632	1550^w	1550
Antiparallel-chain pleated sheet	$\nu(0, \pi)$	1685^w	(1684)	1530^s	(1530)
	$\nu(\pi, 0)$	1632^s	(1632)		1540
	$\nu(\pi, \pi)$		1668	1550^w	1550
Nylon 66		1640^s	(1640)	1540^s	(1540)

* Observed intensities indicated by superscript letters: *s*, strong; *m*, medium; *w*, weak.
† The values in parentheses were used in calculating the interaction constants
‡ *n* is the number of peptide groups per turn of helix.

at *ca.* 1550 cm^{-1} may not necessarily be due to the α-helical conformation.

PARALLEL-CHAIN PLEATED SHEET

For the parallel-chain pleated sheet with the fiber period of 6.5 Å, the internal rotation angles are calculated as $t(C_\alpha—C) = -112°$ and $t(N—C_\alpha) = 118°$ (see Appendix). The polypeptide chain is thus appreciably twisted from the fully extended conformation. Pauling and Corey have suggested that β-keratin is in the parallel-chain pleated sheet (*27*).

A unit cell of the parallel-chain pleated sheet contains two peptide groups from one chain and there are two infrared active amide I (or II) vibrations as schematically shown in Fig. 18.3. Taking into account the vibrational interactions between peptide groups in the chain and through interchain hydrogen bonds, the frequencies of the infrared active vibrations are derived from equation 4 to be

$$\begin{aligned} \nu(0, 0) &= \nu_o + D_1 + D_1' \text{ (parallel band)} \\ \nu(\pi, 0) &= \nu_o - D_1 + D_1' \text{ (perpendicular band)} \end{aligned} \tag{7}$$

Transferring the constants, ν_o, D_1, and D_1' as found for polyglycine I, the two amide I frequencies of the parallel-chain pleated sheet are calculated to be 1648 cm^{-1} and 1632 cm^{-1} as shown in Table 18.3. In fact for steam-stretched horsehair (*33*) a parallel amide I band of medium intensity and a strong perpendicular band have been observed at 1645 cm^{-1} and 1630 cm^{-1}, respectively. The calculated frequencies agree with the observed ones and the parallel band observed at 1645 cm^{-1} may well be due to the $\nu(0, 0)$ amide I vibration of the parallel-chain pleated sheet. Two amide II bands have been observed for stretched horsehair, namely, a strong parallel band at 1530 cm^{-1} and a weak perpendicular band at 1550 cm^{-1}. The frequencies of the two amide II bands are calculated as $\nu(0, 0) = 1530$ cm^{-1} and $\nu(\pi, 0) = 1550$ cm^{-1}, in agreement with the observed frequencies. Accordingly the perpendicular band at 1550 cm^{-1} may well be due to the $\nu(\pi, 0)$ amide II vibration of the parallel-chain pleated sheet.

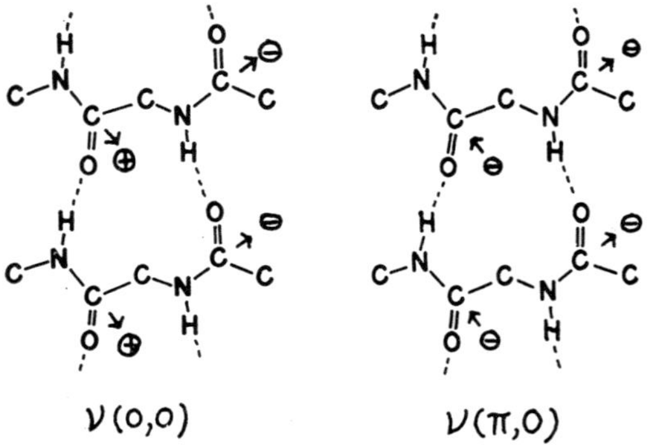

Fig. 18.3. The vibrational modes of the parallel-chain pleated sheet. The arrows represent the components of the transition moments of the peptide groups in the plane of the paper. The plus or minus signs represent the components of the transition moments perpendicular to the plane of the paper, the former pointing upward and the latter pointing downward.

Parallel amide I band of extended conformations.—On the basis of the amide I bands observed in the region 1600–1700 cm^{-1} it is now possible to distinguish between the parallel-chain and antiparallel-chain pleated sheet. Both conformations exhibit the perpendicular amide I band $\nu(\pi, 0)$ at *ca.* 1630 cm^{-1} characteristic to the extended conformation. However, the parallel amide I band $\nu(0, \pi)$ is observed at *ca.* 1645 cm^{-1} for the parallel-chain pleated sheet, whereas the band is observed at *ca.* 1695 cm^{-1} for the antiparallel-chain pleated sheet. Except β-keratin, polypeptides in the extended conformation have been found to show the parallel amide I band at *ca.* 1695 cm^{-1}, as well as the perpendicular amide I band at *ca.* 1630 cm^{-1}, and thus have been found to be in the antiparallel-chain pleated sheet.

The chain packing in the crystalline region may also be deduced from the measurements of the fiber period if the polypeptide in question may be oriented well enough for X-ray diffraction measurements. The present infrared method applies even when the polypeptides are hardly oriented.

Some extended polypeptide chains have been found to be in the cross-β structure (the chain direction being perpendicular to the fiber axis). The polarized infrared spectra of poly-β-propyl-L-aspartate (*34*) show the perpendicular amide I band at 1702 cm^{-1} and the parallel amide I band at 1637 cm^{-1}. Accordingly the extended polypeptide chains, in this case, are considered to be packed similarly as the antiparallel-chain pleated sheet, although the chain direction is perpendicular to the fiber axis.

α-HELIX

In the α-helix there are 3.6 peptide groups per turn of the helix ($\theta = 100°$) and intrachain hydrogen bonds are formed between third neighbors whereas interchain hydrogen bonds are not formed. Taking into account the third-neighbor interaction (D_3) as well as the first-neighbor interaction (D_1), the two infrared active frequencies are derived from equation 4 to be

$$\nu(0) = \nu_o + D_1 + D_3 \quad \text{(parallel band)}$$
$$\nu(\theta) = \nu_o + D_1 \cos\theta + D_3 \cos 3\theta \quad \text{(perpendicular band)} \quad (8)$$

Amide I bands.—For poly-γ-benzyl-L-glutamate in the α-helical conformation the $\nu(0)$ amide I band and the $\nu(\theta)$ amide I band have been observed at 1650 cm^{-1} and 1652 cm^{-1} respectively. Making use of the observed $\nu(0)$ frequency and the values of ν_o and D_3 as found for polyglycine, the first-neighbor interaction constant of the α-helix is calculated to be $D_1 = 10$ cm^{-1} and then the $\nu(\theta)$ amide I frequency is calculated to be 1647 cm^{-1}, in good agreement with the observed value.

For the α-helix the parallel amide I and the perpendicular amide I frequencies are not much different from each other so that the apparent dichroism observed for the amide I band of the α-helix is not too strong. On the other hand, for the antiparallel-chain pleated sheet the perpendicular amide I frequency is separated from the parallel amide I frequency and accordingly the observed dichroism is considerably strong.

Amide II bands.—A weak parallel amide II band and a strong perpendicular amide II band of poly-γ-benzyl-L-glutamate have been observed at 1516 cm^{-1} and 1546 cm^{-1}, respectively. From the observed $\nu(0)$ frequency and the constants $\nu_o = 1535$ cm^{-1} and $D_3 = 2$ cm^{-1} (as discussed before in the section for Nylon 66), the first-neighbor interaction constant for the amide II vibration is calculated to be $D_1 = -21$ cm^{-1} and the $\nu(\theta)$ amide II frequency is calculated to be 1540 cm^{-1} in agreement with the observed value of 1546 cm^{-1}.

The perpendicular amide II band of the α-helix at *ca.* 1550 cm^{-1} is due to the transition moment exactly perpendicular to the helix axis and this band is not overlapped by parallel bands. Accordingly the apparent dichroism of this band may be used as a useful measure of the orientation of the helix axes in the oriented sample.

It should be remarked that the α-helix exhibits the parallel amide II band at 1530 cm^{-1} and accordingly a weak parallel band at 1520 cm^{-1} may not necessarily be due to the extended conformation. The parallel band of natural elephant hair (*33*) at 1515 cm^{-1} is reasonably assigned to the α-helix since the extended conformation is not identified either by the perpendicular amide I band (*ca.* 1630 cm^{-1}) or by the X-ray diffraction measurement.

Infrared spectra of poly-β-benzyl-D-aspartate (PBDA) show the frequency splitting of the amide I band [$\nu(\theta) - \nu(0) = 3$ cm^{-1}] and the amide II band [$\nu(\theta) - \nu(0) = 18$ cm^{-1}] (*35*) similar to those observed for the α-helical conformation of poly-γ-benzyl-L-glutamate (PBLG); however, the amide I and II frequencies of PBDA (*34, 35*) are 10–15 cm^{-1} higher than the corresponding frequencies of PBLG. The conformation of PBDA is considered to be somehow distorted from the α-helical conformation of PBLG.

RANDOM COIL

Solid films of sodium poly-L-glutamate (*36, 37*), poly-L-serine (*38*), and poly-L-lysine hydrochloride (*37*) in the random-coil conformation show the broad amide I band at *ca.* 1655 cm^{-1} and the broad amide II band at *ca.* 1535 cm^{-1}. It is interesting to note that these amide frequencies of the random-coil conformation are close to the corresponding unperturbed frequencies, namely, the amide I frequency of $\nu_o = 1658$ cm^{-1} and the amide II frequency of $\nu_o = 1535$ cm^{-1}. In fact if the conformation is quite random the intrachain interaction or the interchain interactions will average zero, and broad bands centered at ν_o are expected to appear.

It is important to note that the amide I frequency of the random-coil conformation is not much different from the amide I frequency of the α-helix. The apparently nondichroic band of silk gut (*Bombyx mori*) (*31*) at 1660 cm^{-1} may be assigned to the random-coil conformation since the X-ray diffraction diagram of the α-helix has not been observed. Water-soluble silk film (*31*) and poly-DL-alanine (*8*) show the amide I and II bands corresponding to the random-coil conformation.

In aqueous solution sodium poly-L-glutamate and poly-L-lysine hydrochloride are considered to exist in the random-coil conformation (*6*). Two amide I′ bands have been observed for each of these polypeptides in deuterium oxide solution at various temperatures 25°–90° (*37*). The amide I′ band at 1645 cm^{-1} may be assigned to the COND group forming intergroup bonds since the corresponding amide I band of the CONH group (in the random-coil conformation) is observed at 1655 cm^{-1} in the solid film where the CONH groups are expected to form intergroup hydrogen bonds. The amide I′ band observed at 1655 cm^{-1} in deuterium oxide solution is then assigned to the COND group solvated by D_2O molecules. On raising the temperature the intensity of the band at 1655 cm^{-1} increases relative to the band at 1645 cm^{-1}. Accordingly, at higher temperatures the intergroup hydrogen bonds are considered to be broken and then solvated. The temperature effect on the amide II′ bands have also been observed. At higher temperatures the intensity of the amide II′ band at 1440 cm^{-1} increases relative to the amide II′ band at 1460 cm^{-1}. The hydrogen bonds between the peptide group and water are considered to be weaker than the intergroup hydrogen bonds since the solvated amide I band is higher and the solvated amide II′ band is lower than the corresponding frequencies of the peptide group which forms intergroup hydrogen bonds.

Amide III bands of various conformation

The infrared spectra of a number of polypeptides and proteins have been measured in the region 1200–1300 cm^{-1} and two or three peaks have been observed for each case (*39*). This is the region where the bands due to the amide III vibrations are expected to appear. In fact the intensities of the band of polyglycine I (*30*) at 1236 cm^{-1} and the band of poly-DL-alanine (*3*) at 1246 cm^{-1} are reduced on N-deuteration.

The normal mode of the amide III vibration of *N*-methylacetamide is shown in Fig. 18.1. For this vibration the displacement of the α-carbon atoms is negligible as compared with that of the atoms of the peptide group, and the vibrational motions are localized as in the case of the amide I or II vibrations. The amide III bands are also expected to depend upon the helical conformations of polypeptide chains. In fact the main amide III peaks of polyglycine I and II have been observed at 1236 cm^{-1} and 1256 cm^{-1}, respectively (*29*). The amide III bands lie in the region of various C—H bending modes and the splitting of the amide III bands has not been established as clearly as those of the amide I and II vibrations except for a few cases.

ANTIPARALLEL-CHAIN PLEATED SHEET

Infrared spectra of poly-L-alanine in the extended conformation have been measured (*3*). The parallel amide I band has been observed at 1695 cm^{-1} and accordingly the polypeptide chain is considered to be in the antiparallel-chain pleated sheet. The strong parallel band at 1221 cm^{-1} and the weak perpendicular band at *ca.* 1240 cm^{-1} may well be due to the $\nu(0, \pi)$ amide III vibration and the $\nu(\pi, \pi)$ (or $\nu[\pi, 0]$) amide III vibration, respectively. It is interesting to note that the splitting (20 cm^{-1}) of the amide III bands is equal to (but with opposite sign from) the splitting (20 cm^{-1}) of the amide II band.

α-HELIX

For the α-helical conformation of poly-L-alanine (*3*) a weak parallel band and a strong perpendicular band have been observed at *ca.* 1270 cm^{-1} and at 1274 cm^{-1}, respectively. These peaks disappear on N-deuteration (*41*) and may be assigned to the $\nu(0)$ and $\nu(\theta)$ amide III vibrations, respectively, of the α-helix.

Infrared spectra of poly-γ-benzyl-L-glutamate (PBLG), poly-β-benzyl-D-aspartate (PBDA), and their N-deuterated derivatives in the region 1150–1350 cm^{-1} are shown in Fig. 18.4 (*37*). The perpendicular band of PBLG at 1314 cm^{-1} and the perpendicular band of PBDA at 1304 cm^{-1} disappear on N-deuteration and are considered to be due to the $\nu(\theta)$ amide III vibration. For PBDA a weak parallel band is observed at 1335 cm^{-1}. This band also disappears on N-deuteration and may well be due to the $\nu(0)$ amide III vibration.

Amide IV, V, and VI bands of various polypeptides

For monosubstituted amides three characteristic bands (the amide IV, V, and VI bands) of the CONH group are observed in the region 600–800 cm^{-1}. The amide IV band observed at *ca.* 620 cm^{-1} is primarily due to the O=C—N bending mode (Table 18.2) and its frequency does not shift appreciably on N-deuteration. The amide V band observed at *ca.* 750 cm^{-1} is due to the N—H out-of-plane bending mode (*40*) and is shifted to *ca.* 500 cm^{-1} on N-deuteration (Table 18.1 and Fig. 18.1). The amide V frequency changes sensitively with the strengths of the hydrogen bonds, and in the case of *N*-methylacetamide the amide V bands have been observed at 648 cm^{-1} in dilute carbon disulfide solution, at 725 cm^{-1} in the liquid state (40° C.), at 745 cm^{-1} in the crystalline state at 15° C., and at 790 cm^{-1} in the crystalline state at −60° C. (*9*). The amide VI band observed at *ca.* 600 cm^{-1} is considered to

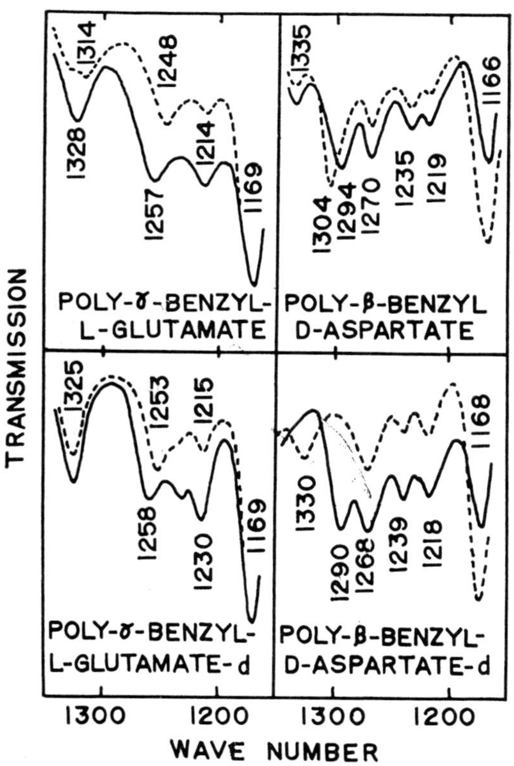

Fig. 18.4. Infrared spectra of poly-γ-benzyl-L-glutamate, N-deuterated poly-γ-benzyl-L-glutamate, poly-β-benzyl-D-aspartate, and N-deuterated poly-β-benzyl-L-glutamate (oriented film cast from chloroform solutions): ———, electric vector parallel to orientation direction; ---------, electric vector perpendicular to orientation direction.

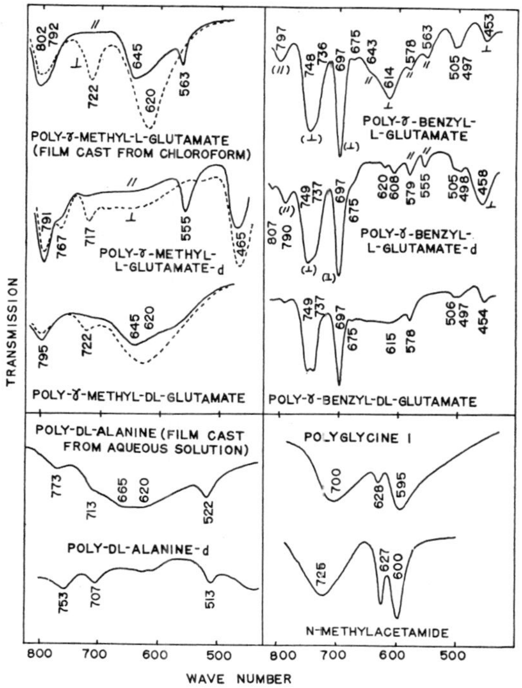

Fig. 18.5. Infrared spectra of polypeptides in the potassium bromide region.

be due to the C=O out-of-plane bending mode (Fig. 18.1). This band does not shift appreciably on N-deuteration.

Polyglycine I

The infrared spectrum observed for polyglycine I (*41*) is shown in Fig. 18.5 along with the spectrum observed for *N*-methylacetamide. The broad bands at *ca.* 700 cm^{-1} and at 595 cm^{-1} correspond to the bands of *N*-methylacetamide at 725 cm^{-1} and 600 cm^{-1}, respectively, and may be assigned to the amide V and VI vibrations.

The normal mode of the amide IV vibration of *N*-methylacetamide is shown in Fig. 18.1. It should be remarked that the displacements of the α-carbon atoms are as large as those of the atoms of the peptide group and thus the vibrational motion is not localized in the peptide group. The interaction between the amide IV vibrations of adjacent groups will be appreciable. In fact the splitting between the parallel and perpendicular amide IV frequencies has been calculated (*41*) to be as large as 150 cm^{-1}.

Referring to the calculated frequencies of 589 and 740 cm^{-1} (Table 18.4), the observed band of polyglycine I at 629 cm^{-1} may be assigned to the parallel amide IV band.

RANDOM COIL

The infrared spectra of poly-DL-alanine and its N-deuterated derivative (*41*) are shown in Fig. 18.5. The broad band at *ca.* 665 cm^{-1} disappears on N-deuteration, and accordingly this band is considered to be primarily due to the N—H out-of-plane bending mode of the random-coil conformation. The corresponding band is also observed at *ca.* 650 cm^{-1} for sodium poly-L-glutamate existing in the random-coil conformation. The peaks of poly-DL-alanine at 773, 713, and 522 cm^{-1} are shifted on N-deuteration and are considered to be due to the peptide group or the polypeptide main chain.

α-HELIX

The infrared spectra of poly-γ-benzyl-L-glutamate (PBLG), its N-deuterated derivative (PBLG-*d*), poly-γ-methyl-L-glutamate (PMLG), and its N-deuterated derivative (PNLG-*d*) in the α-helical conformation have been observed (*42*) as shown in Fig. 18.5. The absorption bands at 748, 736, 697, 675, 578, 505, and 497 cm^{-1} are observed for PBLG and PBLG-*d* but not for PMLG or PMLG-*d*, and most of these bands are presumably due to the benzyl group.

The parallel band at 643 cm^{-1} and the perpendicular band at 620 cm^{-1} are observed for PBLG as well as for PMLG and both these peaks are shifted to *ca.* 460 cm^{-1} on N-deuteration. The frequency ratios of *ca.* 1.35 indicate that these bands are the $\nu(0)$ and $\nu(\theta)$ amide bands primarily due to the N—H out-of-plane bending mode. The peak of poly-DL-alanine at 620 cm^{-1} is considered to be due to the $\nu(\theta)$ N—H out-of-plane bending mode of the α-helical part of this polymer.

For PMLG highly dichroic bands are observed at 722 cm^{-1} (perpendicular dichroism) and at 563 cm^{-1} (parallel dichroism). On N-deuteration these bands are shifted to 717 cm^{-1} and 555 cm^{-1}, respectively, and are considered to be the amide bands of the α-helical conformation. The bands at 563 cm^{-1} and 555 cm^{-1} are also observed for PBLG and PBLG-*d* respectively. The vibrational assignment of these peaks are yet to be made; however, it may be remarked that in the α-helical conformation the planes of adjacent peptide groups make the dihedral angle not much different from 90° and thus the O═C—N bending mode of a peptide group is expected to interact strongly with the N—H and C═O out-of-plane bending modes of adjacent groups.

ANTIPARALLEL-CHAIN PLEATED SHEET

Low molecular weight PMLG has been found to be in the antiparallel-chain pleated sheet and its absorption peak at 700 cm^{-1} has been found to shift to 530 cm^{-1} on N-deuteration (*42*). The corresponding band has been observed at 695 cm^{-1} for low molecular weight poly-L-alanine (*41*) in the antiparallel-chain pleated sheet conformation. The peak is thus considered to be due to the N—H out-of-plane bending mode of the antiparallel-chain pleated sheet.

In concluding the present section it may be noted again that the bands primarily due to the N—H out-of-plane bending mode are observed at *ca.* 700, *ca.* 650, and *ca.* 620 cm^{-1} for the antiparallel-chain pleated sheet, the random-coil, and the α-helix. Thus the conformation dependence of the N—H out-of-plane bending frequency is much greater than the amide I or the amide II frequencies. It is particularly interesting to note that the fraction of the α-helical part may now be estimated by the measurement of the intensity of the perpendicular band of the α-helix at *ca.* 620 cm^{-1} even when the polypeptide in question contains certain fraction of the random-coil conformation. (In such cases the amide I or II bands are not too useful for estimating the fraction of the α-helix.)

Amide VII bands of polypeptides

The torsional vibration about the C—N bond of the CONH group is directly related with the potential barrier hindering internal rotation about this bond. For the molecule of *N*-methylacetamide the amide VII band due to this torsional mode has been observed (*11*) at 206 cm^{-1}, and the twofold potential barrier has been found to be as high as 14 kcal/mole by the normal coördinate treatment of the amide VI and VII vibrations (*11*). Taking the potential barrier of a double bond as 40 kcal/mole, the partial double-bond character of the C—N bond of the CONH group is estimated to be 0.3–0.4, in good agreement with the value estimated from the C—N and C═O bond lengths.

The infrared spectra of the two crystalline forms of polyglycine have been measured in the region 400–190 cm^{-1}, and the bands possibly due to the amide VII vibrations have been observed at 217 cm^{-1} for polyglycine I and at 365 cm^{-1} for polyglycine II (*11*). It is remarkable to note that the frequency observed for polyglycine II is almost as high as twice the frequency observed for polyglycine I. The amide bands primarily due to the torsional vibrations are in fact expected to change sensitively with the helical conformation. The torsional mode about the C—N bond involves large displacements of the α-carbon atoms and accordingly strong vibrational interactions will take place between the amide VII vibrations of adjacent groups. Also for nonplanar conformations of polypeptides, vibrational interactions between the C—N torsional mode and all the valence angle bending modes will be appreciable and all these vibrational interactions are expected to depend sensitively on the helical conformations.

In connection with the studies of the correlations between the low skeletal frequencies and helical conformations of polypeptide chains, the normal vibrations of infinite helical chains,

$$(\text{—CH}_2\text{—})_n,\ (\text{—CH}_2\text{—O—})_n,\ \text{and}\ (\text{—CH}_2\text{—O—CH}_2\text{—})_n,$$

have been calculated for various helical conformations (*43*). It was learned from these calculations that the low skeletal frequencies primarily due to the torsional modes or to the bending modes are very sensitive to the helical conformations of infinite polymer chains (*44*), whereas the high frequencies primarily associated with the stretching modes do not change with the conformation as much as do the low frequencies. The results of these calculations are in accord with the observations that the amide IV, V, VI, and VII bands vary sensitively with the helical conformations of polypeptide chains. It should be useful, then, to establish the correlations between these amide frequencies and the helical conformations.

Normal vibrations of polyglycine I

The in-plane normal vibrations of the fully extended single chain of polyglycine I were calculated (*41*) in order to study the characteristic amide vibrations as well as the skeletal vibrations of polypeptides in the extended conformation. The methylene groups were treated as single dynamic units and the N—H stretching vibration was separated away from the lower frequency vibrations. In Table 18.4 there are shown the calculated and observed frequencies and the approximate description of each normal vibration expressed in terms of the potential energy distributions. It will be seen that the calculated amide frequencies agree satisfactorily with the observed frequencies. The A_1 vibrations are symmetric with respect to the twofold axis (parallel to the fiber axis) and give rise to parallel bands, whereas the B_1 vibrations are antisymmetric with respect to the axis and give rise to perpendicular bands. The vibrational modes of the characteristic amide vibrations were calculated as shown in Fig. 18.6.

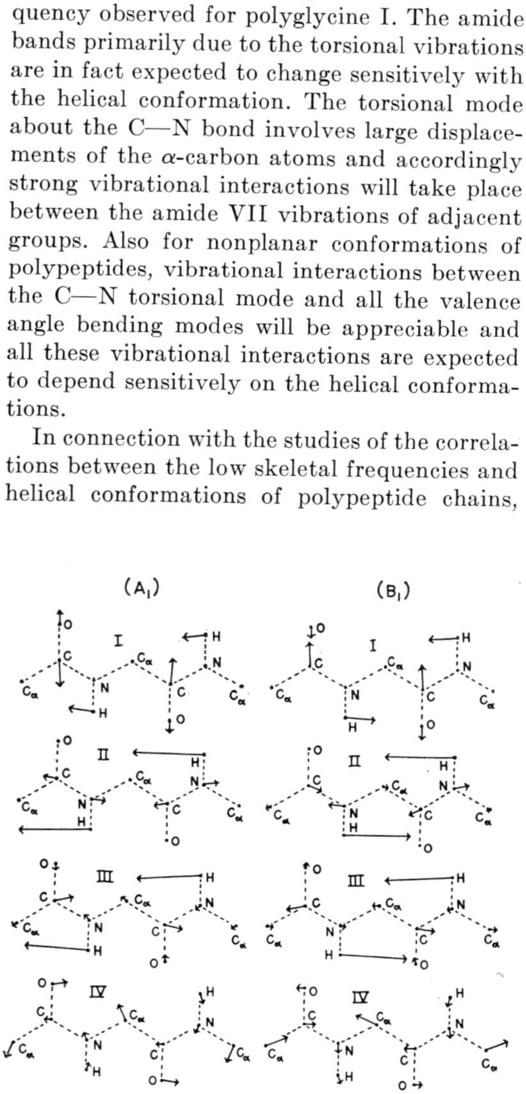

Fig. 18.6. Calculated normal modes of the A_1 and B_1 amide vibrations of polyglycine I.

The present calculation indicates that the amide I, II, and III vibrations of polyglycine are in fact localized in the peptide group (just as expected before from the normal coördinate analysis made of *N*-methylacetamide). Thus, as shown in Fig. 18.6, the displacements of the α-carbon atoms are negligible as compared with

Table 18.4—Calculated and observed skeletal frequencies of polyglycine I and the approximate description of each normal vibration in terms of potential energy distributions

ν_{obs}	ν_{calc}	Assignment*
		A_1 (parallel band)
	180	backbone bending (90%)
	219	backbone bending (100%)
628	589	Amide IV; $\beta_{C=O}$ (75%), $\nu_{C_\alpha-C}$ (15%)
(1015)	989	$\nu_{C_\alpha-C}$ (20%), $\nu_{C=O}$ (15%), backbone bending (40%)
1112	1101	ν_{N-C_α} (70%), $\nu_{C_\alpha-C}$ (20%)
1236	1305	Amide III; ν_{C-N} (35%), β_{N-H} (30%), $\nu_{C_\alpha-C}$ (25%)
1530	1517	Amide II; β_{N-H} (65%), ν_{C-N} (40%)
1688	1651	Amide I; $\nu_{C=O}$ (80%), ν_{C-N} (15%)
		B_1 (perpendicular band)
	279	backbone bending (50%), $\beta_{C=O}$ (40%)
	542	backbone bending (90%)
	740	Amide IV; $\beta_{C=O}$ (30%), backbone bending (35%), $\nu_{C\alpha-C}$ (20%)
	874	$\nu_{C_\alpha-C}$ (45%), $\nu_{N-C\alpha}$ (35%)
(1015)	1001	ν_{N-C_α} (45%), ν_{C-N} (15%)
1260	1294	Amide III; ν_{C-N} (30%), β_{N-H} (30%), $\nu_{C_\alpha-C}$ (20%)
1550	1513	Amide II; β_{N-H} (60%), ν_{C-N} (45%)
1630	1647	Amide I; $\nu_{C=O}$ (80%), ν_{C-N} (10%)

* ν: stretching, β: bending.

those of the atoms of the CONH group. Also it may be noted that the nature of these amide vibrations of polyglycine I (Table 18.4) are nearly the same as the nature of the corresponding vibration of *N*-methylacetamide.

The amide IV vibrations of polyglycine I are not localized in the peptide group and, as shown in Fig. 18.6, the displacements of the α-carbon atoms are prominent. The vibrational interactions between adjacent groups are significant and the splitting ($\nu_{A_1}-\nu_{B_1}$) is calculated to be as large as 150 cm^{-1}. Also the nature of the A_1 and B_1 amide IV vibrations of polyglycine I are different from each other; the A_1 amide IV vibration is essentially due to the C=O bending mode whereas the B_1 amide IV vibration is due to the backbone bending mode as well as to the C=O bending mode.

The strong band of polyglycine I at 1015 cm^{-1} has been considered to be characteristic to the diglycyl group (*30, 39*). This band appears to correspond either to the calculated B_1 frequency of 1001 cm^{-1} or to the calculated A_1 frequency of 989 cm^{-1}. The dichroism of this band has not been determined and it is not clear which one of these vibrations is responsible for the band at 1015 cm^{-1}. There are differences among other bands of polyglycine as reported previously so that the vibrational assignment will not be discussed here in detail.

Infrared spectra of poly-γ-methyl-L-glutamate and poly-γ-methyl-DL-glutamate

The polarized infrared spectra of poly-γ-methyl-L-glutamate (PMLG), N-deuterated poly-γ-methyl-L-glutamate (PMLG-*d*), and poly-γ-methyl-DL-glutamate (PMDLG) have been observed (*42*) as shown in Figs. 18.5, 18.7 and 18.8. PMLG-*d* was prepared by dissolving PMLG in O-deuterated trifluoroacetic acid (*45*). The oriented films were cast from chloroform solutions, and a pair of silver chloride polarizers were used for dichroic measurements.

The observed spectra of PMLG show the amide I and II bands characteristic to the α-helical conformation, namely the parallel amide I band at 1654 cm^{-1}, the perpendicular amide I band at 1656 cm^{-1}, the parallel amide II band at 1520 cm^{-1}, and the perpendicular amide II band at 1550 cm^{-1}. PMLG-*d* shows the unperturbed amide II band (ν_o) at 1538 cm^{-1} and the strong perpendicular amide II′ band at 1441 cm^{-1} just as does PBLG-*d*. The weak perpendicular band of PMLG at 1316 cm^{-1} disappears on N-deuteration and is presumably due to the $\nu(\theta)$ amide III vibration.

The polarized infrared spectra of PMDLG

show weaker dichroism as compared with the spectra of PMLG. The splitting of the parallel and perpendicular amide I bands is almost negligible and their frequencies are not much different from the amide I frequency of the random-coil conformation. The amide II bands of PMDLG, however, are appreciably different from the amide II band of the random-coil conformation. Thus the amide II vibration of PMDLG is clearly split into two components, namely, the parallel band at 1522 cm^{-1} and the perpendicular amide band at 1550 cm^{-1}. This splitting is characteristic to the α-helical conformation. Thus in view of the amide II bands, PMDLG is considered to contain the α-helical conformation although the fraction of the α-helical part cannot be estimated from the amide II bands alone.

The nondichroic bands at 1738 cm^{-1} and 1177 cm^{-1} are due to the C═O and C—O stretching modes of the ester group in the side chain. These bands are also observed for PBLG and PBLG-*d*. There are a number of other infrared bands commonly observed for PMLG, PMLG-*d*, and PMDLG, namely, the bands at 1439 cm^{-1}, 1333 cm^{-1}, 1252 cm^{-1}, 792 cm^{-1}, etc. Most of these are considered to be due to the side chain group $—CH_2—CH_2—COOCH_3$.

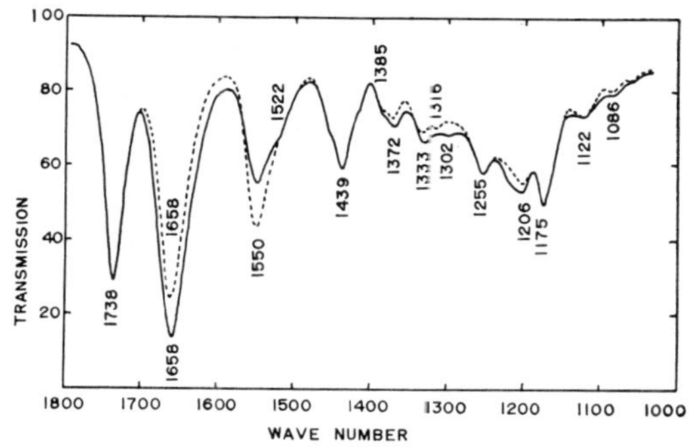

Fig. 18.8. Polarized infrared spectra of poly-γ-methyl-DL-glutamate (film cast from chloroform): ———, electric vector parallel to orientation direction; ---------, electric vector perpendicular to orientation direction.

For PMLG and PMLG-*d* a medium intensity band is observed at 1385 cm^{-1}. For PMDLG a new band appears at 1372 cm^{-1} which is stronger than the band at 1385 cm^{-1}. The relative intensity of these two bands of the copolymers of methyl-D-glutamate and methyl-L-glutamate have been found to vary with the D:L ratios.

Infrared spectra of PMLG in the region 400–800 cm^{-1} have been discussed in a previous section. The highly dichroic bands at 802 cm^{-1}, 722 cm^{-1}, 645 cm^{-1}, 620 cm^{-1}, and 563 cm^{-1} are associated with the peptide group or at least with the backbone chain of polypeptides. These bands are considered to be characteristic to the α-helical conformation since all of these disappear for low molecular weight PMLG which exists in the antiparallel-chain extended conformation (as evidenced by its amide I and II bands). The bands at 802 and 563 cm^{-1} are not observed for PMDLG. The corresponding bands of PBLG are observed at 797 and 563 cm^{-1}; however, these bands are not observed for copolymers of benzyl-L-glutamate and benzyl-D-glutamate containing the D-residue fraction of 20 per cent or more. Thus these bands are very sensitive to the incorporation of the D-form and may well be due to relatively long α-helical segments made of the L-form. On the other hand, the bands of PMLG at *ca.* 720 and 620 cm^{-1} are also observed for PMDLG al-

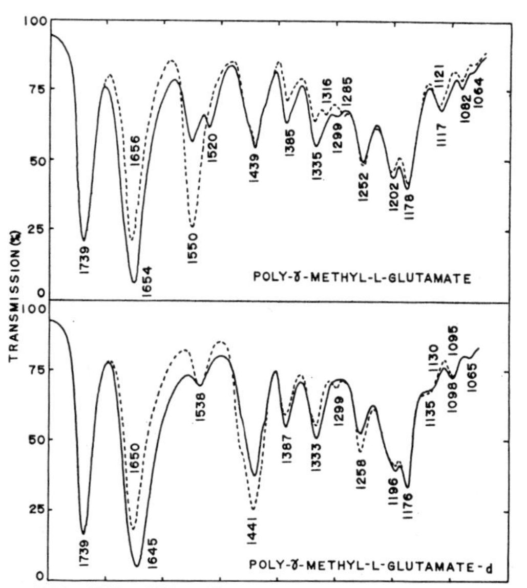

Fig. 18.7. Polarized infrared spectra of poly-γ-methyl-L-glutamate and N-deuterated poly-γ-methyl-L-glutamate (film cast from chloroform solutions): ———, electric vector parallel to orientation direction; ---------, electric vector perpendicular to orientation direction.

though the intensities of these bands of PMDLG are weaker than the intensities of the corresponding bands of PMLG. Also the intensity of the band of PBDLG at 615 cm^{-1} is much weaker than the intensity of the corresponding band of PBLG. These bands at 720 and 620 cm^{-1} as well as the amide II bands appear to be independent of the lengths of the α-helical segments. Also the band at 620 cm^{-1} is well separated from the corresponding bands of the random coil or the extended conformation and will be useful for estimating the fraction of the α-helical part in the copolymers of D- and L-amino acids. For example the fraction of the α-helical part of PBDLG is estimated to be approximately 30 per cent from the intensity of the 615 cm^{-1} band (Fig. 18.5) (*42*). More detailed analyses of the amide bands of various conformations of polypeptides in the potassium bromide region are yet to be made; however, the preliminary results already obtained suggest the usefulness of the low frequency amide bands for detailed conformation studies.

Appendix

THE HELICAL PARAMETERS AND INTERNAL COÖRDINATES OF THE POLYPEPTIDE CHAIN

The helical conformation of the polypeptide chain $(-C_\alpha-C-N-)_n$ may be described by the translation (d) along the helix axis and the angle (θ) of rotation about the axis on passing from a peptide group to the adjacent one. These two parameters are functions of the bond lengths, $r_{12}(C_\alpha-C)$, $r_{23}(C-N)$, and $r_{31}(N-C_\alpha)$, the bond angles, $\phi_1(N-C_\alpha-C)$, $\phi_2(C_\alpha-C-N)$, and $\phi_3(C-N-C_\alpha)$, and the internal rotation angles, $t_{12}(C_\alpha-C)$, $t_{23}(C-N)$, and $t_{31}(N-C_\alpha)$.

The values of the bond lengths and bond angles have been given by Pauling and Corey (*46*) and the internal rotation angle $t_{23}(C-N)$ has been established to be 180° (trans). The internal rotation angles for various helical conformations have been calculated from equations 9 and 10; the α-helix ($d=1.5$ Å and $\theta=100°$): $t(C_\alpha-C)=47°$ and $t(N-C_\alpha)=58°$, polyglycine II ($d=3.1$ Å and $\theta=120°$): $t(C_\alpha-C)=144°$ and $t(N-C_\alpha)=-76°$, the parallel-chain pleated sheet ($d=3.25$ Å and $\theta=180°$): $t(C_\alpha-C)=-112°$ and $t(N-C_\alpha)=118°$, and the antiparallel-chain pleated sheet ($d=3.5$ Å and $\theta=180°$): $t(C_\alpha-C)=-145°$ and $t(N-C_\alpha)=142°$. The position of the C_α, C, or N atom of the main chain relative to the helix axis may also be readily calculated from the bond lengths, bond angles, and internal rotation angles (*47*).

$$\begin{aligned}\cos\tfrac{1}{2}\theta = {} & \cos\tfrac{1}{2}(+t_{12}+t_{23}+t_{31})\sin\tfrac{1}{2}\phi_1\sin\tfrac{1}{2}\phi_2\sin\tfrac{1}{2}\phi_3 \\ & -\cos\tfrac{1}{2}(-t_{12}+t_{23}+t_{31})\cos\tfrac{1}{2}\phi_1\cos\tfrac{1}{2}\phi_2\sin\tfrac{1}{2}\phi_3 \\ & -\cos\tfrac{1}{2}(+t_{12}-t_{23}+t_{31})\sin\tfrac{1}{2}\phi_1\cos\tfrac{1}{2}\phi_2\cos\tfrac{1}{2}\phi_3 \\ & -\cos\tfrac{1}{2}(+t_{12}+t_{23}-t_{31})\cos\tfrac{1}{2}\phi_1\sin\tfrac{1}{2}\phi_2\cos\tfrac{1}{2}\phi_3\end{aligned} \tag{9}$$

$$\begin{aligned}d\sin\tfrac{1}{2}\theta = {} & (+r_{12}+r_{23}+r_{31})\sin\tfrac{1}{2}(+t_{12}+t_{23}+t_{31})\sin\tfrac{1}{2}\phi_1\sin\tfrac{1}{2}\phi_2\sin\tfrac{1}{2}\phi_3 \\ & -(-r_{12}+r_{23}+r_{31})\sin\tfrac{1}{2}(-t_{12}+t_{23}+t_{31})\cos\tfrac{1}{2}\phi_1\cos\tfrac{1}{2}\phi_2\sin\tfrac{1}{2}\phi_3 \\ & -(+r_{12}-r_{23}+r_{31})\sin\tfrac{1}{2}(+t_{12}-t_{23}+t_{31})\sin\tfrac{1}{2}\phi_1\cos\tfrac{1}{2}\phi_2\cos\phi_3 \\ & -(+r_{12}+r_{23}-r_{31})\sin\tfrac{1}{2}(+t_{12}+t_{23}-t_{31})\cos\tfrac{1}{2}\phi_1\sin\tfrac{1}{2}\phi_2\cos\tfrac{1}{2}\phi_3\end{aligned} \tag{10}$$

References

1 C. H. Bamford, L. Brown, E. M. Cant, A. Elliott, W. E. Hanby, and B. R. Malcolm, *Nature* **176**, 396 (1955).
2 F. H. C. Crick and A. Rich, *ibid.* **176,** 780 (1955).
3 A. Elliott, *Proc. Roy. Soc.* **A226,** 408 (1954).
4 L. Pauling, R. B. Corey, and H. R. Branson, *Proc. Nat. Acad. Sci.* **37,** 205 (1951).
5 P. Doty, A. M. Holtzer, J. H. Bradbury, and E. R. Blout, *J. Am. Chem. Soc.* **76,** 4493 (1954).
6 P. Doty, A. Wada, J. T. Yang, and E. R. Blout, *J. Polymer Sci.* **23,** 851 (1957).
7 E. J. Ambrose and A. Elliott, *Proc. Roy. Soc.* **A205,** 47 (1951).
8 A. Elliott, *Proc. Roy. Soc.* **A221,** 104 (1953).
9 T. Miyazawa, T. Shimanouchi, and S. Mizushima, *J. Chem. Phys.* **24,** 408 (1956); T. Miyazawa, *J. Chem. Soc. Japan* **77,** 171, 321, 526, 619 (1956).
10 T. Miyazawa, *J. Mol. Spectroscopy* **4,** 168 (1960).

11 T. Miyazawa, *Bull. Chem. Soc. Japan* **34,** 691 (1961).

12 T. Miyazawa, T. Shimanouchi, and S. Mizushima, *J. Chem. Phys.* **29,** 611 (1958).

13 T. Miyazawa, *J. Chem. Phys.* **32,** 1647 (1960).

14 T. Miyazawa and E. R. Blout, *J. Am. Chem. Soc.* **83,** 712 (1961).

15 M. Beer, G. B. B. M. Sutherland, K. N. Tanner, and D. L. Wood, *Proc. Roy. Soc.* **A249,** 147 (1959).

16 M. Tsuboi, *Bull. Chem. Soc. Japan* **22,** 255 (1949).

17 For example, Ref. *18,* p. 194.

18 C. H. Bamford, A. Elliott, and W. E. Hanby, "Synthetic Polypeptides," Academic Press, Inc., New York, 1956.

19 R. M. Badger and A. D. E. Pullin, *J. Chem. Phys.* **22,** 1142 (1954); see also C. G. Cannon, *Spectrochim. Acta* **16,** 302 (1960).

20 T. Miyazawa, *J. Mol. Spectroscopy* **4,** 155 (1960).

21 R. E. Richards and H. W. Thompson, *J. Chem. Soc.* **1947,** 1248.

22 H. M. Randall, R. G. Fowler, N. Fuson, and J. R. Dangl, "Infrared Determination of Organic Structures," D. van Nostrand Co., Inc., New York, 1949, p. 10.

23 H. Lenormant, *Ann. chim. (Paris)* **12,** 459 (1950).

24 R. D. B. Fraser and W. C. Price, *Nature* **170,** 490 (1952).

25 T. Miyazawa, *J. Chem. Soc. Japan* **76,** 341, 1018 (1955).

26 The methyl groups were treated as single dynamic units.

27 L. Pauling and R. B. Corey, *Proc. Nat. Acad. Sci.* **37,** 729 (1951).

28 Ref. *18,* p. 147.

29 A. Elliott and B. R. Malcolm, *Trans. Faraday Soc.* **52,** 528 (1956).

30 E. R. Blout and S. G. Linsley, *J. Am. Chem. Soc.* **74,** 1946 (1952).

31 E. J. Ambrose and A. Elliott, *Proc. Roy. Soc.* **A206,** 206 (1951).

32 The transition moment of the $\nu(\pi, \pi)$ vibration is perpendicular to the two-dimensional sheet of the antiparallel-chain pleated sheet. If the sheet is perpendicular to the propagation of the incident infrared radiation, this band cannot be observed.

33 K. D. Parker, quoted in Ref. *18,* p. 405.

34 E. M. Bradbury, L. Brown, A. R. Downie, A. Elliott, R. D. B. Fraser, W. E. Hanby, and T. R. R. McDonald, *J. Mol. Biol.* **2,** 276 (1960).

35 K. S. Norland, G. D. Fasman, and E. R. Blout, private communication.

36 A. Elliott, W. E. Hanby, and B. R. Malcolm, *Discussions Faraday Soc.* **25,** 167 (1958).

37 E. R. Blout and T. Miyazawa, unpublished.

38 G. D. Fasman and E. R. Blout, *J. Am. Chem. Soc.* **82,** 2262 (1960).

39 M. Asai, M. Tsuboi, T. Shimanouchi, and S. Mizushima, *J. Phys. Chem.* **59,** 322 (1955).

40 H. K. Kesseler and G. B. B. M. Sutherland, *J. Chem. Phys.* **21,** 570 (1953).

41 K. Fukushima and T. Miyazawa, to be published.

42 Y. Masuda, K. Fukushima, and T. Miyazawa, to be published.

43 T. Miyazawa, *Spectrochim. Acta.* **16,** 1231, 1233 (1960); *J. Chem. Phys.* **35,** 693 (1961).

44 The helical conformations of tetrafluoroethylene, polyoxymethylene, and polyethylene glycol deduced from the normal coördinate analyses are in accord with the conformations deduced from X-ray diffraction studied.

45 O-Deuterated trifluoroacetic acid was originally used by Mr. K. N. Norland for N-deuteration of polypeptides in the α-helical conformation.

46 L. Pauling and R. B. Corey, *Proc. Roy. Soc.* **B141,** 10 (1953).

47 T. Miyazawa, *J. Polymer Sci.* **55,** 215 (1961).

Discussion of Paper 18: Amide Bands and Conformations

A. ELLIOTT

Dr. Miyazawa's paper is a very valuable contribution to our knowledge of the significance of polypeptide infrared spectra, and the new assignments in the low-frequency region will be useful. I should like to comment on the subject of bands arising from the coupling of Amide I modes, however. The band at about 1690 cm^{-1} is I believe correctly interpreted as a (o, π) mode of antiparallel chains, but I am not convinced that the factors D_1 and D_1' have been correctly assessed. According to Miyazawa's calculation, the interaction across the hydrogen bond is much greater than the intrachain interaction. We should therefore expect a considerable splitting of the Amide I band in acetanilide, where the hydrogen-bridged amide groups are arranged in an essentially antiparallel arrangement. In fact, the splitting is only about 1 cm^{-1} (Abbott and Elliott, *Proc. Roy. Soc.* **A234**, 247 [1956]). Although acetanilide is not without objection as a model compound, because of the aromatic ring, there is nothing abnormal about the frequency of the Amide I band (1666 cm^{-1}). The aromatic band at 1600 cm^{-1} (unlike Amide I) is quite unaffected by N-deuteration, and therefore the motions of the aromatic ring should be independent of the motion of the atoms in the amide group. I see no reason why the splitting of Amide I in acetanilide should not be much larger if the interaction across the H-bond is as important as Miyazawa implies.

Again, in Nylon 6 we have an antiparallel chain arrangement, and we have observed a splitting of the Amide I band of about 20 cm^{-1} instead of 36 cm^{-1} as calculated. We are examining the spectra of a number of nylons to see whether the terms D_1 and D_1' are constant. The importance of these observations lies in the doubt they cast on the correctness of the calculation for the (0, 0) band of *parallel* chains, said to be at 1648 cm^{-1}. If the intrachain interaction is as great as, or perhaps greater than the interchain interaction this band would appear at a higher frequency than 1648 cm^{-1} and might not be recognized in the presence of bands from other polypeptide chain conformations. In any case, the peak in Parker's "parallel" spectrum of stretched horsehair (β-keratin) cannot be this band. We have repeated Parker's observations on a number of sections of stretched horsehair. Figure 18.9 shows that this band is at 1657 cm^{-1} and is most reasonably assigned either to a proportion of α-helix or similar fold in the specimen, or more probably to a more or less random, unoriented fraction. Some indication of a weak parallel band at about 1690 cm^{-1} may be seen in Parker's spectrum, and quite clearly in ours. On exposing the stretched hair to D_2O vapor, the Amide I band of the amorphous part (which exchanges its hydrogen rapidly with

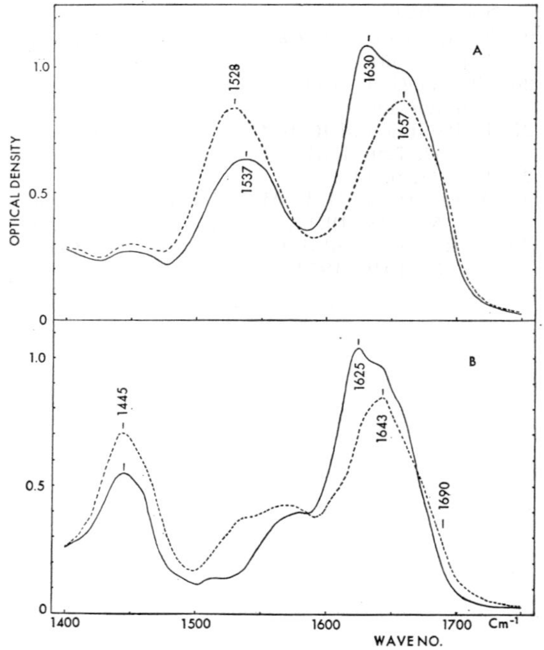

Fig. 18.9. *A:* Infrared spectrum of steam-stretched horsehair (β-keratin). *B:* The same after exposure to D_2O vapor for one week. *Full line,* E vector perpendicular to fiber axis. *Broken line,* E vector parallel to fiber axis.

deuterium) moves to a lower frequency and the band at *ca.* 1690 cm^{-1} is more prominent, though still weak. It may be remarked that contraction of a polypeptide from a repeat length of 7.00 Å to 6.5 Å rotates the amide group in such a way that the intensity of the "parallel" band (either with parallel or with antiparallel chains) drops to about one-third of its initial value. It is possible that the chain arrangement in stretched horsehair is of the antiparallel kind.

W. B. GRATZER

There appears to be conflicting evidence on the conformation of poly-DL-alanine in aqueous solution. We have examined the short wavelength ultraviolet spectrum of low molecular weight poly-DL-alanine and the results here are not entirely consistent with the infrared data presented by Dr. de Lozé. We find a consistent hypochromic effect, which indicates the presence of a periodic structure (Imahori and Tanaka, *J. Mol. Biol.* **9**, 359 [1959]; Rosenbeck and Doty, cited in Doty, *J. Polymer Sci*, **49**, 149 [1961]; Tinoco, Halpern and Simpson in this book, Paper 13). Our experiments with poly-DL-glutamic acid, which shows a hyperchromic increase on ionization, indicate that in the random form the extinction coefficients at 190 mμ of L-polymers and DL-polymers are identical. If the assumption be made that in the hypothetical α-helix made up of equal parts of D- and L-residues the extinction coefficient at this wave length is the same as in an α-helix of L-residues only, the calculated helical content of the poly-DL-alanine is about 30 per cent.

The hypochromicity was essentially unaffected by heat and by 0.1 M sodium dodecyl sulphate. Addition of acetonitrile or isopropanol brought about an appreciable decrease in absorbance, and the presence of these solvents could therefore be said to promote the development of additional helical structure. It was also noted that in D_2O solution the infrared spectrum gave no indication of the presence of the β-form. The curious observation of Berger and Linderstrøm-Lang (1957) that poly-DL-alanine dissolves much less readily in D_2O than H_2O was also confirmed. In D_2O a gelatinous phase, separated by centrifugation, was found to exhibit a band at 1620 cm^{-1}, indicative of the β-structure.

The selectivity between D- and L-residues in the polymerization kinetics referred to by Dr. Bamford and Dr. Wada (see also Lundberg and Doty, *J. Am. Chem. Soc.* **79**, 3961 [1957]) as well as the possibility of specific hydrophobic interactions between D and L side chains might be expected to favor the formation of sections of left- and right-handed α-helix in appreciable degree. In solution the ultraviolet technique here employed certainly seems to offer the best prospect of detecting α-helices under these circumstances. Certainly, however, as Dr. Elliott has pointed out, other types of helix might well be present and could give rise to a hypochromic effect. In this case a precise quantitative evaluation would be precluded.

PAPER 19

The Relative Stabilities of the α-Helix in Synthetic Polypeptides

GERALD D. FASMAN

The stability of the α-helix and the pleated sheet structures proposed for proteins by Pauling and Corey (*1*) depends upon the hydrogen bonding between the carbonyls and the NH in the peptide backbone. In proteins many other forces (*2, 3, 4, 5*) have been proposed which may add stability to native conformations, such as salt linkages, covalent bonds, electrostatic forces, hydrophobic bonds, and dispersion forces. It is often extremely difficult or impossible to differentiate experimentally between these forces. The question has frequently been raised as to whether the stabiziling effect of hydrogen bonding alone would be sufficient to give native proteins their inherent stability (*6, 7, 8*). Of these many factors which may play an important role in protein stabilization, hydrophobic bonding and dispersive forces have recently received serious consideration (*3, 7, 9*).

Synthetic poly-α-amino acids offer unique model systems for studying the contribution of hydrophobic bonding and dispersive forces to the stabilization of the α-helix. Using polymers in which the side chains are either aliphatic, e.g., leucine, or side chains in which the functional groups are blocked, e.g., γ-benzyl-L-glutamate, it is possible to study the effects of the interaction of these side chains and investigate their contribution to stabilization which arises from hydrophobic bonding and dispersive forces. By performing studies in nonaqueous media, such as chloroform, it is possible to overcome the difficulty of dealing with strong competitive hydrogen bonding forces from the solvent as displayed in aqueous solutions. The contribution and importance of these side-chain interactions is assessed by measuring the relative stability of the α-helix for several poly-α-amino acids. The method of "optical titration" used is that first introduced by Doty, Yang, and Blout (*10, 11*). The method consists of dissolving a synthetic polypeptide in a solvent, such as chloroform, in which the α-helical conformation is known to exist (*10, 12, 13, 14, 15*), and examining its optical rotatory dispersion behavior on the addition of a strong hydrogen-bonding solvent such as dichloroacetic acid (DCA) or trifluoroacetic acid (TFA). The parameters calculated are the specific rotation $[\alpha]_\lambda$ or the b_o value (obtained from the Moffitt equation [*16*]) as the solvent compostion is changed from 100 per cent chloroform ($CHCl_3$)→100 per cent DCA→100 per cent TFA (see Fig. 19.1). A sharp transition usually occurs as the helical conformation is destroyed with a concomitant large change in both $[\alpha]_\lambda$ and b_o. The b_o value associated with the helical conformation is of the order of −630 (*16*), while the random-coil

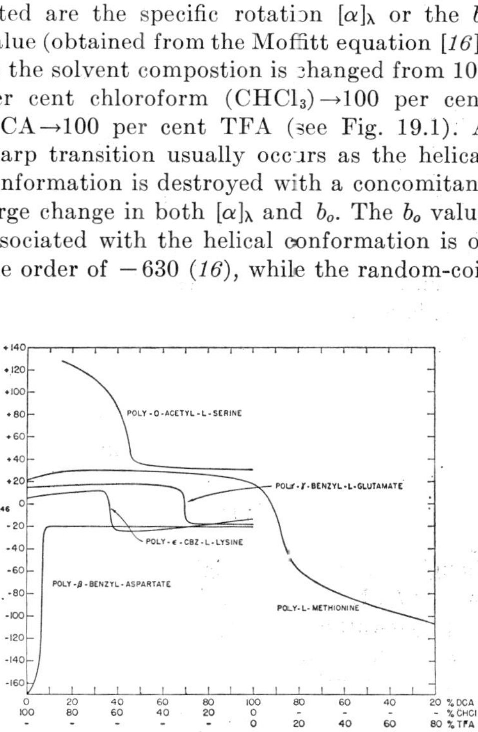

Fig. 19.1. $[\alpha]_{546}$ Values versus solvent composition ($CHCl_3$ — DCA — TFA).

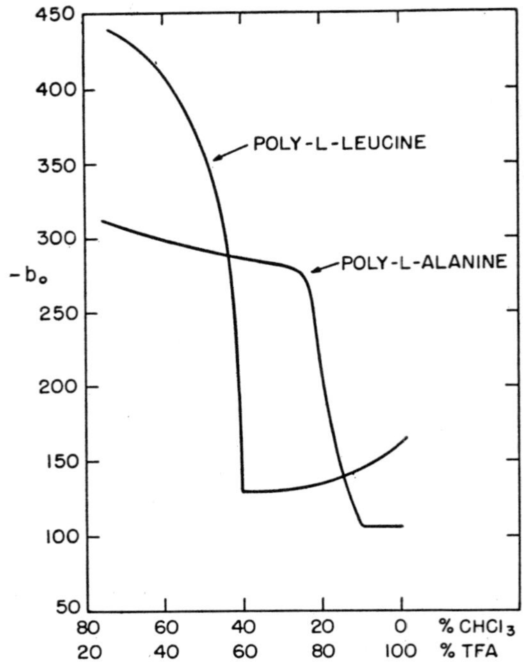

Fig. 19.2. b_o Values versus solvent composition ($CHCL_3$ — DCA — TFA).

value of b_o is usually zero (*7*). With certain polymers it was found that DCA did not disrupt the helix and TFA was necessary to cause a significant change in b_o.

A summation of results is seen in Figure 19.1. It is seen that different amounts of DCA and TFA are required to destroy the helical conformation in these polypeptides. The least stable helix, based on this criterion, is that of poly-β-benzyl-L-aspartate (*18, 19*), which requires as little as 5 per cent to 8 per cent DCA to cause the helix→random-coil transition. This polypeptide is known to have a left-handed sense of twist (*18, 19*). The left-handed helix has been suggested to be weaker (*20*) than the right-handed form, and this suggestion is supported by these data. The other polypeptides are thought to have the right-handed sense of twist (*21*). Next in increasing order of stability is poly-ϵ-carbobenzyloxy-L-lysine (*15*). The breakdown of the helical conformation is seen to occur abruptly between 35 per cent to 40 per cent DCA. The next polypeptide helix, in order of greater stability, was found to be poly-γ-benzyl-L-glutamate (*10, 11, 18*). The helix →random-coil transition for this polymer occurs at 68 per cent DCA. The poly-L-methionine helix was found to be next in order of stability. In 100 per cent DCA a b_o value of -440 (compared to -630 in $CHCl_3$) was found, indicating that the polymer was still about 60 per cent helical in this solvent. The complete breakdown of the poly-L-methionine helix occurs in the presence of about 10 per cent TFA:90 per cent DCA (*22*).

Another polymer, known to exist in the β-structure, poly-*O*-acetyl-L-serine (*23*), also displays the same stability as the helix of poly-ϵ carbobenzyloxy-L-lysine. The β structure is seen to be disrupted at 35 per cent to 40 per cent DCA as measured by a change in $[\alpha]_\lambda$.

Poly-L-alanine (*24*) has a b_o value of -325 in 100 per cent DCA, which would suggest that the polymer is in the transition region and is slightly less stable than poly-L-methionine. However, from the b_o plot versus solvent composition (Fig. 19.2) it is seen that the b_o changes slowly from -425 in 70 per cent $CHCl_3$:30 per cent DCA (minimal amount of DCA required for solubility) to -300 in 40 per cent TFA:60 per cent DCA, whereupon a sharp drop in b_o occurs, finally reaching a minimal value of -115. Thus the sharpest transition in b_o values occurs between 40 per cent and 70 per cent TFA and, if this is truly the helix→coil transition, this polymer is more stable than poly-L-methionine.

The polypeptide which exhibits the greatest stability of all those measured, using the b_o criterion, is poly-L-leucine (*25*). This polymer requires approximately 60 per cent TFA:40 per cent $CHCl_3$ to have the helix destroyed (Fig. 19.3). It will be noticed that this polymer also shows a gradual b_o decrease from -435 in 75 per cent $CHCl_3$:25 per cent TFA (minimal TFA required for solubility) to -360 in 50 per cent $CHCl_3$:50 per cent TFA before showing a rapid decrease to the final value of -125.

From available data on poly-L-alanine and poly-L-leucine, it is not possible to say whether the slow change in b_o represents a gradual loosening of the helix, or whether this represents solvent or other effects. If the first explanation is valid, there would appear to be more than one phase transition during destruction of the helix. To correlate the $CHCl_3$:TFA system with the $CHCl_3$:DCA:TFA system, poly-L-alanine was studied in both solvent systems. It can be seen that its behavior is similar in both systems (Fig. 19.2 and Fig. 19.3). In the DCA:TFA system, the sharpest transition in b_o occurs between 40 per cent and 70 per cent TFA, while in the $CHCl_3$:TFA system the break occurs more sharply at 70 per cent to 90 per cent TFA. Thus, if DCA does not destroy the hydrogen-bonding necessary for helical stability, it operates only in a weak coöperative fashion with TFA. The transition does occur in a more definitive fashion in the $CHCl_3$:TFA system than in the DCA:TFA system. It is interesting to note that both poly-L-alanine and poly-L-leucine do not attain zero b_o values in TFA, where it is assumed they are in the random-coil conformation, but have values about -100.

Thus, a scale of relative stabilities of the α-helix in synthetic poly-α-amino acids soluble in organic solvents has been established. To break down an α-helix, solvent or solute molecules must find their way to the backbone of the polymer and successfully compete in hydrogen-bond formation with the amide links of the polypeptide or protonate the peptide bond. The stronger acid TFA should be a stronger hydrogen-bond breaker, and thus be more efficient in destroying the helical conformation. One other factor may play a role here, namely, the fact that TFA is a smaller molecule than DCA, which may aid in its approach to the backbone. But it is felt that the stronger acidity plays the prominent role.

The different stabilities observed must be due to the different availabilities of the hydrogen-bonded backbone. The solutes used (DCA and TFA) would have greater affinity for the relatively polar environment of the side chain in poly-β-benzyl-L-aspartate, poly-ϵ-carbobenzyloxy-L-lysine, and poly-γ-benzyl-L-glutamate than the relatively less polar side chains of poly-L-methionine. Other factors being equal, the amino acids having less polar side chains should provide a better shield for the internally hydrogen-bonded α-helical backbone than those having polar side chains. A more efficient shield is provided by aliphatic hydrocarbon side chains, such as are present in poly-L-alanine and poly-L-leucine. These methyl and isobutyl groups present a tightly compacted

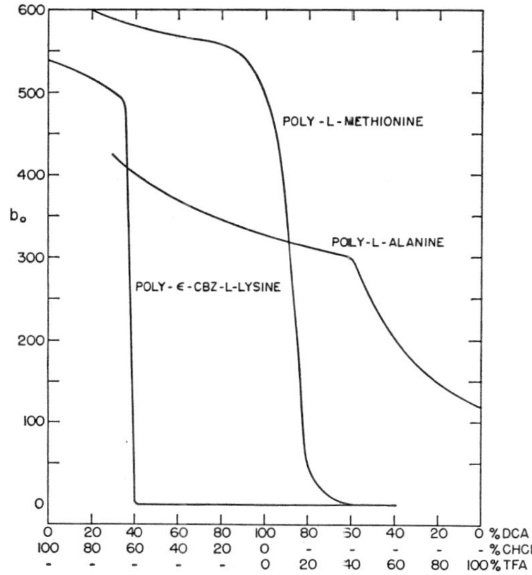

Fig. 19.3. b_o Values versus solvent composition (TFA — chloroform).

hydrocarbon area, due to lyophobic bonding and dispersive forces, which decrease the accessibility of the hydrogen bonds in the backbone to the polar hydrogen-bond breakers.

These observations may also be extended to the probable role these forces play in protein conformation. The α-helix as found in several poly-α-amino acids displays varying stabilities. The stability of the helix increases as the side chains present a more formidable barrier to the polar hydrogen-bond breaker. The protection of the backbone increases as the side chain becomes less polar and as the hydrocarbon lyophobic bonds and dispersive forces combine to exclude the solute responsible for helix destruction. These same forces, when operating in a protein in an aqueous system, would be the hydrophobic bonds and dispersive forces proposed by Kauzmann (*3*, *4*) and recently discussed for several proteins (*7*, *9*). Crick (*26*) had earlier pointed out the possibility of packed α-helices forming coiled coils. The stability of such structures was based on the packing of the nonpolar residues. If the contribution of such forces exert as great an influence on protein conformational stability as they do in synthetic polypeptide helical structures, they may indeed be the major factors determining protein stucture.

Acknowledgment

This work was performed in collaboration with Dr. E. R. Blout and associates.

We gratefully acknowledge the support of the Office of the Surgeon General, Department of the Army, U.S.A., and the invaluable assistance of Miss Carole Lindblow and Miss E. Bodenheimer.

References

1 L. Pauling, R. B. Corey, H. R. Branson, *Proc. Nat. Acad. Sci.* **37,** 205 (1951); L. Pauling, R. B. Corey, *ibid.* **37,** 235, 729, 241 (1951); *Nature* **168,** 500 (1951).

2 K. U. Linderstrøm-Lang and J. A. Schellman, in P. Boyer, H. Lardy, and K. Myrback (eds.), "The Enzymes," Academic Press, Inc., New York, 1959, Vol. 1, p. 443.

3 W. Kauzmann, *Advances in Protein Chem.* **14,** 1 (1959).

4 W. Kauzmann, in W. D. McElroy and B. Glass (eds.), "The Mechanism of Enzyme Action," Johns Hopkins Univ. Press, Baltimore, Md., 1954, p. 70.

5 W. H. Stockmayer, *Rev. Mod. Phys.* **31,** 103 (1959).

6 J. A. Schellman, *Compt. rend. trav. lab. Carlsberg, Sér. Chim.* **29,** 230 (1955).

7 D. T. Warner, *Nature* **190,** 120 (1961).

8 J. T. Edsall and J. Wyman, "Biophysical Chemistry," Academic Press, Inc., New York, 1958, Vol. I, p. 124.

9 C. Tanford, P. K. De, and Valerie G. Taggart, *J. Am. Chem. Soc.* **82,** 6028 (1960).

10 P. Doty and J. T. Yang, *ibid.* **78,** 498 (1956).

11 E. R. Blout, P. Doty, and J. T. Yang, *ibid.* **79,** 749 (1951).

12 P. Doty, A. M. Holtzer, J. H. Bradbury, and E. R. Blout, *ibid.* **76,** 4493 (1954).

13 P. Doty, J. H. Bradbury, and A. M. Holtzer, *ibid.* **78,** 947 (1956).

14 G. R. Bird and E. R. Blout, *ibid.* **81,** 2499 (1959).

15 G. D. Fasman, M. Idelson, and E. R. Blout, *ibid.* **83,** 709 (1961).

16 W. Moffitt and J. T. Yang, *Proc. Nat. Acad. Sci.* **42,** 596 (1956); W. Moffitt, *ibid.* **42,** 736 (1956).

b_o of the second term in the Moffitt equation

$$[m']_\lambda = \frac{a_o\lambda_o^2}{\lambda^2 - \lambda_o^2} + \frac{b_o\lambda_o^4}{(\lambda^2 - \lambda_o^2)^2}$$

where

$$[\alpha]_\lambda = [m']_\lambda \frac{100}{M} \frac{n^2 + 2}{3}$$

17 C. Cohen, A. Szent-Gyorgyi, *J. Am. Chem. Soc.* **79,** 248 (1957).

18 R. H. Karlson, K. S. Norland, G. D. Fasman, and E. R. Blout, *ibid.* **82,** 2268 (1960).

19 E. M. Bradbury, A. R. Downie, A. Elliott, and W. E. Hanby, *Proc. Roy. Soc.* **A259,** 110 (1960).

20 M. Huggins, *J. Am. Chem. Soc.* **74,** 3963 (1952).

21 A. R. Downie, A. Elliott, W. E. Hanby, and B. R. Malcolm, *Proc. Roy. Soc.* **A242,** 325 (1957); D. D. Fitts, J. G. Kirkwood, *Proc. Nat. Acad. Sci.* **43,** 1046 (1957).

22 S. M. Bloom, G. D. Fasman, E. R. Blout, Biophysical Chemistry Society Meeting Abstracts, St. Louis, 1961; G. Perlmann, E. Katchalski, Biophysical Chemistry Society Meeting Abstracts, St. Louis, 1961.

23 G. D. Fasman and E. R. Blout, *J. Am. Chem. Soc.* **82,** 2262 (1960).

24 G. D. Fasman and E. R. Blout, unpublished results.

25 G. D. Fasman and E. R. Blout, unpublished results. A b_o value of −650 in DCA was reported by E. R. Blout in C. Djerassi (ed.), "Optical Rotary Dispersion," McGraw-Hill Book Co., New York, 1960, p. 248.

26 F. A. Crick, *Acta Cryst.* **6,** 689 (1953).

Discussion of Paper 19: Relative Stabilities of the α-Helix

C. H. BAMFORD

Has Dr. Fasman any information about the effect of additions of small quantities of water on the chain conformations in dichloracetic acid or solutions containing this acid? It is known that the presence of a small percentage of water greatly increases the solubility of some β-polypeptides in DCA, presumably by facilitating attack on the hydrogen bonds. By adding controlled quantities of water it may be possible to obtain further information about the polar and shielding effects of the side chains suggested by Dr. Fasman.

SETSUJI SAKURAI

In a manner similar to the method reported for poly-γ-benzyl-L-glutamate (L-PBG) and other synthetic polypeptides, the optical behavior of poly-γ-methyl-L-glutamate (L-PMG) in chloroform solution on the addition of a strong hydrogen bonding solvent dichloroacetic acid (DCA) was studied by Dr. Okuda. The results of these experiments are shown in Fig. 19.4. The abrupt breakdown of the helical conformation is seen to occur between 55 and 60 per cent DCA. Therefore the α-helix in L-PMG appears to be less stable than that of L-PBG, which is known to retain its helical conformation up to 68 per cent DCA. This indicates that even the difference in the γ-ester group provides the different stability of the α-helix. Further, we have synthesized copolymers derived from L-MG and DL-valine and measured their optical rotations in a system chloroform–dichloroacetic acid. Since valine peptide is reported to take no α-helical conformation because of the steric hindrance due to the side chain, the incorporation of valine residue in the polypeptide chain of L-PMG would be expected to cause a reduced helix stability. This, however, was not the case. From Fig. 19.4 it is seen that the breaking occurs at 65 per cent DCA for a copolymer (L-MG:DL-Val = 95:5) and at 70 per cent DCA for a copolymer (90:10). Of particular interest is the fact that with both copolymers transformation points shift towards an increased stability. We consider that this finding may support a suggestion of Dr. Fasman that aliphatic hydrocarbon side chains decrease the accessibility of the hydrogen bonds in the backbone to the polar hydrogen bond breaking solvents.

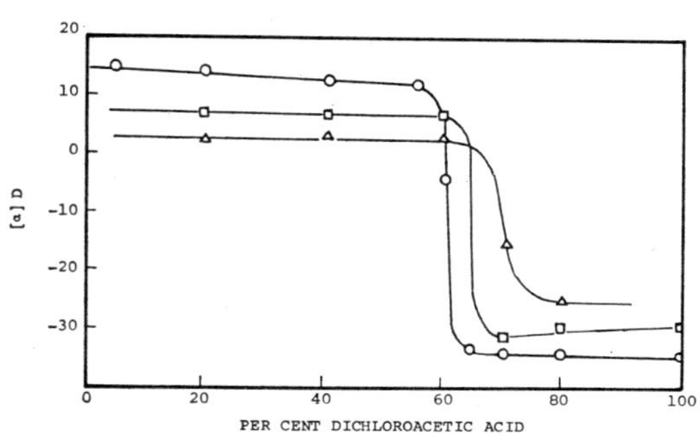

Fig. 19.4. The optical rotation of poly-γ-methyl-L-glutamate and copoly(γ-methyl-L-glutamate-DL-valine) in dichoroacetic acid–chloroform mixtures. L-PMG, ——○——; copoly(L-MG-DL-Val,95:5), ——□——; copoly(90:10), ——△——.

A. ELLIOTT

I should like to mention briefly some measurements made on poly-L-alanines in our laboratory by the late Dr. A. R. Downie. These show that the values of Moffitt's constants a_o and b_o increase with increasing molecular weight of the polymer when the measurements are made in dichloracetic acid. The highest (numerical) values found are: $a_o = 174°$, $b_o = -353°$. It has also been found that the values of these constants (for a sample of given molecular weight) increase with concentration. These results support the suggestion which we put forward some years ago that poly-L-alanine of sufficiently high molecular weight would all be in a right-handed helical form (Downie, Elliott, Hanby and Malcolm, *Proc. Roy. Soc.* **A242,** 325 [1957]). It is very satisfactory that Dr. Fasman's experiments also show the unusual stability of this polypeptide.

JEN TSI YANG

In view of the popular use of the Moffitt equation (W. Moffitt, *J. Chem. Phys* **25**, 467 [1956]; W. Moffitt and J. T. Yang, *Proc. Nat. Acad. Sci.* **42**, 596 [1956])

$$[m'] \equiv \frac{3M_o}{n^2 + 2}[\alpha] = \frac{a_o\lambda_o^2}{\lambda^2 - \lambda_o^2} + \frac{b_o\lambda_o^4}{(\lambda^2 - \lambda_o^2)^2} \quad (1)$$

in the treatment of optical rotatory dispersion data, a few words of caution seem in order:

The choice of the dispersion constant, λ_o (equation 1), is now not firmly supported by theory (W. Moffitt, D. D. Fitts, and J. G. Kirkwood, *Proc. Nat. Acad. Sci.* **43**, 723 [1957]) and accordingly should be used with the caution required by an empirical equation. From our early studies with poly-γ-benzyl-L-glutamates and poly-L-glutamic acids (J. T. Yang and P. Doty, *J. Am. Chem. Soc.* **79**, 76 [1957]) the best value for λ_o was found to be 212 mμ and the mean b_o value about -630 independent of the helix-forming solvents used. Now it is realized that b_o cannot only vary in magnitude but also in sign with different polypeptides, probably because of the complications due to side group–backbone interations (E. R. Blout in "Optical Rotatory Disperson," C. Djerassi [ed.], McGraw-Hill, N.Y., 1960, Chapter 17). There is, however, a tendency among workers in this field to pre-set a common λ_o in the calculation of b_o. Though not disloyal to my own results, I should neverheless point out that 212 mμ is not a magic number and there is still room for refinement. One should never take for granted that any particular λ_o value will automatically be applicable to all polypeptides (J. T. Yang, "International Symposium on Optical Rotation and Rotatory Dispersion," *Tetrahedron,* **13,** 143 [1961]). Now that the ultraviolet range of instrumentation is much improved (our early results were limited down to 350 mμ and occasionally 313 mμ) and new polypeptides are being synthesized, it seems highly desirable to subject the experimental data on various polypeptides to a more careful test. Here we may encounter a dilemma: A precise λ_o value requires measurements over a wide range of wave lengths. (One can obtain almost any positive or negative b_o value by using vastly different λ_o values for a particular set of data if the measurements were limited to a narrow range of wave lengths, say, between 450 and 600 mμ.) On the other hand equation 1 is valid only when λ is far away from any optically active absorption band such as the Cotton effect near 230 mμ (N. S. Simmons, private communication). How low a limit of the wave length should be used in experimental measurements is still an open question. By past experience, 313 mμ or even lower should be included in such a test.

Another point which somewhat worries me is the unconditional use of $\lambda_o = 212$ mμ, even for randomly coiled polypeptide chains in vastly different solvent media. For example, the rotatory dispersion data on poly-γ-benzyl-L-glutamate in dichloroacetic acid was fitted well by a one-term Drude equation

$$[\alpha] = \frac{k}{\lambda^2 - \lambda_c^2} \quad (2)$$

yielding $\lambda_c = 180$ mμ. If the same data is plotted according to equation 1, a small positive b_o value results. In this particular case I would hesitate to consider this an indication of some helices of opposite screw sense present in this solvent. Rather, I suspect that it rests on the fact that the Drude equation can always be explanded into an inverse power series of $(\lambda^2 - \lambda_o^2)$ (see also equation 3 below). The same argument can equally well be applied to some oligopeptides that cannot possibly form helices in various solvents. I am not implying that polypeptides will always exist in the randomly coiled form in solvents such as dichloroacetic acid, but I wonder if there is any advantage in the use of equation 1 under these conditions. There is reason to be wary of any unnecessary confusions that might arise from the uncritical use of equation 1.

At present most workers prefer to estimate the helical content of proteins from the experimental b values, i.e., $b = fb_o$, where f is the fraction of amino-acid residues in the helical conformation. (The $b_o \cong -630$ with λ_o set at 212 mμ, subjected to some possible refinement, probably represents the "normal" value that can be used for proteins. This is based on the belief that the effect of strong side-group interactions which cause abnormal b_o values could very well be only secondary to the total rotation, because of the diversity of amino-acid residues and the absence of a long sequence of one kind of residues in a protein molecule.) Imahori, Klemperer and Doty ("Abstr. Am. Chem. Soc. Meeting," Miami, April, 1957; also P. Doty, *Rev. Mod. Phys.* **31**, 107 [1959]) have further proposed the separation of a_o in equation 1 into $a_o^R + fa_o^H$ for proteins. Here a_o^R and a_o^H are the mean contributions of optical rotation due to the intrinsic residues and the 100 per cent helical backbone. This implies that the protein molecule in its completely nonhelical state would have the same dispersion constant, 212 mμ, as that of the helical backbone. If this is not true, one can rewrite equation 1 into a more general form:

$$[m'] = \frac{a_1^R \lambda_1^2}{\lambda^2 - \lambda_1^2} + \frac{fa_o^H \lambda_o^2}{\lambda^2 - \lambda_o^2} + \frac{fb_o \lambda_o^4}{(\lambda^2 - \lambda_o^2)^2} \quad (3)$$

In any event λ_1 is determined separately from the measurements of denatured proteins. Thus by subtracting this Drude simple dispersion for the residues from equation 3 one can determine both fa_o^H and fb_o simultaneously according to equation 1 in the customary $[m'](\lambda^2-\lambda_o^2)$ versus $1/(\lambda^2-\lambda_o^2)$ plot. The important consequence of this modified treatment is that we can now have two methods for estimating the helical content. The difficulty of course lies in the precise determination of a_o^R and thereby fa_o^H. There is reason to question whether a_o^R remains unchanged when the protein is denatured; for example, portions of the residues which are buried in the interior of the molecules could have a different rotation when exposed to the solvent after denaturation. Doty and coworkers ("Abstr. Am. Chem. Soc. Meeting," Miami, April, 1957; *Rev. Mod. Phys.*, **31,** 107 [1959]; *Proc. Nat. Acad. Sci.* **43,** 213 [1957]) have reported an a_o^H value of about 650–680, but it still must be subjected to extensive tests.

In our original working hypothesis (*J. Am. Chem. Soc.* **79,** 761 [1957]), the protein molecule was assumed to have only two states, helical and nonhelical. Obviously, if more than two conformations are present in the molecule we need independent methods to characterize the various kinds of secondary structures. In the absence of these methods, the determination of fa_o^H provides a double check on the estimate of helical content from fb_o alone. For example, it is now recognized that the rotary dispersion of the β-form can also be fitted with equation 1 with λ_o at 212 mμ in advance. Fasman and Blout (*J. Am. Chem. Soc.* **82**, 2262 [1960]) found that b_o in this case was zero and a_o a large positive number, whereas Imahori (*Biochim. et Biophys. Acta* **37**, 336 [1960]) and Wada (private communication) observed that both were large positive numbers. In either case the estimated helical content from experimental fa_o^H and fb_o should differ if β-form exists in the protein molecule. At present the β-form is still not well characterized and our knowledge of other possible secondary structures that might also contribute to the optical rotations is completely lacking. The use of both a_o^H and b_o, however, should give us more information than either alone, despite some uncertainties.

The recent interesting work by Tanford, De, and Taggart (*J. Am. Chem. Soc.* **82**, 6028 [1960]) on β-lactoglobulin illustrates this point. This protein has an $[\alpha]_D$ of only −28 and λ_c of 242 mμ (as recalculated from their data), both of which seem to indicate the presence of α-helix (J. T. Yang and P. Doty, *J. Am. Chem. Soc.* **79**, 761 [1957]). However, by presetting $\lambda_o = 212$ mμ they obtained the results shown in Table 19.1. These results would indi-

Table 19.1

Solvent		a_o value	b value
(a) 0.01 M $N(CH_3)_4Cl$, pH 2.8–5.7	(mean)	−160	−68
(b) 0.01 M $N(CH_3)_4Cl$, pH 11.8		−623	−77
(c) 8 M urea, pH 5.5		−663	−51

cate the absence of α-helix in β-lactoglobulin (Incidentally, the fact that the b values did not vanish in the denatured states, (b) and (c), perhaps reflects the use of $\lambda_o = 212$ mμ for the denatured protein instead of a somewhat different λ_1.) If one assumes that −623 in (b) represents a_o^R, then the experimental fa_o^H and fb_o simply become +463 and +9, which clearly give quite different estimates of the helical content. This discrepancy could very well be attributed to the presence of more than two conformations in the protein molecule. If one accepts Imahori's and Wada's idea it could mean the coëxistence of both α-helix and β-form (as illustrated in J. T. Yang, "International Symposium on Optical Rotation and Rotatory Dispersion," *Tetrahedron* **13,** 143 [1961]; and in J. A. Schellman and C. G. Shellman, *J. Polymer Sci.* **49,** 129 [1961]). I do not mean to predict the presence of β-form in β-lactoglobulin, which should be easily verified from infrared study. Indeed other conformations, as yet not understood, could also contribute to optical rotations. What I would like to emphasize is that interpretations based on b

values alone do not give the whole picture if more than two states are present in the protein molecule. Of course, one can still argue that the difference of 463 as calculated above actually reflects the solvent effects at different *p*H's, and Tanford and co-workers' conclusion could very well be correct. In this respect it is noted that the use of 8 M ureas does not affect the a_o value very much, and with our present knowledge I can hardly imagine that the solvent effect in this case could amount to 463, especially when all the measurements are carried out in aqueous solution.

At this symposium I have learned that both Drs. Blout and Beychok and also Professor Doty and his co-workers have independently found a fb_o value of about -500 for native sperm whale myoglobin, which amounts to about 80 per cent helices—a very good agreement with Dr. Kendrew's finding from X-ray study. The beautiful work by Drs. Perutz, Kendrew, and their co-workers clearly opens up an ultimate method of determination of the internal structure of the protein molecules. It will be most interesting to find out if the fa_o^H value for myoglobin yields the same helical content as fb_o does. If so, we will certainly have more confidence in the use of equation 3. In a few years the isomorphous replacement method will give us more knowledge of the three-dimensional structure of insulin, ribonuclease, lysozyme, and other proteins. Accordingly the rotatory dispersion method will find new applications. For example, it may provide us with a convenient means for studying the conformational changes, if any, of proteins when they dissolve in water or other nonaqueous media. Carefully used, equation 3 can now provide a reasonable estimate of the helical content and, more important, shed some light on the internal structure of the protein molecule.

Finally, I would like to make a plea about the reporting of experimental data. Understandably, there is now a tendency to follow conformational changes with optical rotations other than the customary $[\alpha]_D$. Some have used 313, 334, 546, 578 mμ, and other wave lengths, while others have simply reported them in terms of equation 1. It becomes increasingly difficult for readers to compare the results in various papers. Failing to reach a general agreement on the treatment of the rotation data, we should at least give enough information so that anyone wanting to can recalculate the raw data.

This work is supported by the USPHS Grants A-4919 and SF-435 and NSF Grant G-15040.

DON WETLAUFER

In connection with Dr. Yang's comments on reporting optical rotatory dispersion data, I agree that before long the constants and perhaps the form of the Moffitt equation will have to be revised. I would suggest that this be deferred until we have some information about the rotational strength and spectral location of the absorption bonds concerned with the optical rotation. Since there is real hope in the next few years of circular dichroism measurements on these materials, which should provide a sound basis for rewriting rotatory dispersion equations, it seems best to continue to use the Moffitt equation as a convenient empirical equation for summarizing experimental findings. To be successful in this respect, it is not enough simply to indicate the value of the Moffitt equation b_o. One must also indicate the value of λ_o employed, the value of a_o, and the wave length range over which the measurements were made. When all of these parameters are specified, it is possible to estimate the actual specific rotations that were measured.

JON APPLEQUIST

Dr. Fasman's work on the copolymer of L-tyrosine and L-glutamic acid is an important piece of information for establishing the helix sense of poly-L-tyrosine. However, I would like to clarify the amount of information that can be gotten from the dependence of b_o on copolymer composition. The nearly linear dependence observed by Dr. Fasman happens to be just what is expected if poly-L-tyrosine and poly-L-glutamic acid are mixed in various proportions, whether they have the same helix sense or not. Similarly, this behavior would be expected of a copolymer consisting of long blocks of either amino acid, such as could arise if the reactivity ratios of the monomers were favorable. On the other hand, if there were some demonstration that the copolymer composition were more or less uniform along the chains, it would be difficult to see how Dr. Fasman's observations could be consistent with opposing helix senses for poly-L-tyrosine and poly-L-glutamic acid.

PAPER 20

The Theory of the Transition between Helix and Random Coil

BRUNO H. ZIMM

All authors seem to agree on the fundamentals of the theory of the transition between helix and random coil. Two parameters are especially important: first, a parameter measuring the free energy of forming a hydrogen bond in a helical sequence; and second, a parameter measuring the free energy of the junction between helical and random sections. The junction free energy (second parameter) is positive, since more amino-acid residues are immoblized than hydrogen bonds are formed in each helical section; the free energy of immobilization of the extra residues appears as the junction free energy and is expected to be positive. In order for the total free energy of helix formation to be negative the number of helix coil junctions must be small, and the helical sections tend to grow long. Their length is limited, however, by occasional thermal fluctuations to the random state. One expects, therefore, that short chains will contain only one helical section, if any, while long chains will have an alternation of helical and random sections. There will also be a critical chain length below which no helices form at all at any given value of the hydrogen bond free energy. The above conclusions agree with the experiments that have been so far reported on polybenzylglutamate polymers of diverse chain lengths. The effect of charges on the molecule, as in salts of polyglutamic acid, is practically one of reducing the hydrogen bond strength as a result of the electrostatic repulsions of the charges.

Introduction

We shall discuss only the case of isolated polypeptide molecules, and shall restrict ourselves to the discussion of equilibrium only. We assume that there is one helical configuration that is much more stable than all other possible helices, and, even more strictly, that if intramolecular hydrogen bonds form at all they are the ones consistent with this helical structure. This is not to deny that other hydrogen bonds may form, but it is to say that the importance of such other bonds is negligible in determining the course of the transition. (There may very well be cases in which this restriction is inappropriate.)

The thermodynamics of the transition involves the balancing of the free energy gains from the formation of the hydrogen bonds against the losses from the concomitant constraints on the chain backbone. These two quantities are involved in significantly different ways. Let us for the moment adopt the crude but perspicuous method that was used by Schellman in the first treatment of the problem (*1*). Let e be the energy gained on forming a hydrogen bond and f be the entropy lost on immobilizing the alpha-carbon of an amino-acid residue by forming the helix. If there are n amino acid residues, then there is a maximum of $n-4$ hydrogen bonds, but $n-2$ alpha-carbons may be immobilized. If we assume that all the possible hydrogen bonds are formed, then the total free energy change, ΔG, for formation of the helix is:

$$\Delta G = T(n-2)f - (n-4)e = 2Tf - (n-4)(e - Tf) \quad (1)$$

The above equation shows us immediately that the helix is not stable (ΔG negative) if the chain length, n, is too small, and that the extra entropy lost at the ends of the helix is the source of the instability. This follows since we have every a priori reason to expect that e and f

should be positive. It is also apparent that when n is large (very long chains) the criterion for stability of the helix is merely that the free energy of formation of a hydrogen bond, $Tf-e$, should be slightly negative.

In addition to the foregoing, a more advanced treatment must consider the number of arrangements of helical and random sections. For example, a hydrogen bond may be broken at any one of $n-4$ places; even if e is very large, there will be a considerable probability of some bond being broken if n is great enough. (In fact, it is probably easier to break three successive bonds rather than one in an alpha helix, since the larger break is free of strain, but the same principle holds.) As a consequence, we expect to find long chains preferentially in the form of several helical sections separated by random regions rather than as one unbroken helix, even under conditions that strongly favor hydrogen bonding.

In many cases we should probably also take account of interactions between the side chains of the amino acids. We are going to hope that these interactions can be lumped together with the hydrogen bonds of the helix into one generalized interaction. Detailed study has shown that this is in fact possible for electrostatic interactions between charged side chains as long as one is not concerned with exceedingly fine details (see "Charged Polypeptides" below).

Fundamental formulas

We shall now give the results of a more exact treatment of the theoretical problem, one which takes account of all the various possible arrangements of hydrogen bonds discussed above. We prefer to introduce two new parameters, s and σ. The parameter s is essentially the equilibrium constant for the addition of an amino-acid residue to a section of helix, the residue being initially at the end of the random portion of the chain adjacent to the helical section. The parameter σ is a factor, much less than unity, that measures the extra free energy lost at the start of a helical section; σs is the equilibrium constant for the formation of the hydrogen bond that initiates the helix. In the simple discussion above we would have put s equal to $\exp\,(e-Tf)/kT$ and σ equal to $\exp\,(-2f/k)$, where k is Boltzmann's constant, but in general the separation of the free energy into parts uniquely related to the energy and entropy of hydrogen bonds and alpha-carbons respectively may not be so clear cut. For example, in the case of polybenzylglutamate, for which experimental data are given below, s has a large entropy part, and the hydrogen bond energy has the wrong sign! This anomaly is certainly the result of changes in the surrounding solvent concomitant with hydrogen bond formation. It seems better, therefore, to work first with the parameters s and σ, which are equilibrium constants that enter naturally into the statistical treatment, and defer analysis of the energy and entropy contributions until later.

There have been given recently several more or less equivalent theoretical descriptions of the transition (*2–6*). In the following we adopt the form given by J. K. Bragg and the author (*2*); this selection is merely for the author's convenience and is not meant to imply any judgment of the ultimate value of the various forms of presentation.

The quantity of primary interest is the fraction of possible hydrogen bonds that are actually formed; we denote this by the symbol θ. For very long chains this fraction is calculated from the expression:

$$\theta = d \ln \lambda / d \ln s, \qquad (2)$$

where λ is the larger value of the expression

$$\lambda = (\tfrac{1}{2})\{1 + s \pm [(1 - s)^2 + 4\sigma s]^{\frac{1}{2}}\}. \qquad (3)$$

The above expression is valid as long as n is greater than a certain minimum value, which seems in practice to be of the order of several hundred amino-acid residues. As long as n is large the course of the transition is independent of chain length, though dependent on the paramater σ, being more abrupt when σ is smaller. The parameter s has a critical value of unity, in the neighborhood of which the transition occurs.

With shorter chains the description is conveniently done differently. Such chains have at the most one helical section per molecule because σ is much smaller than unity. It is then easy to derive a formula for the fraction of hydrogen bonds, θ, by summing the statistical weights of the various configurations; the result for a chain of n residues is:

$$\theta = \frac{(n-4)(s-1) - 2 + [(n-4)(s-1) + 2s]s^{-n+1}}{(n-4)(s-1)\{1 + (s-1)^2 s^{-n}/\sigma - [(n-4)(s-1) + s]s^{-n+1}\}} \tag{4}$$

Close examination of this formula shows that there is a critical value of n at which substantial helix formation first appears for any given value of s (greater than unity) and σ. The value is approximately that at which the following relation is satisfied:

$$(s-1)^2 s^{-n} = \sigma. \tag{5}$$

(We should repeat that n in these formulas is the number of amino-acid residues, which is one greater than the number of amide residues; the latter number was used in a previous paper [*2*].)

The calculated behavior of the fraction of hydrogen bonds, θ, for the cases of long chains and short chains is illustrated in Figures 20.1 and 20.2 respectively. Figure 20.1 shows how the transition sharpens as σ is decreased. Figure 20.2 emphasizes the importance of the chain length, n, in determining the behavior of short chains. Also shown in Figure 20.2 (dashed curves) is the probability that the end amide groups be hydrogen bonded. Since the end amide groups can dissociate from the helix without disrupting it, in contrast to groups in the middle, the presence or absence of bonding at these points makes little difference in the free energy of the molecule as a whole, and the end groups show a strong tendency to come loose from the helix. The effect is evident in Figure 20.2, where the plateaus of the dashed curves at right have levels much less than unity. This effect may be significant in the kinetics of polymerization (*7*).

The quantities, stability constant and chain length, make suitable coördinates on which to display a "phase diagram" of the molecule. Such a diagram is shown in Figure 20.3 for the case of σ equal to 10^{-4}. The use of the term, "phase diagram," represents an extension of the traditional concept of phase in chemical systems and one which may be objectionable to some, although respectable reasons can be advanced for this extension. The points of difference lie in the diffuseness of the transition and in the impossibility of separating the coexistent phases on a macroscopic scale. Both points arise from the same cause, the very small, frequently submolecular, sizes of the helices which are one of the phases.

Turning to the diagram itself, Figure 20.3, we can see that for small values of n or s the chains are in the random, unbonded configuration, while at larger values of n or s the helical configurations dominate. The shaded region

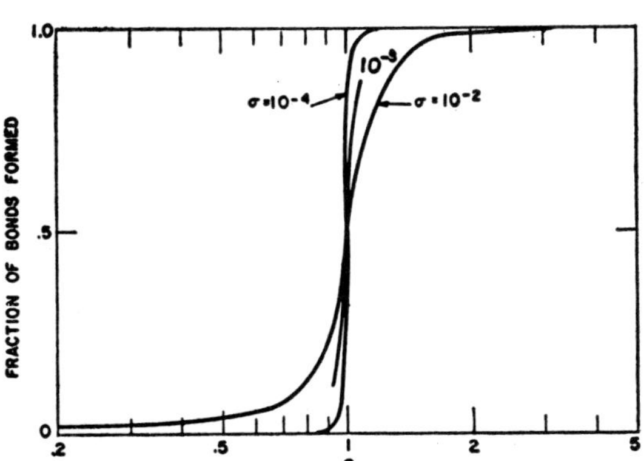

Fig. 20.1. Fraction of intersegment hydrogen bonds θ as a function of the equilibrium constant s for various values of the initiation parameter σ.

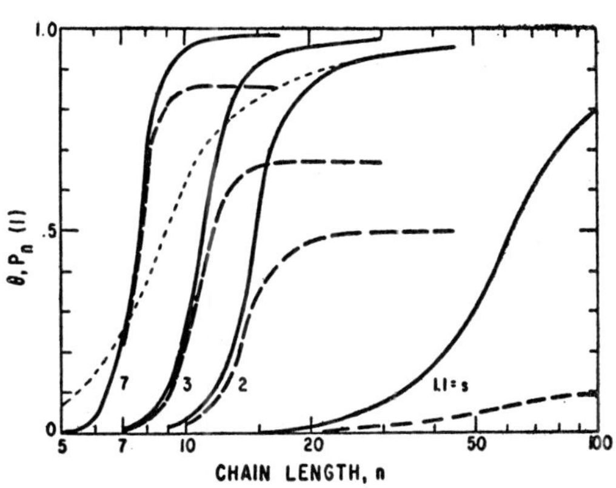

Fig. 20.2. Fraction of intersegment hydrogen bonds θ and probability that the end segment be hydrogen-bonded, P_n (1), as functions of the number of segments n at various values of s and σ. Solid curves θ at $\sigma = 10^{-4}$ and the indicated values of s; thin dashed curve, θ for $s = 2$ and $\sigma = 10^{-2}$; heavy dashed curves, $P_n(1)$ at $\sigma = 10^{-4}$ and the indicated values of s.

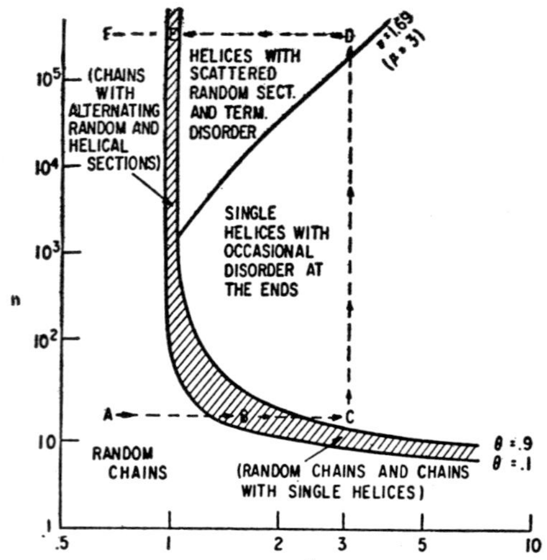

Fig. 20.3. The "phase diagram" in the n-s plane, calculated for $\sigma = 10^{-4}$, and showing the characteristics of the chains in the various regions.

between is the transition region, with boundaries represented, rather arbitrarily, by two curves of constant θ. The sector of moderate n and large s differs from that of large n and moderate s; in the former each chain contains only one unbroken and necessarily short helical section, while in the latter each chain contains several helical sections. The boundary separating the two sectors has been drawn along the line where half the chains contain one unbroken helical section. A more extensive discussion was given in the original paper (*2*).

Comparison with experiment

Systems in which the helix-coil transition can be seen in its pure form without complications of aggregation, insolubility, or more poorly defined complications seem to be rather few in number, so that comparison of the theory with experiment has not been as extensive as one would wish. The only critical test is afforded by data on solutions of poly-γ-benzyl-L-glutamate in mixtures of dichlorethane and dichloracetic acid in which the transition can be followed by the change in optical rotation with varying temperature. The optical rotation is assumed to be proportional to the fraction of helices. The most recent data (*8*) are shown in Figure 20.4, in which are also included relevant data from earlier work by Doty and Yang (*9*). There are essentially three parameters to be determined: (1) one value of s, (2) the rate of change of s with temperature, and (3) σ. Since s is an equilibrium constant, it is assumed that log s varies linearly with $1/T$.

It is immediately found that data on one sample are not enough to establish all these parameters; data on at least two samples are required, and one of these must be of short enough chain length to lie in the lower part of the phase diagram (Figure 20.3). In Figure 20.4 are shown the results for three samples, one more than necessary, in order that the theory should get a serious test. The curves are the theoretical ones, as calculated from the equations above. It can be seen that the agreement is quite satisfactory. The initiation parameter, σ, comes out to be 2×10^{-4}, a value which is reasonable enough in the light of our present imperfect knowledge about the rotational characteristics of the bonds of the peptide backbone. The heat of hydrogen bonding, as determined from the temperature coefficient of s, has a positive sign, and corresponds to heat *absorbed* on bonding in this case.

The interpretation of the data by the theory thus appears to be reasonably successful. However, further studies would be most desirable, particularly with samples of short chain length

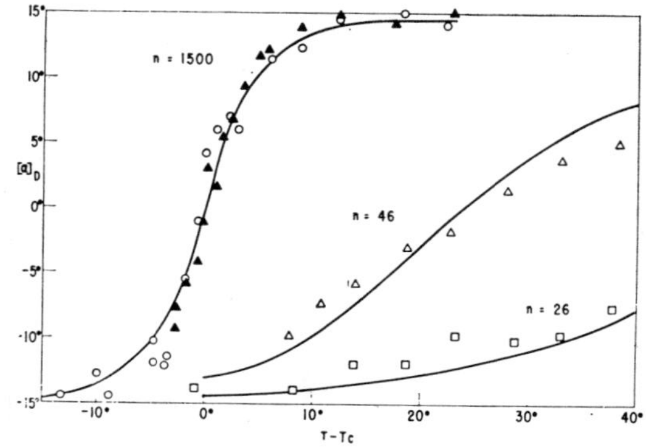

Fig. 20.4. Experimental results for the optical rotation, $[\alpha]_D$, as a function of temperature, T, minus critical temperature, T_c, for samples of various degrees of polymerization, n. ○, data of Ref. *9;* other symbols, data of Ref. *8.* T_c for ○, 28.7° C.; T_c for other data, 11.8° C. Solvent for ○, dichloracetic acid, 80 parts, ethylene dichloride, 20 parts; for the other data, the composition was 70:30.

in various solvents. The questions of the constancy of σ, and of the dependence of s on solvent and temperature, are in particular need of further study.

Charged polypeptides

To go far into the detailed theory of charged polypeptides would involve us in all the poorly resolved complications of polyelectrolyte theory. Fortunately, the main result is simple enough: the electrostatic interactions affect the value of the parameter s. As the helix becomes charged, as during a titration for example, the electrostatic repulsions weaken the bonding between successive coils and reduce the effective value of s. If s drops below unity the helix melts. This effect is seen with polyglutamic acid, which was shown by Doty, Wada, Yang, and Blout (*10*) and by Wada (*11*) to be helical in aqueous solution in the acid form but random in the charged basic form.

Rice and I (*12*) were able to predict this behavior quantitatively from polyelectrolyte theory, though there are reasons for thinking that the agreement with experiment was partly fortuitous. Nevertheless, the following interesting point deserves to be brought out. Since it is possible to carry out the transition reversibly and measure the necessary work, for example by using a hydrogen electrode as the source of hydrogen ions, the free-energy change is directly accessible. In effect, this makes it possible to obtain s directly, instead of having to infer it from the fitting of curves to temperature-transition data. The only remaining parameter then is σ; it is determined by the sharpness of the transition. The data on polyglutamic acid are in accord with a value of σ in the neighborhood of 10^{-4}, as with the benzyl ester, except for some unexplained discrepancies at low ionic strength.

References

1 J. A. Schellmann, *Compt. rend. trav. lab. Carlsberg, Sér. chim.* **29**, No. 15 (1955).
2 B. H. Zimm and J. K. Bragg, *J. Chem. Phys.* **28**, 1246 (1958); **31**, 526 (1959).
3 J. H. Gibbs and E. A. DiMarzio, *ibid.* **28**, 1247 (1958); **30**, 271 (1959).
4 S. A. Rice, A. Wada, and E. P. Geiduschek, *Discussions Faraday Soc.* **25**, 130 (1958).
5 T. L. Hill, *J. Chem. Phys.* **30**, 383 (1959).
6 L. Peller, *J. Phys. Chem.* **63**, 1194 (1959).
7 P. Doty and R. D. Lundberg, *J. Am. Chem. Soc.* **78**, 4810 (1956).
8 B. H. Zimm, P. Doty, and K. Iso, *Proc. Nat. Acad. Sci.* **45**, 1601 (1959).
9 P. Doty and J. T. Yang, *J. Am. Chem. Soc.* **78**, 498 (1956).
10 P. Doty, A. Wada, J. T. Yang, and E. Blout, *J. Polymer Sci.* **23**, 851 (1957).
11 A. Wada, *Mol. Physics* **3**, 409 (1960).
12 B. H. Zimm and S. A. Rice, *ibid.* **3**, 391 (1960)

PART IV

Protein and Polypeptide Structure

PAPER 21

The Structure of Globular Proteins

J. C. KENDREW

The study of the structure of globular proteins by X-ray methods began in the late 1930's when the first X-ray photographs of protein crystals were obtained by Bernal, Crowfoot, Fankuchen and Perutz. Proteins are so much more complex than the molecules which were within the scope of X-ray methods at that time and for many years afterwards that little progress was made in terms of results, though much in terms of techniques, for about fifteen years. The decisive advance came in 1953, when Perutz (*1*) showed that it was possible to prepare heavy-atom derivatives of hemoglobin which crystallized isomorphously with the unsubstituted protein and whose crystals gave appreciably different diffraction patterns; he was then able to use the so-called method of isomorphous replacement, which in principle enables an unambiguous structure determination to be carried out. During more recent years a number of crystalline proteins have been studied intensively with the object of applying this method, especially hemoglobin, myoglobin, insulin, ribonuclease, and lysozyme. Always the difficulty has been to obtain suitable isomorphous heavy-atom derivatives; only one chemically rational method—the attachment of mercury or silver atoms to free sulphydryl groups, as originally used by Perutz—is known even today, and this is more often than not inapplicable since most of the proteins suitable for analysis on other grounds (e.g., low molecular weight, ease of crystallization) do not contain free sulphydryl groups. For the most part it has been necessary to search *ad hoc* for methods of attaching heavy atoms, those used successfully for one crystalline protein generally not being applicable to others. Only in hemoglobin and myoglobin has the isomorphous replacement method so far been applied with complete success. Perutz's early results with hemoglobin took the form of a projection of a row of hemoglobin molecules onto a plane: they were the first pictures of a protein molecule which were demonstrably correct, but the amount of structural information which could be derived from what was in effect a silhouette of a very complicated molecule was small. It was clear that if the method was to be exploited it was necessary to extend it into the third dimension, and this was first achieved in 1957 in the analysis of the simpler protein myoglobin (*2*).

The first stage of the X-ray analysis of myoglobin resulted in a three-dimensional representation of the molecule at a resolution of 6 Å, which was insufficient to reveal the side chains or even the configuration of the polypeptide chain itself, but which did give a picture of the general arrangement of the polypeptide chain and of the haem group—in other words of the tertiary structure of the molecule. See Plate 21.1. This tertiary structure proved to be unexpectedly irregular and complex: clearly the simple geometrical arrangements proposed in many earlier theories of protein structure could not be correct. More recently the resolution of the analysis has been increased to 2 Å (*3*). See Plate 21.2. Although neighboring covalently bonded atoms are still not distinctly resolved, it is now possible to separate atoms which are hydrogen-bonded or in Van der Waals contact, with the result that the atomic arrangement of most of the molecule can be inferred. At this resolution the appearance of the haem group with its central iron atom corresponds closely with theoretical expectation, thus providing a valuable check on the correctness of the analysis, and it can be seen that the iron atom is attached to a neighboring polypeptide chain by means of a group which is almost certainly the imidazole ring of a histidine residue. Most of the polypeptide chain consists of a series of segments of right-handed α-helix, with parameters closely corresponding to those proposed by Pauling and Corey in 1951 (*4*). This was the first time that the α-helix had been directly seen in a protein molecule; the results confirmed earlier plausible evidence that this configuration was important in the structure of globular proteins as well as in those of many fibrous proteins. It turns out that the myoglobin molecule consists of eight segments of α-helix joined by irregular regions of varying length: the helical segments comprise 75 per cent of the amino-acid residues, in agreement with estimates made on the basis of optical rotation and deuterium exchange studies.

One reason for the choice of 2 Å as the resolution of the second stage of the myoglobin analysis was that the secondary structure (or polypeptide chain configuration) should be clearly revealed at this resolution, as indeed proved to be the case; it was not anticipated that side chains would be identifiable, so that progress could be made towards a direct determination of the amino-acid sequence by X-ray methods alone. In the event expectation was much exceeded, for it was possible to see the individual side chains as dense regions emerging from the helical main chain at intervals corresponding to the parameters of the α-helix, and a close examination of these regions often made it possible to identify side chains with certainty; in other cases some ambiguity remained but the choice of side chain could be reduced to two or three. At the present resolution about one-third of the side chains can be identified with certainty, and another third with fairly high probability (*5*). See Plate 21.3. It now becomes possible to correlate these X-ray results with the preliminary data obtained by Dr. A. Edmundson, who is engaged in working out the amino-acid sequence of myoglobin by chemical methods (*6*). He has broken down the molecule into short peptides by means of tryptic digestion, and he has determined the composition, and in a few cases the internal sequence, of these peptides. By comparing his results with our own it has proved possible to place almost all the tryptic peptides along the polypeptide chain, and the order of peptides thus ascribed corresponds with the order suggested in a few cases by Edmundson on the basis of preliminary work using chymotryptic digestion. Although some discrepancies remain and although the amino-acid sequence has not yet been completely determined, its main features are now beyond doubt. We are at present engaged in an attempt to increase the resolution of the X-ray results still further, and we hope that the remaining ambiguities will then be removed.

Meanwhile Perutz and his collaborators have obtained a three-dimensional representation of the hemoglobin molecule at a resolution of 5.5 Å (*7*). See Plate 21.4. Hemoglobin is known to be made up of four subunits, each consisting of a single polypeptide chain and an associated haem group. Larger protein molecules generally consist of several subunits, whose spatial arrangement is termed the quaternary structure; Perutz's results show that in hemoglobin the quaternary structure is of a very simple and symmetrical kind, the four subunits being arranged at the vertices of a nearly regular tetrahedron. The most astonishing result of the analysis, however, is that the tertiary structure of each of the four subunits very closely resembles that of myoglobin; by implication it follows that the secondary structure must also be very similar. See Plate 21.5. This remarkable finding indicates that the irregular arrangement of the polypeptide chain now found in two distinct proteins, derived from very different species—namely, horse and sperm whale—has some general significance which we cannot at present understand. A further extension of the range of this structure has been provided by the work of Scouloudi (*8*), who has shown that the molecule of seal myoglobin very closely resembles that of horse myoglobin.

The structures of virus particles lie outside the scope of the present paper; it is, however, relevant to note that in these still more complex molecules protein subunits are also arranged in a highly regular manner. For example, in tobacco mosaic virus the 2,000 identical protein subunits, each as large as a myoglobin molecule, are arranged in the form of a helix; in the spherical viruses the subunits are disposed symmetrically on the surface of regular or semiregular polyhedra.

Thus we are now beginning to get a first glimpse at the general nature of protein structure at all levels of complexity. Hitherto the

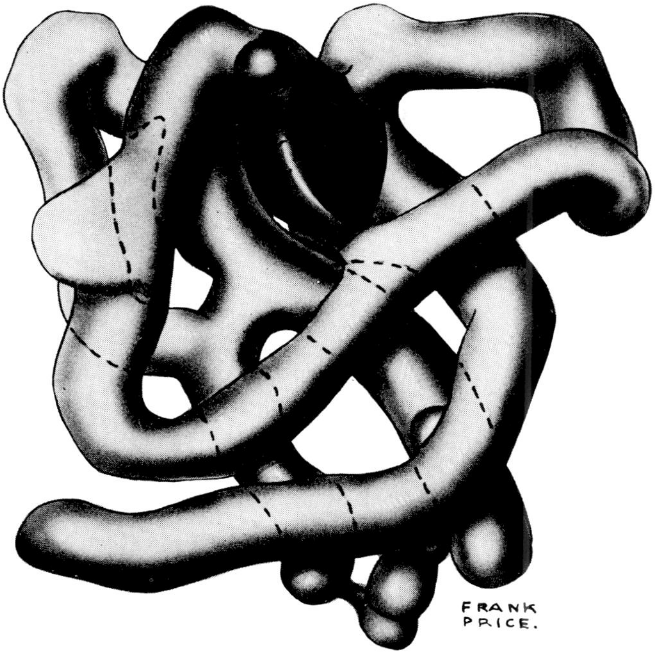

Plate 21.1. Model of the myoglobin molecule derived from the 6 Å Fourier synthesis.

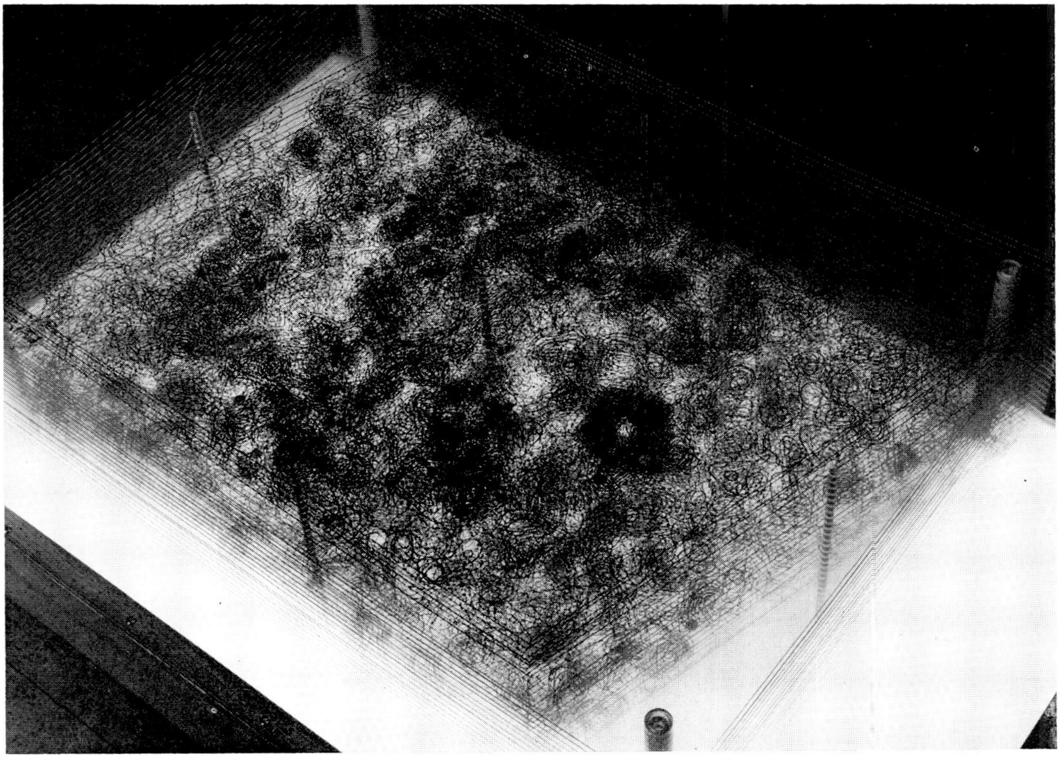

Plate 21.2. End-on view of a helical segment of polypeptide chain in the 2 Å Fourier synthesis of hemoglobin, showing the hole down the center.

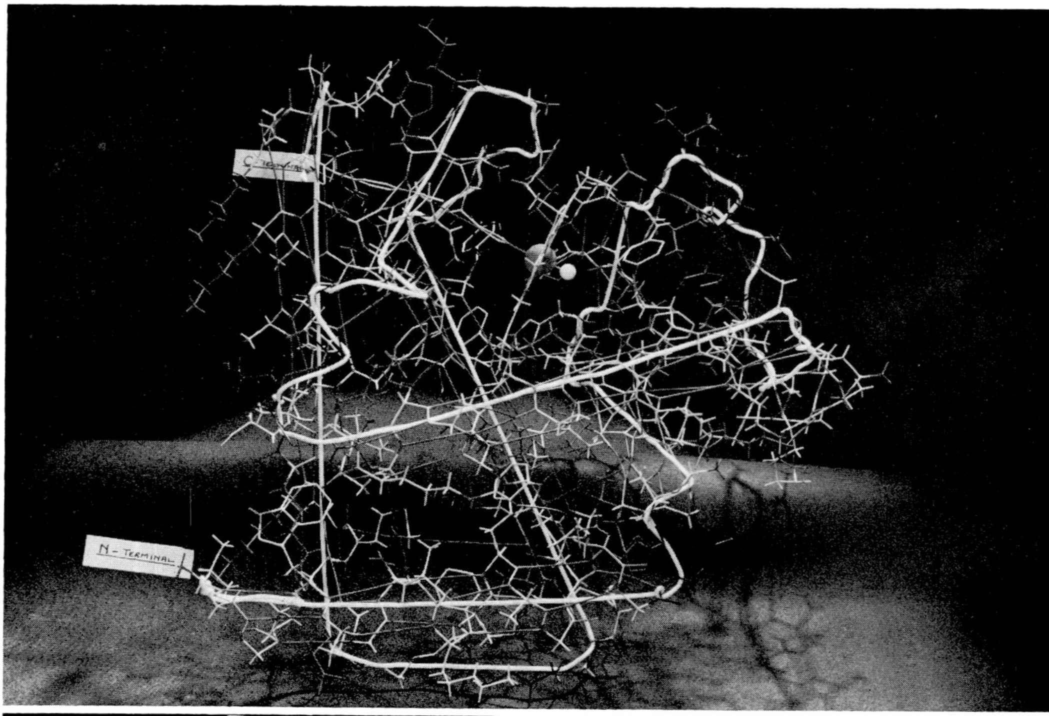

Plate 21.3. Model of the myoglobin molecule derived from the 2 Å. Fourier synthesis, including most of the side chains.

Plate 21.4. The molecule of hemoglobin derived from a 5.5 Å Fourier synthesis.

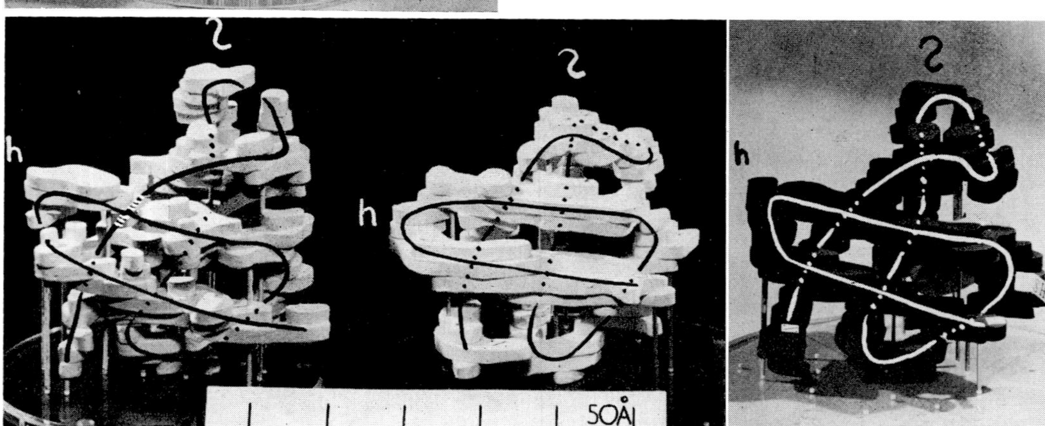

Plate 21.5. Comparison between the subunits of hemoglobin and the molecule of myoglobin. *Left,* myoglobin; *center and right,* hemoglobin chains.

chief difficulty in applying to biological problems the results of the exhaustive physicochemical and chemical studies of proteins carried out in many laboratories during the last half-century has been our lack of knowledge of the three-dimensional arrangement of the atoms of which protein molecules are composed. Such knowledge is an essential complement to an understanding of the chemical interconnections, because it is the spatial relations between the side chains which determine the chemical behavior and biological specificity of the protein molecule as a whole, and these relations cannot be determined, except in a fragmentary manner, by purely chemical techniques. The rate of progress on all these fronts is now so great that we are undoubtedly on the eve of a major advance in our understanding of one of the central problems of biology—the relation between the structure of the molecules making up living cells and their biological functioning.

References

1 D. W. Green, V. M. Ingram, and M. F. Perutz, *Proc. Roy. Soc.* **A225**, 315 (1954).

2 G. Bodo, H. M. Dintzis, J. C. Kendrew, and H. W. Wyckoff, *ibid.* **A253**, 70 (1959).

3 J. C. Kendrew, R. G. Dickerson, B. E. Strandberg, R. G. Hart, D. R. Davies, D. C. Phillips, and V. C. Shore, *Nature* **185**, 422 (1960).

4 L. Pauling, R. B. Corey, and H. R. Branson, *Proc. Nat. Acad. Sci., Wash.* **37**, 205 (1951).

5 J. C. Kendrew, H. C. Watson, B. E. Strandberg, R. E. Dickenson, D. C. Phillips, and V. C. Shore, *Nature* **190**, 666 (1961).

6 A. B. Edmundson and C. H. W. Hirs, *ibid.* **190**, 663 (1961).

7 M. F. Perutz, M. G. Rossmann, A. F. Cullis, H. Muirhead, G. Will, and A. C. T. North, *ibid.* **185**, 416 (1960).

8 H. Scouloudi, *Proc. Roy. Soc.* **A258**, 181 (1960).

PAPER 22

Effect of Side-Chain Hydrogen Bonds, Hydrophobic Bonds, and Covalent Cross-Links on Protein Conformation in Solution

HAROLD A. SCHERAGA

While our primary concern here is with the effect of side-chain hydrogen bonds, hydrophobic bonds, and covalent cross-links such as disulfide bonds on protein conformation in solution, it is appropriate to begin by recalling that the problem of protein and polypeptide conformation in solution has recently been viewed in terms of polymer crystallinity. Consequently, reversible changes in conformation have been regarded as phase transitions. If we are to consider the effect of side-chain interactions on these phase transitions, cognizance must first be taken of the fact that the stable conformation of a protein or polypeptide can be quite strongly dependent on the protein concentration. Flory has pointed this out most clearly in Fig. 1, and the related discussion, of his recent paper entitled "Phase Changes in Proteins and Polypeptides" (*1*). In very dilute solution we may have equilibrium between dispersed helices and dispersed random coils. However, at higher concentrations, the helices may associate into larger, dense crystalline regions. Such further crystallization would deprive the random state of some of its statistical advantage. In other words, crystallization at high concentration can lead to a change in the conformation of the molecule in order that it can be accommodated in the crystal lattice. As a consequence of these considerations, we should be very careful to recognize that the conformation of a protein molecule in dilute solution may be considerably different from that in the wet crystalline form which the X-ray crystallographer investigates. The principles of polymer crystallinity and the thermodynamics of phase transitions are applicable to both forms, even though the conformations may differ.

We shall apply these same principles to conformational changes both in dilute solutions and in more concentrated systems, namely protein fibers. The applicability of these principles to both ranges of concentration gives us additional confidence in their validity. Fortunately, a variety of methods are available for investigating conformational changes in dilute solutions and in fibers.

It is worth noting that the natural habitat of proteins in some cells and tissues may give rise to rather high protein concentrations in which the dense crystalline phase or tactoidal anisotropic solution phase, discussed by Flory (*1*), may be stable.

Adopting the point of view of Flory (*1*), we have carried out a series of investigations on reversible protein denaturation to test current theories of crystalline-amorphous (e.g., helix–random coil) equilibrium and the effects thereon of side-chain hydrogen and hydrophobic bonds, covalent cross-links, and cooperative effects between helices. The investigations of fibers have contributed to the establishment of the general validity of the principles involved* and have given confidence in their applicability to reversible conformation changes in dilute solutions. Thus, we are able to obtain information about side-chain interactions in the native molecule in dilute solutions. Some aspects of the theory and experimental results will be discussed here. We are

* The validity of these principles has been extensively verified in experiments of Flory (*2*, *3*) and Mandelkern (*4*).

also applying these principles to kinetic studies but space limitation precludes their consideration at this time.

Conformational problem in dilute solution

It is now recognized that a globular protein can be made up of several regions, each of which may exist in a different conformation. The various conformations currently considered are: left- and right-handed α-helices, random coils, and extended chains (e.g., β-structures). Conformational changes can occur in which, for example, helical regions of a molecule are converted to randomly coiled ones. Interconversion between other forms are, of course, also possible. In order to simplify the discussion we shall, *for illustrative purposes only,* consider the molecule as consisting of only two regions, namely right-handed α-helices and random coils. Denaturation will then be regarded as the conversion of part or all of the helical region of a molecule to a randomly coiled form, without the rupture of any covalent bonds. Such a transformation of a helical region is shown in Fig. 22.1. The process is divided into two steps; in the first step, the side-chain interactions are disrupted, and in the second step the NH$\cdots$OC hydrogen bonds of the helix are disrupted.

If the various forms in Fig. 22.1 are regarded as being in equilibrium with each other, we can shift this equilibrium in either direction by changing the chemical potentials of the various species in the various phases, e.g., by changing the temperature, pH, solvent, etc. These transformations can be observed in dilute solutions by measurements of optical rotation, ultraviolet light absorption, etc. They can also be observed in fibers by measurements of elastic properties, birefringence, X-ray diffraction, etc. Our concern here is with the thermodynamics of these reversible changes in conformation.

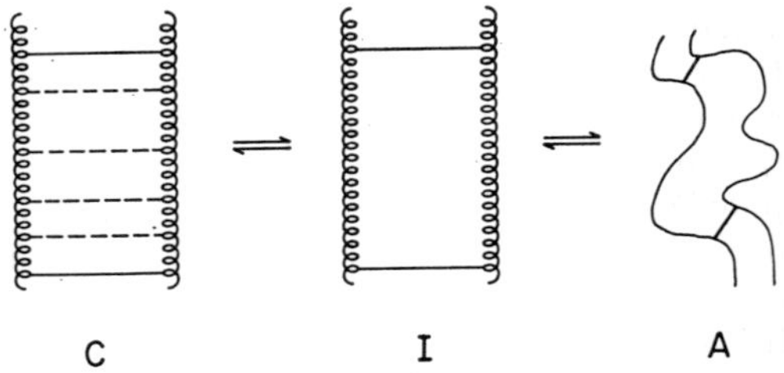

Fig. 22.1. Representation of equilibria between helix and random-coil forms. The solid lines represent covalent crosslinks and the dashed lines side-chain hydrogen or hydrophobic bonds. Species C, I, and A are crystalline (helix), intermediate, and amorphous (random coil) forms, respectively. Forms C and A may also be regarded as native and denatured forms, respectively. In any given protein molecule the native conformation may actually consist of both helical and randomly coiled regions, i.e., the helices may not be perfect (*5*).

*Thermodynamics of protein denaturation**

If a protein, conforming to the model of Fig. 22.1, is denatured, the observed standard free energy of denaturation, ΔF_{obs}, may be written as

$$\begin{aligned}\Delta F_{\text{obs}} &= \sum \Delta F_i \\ &= \Delta F_b + \Delta F_{\text{H}} + \Delta F_{NP} + \Delta F_e + \Delta F_m \\ &\quad + \Delta F_x + \cdots \end{aligned} \tag{1}$$

where the summation is taken over all possible contributions to ΔF_{obs}; these contributions arise from a variety of sources.

The term ΔF_b is the free energy change for unfolding a helix containing n amino acid residues (transformation I$\rightarrow$A of Fig. 22.1). This problem has been treated by a number of investigators (*7–13*). For simplicity, and to show the essential features of the problem, we shall use Schellman's original result (*7*), which corresponds to the all-or-none transition of a helix to a random coil. The helical segments in a protein are probably short enough for this to be a reasonable assumption.

$$\Delta F_b = (n - 4)\Delta H_{\text{res}} - (n - 1)T\Delta S_{\text{res}} \tag{2}$$

where ΔH_{res} and ΔS_{res} are the changes in enthalpy and entropy, respectively, per amino-acid residue, for converting an infinitely long helix to a random coil.

The term ΔF_{H} is the free energy change for rupturing the side-chain hydrogen bonds between the helices† (transformation C$\rightarrow$I of Fig. 22.1). This quantity has been evaluated for a number of different kinds of side-chain hydro-

* Further details may be found in a recent review of denaturation (*6*).

† The side-chain interactions discussed throughout this paper can exist between separate helices or between groups on the same helix.

gen bonds (*14*). For a heterologous single bond between a donor in the i position and an acceptor in the j position it is

$$\Delta F_H = -RT \sum \ln (1 - x_{ij}) \qquad (3)$$

where x_{ij} is the fraction of the molecules having a side-chain hydrogen bond between the given donor and acceptor groups, and the sum is taken over all such bonds. The value of x_{ij} is

$$x_{ij} = \frac{K_{ij}}{1 + K_{ij} + K_1/[H^+] + [H^+]/K_2} \qquad (4)$$

where K_{ij} is the equilibrium constant for formation of the ij bond, K_1 and K_2 are the ionization constants of the nonhydrogen bonded donor and acceptor groups, respectively, and $[H^+]$ is the hydrogen ion activity. Similar expressions can be written for other types of side-chain hydrogen bonds (*14*).

The term ΔF_{NP} is the free energy change for rupturing the side-chain nonpolar (or hydrophobic) bonds between the helices. The evaluation of this quantity was based on a statistical mechanical treatment of liquid water and of aqueous hydrocarbon solutions (*15*).

The term ΔF_e is the free energy change arising from the difference in electrostatic free energy between the two conformations involved in the transition. While some estimates of this quantity have been made in the reversible denaturation of ribonuclease (see below), we have thus far tried to bypass this problem by using solutions of high ionic strength to minimize and, thereby, neglect the term ΔF_e.

The term ΔF_m is the free energy change for the mixing of the amorphous or randomly coiled chain with the solvent. In the derivations of ΔF_b, ΔF_H and ΔF_{NP}, the term ΔF_m, arising from water, has usually been included already and, therefore, need not be explicitly introduced into equation 1. However, where the solvent contains an agent like urea, then this contribution to ΔF_m must be included. According to Schellman (*7*), the contribution from urea is

$$\Delta F_m = -2pRT \ln (1 + Kc) \qquad (5)$$

where K is the equilibrium constant for the binding of urea to the peptide NH or CO group, c is the molarity of urea, and p is the number of residues in the crystalline chain which combine with urea after unfolding. At low values of c, equation 5 may be approximated by

$$\Delta F_m = -2pRTKc \qquad (6)$$

The term ΔF_x is the free energy change arising, according to Flory's original treatment of the cross-linking as an athermal process (*17*), from the entropy decrease of the random coil due to the presence of cross-links between helices in the *crystalline* phase. Therefore,

$$\Delta F_x = -T\Delta S_x \qquad (7)$$

where

$$\Delta S_x = -(3R\nu/4)[\ln n' + 3] \qquad (8)$$

The quantity ν is the number of cross-linked helices, and n' is the number of statistical elements (assumed here to be equal to the number of peptide residues) between cross-links.

The quantities ΔF_b and $-T\Delta S_x$ in equation 1 are independent of pH since they pertain to processes which do not involve ionizable groups. The other quantities in equation 1 will depend on pH. For example, if the donor and/or acceptor groups in the side-chain hydrogen bond are ionizable, then the bond will exist only in the pH range where the donor has a proton and the acceptor is lacking one. Outside of this pH range ΔF_H will be zero. Alternatively, if an ionizable group is embedded among nonpolar side-chains in a native protein but is accessible to water in the denatured form, then the pK's of the group in the native and denatured forms will differ. This will give rise to a pH dependence of ΔF_{obs}. Similarly, the pH can affect ΔF_e and ΔF_m and, thereby, ΔF_{obs}. The pH dependence of ΔF_{obs} has heretofore been treated explicitly in terms of the pH dependence of ΔF_H (*5*). However, more recently a qualitative discussion of the pH dependence of the other terms of equation 1 has been given (*18*), and attempts are now being made to treat these parameters quantitatively.

At the present time we have less detailed information about ΔF_{NP} and ΔF_e than about the other terms in equation 1. Therefore, we shall deal here with equation 1 in the form

$$\Delta F_{obs} = (n-4)\Delta H_{res} - (n-1)T\Delta S_{res} + \Delta F_H - T\Delta S_x \qquad (9)$$

and will add in the term ΔF_m when we are dealing with urea. The presence of nonpolar groups in the neighborhood of the polar ones can be expected to provide additional stability to the side-chain hydrogen bonds. It is expected that current efforts in the evaluation of

ΔF_{NP} and ΔF_e will permit us to add these terms to the right-hand side of equation 9, thereby giving us a more complete theory for protein denaturation. For the present, we shall make only passing reference to ΔF_{NP} and ΔF_e.

Magnitude of free-energy terms

In order to discuss the magnitude of the contribution of the various free-energy terms to ΔF_{obs} it will be helpful to imagine a hypothetical protein with specific values of the conformational parameters. For illustrative purposes we shall take $n = 14$; also we shall use Schellman's values (*7*) of $\Delta H_{\text{res}} = 1500$ cal./mol and $\Delta S_{\text{res}} = 4.2$ e.u. The backbone then makes a contribution of $(n-4)\Delta H_{\text{res}} = 15{,}000$ cal. and of $(n-1)T\Delta S_{\text{res}} = 16{,}380$ at 300° K., or $\Delta F_b = -1380$ cal. at 300° K.

If $K_{ij} = 1$, the maximum value of ΔF_{H} is 414 cal. at 300° K. for one side-chain hydrogen bond per helix of 14 residues. This contribution would be made at a pH intermediate between $p\text{K}_1$ and $p\text{K}_2$ (see equations 3 and 4). At low and high pH's the value of ΔF_{H} would fall to zero. Thus, the protein would be most stable in the pH region between $p\text{K}_1$ and $p\text{K}_2$. If K_{ij} were greater than unity, i.e., if the hydrogen bond were stronger, then the value of ΔF_{H} would be greater than 414 cal.—e.g., if $K_{ij} = 100$, the maximum value of ΔF_{H} would be 2760 cal. Previous estimates of K_{ij} place it in the range of unity (*14*). However, these estimates were made on the assumption that the donor and acceptor groups are in an aqueous environment. If, on the other hand, the donor and acceptor groups are surrounded by hydrophobic regions which hamper the motion of the free polar side-chains, then K_{ij} can be much larger (*19*), making the magnitude of ΔF_{H} much larger. Thus, the side-chain hydrogen bond (in coöperation with the surrounding nonpolar groups), can play a very important role in stabilizing the native protein. A calculation of this effect is now in progress (*18*). The previous success of the hydrogen-bonding theory probably arises from the partially nonpolar environment of the polar side-chains.

Preliminary estimates (*15*) of ΔF_{NP} range from 500 to 1500 cal. per mole. These values for the hydrophobic bond are comparable to those for ΔF_{H} for a heterologous single, side-chain hydrogen bond. Therefore, side-chain hydrophobic bonds can also stabilize the native conformation. A generalized electrostatic effect, contributing to ΔF_e, of about the same size may also be expected (*20*).

The effect of urea (*21*), according to equation 6, can lead to a value of $\Delta F_m = -300$ cal. for $c = 1$ M, $p = 10$, $T = 300°$ K., and $K = 0.025$. Thus, urea decreases ΔF_{obs}, rendering the protein less stable.

For $n' = 14$ and $\nu = 1$, $\Delta S_x = -8.5$ e.u. (equation 8); therefore $-T\Delta S_x = 2550$ cal. at $T = 300°$ K. This contribution from cross-linking stabilizes the native protein. The larger the degree of cross-linking, the more stable is the native protein—i.e., since ΔS_x is negative, ΔF_{obs} becomes more positive as the degree of cross-linking increases.

Transition temperature

The transition temperature, T_{tr}, for the transformation is the temperature at which ΔF_{obs} is zero. If experiments are carried out in urea at low pH and high salt concentration, where ΔF_{H}, and ΔF_e may be taken equal to zero, then it is possible (*7*, *21*) to evaluate ΔH_{res} and ΔS_{res} from the equation

$$\frac{1}{T_{tr}} = \frac{(n-1)\Delta S_{\text{res}} + \Delta S_x}{(n-4)\Delta H_{\text{res}}} + \frac{2pRKc}{(n-4)\Delta H_{\text{res}}} \tag{10}$$

For this evaluation K may be considered as a constant (or its variation with temperature can be taken into account [*21*]) and $1/T_{\text{tr}}$ plotted against c. The values of ΔH_{res} and ΔS_{res} can be obtained from the slope and intercept if a model is assumed for the crystalline region. This model is specified in terms of n, p, and ΔS_x.

Using these values of ΔH_{res} and ΔS_{res}, the pH dependence of T_{tr} in the absence of urea (i.e., $\Delta F_m = 0$) may then be computed, if the pH dependence is attributed to the term ΔF_{H}. The results of such a calculation are shown in Fig. 22.2 for several different kinds of side-chain hydrogen bonds.

It is easiest to see the effect of cross-linking by considering a protein which contains no side-chain hydrogen or hydrophobic bonds. Then

$$\Delta F_{\text{obs}} = (n-4)\Delta H_{\text{res}} - (n-1)T\Delta S_{\text{res}} - T\Delta S_x \tag{11}$$

Since $\Delta F_{obs} = 0$ at $T = T_{tr}$ we obtain

$$T_{tr} = \frac{(n-4)\Delta H_{res}}{(n-1)\Delta S_{res} + \Delta S_x} \tag{12}$$

Since ΔS_x is negative, an increasing degree of cross-linking leads to increasing values of T_{tr}.

It should be noted that ΔF_{obs} is an extensive quantity. Therefore, if r helices coöperate* in an all-or-none helix–random coil transition, the value of T_{tr} would be unaffected, i.e., a factor r would appear in both the numerator and denominator of equation 12 and would cancel. However, as will be discussed below, the sharpness of the transition *will* be affected.

Sharpness of transition

(1, 23, 24, 25)

If the value of ΔF_{obs} is known as a function of T, pH, etc., then the fraction of material denatured, α, may be defined (*7*) by the equation

$$\alpha = [1 + \exp(\Delta F_{obs}/RT)]^{-1} \tag{13}$$

The sharpness of the transition may be taken as the slope at T_{tr}, i.e., as $(d\alpha/dT)_{T_{tr}}$.

$$\left(\frac{d\alpha}{dT}\right)_{T_{tr}} = \frac{\Delta H_{obs}}{4RT_{tr}^2} \tag{14}$$

We may write the following expression for ΔH_{obs}.

$$\Delta H_{obs} = (n-4)\Delta H_{res} + \Delta H_H \tag{15}$$

where

$$\Delta H_H = \sum x_{ij}\left[-\Delta H_{ij} + \frac{(K_1/[H^+])\Delta H_1 - ([H^+]/K_2)\Delta H_2}{1 + K_1/[H^+] + [H^+]/K_2}\right] \tag{16}$$

Here x_{ij} is defined in equation 4, ΔH_{ij} is the heat of formation of the ij hydrogen bond, ΔH_1 and ΔH_2 are the heats of ionization corresponding to the equilibrium constants K_1 and K_2, respectively, and the sum is taken over all the side-chain hydrogen bonds in the crystalline unit. The pH dependence of $(d\alpha/dT)_{T_{tr}}$ arises from the pH dependence of ΔH_H; the sharpness of the transition, like the transition temperature itself, is greatest in the pH range where the degree of hydrogen bonding is maximal.

* This coöperation between helices has been referred to as the quaternary structure by Bernal (*22*).

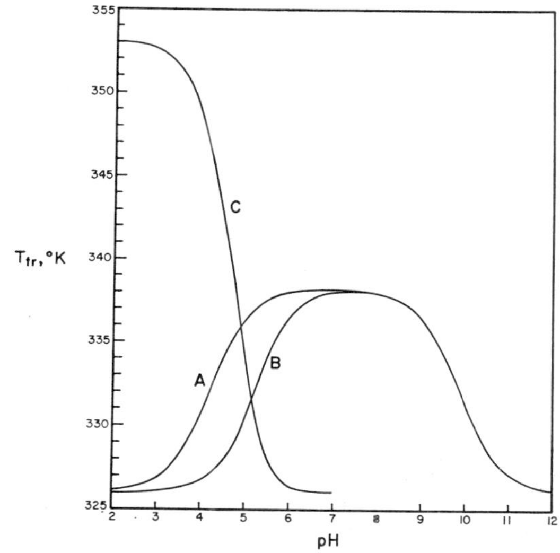

Fig. 22.2. Transition temperature as a function of pH for a hypothetical protein stabilized by: A, tyrosyl-carboxylate ion; B, tyrosyl-histidine; and C, carboxyl-carboxyl hydrogen bonds (*5*).

Hydrophobic bonds will also affect the sharpness according to their effect on ΔH_{obs} and T_{tr}.

In order to see the effect of cross-linking on the sharpness we shall assume that the pH is sufficiently low so that no heterologous single side-chain hydrogen bonds exist. For such a system, T_{tr} is given by equation 12 and $(d\alpha/dT)_{T_{tr}}$ by

$$\left(\frac{d\alpha}{dT}\right)_{T_{tr}} = \frac{[(n-1)\Delta S_{res} + \Delta S_x]^2}{4R(n-4)\Delta H_{res}} \tag{17}$$

Since ΔS_x is negative, the sharpness decreases with increasing cross-linking, even though T_{tr} itself increases.

While T_{tr} is unaffected by the coöperation of r helices in an all-or-none transition, there is an effect on the sharpness of the transition. For such a case equation 14 is replaced by

$$\left(\frac{d\alpha}{dT}\right)_{T_{tr}} = \frac{r\Delta H_{obs}}{4RT_{tr}^2} \tag{18}$$

Thus, if r is sufficiently greater than unity, the protein can exhibit a very sharp transition (*1, 23, 25*).

Some experimental results

Typical experimental results can be cited to verify the various aspects of the theory presented here. These results fall into two categories, viz., (*a*) films and fibers and (*b*) dilute solutions. The conformational changes in films and fibers were investigated by observing their effect on the dimensions and elastic properties of the films and fibers; the conformational changes in dilute solutions were investigated by means of optical rotation and ultraviolet-difference spectra measurements. There are several advantages in using films and fibers to establish the principles involved in conformational changes. These include the possibility of applying X-ray diffraction and rubber elasticity theories to ascertain the presence of crystalline and amorphous forms; also films and fibers lend themselves easily to swelling measurements to provide information about ΔF_m, and to cross-linking experiments to provide information about ΔS_x. The systems to be discussed are: ribonuclease films (*21*), insulin fibers (*26*), fibrin films and fibers (*25*), and dilute solutions of ribonuclease (*20, 27*). Details for the preparation of the films and fibers, as well as other experimental aspects, may be found in the references cited.

We can immediately provide an experimental verification of Schellman's estimates (*7*) of the values of ΔH_{res} and ΔS_{res}. For this purpose, we shall use data for the effect of urea on the transition temperature in ribonuclease films (*21*), and assume that the crystalline unit of the film is composed of two α-helices, each containing 12 residues and cross-linked at both ends. The urea-denaturation experiments were carried out at pH 2.35, to eliminate side-chain hydrogen bonds, and in 2 M KCl, to eliminate electrostatic effects. For such a model, equation 1 (with the aid of equations 2, 6, 7, 8) becomes

$$\Delta F_{obs} = 2\{(n-4)\Delta H_{res} - (n-1)T\Delta S_{res}\} - T\Delta S_x - 2pRTKc \quad (19)$$

where $n = n' = 12$, $\nu = 2$, $p = 24$, and $K = 0.025$.* If ΔF_{obs} is set equal to zero at the transition temperature, we obtain the following analog of equation 10.

$$\frac{1}{T_{tr}} = \frac{11\Delta S_{res} - 8.25}{8\Delta H_{res}} + \frac{3RKc}{\Delta H_{res}} \quad (20)$$

From the resulting plot of $1/T_{tr}$ against c, the values obtained from the slope and intercept are: $\Delta H_{res} = 1700$ cal./mol and $\Delta S_{res} = 4.83$ e.u. These values are in good agreement with Schellman's estimates (*7*), and will be used in further applications of equation 1 to experimental data. It should be emphasized that other assumptions about the crystalline unit could lead to similar values of ΔH_{res} and ΔS_{res}. However, these assumptions do not affect our considerations of the effect of the *side-chains* on the stability, which is our primary concern here. Further, the values of ΔH_{res} and ΔS_{res} determine only the range of values of T_{tr} but not the pH-dependence of T_{tr}.

We shall next consider pH-dependent denaturation of ribonuclease films in the absence of urea. For this case, the last term of equation 19 is omitted, and the term from equation 3 is included instead, i.e.,

$$\Delta F_{obs} = 16\Delta H_{res} - 22T\Delta S_{res} + 16.5T - RT\sum \ln(1 - x_{ij}) \quad (21)$$

The transition temperature is obtained as a function of pH from equation 21 by setting ΔF_{obs} equal to zero. Details of such calculations have been described previously (*5*). The experimental results, in the presence and absence of 2 M KCl, are shown in Fig. 22.3. It can be seen

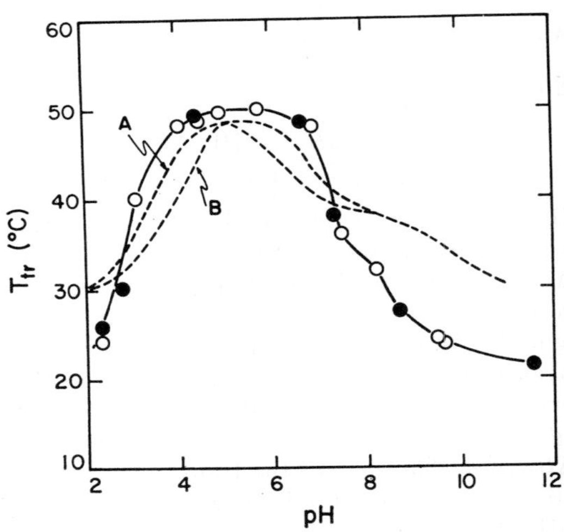

Fig. 22.3. Experimental (solid line) and calculated (broken lines) results for pH dependence of transition temperature, T_{tr}, for ribonuclease films. Experimental values in 0.07 M buffer solutions (open circles); in the same buffer with added 2 M KCl (solid circles) (*21*).

* Actually the temperature dependence of K was taken into account (*21*) in Equation 19, but it is omitted here for simplicity since the resulting values of ΔH_{res} and ΔS_{res} are not very sensitive to the temperature variation of K.

that T_{tr} is unaffected by the excess salt. Thus, ΔF_e is negligible *in this particular case.* Also, the degree of swelling is independent of pH in this system (*21*). Thus ΔF_m does not contribute to the pH dependence of ΔF_{obs} *in this particular case.* The two dashed curves in Fig. 22.3 are theoretical ones, calculated from equation 21, using the values of ΔH_{res} and ΔS_{res} obtained from the urea-denaturation experiments, and assuming two side-chain hydrogen bonds per crystalline unit to be contributing to the sum in equation 21. The two side-chain bonds are assumed to be one tyrosyl-carboxylate ion and one histidyl-carboxylate ion hydrogen bond; curves A and B differ only in the values of the parameters used to compute x_{ij} from equation 4. The theoretical curves fit the experimental data fairly well, especially in the region of maximum stability and at low pH. The discrepancy at high pH may be due to electrostatic interactions which were neglected here or, of course, to inadequacies of the model. These theoretical curves are similar to those represented in Fig. 22.2.

Similar results for insulin fibers are shown in Fig. 22.4. The crystalline unit of the insulin fibers was assumed to contain two loops (*26*),

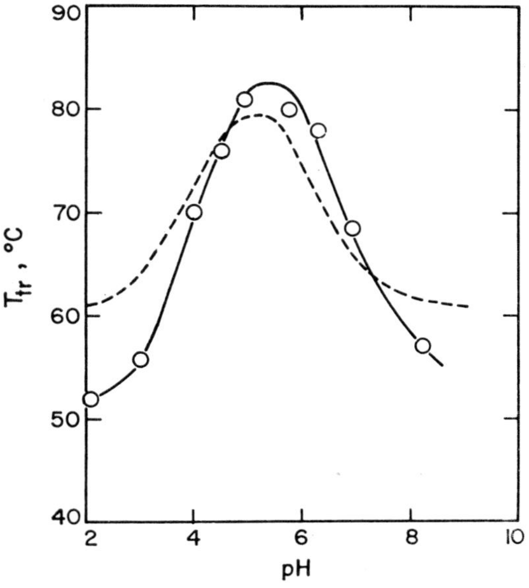

Fig. 22.4. Relation between the transition temperature and pH of the diluent for insulin fibers. The solid curve is experimental and the dashed one theoretical (*26*).

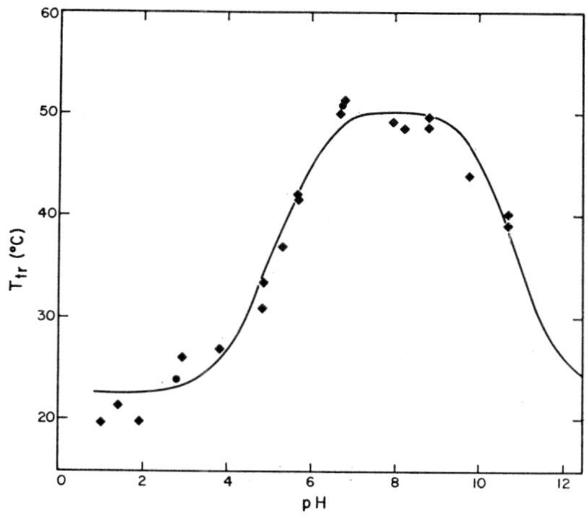

Fig. 22.5. Relation between the transition temperature and pH of the diluent for fibrin fibers. The points are experimental and the curve theoretical (*25*).

in accordance with the Sanger structure of insulin (*28*). It was also assumed to contain one side-chain hydrogen bond between a histidyl donor and a carboxylate ion acceptor. Here again the agreement is reasonable.

In the case of fibrin fibers (Fig. 22.5), the agreement between theory and experiment is even better. The model was assumed to consist of chains of 40 amino-acid residues each between disulfide cross-links. Thirty-five per cent of the residues were assumed to be in a helical conformation, each helix being stabilized by two side-chain hydrogen bonds (one tyrosyl-carboxylate ion bond and one tyrosy-histidine bond; ref. *25*).

In the case of fibrin the transitions are very sharp, occurring over a range of 2° C. or less at pH 7. Therefore $(d\alpha/dT)_{T_{tr}}$ has a minimum value of 0.5 deg^{-1}. Attributing the excessive sharpness to a coöperative effect, and applying equation 18 to the data, a value of $r = 18$ was obtained (*25*), i.e., at least 18 chains are involved in a coöperative transition.

An illustration of the effect of cross-linking is shown in Fig. 22.6. It can be seen that the introduction of cross-links into (presumably the crystalline portions of) insulin fibers by tanning with formaldehyde and p-benzoquinone raises the transition temperature, as implied by the ΔS_x term of Equation 8.

Accepting the validity of the theory on the basis of the results on fibers and films we can

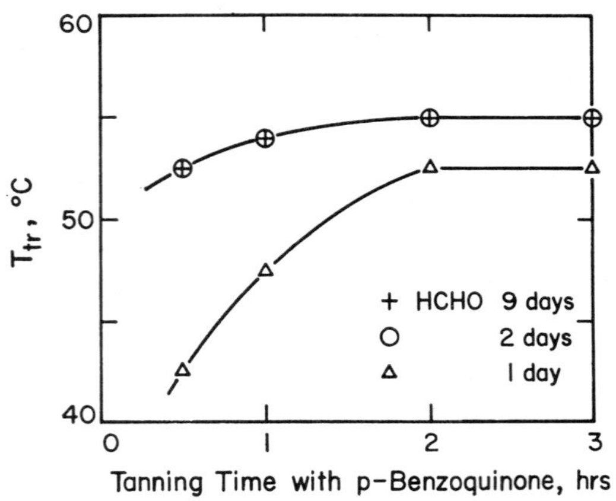

Fig. 22.6. Effect of tanning with formaldehyde and *p*-benzoquinone on the transition temperature of insulin fibers in a buffer solution of *p*H 6.92. The fibers, after treatment with formaldehyde for the period shown, were tanned in a 0.5 per cent quinone solution of *p*H 8.86 for various times, and then stretched 200 per cent (*26*).

consider an example of *p*H-dependent denaturation in dilute solution, namely ribonuclease (*20, 27*). The *p*H dependence of the transition temperature is shown in Fig. 22.7. The points are experimental and the curves are theoretical for two alternative models, the distinction between which it is difficult to make, although model A seems to be somewhat preferable to model B. In model A, the *p*H dependence of ΔF_{H} arises solely from local interactions of two carboxyl groups (presumably two tryrosyl-carboxyl hydrogen bonds imbedded in nonpolar regions). In model B, only one such local interaction contributes to the *p*H dependence of ΔF_{H}, with an additional contribution from a generalized electrostatic interaction, ΔF_e, due to different conformations of the native and denatured molecules. These experiments on ribonuclease could not be carried into the alkaline region because of irreversible effects. However, since the transition was observed by means of a perturbation of the tyrosyl ultraviolet absorption spectrum, the involvement of a tyrosyl donor(s) seems probable. As in the case of films and fibers, the *p*H dependence of T_{tr}, shown in Fig. 22.7, resembles that predicted from theory (Fig. 22.2).

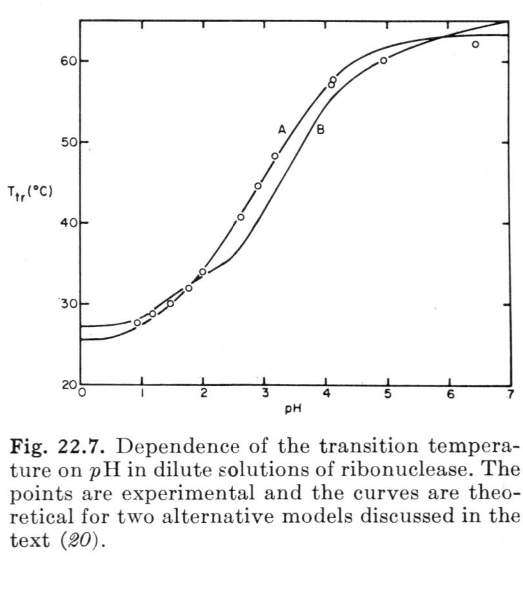

Fig. 22.7. Dependence of the transition temperature on *p*H in dilute solutions of ribonuclease. The points are experimental and the curves are theoretical for two alternative models discussed in the text (*20*).

Discussion

The experimental results cited here provide confidence in the model used for accounting for reversible denaturation (*5*). In this model the tertiary structure (side-chain hydrogen bonds, hydrophobic bonds, and disulfide cross-links) plays an important role in determining the stability of helical portions (secondary structure) of the protein molecule. Further studies of reversible denaturation in dilute solution should provide us with information about the nature of side-chain interactions, as they already have done in the case of ribonuclease (*20*). Such investigations are now being carried out on soybean trypsin inhibitor (*29*) and paramyosin (*30*). In this way, we should obtain information about protein conformation in solution.

There are still some unsolved problems which are presently under investigation (*18*). First of all, it will be necessary to evaluate ΔF_{NP} and ΔF_e, and to compute more precisely the effect of a nonpolar environment on ΔF_{H}. In addition, we need further information about ΔF_m. Probably, the theories of Flory (*1*) and Oplatka and others (*31*) for the change in composition of the amorphous phase of a fiber upon melting of the crystalline form will be applicable for the evaluation of ΔF_m for denaturation in dilute solutions. Finally, we need more experimental data in dilute solutions to verify the calculations of ΔF_{H}, ΔF_{NP}, ΔF_e and ΔF_m. Denaturation studies in various solvents, ionic strengths, and *p*H's on reversible systems will be helpful.

Also, it will be useful to have more experimental data on the mixing of randomly coiled polypeptide chains with solvent. Here again, the thermodynamic methods used in the investigations of solutions of synthetic polymers may be useful—e.g., measurements of hydrodynamic properties and of virial coefficients from light scattering for random coils in a variety of solvents at various temperatures will provide information on the mixing process. When these problems are solved we may expect to have a more complete theory of protein denaturation. We seem to have a correct qualitative picture, and we should be able to improve the quantitative aspects of the theory in the near future.

Added in proof.—The evaluation of ΔF_{NP}, referred to here, has now been completed (*32, 33*). Also, the effect of hydrophobic bonds on the thermal stability of protein conformations has been computed (*34*). In addition, calculations have been carried out on the effect of a nonpolar environment on ΔF_{H} (*35*). Finally, some preliminary experimental data, which are compatible with the computed values of ΔF_{NP}, have been obtained (*36*).

Acknowledgment

The preparation of this manuscript and the work of the author and his collaborators, to which reference is made here, were supported by research grant E-1473 from the National Institute of Allergy and Infectious Diseases, Public Health Service, Bethesda, Md., and by research grant G-14792 from the National Science Foundation, Washington, D.C.

References

1 P. J. Flory, *J. Polymer Sci.* **49,** 105 (1961).
2 P. J. Flory, *J. Cellular Comp. Physiol.* **49,** Sup. 1, 175 (1957).
3 P. J. Flory and O. K. Spurr, Jr., *J. Am. Chem. Soc.* **83,** 1308 (1961).
4 L. Mandelkern, *Rubber Chem. and Technol.* **32,** 1392 (1959).
5 H. A. Scheraga, *J. Phys. Chem.* **64,** 1917 (1960).
6 H. A. Scheraga, "Protein Structure," Academic Press, New York, 1961, Chap. IV.
7 J. A. Schellman, *Compt. rend. trav. lab. Carlsberg, Sér. Chim.* **29,** 223, 230 (1955).
8 B. H. Zimm and J. K. Bragg, *J. Phys. Chem.* **28,** 1246 (1958); **31,** 526 (1959).
9 J. H. Gibbs and E. A. DiMarzio, *ibid.* **28,** 1247 (1958); **30,** 271 (1959).
10 T. L. Hill, *J. Polymer Sci.* **23,** 549 (1957); *J. Chem. Phys.* **30,** 383 (1959).
11 L. Peller, *J. Phys. Chem.* **63,** 1194, 1199 (1959).
12 S. A. Rice and A. Wada, Abstracts of the 134th meeting of the Am. Chem. Soc., Chicago, Ill., p. 41S, Sept., 1958; *J. Chem. Phys.* **29,** 233 (1958).
13 J. A. Schellman, *J. Phys. Chem.* **62,** 1485 (1958).
14 M. Laskowski, Jr., and H. A. Scheraga, *J. Am. Chem. Soc.* **76,** 6305 (1954); **78,** 5793 (1956).
15 G. Nemethy and H. A. Scheraga, Abstracts of the 138th meeting of the Am. Chem. Soc., New York, N. Y., Sept., 1960, p. 4C; see also the discussion at the end of Klotz's paper (ref. *16*).
16 I. M. Klotz, *Brookhaven Symposia in Biol.* **13,** 25 (1960).
17 P. J. Flory, *J. Am. Chem. Soc.* **78,** 5222 (1956).
18 H. A. Scheraga, *J. Phys. Chem.* **65,** 1071 (1961).
19 Ref. *6,* Chap. II.
20 J. Hermans, Jr., and H. A. Scheraga, *J. Am. Chem. Soc.* **83,** 3283, 3293 (1961).
21 A. Nakajima and H. A. Scheraga, *ibid.* **83,** 1575 (1961).
22 J. D. Bernal, *Discussions Faraday Soc.* **25,** 7 (1958).
23 B. H. Zimm, P. Doty, and K. Iso, *Proc. Nat. Acad. Sci.* **45,** 1601 (1959).
24 H. A. Scheraga, R. A. Scott, G. I. Loeb, A. Nakajima, and J. Hermans, Jr., *J. Phys. Chem.* **65,** 699 (1961).
25 G. Loeb and H. A. Scheraga, *J. Am. Chem. Soc.* **84,** 134 (1962).
26 A. Nakajima and H. A. Scheraga, *ibid.* **83,** 1585 (1961).
27 H. A. Scheraga, C. Y. Cha, J. Hermans, Jr., and C. L. Schildkraut, in "Amino Acids, Proteins and Cancer Biochemistry," Academic Press, N. Y., 1960, p. 31.
28 A. P. Ryle, F. Sanger, L. F. Smith, and R. Kitai, *Biochem. J.* **60,** 541 (1955).
29 Y. V. Wu and H. A. Scheraga, submitted to *Biochemistry.*
30 L. Riddiford and H. A. Scheraga, *Biochemistry* **1,** 95, 108 (1962).
31 A. Oplatka, I. Michaeli, and A. Katchalsky, *J. Polymer Sci.* **46,** 365 (1960).
32 G. Nemethy and H. A. Scheraga, *J. Chem. Phys.,* in press.
33 G. Nemethy and H. A. Scheraga, *J. Chem. Phys.,* in press.
34 H. A. Scheraga, G. Nemethy, and I. Z. Steinberg, *J. Biol. Chem.,* in press.
35 G. Nemethy, I. Z. Steinberg, and H. A. Scheraga, to be submitted to *J. Am. Chem. Soc.*
36 I. Z. Steinberg and H. A. Scheraga, *J. Am. Chem. Soc.,* in press.

PAPER 23

Structural Studies with Small-Angle X-Ray Scattering

W. W. BEEMAN

Because equipment is not readily available commercially and is relatively complex, only a few laboratories are doing experiments of the kind I shall discuss. The methods and applications are not as well known as they deserve to be and for this reason a fairly general discussion is appropriate.

The "small angle" aspect of the work is an unavoidable nuisance rather than a virtue. It results from the circumstance that the most available and useful X-ray wave lengths are in the one to two Ångstrom region while molecules of biological interest have linear dimensions a couple of orders of magnitude greater. The same considerations sometimes force crystallographers to take data at very small scattering angles. However what is usually meant by small-angle X-ray scattering is the diffuse scattering from a dilute system of randomly arranged large particles, for instance a solution of macromolecules. I shall confine my remarks to this area. In many ways the field is more closely related to light scattering than to X-ray crystallography.

The scattering function and the radius of gyration

Unless a particle has spherical symmetry the angular distribution of scattered X rays is a function of its orientation with respect to the incident X-ray beam. By the scattering function we mean the average of this angular distribution over all orientations of the particle or, what is equivalent, the average over a population of identical, noninteracting, randomly oriented particles. Usually the function is normalized to the value unity for forward scattering. A useful approximate form for the scattered intensity is a gaussian; first suggested by Guinier (*1*, *2*)

$$I = I_o N(n - n_o)^2 e^{-\frac{1}{3}(2\pi R\theta/\lambda)^2}$$

Here I is the scattered intensity in some experimental scattering geometry; I_o is the intensity scattered by a single electron in the same geometry; N is the number of particles illuminated by the incident beam; n and n_o are respectively the number of electrons in the particle and in an equal volume of solvent; θ is the scattering angle; λ, the X-ray wave length; and R, the electronic radius of gyration of the particle. R is defined by

$$R^2 = \int_o^V \rho(r) r^2 dV \Big/ \int_o^V \rho(r) dV$$

where $\rho(r)$ is the electron density (above that of the solvent) at a distance r from the center of charge of the particle and V is the volume of the particle.

The gaussian approximation to the scattering function is best at the smallest angles. It is good enough over a sufficiently great angular region to permit quite accurate (2 per cent or 3 per cent) experimental determinations of R and these have been made for a number of macromolecules. No assumption of uniform electron density within the particle is made, although if the electron density is not uniform the radius of gyration is a function not only of the structure of the particle but also of the electron density of the solvent. This is easily observed with the spherical plant viruses (Geil, ref. *3*) where the protein and nucleic-acid components have quite different average electron densities. The same virus has a smaller radius of gyration in a water-sucrose solvent than in pure water. This is expected since the denser component (RNA) is at the center of the particle.

By matching the solvent density to the protein shell one can arrange that the X rays "see" only the denser component. This has been done with ferritin (Kratky, ref. *4*; Fischbach and Anderegg, ref. *5*).

It should be noted that if appreciable regions of the particle have an electron density less than that of the solvent both the definition of the radius of gyration and the gaussian approximation may fail.

Thus the radius of gyration is under most circumstances a well-defined linear parameter of a macromolecule which can be measured with accuracy and relative ease. It does not, however, have the immediate chemical and structural implications of a molecular weight, for instance, and it appears unlikely that anyone will bother to measure and catalog, for their own sake, an appreciable number of radii of gyration. The measurement unlike a sedimentation constant gives no test of the homogeneity of the preparation. It appears that many small disagreements in measured radii of gyration reflect small variations in the amount of dimerization present. As a probe of changes in molecular configuration or a test of a structural hypothesis the radius of gyration can be important. Echols and Anderegg (*6*) have observed large changes as serum albumin unfolds in urea. The original value returns when the urea is removed. Since radii of gyration can be measured under conditions of solution which approximate those of the biologically functioning molecule, it will be of great interest to see if molecules have, in solution, the radii of gyration predicted by the detailed crystallography on single crystals. We are attempting accurate measurements on myoglobin and hemoglobin at this laboratory.

At angles greater than roughly 0.2 λ/R the gaussian approximation becomes quite inaccurate and the scattering function depends on details of the size and structure of the particle. Many calculations of scattering functions have been made (Guinier, *7*; Fournet, *8*; Porod, *9*; Malmon, *10*) particularly for ellipsoids of revolution and cylinders of constant electron density. For two reasons only very qualitative information can be obtained by matching observed scattering curves with calculated scattering functions: First and most important, the experimental scattering curve determines only the electron density correlation function and not the actual charge distribution. Many different particles may have the same scattering function. Secondly, the assumption of uniform electron density is unjustified at angles greater than λ/D where D is the length of some important internal periodicity. In protein macromolecules D will be at least as large as 10 Å or 12 Å, the diameter of an α-helix. With the exception of certain kinds of measurements on long rodlike particles, results which depend on the larger angle part of the scattering curve should be viewed with caution.

Near spherical particles are another important exception. In this case the scattering curve has a sequence of subsidiary maxima and minima not shown by asymmetrical particles, and an inversion of the experimental curve gives the radial electron density distribution. Interesting work has been done on the spherical plant viruses by Leonard (*11*), Schmidt (*12*), Geil, Anderegg, Kaesberg (*13*), and others.

Finally we remark that, as with light scattering, interparticle interference effects are important even at low concentrations. These are most important at the smallest angles. For accurate results a radius of gyration must be measured at a sequence of concentrations and extrapolated to zero concentration. Weight concentrations between 5 per cent and 0.5 per cent are usually sufficient.

Fournet (*14*) has shown that careful measurements of interparticle interference effects can lead to an evaluation of the interparticle potential. Such measurements are quite feasible but almost no work has been done to date (see Schmidt, Kaesberg, and Beeman, ref. *12*).

Solvent density studies and the excluded volume

Referring to the Guinier expression for the scattered intensity, we note that the square root of the scattered intensity (extrapolated to zero angle) is proportional to the difference between the average electron density of the solvent and the particle. Some very interesting experiments are possible if one deliberately varies the average density of the solvent by the addition of a small molecule, for instance, sucrose. It is then possible to determine the average electron density of the particle from a plot of the square root of the zero angle intensity versus the solvent electron density. Experiments with serum albumin, catalase (Geil, *3*; Shurman, *5*; Malmon, *15*), and a number of other molecules have shown that this plot is generally an excellent straight line. This indicates that the region around each macromolecule from which the added small molecule is excluded does not change in volume or content as the solvent electron density is varied.

The "particle" in the sense of these experiments is the entire volume from which the added small molecule is excluded. It may include an external hydration layer if this layer is so tightly bound to the macromolecule that sucrose, for instance, cannot enter. At least in the case of serum albumin there is very little experimental evidence for such an external hydration layer. The measured radius of gyration changes only slightly as sugar is added to a serum albumin solution. If there were an external hydration layer which excluded sugar, one would expect the radius of gyration to decrease. This is clear from the fact that the hydration layer will have an electron density less than that of the sugar-water solvent, while an immediately adjacent amino-acid layer will have a density greater than that of the solvent. Thus the hydration layer effectively cancels the scattering from an outer layer of the serum albumin leaving a smaller structure to determine the radius of gyration.

The measured average electron density of the excluded volume of serum albumin gives a total hydration of about 0.4 gm of H_2O per gram of protein. From the constancy of the radius of gyration one infers that this is mainly internal hydration. This value of internal hydration is in good agreement with that obtained earlier from size and shape estimates (Anderegg, Beeman, Shulman, and Kaesberg, ref. *16*) not involving variation of the solvent density. The average electron density measured for serum albumin in dilute solution is also very close to that obtained by Perutz (*17*) for the hemoglobin molecule in the wet crystal. Perutz defined an excluded volume about each molecule by diffusing salt into the crystal.

Absolute intensity measurements

None of the measurements we have described thus far involves more than the shape of a scattering curve or perhaps the relative shapes and intensities of several different curves. If in addition the incident and scattered intensities can be compared, new results are possible; in particular, molecular weights can be determined. The ratio of incident to scattered intensities is of the order of 10^5 to 10^6. The accurate determination of so great a ratio is a matter of some difficulty but the problem has been attacked recently by Kratky (*4*), by Luzzati (*18*), and in this laboratory. Luzzati has had considerable success using a crystal monochromated beam which can be attenuated with nickel filters. In this laboratory we calibrate a given beam and scattering geometry using a sample whose scattering cross section can be calculated, for instance, a gas.

To illustrate what must be measured to determine a molecular weight we recast the Guinier scattering formula in the following form

$$I = I_o tc \frac{M}{A} (1 - v\rho_s l_s)^2 \epsilon^{-1/3(2\pi R\theta/\lambda)^2}$$

I, I_o and the gaussian have their earlier meanings, t is the thickness of the scattering sample (used in transmission), c the weight concentration of the dissolved substance, M its molecular weight, A is Avogadro's number, l and l_s are the number of electrons per gram of the dry substance and of the solvent, respectively, ρ_s is the mass density of the solvent, and v is the dry partial specific volume of the substance. Everything is available for the determination of M except I_o, and this necessitates a measurement of the incident intensity. It is perhaps obvious that the method is essentially the same as a light-scattering determination of molecular weight. There appears to be no good reason why the X-ray method should not be as accurate as light scattering. No assumptions about the structure of the molecule need be made but of course interparticle interference effects must be avoided.

Rodlike structures

Porod (*9*) was the first to emphasize a characteristic of the scattering functions of very long particles which has led to extremely interesting applications. It is simply that there exists an appreciable region of the scattering curve which is determined by the cross section of the rod even when the rods are randomly oriented in the sample. More precisely there is a part of the scattering curve which varies as the reciprocal of the scattering angle times a gaussian which contains as a parameter the radius of gyration of the cross section. All of the remarks and methods we have described may be applied to the determination of quantities related to the cross section. In particular, absolute intensity measurements furnish a mass per unit length along the rod. Because of the great interest in linear polypeptide and nucleic-acid structures, such measurements have been most fruitful. Professor Luzzati (*19*) has done much of the recent work which I will not discuss further since it will be covered in Luzzati's contribution to this book (Paper *11*).

Conclusion

We have described briefly some of the measurements possible on macromolecular systems using small angle X-ray scattering techniques. The measurements can be made on dilute solutions under a wide variety of conditions of solution and temperature. In most cases analogous things can be done with light scattering but the much smaller X-ray wave length permits much greater spatial resolution. Only small samples are needed, between 0.1 and 0.01 cm^3.

As a probe of structural changes induced by a wide variety of chemical and physical means small-angle X-ray scattering often has unique advantages.

References

1 A. Guinier and G. Fournet, "Small Angle Scattering of X-Rays," John Wiley and Sons, New York and London, 1955.

2 W. W. Beeman, P. Kaesberg, J. W. Anderegg, and M. B. Webb, in "Handbuch der Physik," Volume 32, S. Flugge (ed.), Springer-Verlag, Berlin, 1957.

3 P. H. Geil, Ph.D. thesis, University of Wisconsin, Department of Physics (1956).

4 O. Kratky, *Makromol. Chem.* **35A,** 12 (1959).

5 Unpublished research (University of Wisconsin).

6 G. H. Echols and J. W. Anderegg, *J. Am. Chem. Soc.* **82,** 5085 (1960).

7 A. Guinier, *Ann. phys.* **12,** 161 (1939).

8 G. Fournet, *Bull. soc. franç. minéral. et crist.* **74,** 39 (1951).

9 G. Porod, *Acta Phys. Austriaca* **2,** 255 (1948).

10 A. G. Malmon, *Acta Cryst.* **10,** 639 (1957).

11 B. R. Leonard, J. W. Anderegg, S. Shulman, P. Kaesberg, and W. W. Beeman, *Biochim. et Biophys. Acta* **12,** 499 (1953).

12 P. Schmidt, P. Kaesberg, and W. W. Beeman, *ibid.* **14,** 1 (1954).

13 J. W. Anderegg, P. H. Geil, W. W. Beeman, and P. Kaesberg, *Biophysical J.* **1,** 657 (1961).

14 G. Fournet, *Compt. rend.* **228,** 1421 (1949a).

15 A. G. Malmon, *Biochim. et Biophys. Acta* **26,** 233 (1957).

16 J. W. Anderegg, W. W. Beeman, S. Shulman, and P. Kaesberg, *J. Am. Chem. Soc.* **77,** 2927 (1955).

17 W. L. Bragg and M. F. Perutz, *Acta Cryst.* **5,** 277 (1952).

18 V. Luzzati, J. Witz, and A. Nicolaieff, *J. Mol. Biol.* **3,** 367 and 379 (1961).

19 V. Luzzati, A. Nicolaieff, and F. Masson, *ibid.* **3,** 185 (1961).

PAPER 24

Conformations of Polypeptide Chains in the Solid State

A. ELLIOTT, E. M. BRADBURY, (the late) A. R. DOWNIE, and W. E. HANBY

It is now well known that polypeptide chains may in principle fold in several ways, and that one of the important factors in this folding is the formation of hydrogen bonds between the NH and C=O groups of the peptide backbone. We shall not here attempt a complete historical account of the subject, which has been given elsewhere (*11*), but will review the evidence for the conformations which have been proposed for a number of actual polymers. We shall also try to assess the factors determining which of several possible conformations is adopted in a given set of circumstances. This will lead us to consider whether or not a system is in equilibrium, and the ease of transition from one form to another. We shall see that the nature of the side chains is important, and we shall try to indicate in what regions knowledge is lacking, and to indicate where possible a line of investigation.

The evidence for various conformations

TECHNIQUES

The direct experimental evidence for conformations in the solid state is chiefly from X-ray diffraction, with infrared spectroscopy (especially of oriented specimens with polarized radiation) as a very useful adjunct. Optical rotation has also afforded valuable evidence, and deuterium exchange between NH groups and a suitable deuterating agent has sometimes provided a clue. The use of X-ray diffraction focuses attention on those parts of the polymer which are crystalline, and this technique reveals little of the properties of other parts. The other techniques to which we refer are particularly valuable since they give information about the whole polymer. By comparing other results with those of X-ray diffraction, something can be inferred about the less ordered portions.

For the most part, the diffraction pattern for Cu K_α radiation from oriented fibers has been observed and compared with the pattern calculated for the proposed model structure. This model is usually one in which the bond dimensions (including those of the hydrogen bond) have been derived from independent X-ray studies of simple compounds (*26*). In this way additional information is supplied to supplement the rather meager amount contained in the usual polymer fiber diffraction pattern; but the procedure has obvious dangers. The specimens used are generally polycrystalline, oriented fibers in which only one of the crystal axes has a preferred direction, usually the direction of stretching or rolling, etc. (This is known as fiber orientation, the preferred axis being referred to as the fiber axis; it can, however, occur also in films.) In consequence, the information is far less complete than that which could be obtained from a single crystal. In some cases (especially extended chain or β-structures) double orientation may be produced by rolling. This causes another of the crystal axes to have a preferred orientation in the plane of the rolled sheet. Such orientation helps in separating *hol* reflections from *okl*, but makes it even more difficult satisfactorily to obtain a set of relative intensities of the reflections. We shall consider X-ray diffraction in some individual cases later.

Films rather than fibers must usually be used with polarized infrared radiation in the region of fundamental frequencies, because the extinction coefficients of the amide bands are so high that specimens should not be thicker than one or two microns. Direct evidence as to the approximate orientation of the amide group can be obtained from the dichroic effects which are seen with oriented specimens, and it is readily inferred that in structures of the α-helix type the C=O and NH bonds are approximately along the direction of stretching, whereas with the normal extended chain of a β-structure these bonds are approximately at right angles to this direction. The uncertainties in the interpretation of dichroic effects are such that one could not expect to distinguish between several helices which have been proposed (α, γ, π, etc.) nor between various near-extended chain conformations. These are partly caused by imperfect orientation of the molecules, so that the observed dichroism is not directly related to the orientation of the peptide group within the molecule in a simple manner. This difficulty may be overcome by making observations on several bands associated with the peptide group, whose transition moments have different but known directions within the group (*15*). (The transition moment is a vector whose direction is that of the changing dipole moment associated with a mode of vibration; it is the direction which the ***E*** vector of the infrared radiation must have for maximum absorption.) A more fundamental difficulty is that the directions of the transition moments for various vibrational modes may vary somewhat from one compound to another, or from one chain conformation to another. The directions of the transition moments for the chief amide bands have been measured in model compounds (*1*, *74*), but it is possible that they may be slightly different in some polymers.

The bands on which most reliance can be placed for structural investigations are those due to the NH stretch, the NC=O out-of-phase mode (often called Amide I, or C=O stretch, since most of the energy associated with this mode is due to the stretching of the C=O bond) and the NH out-of-plane deformation. The two NH modes are found, respectively, at about 3310 cm^{-1} and 700 cm^{-1}. The former frequency changes but little with chain conformation, and not in a regular way. The latter band is very broad and the maximum is ill defined. The C=O stretching mode is, however, sharp and its frequency changes considerably with chain conformation. The correlation will be discussed in the sections on individual chain conformations below. The Amide II band at 1530–60 cm^{-1} is sensitive to chain conformation. It arises because of an accidental coupling between NH deformation and CN stretching (*42*). Other vibrational modes of the amide group have been listed by Miyazawa (*59*).

Optical rotation has been used mainly to study structural properties of optically active polypeptides in solution (see reviews of the subject, e.g., references *80* and *77*). It has, however, revealed the existence of an apparently random form in the solid (*40*). The property examined is the specific rotation as a function of wave length. As Moffitt (*61*) has shown, when chromophoric groups such as the peptide groups are situated on a helix, the optical rotation as a function of wave length is represented by a two-term formula.

$$[R_{vac}]_\lambda = a_o\lambda_o^2/(\lambda^2 - \lambda_o^2) + b_o\lambda_o^4/(\lambda^2 - \lambda_o^2)^2$$

Here $[R_{vac}]_\lambda$ is the specific rotation $[\alpha]_\lambda$ reduced to the value which it would have in a medium of unit refractive index and expressed on a molar basis.

$$[R_{vac}] = \frac{\alpha R}{100}\,\frac{3}{n^2 + 2}$$

where R is the residue weight.

The constant a_o is expected to vary with the nature of the side chain and the solvent while b_o and λ_o should be intrinsic properties of the helical skeleton provided the interactions of chromophores in the side chains can be neglected. The sign of b_o is reversed when the screw sense of the helix is changed. Although it is now believed that Moffitt's calculation of b_o and λ_o is based on an assumption which cannot be justified theoretically (*63*) the predictions it makes about the rotatory dispersion of the α-helix seem to be in substantial agreement with experiment.

The results are displayed by plotting $[R_{vac}]_\lambda(\lambda^2-\lambda_o^2)$ against $(1/\lambda^2-\lambda_o^2)$. When accurate results over a sufficient wave-length range are available, an independent determination of λ_o is possible by trial, that value of λ_o for which the plot is a straight line being adopted as the correct one. More often, it is necessary to assume that λ_o will have the value (2120 Å) which has been found for a polypeptide in solution (*62*). In solution, b_o has been found to lie in the range −400 to −700 deg cm²/decimole for a variety of L-polypeptides in solvents which are not too polar to allow the formation of helices. For D-polypeptides positive values of b_o are found. Low values of b_o (<100°) are found in polar solvents which produce a state of low order with the amide groups (or at least a proportion of them) solvated.

With films, it is essential to avoid birefringence, and this is done by making the films quite thin, and using a sufficient number to give a measurable rotation. In consequence, the accuracy with which the mass of polymer per unit area can be measured is low, for the total mass is only a few milligrams. In solution, the freedom of rotation around single bonds will usually prevent side-chain chromophores from being ordered, but in solids this will not always be the case; and if the side chains are ordered the value of b_o may be affected (see below, section on α-helix). For this reason, caution must be used in interpreting the results of optical rotation measurements on solid films.

Observation of the rate of deuteration of the NH groups in polypeptides may reveal the presence of more than one structure, and may distinguish between the random and a regular form. When deuteration reactions are carried out in solutions, the deuterating agent has free access to the polypeptide molecules (though not necessarily to all parts of them). A striking difference in the rate of exchange of the NH groups in solutions of poly-γ-benzyl-L-glutamate according to whether the polymer is in the solvated coil or in the helix form has been found (*36*). In this case, complete exchange occurs in a few minutes for the random coil, but with the helix form little exchange has occurred after many weeks. Linderstrøm-Lang and his school have found examples of globular protein solutions in D_2O in which a fraction of the NH groups are slow to exchange. With solid films, the molecular chains may not be accessible to a deuterating agent to which the polymer is exposed and it may be difficult to interpret the results. However, interesting observations have been made on fibrous proteins which show that parts of the material exchange rapidly when exposed to D_2O vapor (*78, 64, 19*). The part which does not exchange is the more ordered and more highly oriented part, for the dichroism of the remaining NH stretching band is much higher than in the original fiber, whereas the ND band shows little dichroism.

If there is apparently only one rate of reaction when a solid film is exposed to D_2O, it is more difficult to relate it to the structure of the film. However, a reaction in which the half-life of the NH groups is of the order of minutes is indicative of a random structure. When the half-life is many hours or days, an ordered structure is probably present. Deuteration rates by themselves cannot enable a distinction to be made between one ordered form and another.

Specific conformations

THE α-HELIX

The α-helix was proposed by Pauling, Corey and Branson and considered in more detail by Pauling and Corey (*66*). Evidence for one of its characteristic features, the residue translation 1.5 Å along the helix axis, was found by Perutz (*69*), who observed a strong meridional reflection of 1.5 Å in the X-ray diffraction pattern of poly-γ-benzyl-L-glutamate. Again, the layer lines which appeared most strongly in the diffraction pattern of poly-γ-methyl-L-glutamate (*12*) were shown by Cochran, Crick, and Vand (*25*) to be those expected for a structure such as the α-helix with a pitch of 5.4 Å and a residue translation of 1.5 Å. It was therefore surprising that, when well-oriented diffraction photographs of poly-L-alanine (side-chain CH_3) in the α-form became available (*13*), Brown and Trotter (*24*) found large discrepancies between observed and calculated intensities, both for right- and for left-handed screw senses of the helix. The situation was clarified when it was found that all the features of the diffraction pattern (which contains layer-line streaks as well as sharp arcs) could be accounted for by a hexagonal arrangement of right-handed helices in which the chain sense (i.e., the sequence CO·NH·CHR) was random (*39*). The agreement between observed and calculated intensities was reasonably good for a right-handed helix, but quite unacceptable for the left-handed one. This work establishes the right-handed α-helix for the L-enantiomorph of polyalanine. No similarly detailed analysis of the structure of any other synthetic polypeptide has yet been made, but the X-ray diffraction patterns of poly-γ-methyl-L-glutamate and of poly-γ-benzyl-L-glutamate are so similar to those of poly-L-alanine that it cannot be doubted that these polymers also are α-helical. The details of their structures remain unknown, however, because the information available does not serve to determine the conformation of the bulky, flexible side chains. In cases where a layer line corresponding to 5.4 Å and a meridional streak of 1.47–1.50 Å are observed it is reasonable to suppose that α-helices are present.

The screw sense of other L-polypeptides has been linked with that of poly-L-alanine by optical rotation measurements (*40*). The value of b_o for films of poly-L-alanine is $-475°$, showing that the negative value is associated with the right-handed screw sense. It is of interest to note that the value of b_o is changed but little if an appreciable fraction of the L-residues are replaced by D-groups, demonstrating rather strikingly the relative independence of b_o on the side groups (this has also been found in solutions of polyleucines [*34*]). Most synthetic polypeptides so far examined have negative values of b_o for the L-enantiomorph when dissolved in not too polar solvents and must be presumed to be right-handed in solution; it is unlikely that the sense is reversed in the solid state.

In films of poly-L-alanine, the methyl side groups must share the orientation and regular arrangement of the main chains. Because they have no absorption bands in the near ultraviolet region, they do not affect the dispersion of the optical rotation, and b_o has a value similar to that found in some solutions. Films of poly-γ-benzyl-L-glutamate give also a near-normal value of b_o for an α-helix (about -800 deg cm²/decimole [Elliott, unpublished]) and here, since the side chain is chromophoric we must suppose that it is to a considerable extent disordered. With films of poly-β-benzyl-L-aspartate, however, b_o is about $-3560°$ (compare the value $+600°$ in solution, where the helix is thought to be left-handed) when the polymer is cast from chloroform (*21*), and is in the α-helix form. In this case, the chromophoric groups clearly must be sufficiently ordered to affect the value of b_o to an enormous extent. It is evident from this example that such effects may vitiate conclusions drawn from measurements of b_o in the solid state.

The polarized infrared spectra of polypeptides which give the X-ray diffraction features typical of α-helices are as would be expected for such structures; the dichroism of NH stretching bands is parallel, as is that of the C=O stretching band, though the dichroism of the latter band is not so high. This, as was first pointed out by Fraser and Price (*42*), is to be expected, since the transition moment of the C=O frequency is inclined at about 20° to the C=O bond.

The frequency of the C=O band for polypeptides in the α-helix form lies in the narrow range 1652–1665 cm^{-1} (*11*). *Meso* forms, which are sterically strained because of contacts between D and L residues and because of a short distance between βC and O atoms (*45*), have their C=O bands at the higher end of the range. As will appear in the section on random-coil forms, a C=O band in this range does not necessarily indicate the presence of α-helices, unless supported by dichroism.

The β-conformation

The extended or near-extended conformation of the polypeptide chain was proposed many years ago for the silk produced by the larvae of *Bombyx mori* (*57*) and for hair keratin which had been steam-stretched (*6, 7*). In this, the β-form, hydrogen-bonding takes place between NH groups of one chain and the C=O groups of the chains on either side, so making a sheet held together by hydrogen bonds. Neighboring sheets are then held together by van der Waals' forces. This interpretation of the β-form has never been seriously questioned, and later work has concentrated on details such as whether the chains in a sheet are all parallel, with the CO·NH·CHR sequence always in the same direction, or whether there is an alternation of this sequence ("antiparallel chains"; references *44, 66, 67, 68*). The possibility of small contractions of the polypeptide chain has also been examined, and precise conformations for parallel and for antiparallel forms have been proposed by Pauling and Corey.

The structural features common to all β-structures (with whatever side chains) are these: (*a*) a repeat of pattern of the polypeptide backbone along the chain axis of 7 Å or somewhat less; (*b*) an interchain distance of about 4.7 Å in the direction of the hydrogen bond; (*c*) C=O and NH bonds nearly at right angles to the chain axis; (*d*) in all probability, a two-fold screw axis (for the polypeptide backbone) along the chain axis.

It happens that with a near-extended chain the atoms, projected onto the chain axes, give six nearly equi-spaced intervals within the repeat of pattern. This gives rise to a strong reflection corresponding to the sixth order of the repeat, about 1.16 Å or less.

Feature (*b*) is responsible for a strong reflection corresponding to the hydrogen-bonding spacing. The spacing observed depends on the shape of the base of the unit cell, and varies from about 4.36 Å for the shortest side chains (in polyglycine) to nearly 4.8 Å when the side chains are large (*5*). High perpendicular dichroism can be seen in the NH and C=O stretching bands of the spectrum of well-oriented silk (*2*).

Marsh, Corey, and Pauling (*54, 55*) have proposed structures for the fibroin of *Bombyx mori* and of tussah silk (*Antherea perneyi*) which give a reasonable agreement between observed and calculated intensities in the X-ray diffraction pattern. They assume that the hydrogen-bonded sheets have the anti-parallel arrangement, since this gives the most linear hydrogen bond. No reflections have been recorded in these diffraction patterns which compel the assumption of the antiparallel arrangement, however, and positive evidence from X-ray diffraction for this feature of the structure is lacking. It might be thought that the structure of the β-form of poly-L-alanine could be determined from its diffraction pattern, which is quite well-developed (*13, 24*), but this has not yet been accomplished. The unit cell is not known with certainty, and there are layer-line streaks indicative of some disorder. It is possible that, as in the α-form of the same material, no unit cell exists in the proper sense of the term.

The infrared spectra of synthetic polypeptides and fibrous proteins in the β-form have some interesting features. The dichroism has been referred to above. The C=O stretching band (Amide I) (which in the α-helix form is near 1660 cm^{-1}) is near 1630 cm^{-1} in all β-forms hitherto examined. The frequency of the Amide II band is also lowered by about the same amount compared with that of the α-helix.

These empirical observations enable the β-structure to be recognized in polypeptides to which they have been shown to apply, even in unoriented specimens or with unpolarized radiation. There is, of course, the possibility that abnormal frequencies may be encountered in new polypeptides or proteins, or that some other conformation may be found to give β-type frequencies.

Recently an attempt has been made to connect the frequencies of the amide bands I and II with chain conformation in a more fundamental way (*58, 59, 60*). The papers contain much else which is of interest, but we shall consider here only the claim which the authors make to distinguish between parallel and anti-parallel arrangements of near-extended chains (pleated sheets). In a crystalline, polymer-like Nylon 66 there is only one chain in the unit cell and it is known that normal modes which are potentially infrared active are ones in which all unit cells in the crystal are in phase. No possibility exists of coupling of vibrations in neighboring chains to give a splitting into in-phase and out-of-phase modes. The separation of amide groups along a chain is too large for intramolecular coupling to be appreciable, and hence the amide bands in Nylon 66 should be single, as is observed. In an extended polypeptide an anti-parallel arrangement is expected, with two chains in the unit cell. The vibrations in these chains may be either in-phase or out-of-phase. Moreover, successive amide groups along a chain may be coupled. In arrangements where there is a twofold screw axis (as in β-chains) there are two amide groups in the repeat length, and these also may be in-phase or out-of-phase. These possibilities lead to an anti-parallel chain having four modes for every mode of a single amide group provided that the transition moment of the mode in the single group is neither exactly parallel nor perpendicular to the chain axis. The C=O or Amide I band (transition moment at about 70° to the chain axis in a fully extended chain and almost in the plane of the hydrogen-bonded sheets) should be split into four modes, one inactive, one very weak, the strongest a perpendicular mode, and also a considerably weaker parallel mode. These two latter Miyazawa identifies respectively with the strong band at about 1630 cm^{-1} (perpendicular) and a much weaker band at 1685–1700 cm^{-1} (parallel). The latter band has been observed in many β-polypeptides. On the basis of this assignment he and Blout use the observed frequencies of polyglycine I and Nylon 66 to calculate the intermolecular coupling constant and the intramolecular constant for β-polypeptide chains; the intramolecular constant for Nylon 66 is assumed to be negligible.

If the two bands just referred to are correctly identified, we should expect a similarly related pair in the spectrum of Nylon 6, in which a thorough X-ray investigation has shown that the chains are anti-parallel (*43*). The intra-chain coupling should be small, and from the interaction constants of Miyazawa we should expect a splitting of 36 cm^{-1}. We have found that the C=O bond in Nylon 6 does appear double when an oriented specimen is examined with polarized radiation. The splitting is difficult to measure but it appears to be about 25 cm^{-1}, appreciably less than the predicted value. Again, in acetanilide crystals the arrangement of those molecules which are connected by hydrogen bonds is such that the amide groups are related in similar fashion to those in an anti-parallel polypeptide sheet. The C=O band is split into three components with transition moments along each of the three orthorhombic axes (*1*) but the frequencies of these components differ only by a very few wave numbers. Evidently the interaction constants must vary considerably from one compound to another, and it is not admissible to calculate them by using frequencies culled from the spectra of materials as different as Nylon 66 and polypeptides. Until these points are cleared up we consider that caution should be used in using the presence or absence of a band at about 1690 cm^{-1} (the range observed covers 1685 for polyglycine I to 1702 cm^{-1} for poly-β-*n*-propyl-L-aspartate) as an indication of anti-parallel or parallel chains, respectively. Nevertheless, Miyazawa has rightly indicated the importance of examining the interactions between amide groups.

As mentioned above, nearly-extended chains are expected to have an anti-parallel arrangement. When, however, the chain repeat distance contracts to 6.6 Å as in steam-stretched keratin (*6*), the hydrogen bond may perhaps be better formed by a parallel arrangement (*68*). The X-ray diffraction photograph of steam-stretched hair contains a reflection which shows that the repeat distance in the direction of the hydrogen bonds is not 4.65 Å but 2×4.65 Å. This shows that successive chains in the hydrogen-bonded sheet differ in some may. This may indicate anti-parallel sheets, but Astbury, Beighton and Parker (*8*) have suggested that a parallel-chain arrangement is possible if the side chains have different conformations in alternate chains. Evidence of variable side-chain conformation has been found in synthetic polypeptides (see next section). Some remarks on the spectrum of β-keratin will be found in the discussion on Dr. Miyazawa's paper.

THE CROSS-β CONFORMATION

A variety of the near-extended chain in which the chain axis is at right angles to the fiber axis was reported in proteins by Astbury, Dickinson, and Bailey (*9*). Rudall (*73*) has described the occurrence of this form in epidermin. The evidence in this earlier work was the observation of an X-ray reflection on or near the meridian of the photograph, corresponding to 4.7 Å. Confirmation of this structure is obtained from polarized infrared spectra of fibrous insulin (*3*) and of epidermin (*70*). The dichroism is similar to that of an oriented arrangement of α-helices with the helix axes along the fiber axis, but the frequencies are those of the β-form. Much more complete evidence of a cross-β structure has been obtained by Parker and Rudall (*65*) from the silk of the egg-stalk of the green lace-wing fly, *Chrysopa flava*. The X-ray diffraction pattern is well-developed and shows that the chain axis is perpendicular to the fiber axis. On drawing the fiber, the cross-β diffraction pattern disappears and is replaced by a normal β-pattern. Intermediate orientations are not observed during stretching, suggesting an intramolecular change rather than a re-orientation of the crystallites. The infrared observations support the picture obtained from X-ray diffraction. It is possible, though not proven, that the fibroin molecules in the egg-stalk are folded backwards and forwards like a Chinese cracker, and that the folds pull out on stretching. The fibroin of *Chrysopa* silk contains an unusually large proportion of serine residues (*53*). Poly-L-serine has, however, been reported as giving the random form on casting from an unspecified solvent (*18*).

A detailed examination of some synthetic materials of rather low molecular weight has shown that the cross-β form is obtained by suitable treatment of poly-γ-benzyl-L-glutamate, poly-β-benzyl-L-aspartate, and poly-β-*n*-propyl-L-aspartate (*22*). Well-developed X-ray diffraction photographs of the propyl aspartate polymer showed that there are not less than four chains in the unit cell. At least two of these chains must have different conformations; the differences may be restricted to the side chains. An anti-parallel arrangement within a hydrogen-bonded sheet appears likely, but a diffuseness of some of the X-ray reflections suggests that a strict alternation of chain sense may not obtain.

THE RANDOM COIL OR STATE OF LOW ORDER IN THE SOLID STATE

The random, solvated polypeptide form in solution has been established by light-scattering, viscometric and optical rotation techniques (*32, 31, 62, 30*). A form of the polypeptide chain of such low order that it may perhaps be referred to as random is also known in a few synthetic polypeptides and proteins (both fibrous and globular) in solid films. In these cases, substantially all the hydrogen bonds appear to be formed.

Evidence for a random arrangement comes from measurements of the dispersion of optical rotation produced by thin films of the material cast on glass (*40*). Films of polyalanine containing from 80 per cent to 100 per cent of the L-enantiomorph give dispersions for which a Moffitt plot indicates a helical arrangement, with b_o equal to about -500 deg cm^2 per decimole. This falls within the range found for solutions of polypeptides in the helix form. Films of water-soluble fibroin prepared from *Bombyx mori* silk dissolved in aqueous lithium bromide (*4*), as well as films of the sodium and potassium salts of poly-glutamic acid, and of lysozyme cast from aqueous solution gave Moffitt plots with very low b_o indicating the absence of the helix form. Films prepared in the same way for infrared examination all show a single peak at about 1660 cm^{-1} with no indication of a peak at 1630 cm^{-1}. It is therefore clear that the structure is neither of the α-helix nor of the β-type. The evidence for a random form is not conclusive, but is supported by the amorphous character of the X-ray diffraction pattern produced by water-soluble silk. In any case, it is clear that a C=O band at 1660 cm^{-1} is not necessarily an indication of an α-helix form, as was pointed out in connection with the structure of silk by Elliott and Malcolm (*37*). Thus, although silk gut (mainly β-form) has a spectrum with a distinct peak at 1660 cm^{-1} when the radiation has its electric vector parallel to the fiber axis, no reflections corresponding to an α-helix are seen in the X-ray diffraction pattern. That part of the protein from which the 1660 cm^{-1} peak arises must be amorphous (*2, 35*). It is reasonable, in view of what has been said above, to suppose that this material is in a random form. It is interesting to note that the vapor of D_2O exchanges with the NH groups of water-soluble silk and of lyzozyme fairly rapidly. It is known that, in solution, poly-γ-benzyl-L-glutamate exchanges rapidly with a deuterating agent when in the random form, but scarcely at all when in a helix (*36*).

THE ω-FORM

When poly-β-benzyl-L-aspartate is cast from chloroform and stroked during drying, an oriented film is obtained whose X-ray diffraction photograph and infrared spectrum show all the characteristic features of an α-helix form, although the crystallinity, judged by the breadth of the X-ray reflection, is very low. When the film is heated in vacuo a new form (the ω-form) is obtained, without apparent loss of orientation and with high crystallinity. The equatorial reflections fit a cell with a tetragonal base (α-helix forms usually pack hexagonally) and the meridional reflection of spacing 1.5 Å characteristic of the α-helix has disappeared. A new meridional reflection of spacing 1.3 Å is found, and the layer lines are orders of 5.3 Å, which appears to be the chain repeat length. It has been suggested that the chain conformation corresponds to the 4_{13} helix proposed by Bragg, Kendrew and Perutz (*23*; also *21*). This helix is in effect an α-helix distorted (with nonplanar amide groups or nonlinear hydrogen bonds, or a combination of both) so that the residue rotation is exactly 90° instead of about 100°; it has a four-fold screw axis and a residue translation of about 1.3 Å. The infrared spectra of the α and ω forms are shown in Fig. 24.1.

The structure of the ω-form has now been

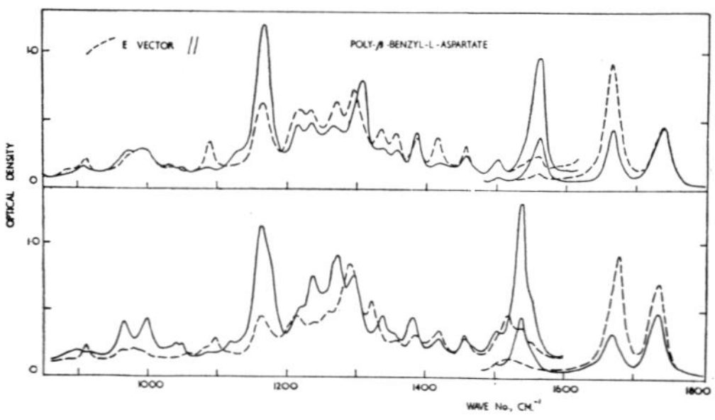

Fig. 24.1. Infrared spectra of films of oriented poly-β-benzyl-L-aspartate. *Full line:* **E** vector perpendicular to fiber axis. *Broken line:* **E** vector parallel to fiber axis. *Upper:* polymer in α-helix form. *Lower:* polymer in ω-form.

examined in detail in the Courtauld Laboratories and we are of the opinion that it is indeed the 4_{13} helix. That poly-β-benzyl-L-aspartate should be different from most synthetic polypeptides in the solid state is not surprising, for it has been found that in solution the helix form is left-handed and is less stable compared with the random-coil form than, for instance, poly-γ-benzyl-L-glutamate (*46, 20*). Infrared spectra obtained with polarized radiation have been of great help in limiting the possible conformations of the side chains, and the X-ray diffraction patterns of the resulting models have been compared with that of the polymer. It has been found necessary to assume that there is a random arrangement of chain sequence, as in poly-L-alanine (above), so that no crystallographic unit cell exists. The chosen conformation has been confirmed by a two-dimensional Fourier synthesis of the projection on a plane normal to the chain axis. This synthesis, although of low resolution, shows the main peptide chain, side chain, and benzene rings. The chain conformation is shown in Fig. 24.2.

Isolated molecules are unlikely to occur in the ω-form, as the isolated α-helix clearly has lower free energy. The ω-form will therefore be produced by re-arrangement of the α-form in the solid state and this will be easy only when it does not involve the very far-reaching reorganization necessary to bring about a reversal of the screw sense. Examination of scale models suggests that a 4_{13} helix can be formed only when the helix is left-handed for L-residues, for steric reasons. Since α-helices with this screw sense are rare, the ω-form is also expected to be found very rarely. The examination of model structures has emphasized the difficulty of taking account of steric interactions, and so far no very convincing reason has emerged for the apparently greater stability of the ω-form compared with the α-form in poly-β-benzyl-L-aspartate, though it is clearly connected with intermolecular interactions.

Poly-L-proline

Poly-L-proline differs from the other synthetic polypeptides considered above in having no NH groups and hence no possibility of forming NH · · · O bonds. It is also peculiar in that, for steric reasons, it cannot form a fully extended chain, but with planar amide groups in the *trans* form it is constrained to fold into a left-handed helix. In the crystal, the molecule has a three-fold screw axis and a chain repeat distance of 9.36 Å (*28*). X-ray diffraction photographs of oriented specimens of poly-L-proline have been published by Cowan and McGavin in a preliminary report on the structure. A further remarkable property of poly-L-proline in solution is the occurrence of a spontaneous muta-rotation, as was found by Kurtz, Berger, and Katchalski (*48, 49*), who suggested that this might be caused by a transition from *cis* to *trans* amide groups. The

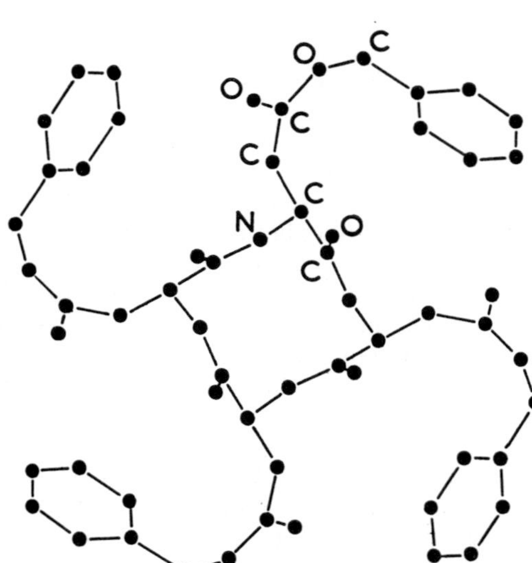

Fig. 24.2. ω-Form of poly-β-benzyl-L-aspartate. Atoms projected on plane normal to 4_{13} helix axis.

evidence for this is quite strong (*33*) for the energy of activation of the reaction is almost exactly equal to the resonance energy of the amide group. The two forms of poly-L-proline are known, respectively, as I and II, the latter being the *trans* form. The structure of this form in the solid has been examined more recently by Sasisekharan (*75*), who has considered the effect on the calculated intensities of the X-ray reflections of rotating the chain in the unit cell around the chain axis. The chain conformation is assumed to be the same as that proposed by Cowan and McGavin. The best fit of observed and calculated intensities occurs at the position where steric interference between chains is a minimum. There is, however, a short contact (CH_2 to O = 3.00 Å) and this, it is suggested, is a CH · · · O bond. In the structure which is put forward for poly-L-proline II, it is assumed that all chains are "parallel," since no X-ray reflections occur which require that there should be more than one chain in the unit cell. The possibility of a random arrangement of chain sequence should, however, be considered; this would appear statistically much more probable in a crystallite containing long polymer chains. The quality of the X-ray diffraction photographs so far published would hardly allow such details to be decided; Sasisekharan's work is based on a powder photograph and we consider that diffraction photographs from well-oriented fibers are needed before it can be claimed that the structure is completely known.

Films of polyproline I so far reported give only two or three poorly oriented and diffuse reflections (*27*). Their intensity distribution is compatible with a right-handed arrangement of *cis* amide groups.

The infrared spectra of oriented films produced from evaporated solutions of forms I and II have been given by Blout and Fasman (*16*). There appears to be little difference in the frequency or dichroic properties of the C=O band at 1650 cm^{-1}. A band only slightly weaker than the C=O band, of perpendicular type like the latter, appears in both spectra at about 1435 cm^{-1}; this probably corresponds to the Amide II′ band of N-deuterated polypeptides. The spectrum of Form I shows all the bands of Form II with a few additional weak bands. The spectra do not enable any conclusions to be drawn as to the presence of *cis* or of *trans* amide groups, the spectra of which in the case of a tertiary amide are unlikely to show profound or predictable differences.

Poly-L-hydroxyproline

The structure of poly-L-hydroxyproline has been determined by Sasisekharan from X-ray diffraction photographs (*76*). Powder photographs were chiefly employed, but a film with low orientation gave confirmation of the assignments made from powder photographs. The unit cell is hexagonal, with $a = 12.3$ Å, $c = 9.15$ Å, and contains three prolyl residues on each of three chains. The three residues on one chain are not exactly related by a threefold screw axis, though this symmetry is approximately attained. The three chains are, however, related exactly by a threefold axis. The conformation of each chain is not very different from that of poly-L-proline, but by making slight distortions (which destroy the threefold screw symmetry of each chain) a closer packing is achieved than when the chains have threefold screw axes. Hydrogen bonds of lengths 2.50–2.56 Å are formed between all the carbonyl oxygen atoms and the nearest OH groups. The agreement between calculated and observed intensities is reasonable, but since the observed intensities are graded only into four categories the test of the structure is not searching. It will be noticed that as in the poly-L-proline II structure all the polypeptide chains are assumed to be "parallel"; the possibility of anti-parallel chains has not been considered. This should be done because such an arrangement is statistically likely. It may require the production of well-oriented fibers before the structure can be decided satisfactorily.

The infrared spectrum of poly-L-hydroxyproline has the C=O band at the low frequency of 1620 cm^{-1} according to Kurtz, Fasman, Berger, and Katchalski (*50*), who ascribe this to the formation of hydrogen bonds with OH groups. This is in agreement with the structure proposed by Sasiskeharan; the frequency is, however, unusually low.

Polyglycine II

Two different X-ray powder diffraction photographs of polyglycine have been described in the literature (*56*, *10*) and it was not clear whether in these forms the chain conformation was the same or not. When it was found that different infrared spectra were associated with the two X-ray diffraction patterns it appeared likely that two conformations were in fact involved (*14, 38*). Form I, obtained by casting from solution in dichloracetic acid, has X-ray reflections and infrared band frequencies characteristic of β-structures. The other form is obtained by precipitating from solution (either in saturated aqueous lithium bromide or dichloracetic acid) by means of water. In this form the frequency of the Amide I band is 1648 cm^{-1} intermediate between that of α-helix and β-structures.

Crick and Rich (*29*) have put forward a structure for polyglycine II. In this, each chain has a threefold screw axis with a residue translation of 3.1 Å, and the chains are arranged with their helix axes on a hexagonal net. The interchain distance is 4.8 Å and each chain forms hydrogen bonds with its six neighbors. The intensities of the reflections from such an arrangement (with parallel chains) are in approximate agreement with those observed by Bamford *et al.* However, as Crick and Rich point out, the details of the structure (screw sense of helix and occurrence of parallel or antiparallel chains) remain unknown. Until an oriented fiber is obtained they are likely to remain so. There can be little doubt that the conformation is very close to that suggested by Crick and Rich.

Factors influencing the stability and ease of transformation of α- and β-conformations in polypeptides

It is well known that many synthetic polypeptides can exist in the solid form as α-helix or as β-structures, or more frequently as mixtures of those forms (*11*). Other forms of less frequent occurrence are referred to above. With a single sample of polymer, the proportions of α- and β-forms depend on the method of preparation of the solid specimen, and transformations from one form to another have been observed. In some polypeptides which can form α-helices, specimens of low molecular weight tend to form β-structures and the degree of polymerization beyond which it is difficult to prepare specimens in the β-form by simple casting varies greatly from one polypeptide to another. An example is given in Table 24.1,

Table 24.1

% β in film	Reduced viscosity* dl/gm	
	PBLG	PMLG
25	0.061	0.22
50	0.054	0.12

* Concentration for all results in table and text is 0.5 per cent *w/v* in dichloracetic acid.

which shows some unpublished results obtained by Bamford and Hanby in this laboratory. Films of poly-γ-methyl- and poly-γ-benzyl-L-glutamate were cast from *m*-cresol solutions and their β-contents were estimated by comparing the intensities of the bands at *ca.* 1630 and 1654 cm^{-1}. In the case of the methyl ester the β-content diminishes rather slowly with increasing molecular weight and is about 10 per cent even at a reduced viscosity of 0.32 dl/gm. The β-content falls off much more rapidly in the films of the benzyl ester. Although the viscosity–molecular weight relation cannot be assumed to be the same for the two polypeptides, it seems safe to conclude that for a given fraction of α-material the methyl polymer has a much higher molecular weight than the benzyl polymer.

The effect of molecular weight on conformation may also be seen in poly-β-benzyl-L-aspartate (*20, 22*). When films are cast from a polymer of fairly high molecular weight ($\eta_{sp}/c=0.54$ dl/gm) dissolved in *m*-cresol and stroked out at about 60° C. during drying, an α-helix form is obtained which when heated to 160° C. is converted to the ω-form (see above). In neither case is any β-peak at 1630 cm^{-1} to be seen. With a specimen of the lower molecular weight ($\eta_{sp}/c=0.121$ dl/gm), however, the α-helix form obtained by casting in the same way is changed on heating to a mixture of ω- and cross-β forms.

In a recent note, Blout, de Lozé, Bloom, and Fasman (*18*) have reported some interesting new results on the chain conformation in poly-L-serine and other synthetic polypeptides. By consideration of the infrared spectrum and dichroism of oriented films (the method of preparation of which they do not specify) they have concluded that there are two classes of polypeptides derived from α-amino acids, those which form helical (α) structures and those which form either random or extended

(β) structures. Presumably they mean by this that, whereas the first class may form α-helices or other structures, the second class does not form α-helices. Blout *et al.* infer from their results that the conformation does not depend on the degree of polymerization in the samples examined. Their figures do indeed show that the conformation is not determined by molecular weight alone; however, their observations, being confined to a single molecular weight for each polypeptide species, give no indication of whether the chain conformation is influenced by molecular weight in a given species. It is not possible to conclude that a given polypeptide cannot form an α-helix because this form has not been observed; it is always possible that different methods of specimen preparation or samples of higher molecular weight would give the helical form. This should be kept in mind particularly when considering polypeptides which are soluble only in polar liquids. When the solution from which the specimen is prepared contains α-helices, we may expect to find them in the solid film. When, however, the solution contains a solvated random-coil form, opportunity is afforded for the formation of the β-structure. It may happen that as the concentration is increased or as precipitant is added a change in the conformation in solution will occur; thus poly-γ-benzyl-L-glutamate in dilute solution in dichloracetic acid has the random solvated coil form, but at about 20 per cent *w/v* this goes over to the helix form (*34*). Again, we have found from measurements of optical rotatory dispersion that when acetone is added to a solution of poly-γ-methyl-L-glutamate in trifluoracetic acid, complete conversion to the helix form occurs before sufficient acetone has been added to precipitate the polymer. The situation is further complicated by the fact that the solid specimens may not be in their equilibrium state. The change from α-helix to β-structure must be hindered by a high potential barrier and the ease of transformation depends upon the height of this barrier rather than on the difference in the free energy of the two states, about which little is known.

The effect of the solvent and the fact that equilibrium is not easily attained is shown by the following observations (*20, 22*). A preparation of poly-β-*n*-propyl-L-aspartate ($\eta_{sp}/c = 0.095$ dl/gm) when cast by evaporation at 60° C. from dichloracetic acid gives films which are in the cross-β form. When cast from chloroform at room temperature, an oriented α-helix form is obtained. On heating to 160° C. the α-helix form is converted to the β-form, which persists after cooling. The polymer may be reconverted to the α-helix form, however, by dissolving in chloroform and recasting. In this case the existence of suitable solvents allows such a cycle to be completed.

When poly-β-benzyl-L-aspartate of fairly high molecular weight ($\eta_{sp}/c = 0.54$ dl/gm) is cast from *m*-cresol at about 60° C. an α-helix form is obtained which when heated to 160° C. is converted to the ω-form. In neither case is any β-peak at 1630 cm^{-1} to be seen. With a specimen of lower molecular weight ($\eta_{sp}/c = 0.121$ dl/gm), however, the α-helix form obtained by casting in the same way is changed on heating to a mixture of ω- and cross-β forms. It would be expected that the height of the potential barrier associated with the change $\alpha \rightleftharpoons \omega$ (which presumably does not involve the rupture of hydrogen bonds) would be much less than for the change $\alpha \rightleftharpoons \beta$. However, no conversion of α to ω or vice versa is observed at room temperature.

Further examples of polypeptides which are not in their equilibrium states are furnished by the so-called water-soluble silks and cold-water-soluble gelatin (*4, 71*). These substances will disperse in cold water, but after a longer or shorter time at room temperature a gel is produced. Various lines of evidence suggest that there is a tendency, in silks, to form a β-structure, and in gelatine to form the normal collagen structure. In the case of the water-soluble silks, the ability to disperse completely in water is lost if films of the dispersable material are kept for some months at room temperature. It is not always possible to detect a β-structure in silk which has become insoluble, but it appears likely that very little of this cross-linked form is needed to prevent complete dispersion. Likewise, poly-DL-alanine film or powder when stored for some time at room temperature may become insoluble in water, probably for the same reason. Poly-DL-alanine rapidly comes out of solution in water when the temperature is raised, and films made from the resulting gel contain a large fraction in the β-form (*11*, p. 333). The conformation of poly-DL-alanine in aqueous solution is considered in the discussion on Professor Doty's paper.

Blout *et al.* suggest that the non-helix-forming class comprises those polypeptides in which the side chains contain two atoms other than hydrogen directly attached to the β-carbon atom, or in which an oxygen or a sulphur atom is attached to the β-carbon atom. They ascribe the absence of an α-form in poly-L-valine to steric factors. We have built models of this polypeptide and find that there is indeed some interference between the carbonyl carbon atom and one of the methyl groups in the right-handed α-helical form. However, this interference appears to be no more serious than that encountered in the left-handed helical form of poly-β-benzyl-L-aspartate, whose existence seems to be established (*46, 20*). Similar or perhaps worse conditions are encountered in an α-helix of poly-α-amino-*iso*butyric acid, where there is an oxygen-β-carbon contact. (This is the contact which prevents most L-polypeptides from occurring in a left-handed form.) There is in addition a short β-carbon to β-carbon distance of 3.4 Å between residues 1 and 4, and between residues 4 and 7. Nevertheless, the X-ray diffraction pattern of an unoriented film has been obtained and resembles that of an α-helix, not a β-form (*38*). Drs. Blout and Fasman have pointed out to us that in this polymer steric hindrance would be much less in a 3_{10} helix than in an α-helix. The possibility that 3_{10} helices are present in significant amount seems to be excluded by consideration of the density of such a film. The measured density is in excess of 1.3 gm/ml; if the very strong reflection corresponding to 7.54 Å on the X-ray photograph is $10\bar{1}0$, the density of the crystalline regions would be 1.45 and 1.08 gm/ml for α-helices and 3_{10} helices, respectively. It is inconceivable that a crystalline density would be much lower than the measured density of the whole polymer.

Of all the polypeptides containing β-carbon atoms, poly-L-alanine in a right-handed helix would be expected to have the most stable α-form, since steric interference between side chain and backbone is at a minimum. This expectation is borne out by the fact that, of the polypeptides so far examined, poly-L-alanine is the only one which in dilute solution in dichloracetic acid exhibits the helix form (*34*). The transition in the solid from the α- to the β-form is, however, readily induced by stretching, even in the cold form. Presumably the small side chains are an advantage in the β-form also and may reduce the potential barrier between the two forms. The next member in the series of polypeptides with straight hydrocarbon side chains, poly-α-amino-*n*-butyric acid, is much more difficult to convert to the β-form (Hanby and Malcolm, unpublished). Similarly, poly-L-glutamic esters form a series in which the ease of obtaining a β-structure diminishes as we go from the methyl to the ethyl and to the benzyl esters. However, it would be premature to make the generalization that ease of transformation decreases with length of side chain.

That other influences besides steric factors consequent upon side-chain size are important is shown by the effect of glycine residues, which when copolymerized with other amino acids have a very strong tendency to produce β-structures. This is shown by the silk fibroins, all of which contain a substantial proportion of glycine and which are of rather high molecular weight. A number of fibroins when cast from dichloracetic acid give films predominantly in the β-form (*37*). A striking example is furnished by the fibroin of *Anaphe moloneyi* silk, which contains alanine and glycine in almost equal amounts, and little else (*52*). Although poly-L-alanine when cast from dichloracetic acid gives the α-helix form, with *A. moloneyi* fibroin the β-form is obtained.

In considering the relative stability of α-helix and β-forms, we must take account of at least three sources of steric interaction. There is the interaction between the side chains and α-helix backbone; the interactions between the side chains on neighboring molecules which largely determine the packing of the α-helices and therefore influence the energy; and lastly the way in which the side chains can be accommodated within the β-structure. We must in all these cases remember the stabilizing effect of van der Waals forces as well as the disrupting effect of steric interference. Also it must be remembered that it is the *free* energy which determines the stability of a structure and the entropy even of a solid system is to some extent dependent on the amount of intramolecular motion possible.

In the case of DL-polypeptides the position is even more complicated and the stability of both α- and β-forms is affected by additional possibilities of steric interference. A proportion of the residues will in this case be carried by helices of the less preferred sense, and there may also be steric interference between residues on the same helix. On the other hand, in the β-structure there will be difficulty in accommodating L and D residues in corresponding positions on adjacent chains. The importance of these effects will depend on the distribution of L and D residues along the polymer chain, about which nothing is known. The factors peculiar to DL-polypeptides will be of even greater importance in polypeptides containing two β-carbon atoms such as poly-α-amino-*iso*butyric acid to which we have referred above.

Very remarkable transformations from α-helix to β-form and vice versa are found in films of poly-L-lysine hydrohalides (*17*, *41*) and in the sodium salt of poly-L-glutamic acid (*51*). Oriented films of these polypeptides are found to be in the α-helix form when surrounded by air with a high relative humidity (70 per cent) and go over into the β-form when the humidity is reduced, with little loss of orientation. The change appears to be reversible and has been followed by infrared spectroscopy with polarized radiation, which allows the change in dichroism of the amide bands to be seen.

In many cases there are indications of a conflict between main and side chains. For instance, in poly-β-benzyl-L-aspartate the hydrogen bonds of the ω-helix are by no means linear and hence are weaker than in the α-helix. The 4_{13} helix does however allow the structure as a whole to pack in the crystal in such a way that side-chain interactions are more favorable than they would be in an α-helix. Although an α-helix arrangement in the solid form is possible, it is clearly unstable for with a rise in the temperature the ω-form is obtained. This form persists on cooling.

In β-structures the hydrogen-bonding forces fix the separation of the β-carbon atoms in the plane of the hydrogen-bonded sheets. As pointed out by Bradbury *et al.* (*22*), identity of all the side-chain conformations might cause the interatomic distances to be greater, on the whole, than those corresponding to normal van der Waals forces. In such a case minimum potential energy might be achieved with more than one side-chain conformation. There is very good evidence that the chains in the cross-β form of poly-β-*n*-propyl-L-aspartate take up more than one configuration in the unit cell (*22*).

When there is a random arrangement of chain sense (which is known to occur in the α-helix form of poly-L-alanine and which almost certainly occurs in the ω-form of poly-β-benzyl-L-aspartate) the several molecules will have a variety of environments. The conformation of the side chains may then be expected to vary according to the environment. In poly-L-alanine this variation can affect only the position of the hydrogen atoms of the methyl groups. With a flexible side chain the differences in conformation may be much greater. In the ω-form of poly-β-benzyl-L-aspartate it appears necessary to postulate more than one conformation for steric reasons.

Concluding remarks

It will appear from the foregoing account that the main features of several quite distinct chain conformations in polypeptides are now known, and in a few instances something is also known concerning the way in which the chains pack in crystalline regions. The details remain unknown in many cases, however, and there is need for X-ray diffraction work on specimens of good orientation and crystallinity. All the synthetic materials to which reference has been made are heterogeneous as regards molecular weight; this fact may contribute to the difficulty in establishing a state of true equilibrium, and it would be of considerable interest to examine polypeptides monodisperse with respect to molecular weight.

The extremely interesting paracyrstalline phenomena which Robinson (*72*) has examined in solutions of poly-γ-benzyl-L-glutamate have not been discussed, for they do not properly belong to an account of the solid state. Since, however, the solid state is reached by evaporation of solutions, we may expect that some trace of such paracrystalline arrangements will be found if the structure of the solid film were examined in detail.

Lastly, there is the possibility that even in polymers polydisperse with respect to molecular weight single crystals may be found (as, for instance, in polyethylene [*79*, *47*]). Such forms would enable much more information to be obtained from X-ray or electron diffraction.

References

1 N. B. Abbott and A. Elliott, *Proc. Roy. Soc.* **A234,** 247 (1956).
2 E. J. Ambrose and A. Elliott, *ibid.* **A206,** 206 (1951).
3 E. J. Ambrose and A. Elliott, *ibid.* **A208,** 75 (1951).
4 E. J. Ambrose, C. H. Bamford, A. Elliott, and W. E. Hanby, *Nature* **167,** 264 (1951).
5 W. T. Astbury, *ibid.* **163,** 722 (1949).
6 W. T. Astbury and A. Street, *Phil. Trans. Roy. Soc.* **A230,** 75 (1931).
7 W. T. Astbury and H. J. Woods, *ibid.* **A232,** 333 (1933).
8 W. T. Astbury, E. Beighton, and K. D. Parker, *Biochim. et Biophys. Acta* **35,** 17 (1959).
9 W. T. Astbury, S. Dickinson, and K. Bailey, *Biochem. J.* **29,** 2351 (1935).
10 W. T. Astbury, C. E. Dalgleish, S. E. Darmon, and G. B. B. M. Sutherland, *Nature* **162,** 596 (1948).
11 C. H. Bamford, A. Elliott, and W. E. Hanby, "Synthetic Polypeptides," Academic Press, Inc., New York, 1956.
12 C. H. Bamford, L. Brown, A. Elliott, W. E. Hanby, and I. F. Trotter, *Nature* **169,** 357 (1952).
13 C. H. Bamford, L. Brown, A. Elliott, W. E. Hanby, and I. F. Trotter, *ibid.,* **173,** 27 (1954).
14 C. H. Bamford, L. Brown, E. M. Cant, A. Elliott, W. E. Hanby, and B. R. Malcolm, *ibid.* **176,** 396 (1955).
15 M. Beer, G. B. B. M. Sutherland, K. N. Tanner, and D. L. Wood, *Proc. Roy. Soc.* **A249,** 147 (1958).
16 E. R. Blout and G. D. Fasman, "Recent Advances in Gelatin and Glue," Pergamon Press, London, 1958, p. 122.
17 E. R. Blout and H. Lenormant, *Nature* **179,** 960 (1957).
18 E. R. Blout, C. de Lozé, S. M. Bloom, and G. D. Fasman, *J. Am. Chem. Soc.* **82,** 3787 (1960).
19 E. M. Bradbury, R. E. Burge, J. T. Randall, and G. R. Wilkinson, *Discussions Faraday Soc.* **25,** 173 (1958).
20 E. M. Bradbury, A. R. Downie, A. Elliott, and W. E. Hanby, *Proc. Roy. Soc.* **A259,** 110 (1960).
21 E. M. Bradbury, L. Brown, A. R. Downie, A. Elliott, W. E. Hanby, and T. R. R. McDonald, *Nature* **183,** 1736 (1959).
22 E. M. Bradbury, L. Brown, A. R. Downie, A. Elliott, R. D. B. Fraser, W. E. Hanby, and T. R. R. McDonald, *J. Mol. Biol.* **2,** 276 (1960).
23 W. L. Bragg, J. C. Kendrew, and M. F. Perutz, *Proc. Roy. Soc.* **A203,** 321 (1950).
24 L. Brown and I. F. Trotter, *Trans. Faraday Soc.* **52,** 537 (1956).
25 W. Cochran, F. H. C. Crick, and V. Vand, *Acta Cryst.* **5,** 581 (1952).
26 R. B. Corey and L. Pauling, *Proc. Roy. Soc.* **B141,** 10 (1953).
27 P. M. Cowan and R. E. Burge, "Recent Advances in Gelatin and Glue Research," B.G.G.R.A., London, 1958, Discussion, p. 16.
28 P. M. Cowan and S. McGavin, *Nature* **176,** 501 (1955).
29 F. H. C. Crick and A. Rich, *ibid.* **176,** 780 (1955).
30 P. Doty and J. T. Yang, *J. Am. Chem. Soc.* **78,** 498 (1956).
31 P. Doty, J. H. Bradbury, and A. M. Holtzer, *ibid.* **78,** 947 (1956).
32 P. Doty, A. M. Holtzer, J. H. Bradbury, and E. R. Blout, *ibid.* **76,** 4493 (1954).
33 A. R. Downie and A. A. Randall, *Trans. Faraday Soc.* **55,** 2132 (1959).
34 A. R. Downie, A. Elliott, W. E. Hanby, and B. R. Malcolm, *Proc. Roy. Soc.* **A242,** 325 (1957).
35 A. Elliott, *J. Appl. Chem.* **6,** 341 (1956).
36 A. Elliott and W. E. Hanby, *Nature* **182,** 654 (1958).
37 A. Elliott and B. R. Malcolm, *Biochim. et Biophys. Acta,* **21,** 466 (1956).
38 A. Elliott and B. R. Malcolm, *Trans. Faraday Soc.* **52,** 528 (1956).
39 A. Elliott and B. R. Malcolm, *Proc. Roy. Soc.* **249A,** 30 (1959).
40 A. Elliott, W. E. Hanby, and B. R. Malcolm, *Discussions Faraday Soc.* **25,** 167 (1958).
41 A. Elliott, B. R. Malcolm, and W. E. Hanby, *Nature* **179,** 960 (1957).
42 R. D. B. Fraser and W. C. Price, *ibid.* **170,** 490 (1952).
43 D. R. Holmes, C. W. Bunn, and D. J. Smith, *J. Polymer Sci.* **17,** 159 (1955).
44 M. L. Huggins, *Chem. Revs.* **32,** 195 (1943).
45 M. L. Huggins, *J. Am. Chem. Soc.* **74,** 3963 (1952).
46 R. H. Karlson, K. S. Norland, G. D. Fasman, and E. R. Blout, *ibid.* **82,** 2268 (1960).
47 A. Keller, *Phil. Mag.* **2,** 1171 (1957).
48 J. Kurtz, A. Berger, and E. Katchalski, *Nature* **178,** 1066 (1956).
49 J. Kurtz, A. Berger, and E. Katchalski, "Recent Advances in Gelatin and Glue Research," Pergamon Press, London, 1958, p. 131.
50 J. Kurtz, G. D. Fasman, A. Berger, and E. Katchalski, *ibid.,* p. 270.
51 H. Lenormant, A. Baudras, and E. R. Blout, *J. Am. Chem. Soc.* **80,** 6191 (1958).
52 F. Lucas, J. T. B. Shaw, and S. G. Smith, *Shirley Inst. Mem.* **28,** 77 (1955).
53 F. Lucas, J. T. B. Shaw, and S. G. Smith, *Nature* **179,** 905 (1957).

54 R. E. Marsh, R. B. Corey, and L. Pauling, *Acta Cryst.* **8**, 62 (1955).
55 R. E. Marsh, R. B. Corey, and L. Pauling, *ibid.* **8**, 710 (1955).
56 K. H. Meyer and Y. Go, *Helv. Chim. Acta* **17**, 1488 (1934).
57 K. H. Meyer and H. Mark, *Ber. deut. chem. Ges.* **61**, 1932 (1928).
58 T. Miyazawa, *J. Chem. Phys.* **32**, 1647 (1960).
59 T. Miyazawa, this book, Paper 18 (1961).
60 T. Miyazawa and E. R. Blout, *J. Am. Chem. Soc.* **83**, 712 (1961).
61 W. Moffitt, *Proc. Nat. Acad. Sci.* **42**, 736 (1956).
62 W. Moffitt and J. T. Yang, *ibid.* **42**, 596 (1956).
63 W. Moffitt, D. D. Fitts, and J. G. Kirkwood, *ibid.* **43**, 723 (1957).
64 K. D. Parker, *Biochim. et Biophys. Acta* **17**, 148 (1955).
65 K. D. Parker and K. M. Rudall, *Nature* **179**, 905 (1957).
66 L. Pauling and R. B. Corey, *Proc. Nat. Acad. Sci.* **37**, 235 (1951).
67 L. Pauling and R. B. Corey, *ibid.* **37**, 729 (1951).
68 L. Pauling and R. B. Corey, *ibid.* **39**, 253 (1953).
69 M. F. Perutz, *Nature* **167**, 1053 (1951).
70 K. M. Rudall, *Advances in Protein Chem.* **7**, 253 (1952).
71 C. Robinson, in J. T. Randall (ed.), "Nature and Structure of Collagen," Butterworths, London, 1953, p. 96.
72 C. Robinson, *Trans. Faraday Soc.* **52**, 571 (1956).
73 K. M. Rudall, "Symposium on Fibrous Proteins," Soc. Dyers Colourists, London, 1946, p. 15.
74 I. Sandeman, *Proc. Roy. Soc.* **A232**, 105 (1955).
75 V. Sasisekharan, *Acta Cryst.* **12**, 897 (1959).
76 V. Sasisekharan, *ibid.* **12**, 903 (1959).
77 E. M. Shooter, *Prog. in Biophys. and Biophys. Chem.* **10**, 195 (1960).
78 G. B. B. M. Sutherland, *Rend. inst. lombardo sci.* **89**, 67 (1955).
79 P. H. Till, *J. Polymer Sci.* **24**, 301 (1957).
80 A. Todd, in P. Alexander (ed.), "Laboratory Manual of Analytical Methods of Protein Chemistry," Pergamon Press, London, Vol. II, p. 245 (1960).

Discussion of Paper 24: Solid State Conformations

AKIYOSHI WADA

In connection with the problem of the side-chain–side-chain interaction, I would like to mention a recent work on a mixture of PBLG and PBDG made by Tsuboi and myself. Infrared absorption spectrum of a film cast from a dioxane solution of a 50%–50% mixture of high molecular weight poly-γ-benzyl-L-glutamate (PBLG) and poly-γ-benzyl-D-glutamate (PBDG) has been found to be somewhat different from the spectrum of a film of the pure PBLG or pure PBDG. The differences are observed mostly in the relative intensities and dichroic properties of the side-chain bands. The positions, intensities, and dichroic properties of the NH stretching, Amide I, Amide II, Amide III and Amide V bands (respectively at about 3290, 1650, 1550, 1315, and 613 cm^{-1}) were found to be similar. While, for example, the dichroic ratios ($\perp/\|$) of the phenyl B_2 bands at about 750 and 700 cm^{-1} greatly increase, the band shapes of the phenyl B_2 band at about 750 cm^{-1} and the ester C—O band at about 1160 cm^{-1} are altered, and the relative intensities of the two bands at 1360 and 1210 cm^{-1} (which also are probably ascribed to some side-chain vibrations) greatly increase on going from the pure L (or D) film to the DL mixture. In the X-ray diffraction diagram obtained for a fiber drawn from a viscous dioxane solution of a 50%–50% mixture of PBLG and PBDG, there is a strong meridional arc of 10.4 Å which is absent in the fiber diagram of pure PBLG or PBDG.

From these facts, it may be concluded that there is an appreciable interaction between the side chains of adjacent polypeptide chains and that the interaction between the D- and L-chains is somewhat different from that between two L-chains or between two D-chains. One of the possible explanations of the appearance of the 10.4 Å fiber period in the DL mixture fiber is given by assuming a special side-chain–side-chain interaction between the L- and D-chains running antiparallel to each other and forming a pair with a center of symmetry. If, in a pitch of α-helix, 3.5 residues (instead of 3.6 residues) are involved, such a side-chain–side-chain interaction may take place once every two pitches of the helix. The pair of α-helices with such a side-chain–side-chain interaction could be constructed by only a slight deformation of "natural" 5-turn 18-residue α-helices which have 3.6 residues per turn. This pair of α-helices (in a crystalline arrangement) would show the 10.4 Å fiber period, just as is actually observed.

TATSUO MIYAZAWA

In discussing the vibrational interactions among the amide vibrations of peptide groups, Dr. Elliott *et al.* refer to their experimental observations made on acetanilide. Apparently, however, polypeptides may not be compared with acetanilide, since an acetanilide molecule contains a benzene ring and its ring vibration at *ca.* 1600 cm^{-1} will interact strongly with the Amide I and II vibrations of the peptide group.

The 1690 cm^{-1} band in question has been assigned before to some impurities or to a combination band by Dr. Elliott *et al.* However, the experimental observations are not in accord with their assignments. The dichroic behavior as well as the shift on N-deuteration of the 1690 cm^{-1} band indicate that this band is due to the $\nu_{\|}(0, \pi)$ Amide I vibration of the antiparallel-chain extended conformation.

E. R. BLOUT

Dr. Elliott has indicated in his paper that our recent publication (E. R. Blout, C. de Lozé, S. M. Bloom, and G. D. Fasman, *J. Am. Chem. Soc.* **82**, 3787 [1960]) infers that the conformation of a polypeptide does not depend on the degree of polymerization. This interpretation of our results is not correct. Our evidence indicates that, once a polypeptide has reached a certain degree of polymerization, its conformation is determined by its amino-acid composition. In our work we have been careful to use synthetic polypeptides whose degree of poly-

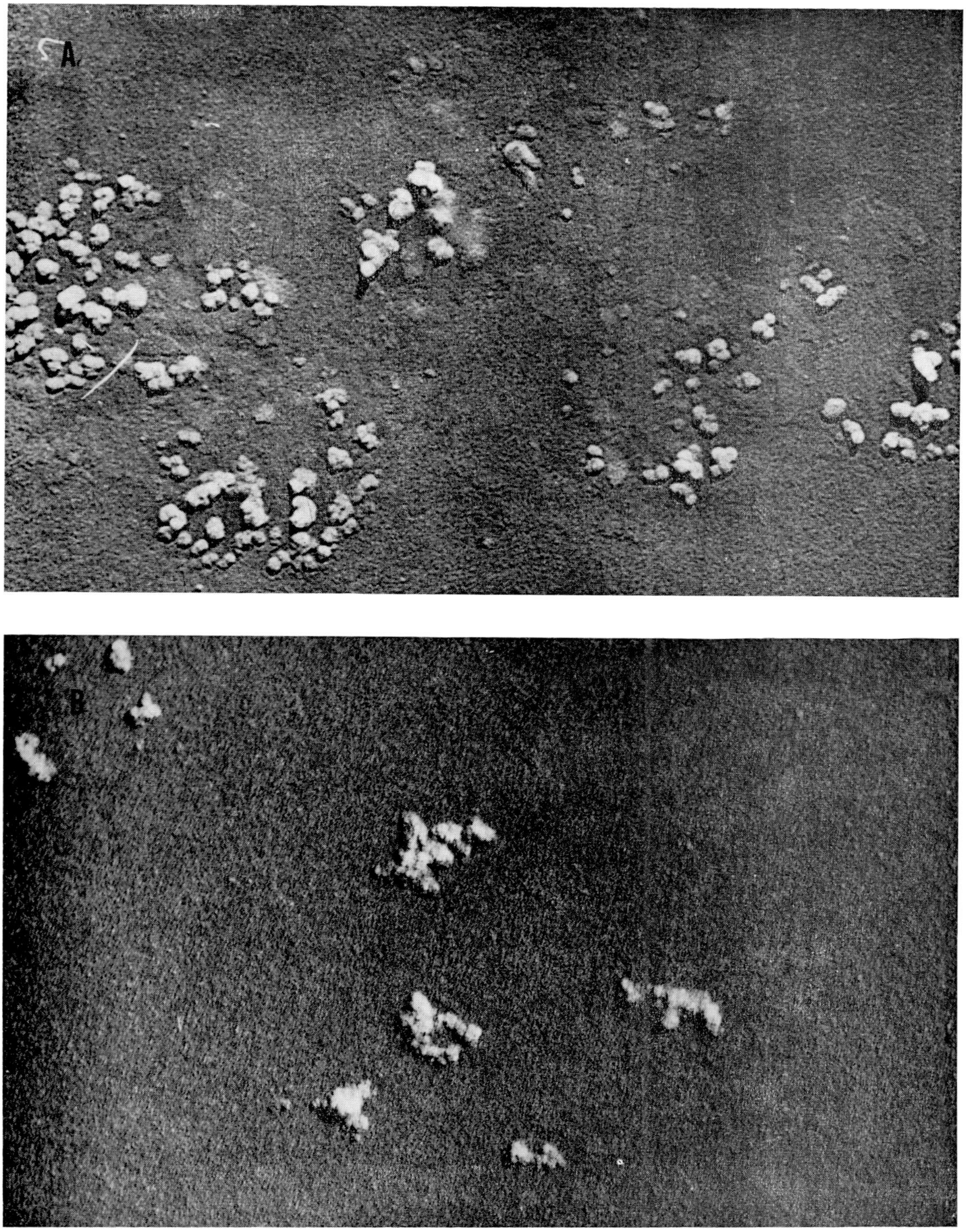

Plate 24.1. *A:* A uranium shadowed, air dried preparation of a 1 per cent solution of poly-γ-benzyl-L-glutamate 2H2 in dioxane which had been sprayed on a carbon-coated, collodion-covered electron microscope grid. The plate shows droplet pattern in which poly-γ-benzyl-L-glutamate is aggregated into irregular shaped particles. Mag. 49,000 ×.

B: A uranium-shadowed, frozen-dried preparation of a 1 per cent solution of poly-γ-benzyl-L-glutamate (2H2) in dioxane. The plate shows droplet pattern somewhat less closely aggregated than in Plate 24.1A. Mag. 47,500 ×.

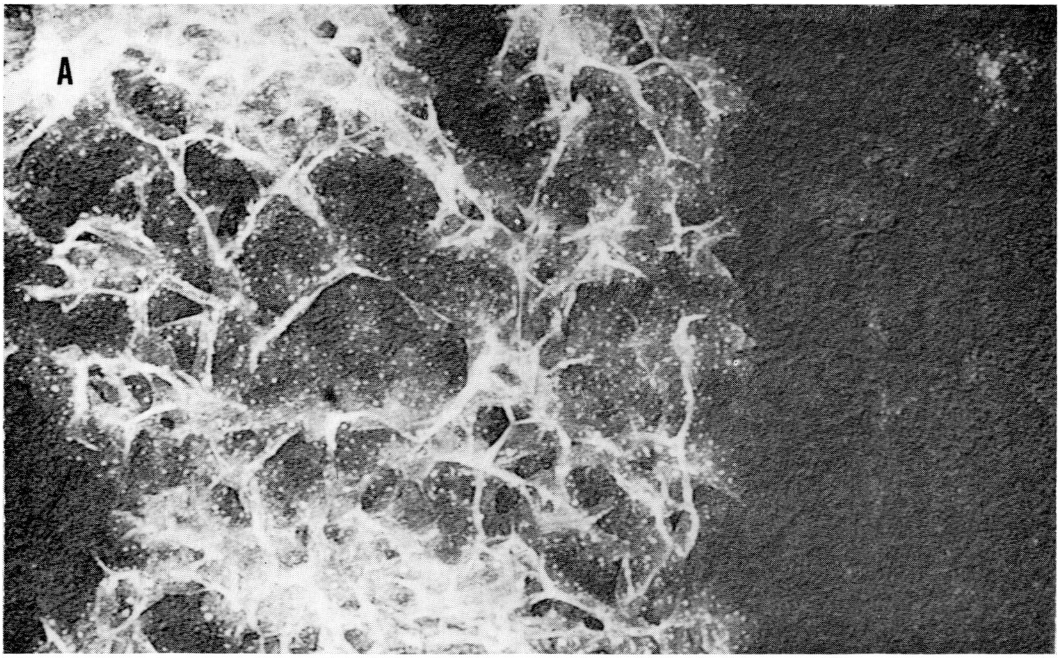

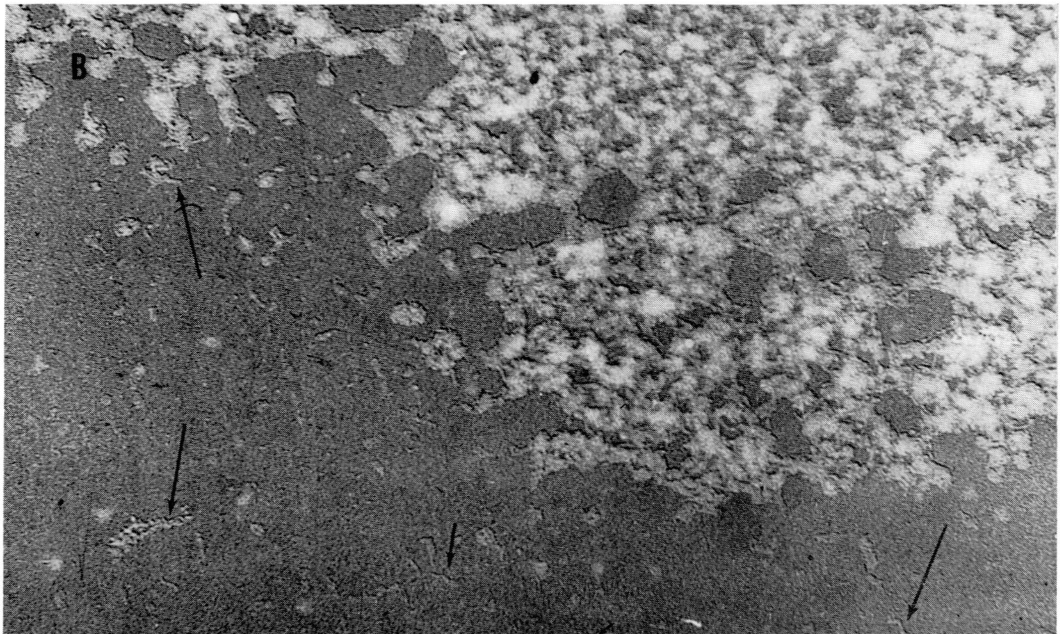

Plate 24.2. *A:* A platinum replica of an air-dried droplet of a 0.02 per cent solution of poly-γ-benzyl-L-glutamate 2N4 in *m*-cresol. The plate shows droplet edge in which some of the high molecular weight polymer is entangled in fibrous strands and some appears as unaggregated particles. Mag. 66,500 ×.

B: A uranium-shadowed preparation of a 1 per cent solution of poly-γ-benzyl-L-glutamate (2H2) in dimethylformamide which had been sprayed on a carboned collodion grid. This plate shows some particles in a linear configuration (indicated by arrows) and masses of small particles. Mag. 18,000 ×.

merization was 100 or more so that the obvious consequences of polydispersity and low molecular weight (in the range of degree of polymerization 3 to 15) would be minimized or eliminated.

I would now like to consider Dr. Elliott's result with poly-α-amino-isobutyric acid. According to the published reference (A. Elliott and B. R. Malcolm, *Trans. Faraday Soc.* **52,** 528 [1956]), unoriented films of this polypeptide show a 1.5 Å spacing and it is therefore concluded that the X-ray diffraction pattern is "undoubtedly that of an α-helix." Since poly-α-amino-isobutyric acid would fall in our classification of non-helix-forming amino acids, we consider the determination of the conformation of this synthetic polypeptide to be an important test of our hypothesis relating the amino-acid composition and conformation of polypeptides. We therefore look forward with interest to further evidence such as X-ray diffraction data on oriented fibers or infrared dichroism of oriented films of poly-α-amino-isobutyric acid which would allow the unequivocal identification of its conformation in the solid state. If, indeed, it is shown that poly-α-amino-isobutyric acid may exist as an α-helix, we would have to modify our hypothesis.

G. FASMAN

Positive b_o values in the Moffitt equation have been reported for a number of poly-α-amino acids in contrast to the negative value (−630) usually associated with the helical conformation of poly-L-glutamic acid, poly-γ-benzyl-L-glutamate, and for poly-L-alanine (W. Moffitt, *Proc. Nat. Acad. Sci.* **42,** 736 [1956]; W. Moffitt and J. T. Yang, *ibid.* **42,** 596 [1956]; R. H. Karlson, K. S. Norland, G. D. Fasman, and E. R. Blout, *J. Am. Chem. Soc.* **82,** 2268 [1960]; A. Elliott, W. E. Hanby, and B. R. Malcolm, *Nature* **180,** 1340 [1957]). The latter has been shown to exist as a right-handed helix (A. R. Downie, A. Elliott, W. E. Hanby, and B. R. Malcolm, *Proc. Roy. Soc.* **A242,** 325 [1957]). The first report of a positive b_o value for poly-L-tyrosine (A. Elliott, W. E. Hanby, and B. R. Malcolm, *Nature* **180,** 1340 [1957]) and two recent papers (A. R. Downie, A. Elliott, and W. E. Hanby, *ibid.* **183,** 110 [1959]; J. D. Coombes, E. Katchalski, and P. Doty, *ibid.* **185,** 534 [1960]) have discussed the conformation of this polymer in various solvents. The positive b_o value reported for poly-β-benzy-L-aspartate was shown to be due to a different sense of twist than that found for poly-γ-benzyl-L-glutamate, presumably a left-handed helix (E. R. Blout and R. H. Karlson, *J. Am. Chem. Soc.* **80,** 1259 [1958]; R. H. Karlson, K. S. Norland, G. D. Fasman, and E. R. Blout, *ibid.* **82,** 2268 [1960]; E. M. Bradbury, A. R. Downie, A. Elliott, and W. E. Hanby, *Proc. Roy. Soc.* **A259,** 110 [1960]).

The following data strongly indicate that the sense of helix of poly-L-tyrosine is the same as that found for poly-α-L-glutamic acid, despite its opposite b_o value.

Poly-L-tyrosine and a series of copolymers of L-tyrosine and L-glutamic acid with various mole ratios were prepared and their optical rotatory dispersions were measured in dimethylsulfoxide. The b_o for each was calculated using $\lambda_o = 212$ mμ. Fig. 24.3 shows that b_o changes linearly from −560, found for poly-α-L-glutamic acid, to a value of +494, that found for poly-L-tyrosine. The b_o found for sodium poly-L-tyrosinate in this same solvent was +263, presumably the random coil, and the

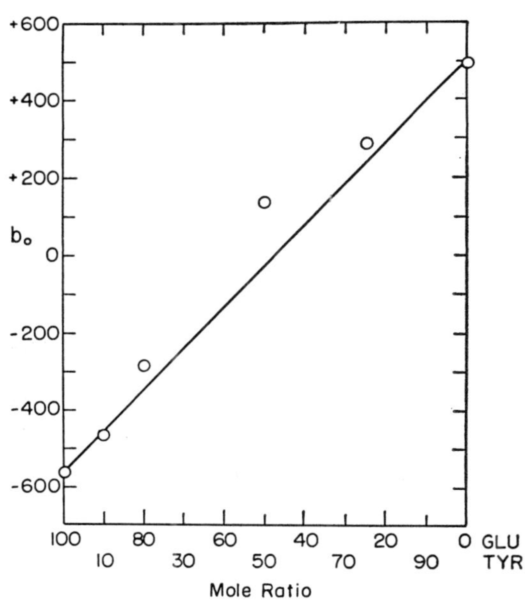

Fig. 24.3. The dependence of b_o (assuming $\lambda = 212$ mμ) on the composition of copolymers of L-glutamic acid and L-tyrosine. Measurements were made in the spectral range 365 to 578 mμ using dimethylsulfoxide as solvent.

higher value of +494 that associated with the helix. From the linear relationship between the b_o values and composition it may be concluded that the same sense of twist has been maintained throughout the series. If poly-L-tyrosine and poly-α-L-glutamic acid were helices of the opposite sense of twist, the b_o of their copolymers should change markedly over a relatively narrow range of copolymers, representing a region of transition from one helix sense to the other, as found for copolymers of β-benzyl-L-aspartate and γ-benzyl-L-glutamate (R. H. Karlson, K. S. Norland, G. D. Fasman, and E. R. Blout, *J. Am. Chem. Soc.* **82,** 2268 [1960]; E. M. Bradbury, A. R. Downie, A. Elliott, and W. E. Hanby, *Proc. Roy. Soc.* **A259,** 110 [1960]). Thus poly-L-tyrosine and poly-γ-benzyl-L-glutamate have the same sense of helix despite their opposite b_o values.

C. DE LOZÉ and E. R. BLOUT

Our results on the determination of the infrared spectra and infrared dichroisms of polypeptide films cast from various solvents have led us to the conclusion that the conformations of polypeptides (having a weight average degree of polymerization greater than 100) do not depend on the solvent from which the sample is cast. For example, oriented films of poly-γ-benzyl-L-glutamate and poly-β-benzyl-L-aspartate show exactly the same absorption frequencies and dichroisms when cast either from chloroform or from trifluoroacetic acid or dichloroacetic acid; the frequencies and dichroisms are characteristic of helical conformations. These examples were chosen because chloroform is known to interact weakly with the peptide hydrogen bonds which stabilize the helix, while trifluoroacetic acid and dichloroacetic acid interact with peptide hydrogen bonds and transform these two polypeptides into random-coil conformations in solution. Our results indicate that when the solvent is evaporated the polypeptides regain their "natural" conformation.

Preliminary results with other polypeptides indicate that this phenomenon is general and that when a polypeptide of high enough molecular weight is soluble in several solvents, its conformation in the oriented solid state is always the same when the solvent has completely evaporated.

DONALD E. SLAGEL

There has been some interest expressed as to what contribution electron microscopy could make to the study of polypeptide configuration. I would like to describe some of the results which we have obtained in Dr. Stahmann's laboratory. We have worked with poly-γ-benzyl-L-glutamate deposited from solvents that cause an association of the polypeptide and solvents in which it is not associated. These included the four solvents, dioxane, chloroform, *m*-cresol, and dimethylformamide. When 1 per cent solutions of poly-γ-benzyl-L-glutamate in dioxane or chloroform (associating solvents) were sprayed on the electron microscope grid and air dried, the polymer material was aggregated into spherical or irregularly shaped particles with very few if any small particles lying between the aggregates. This is illustrated in Plate 24.1A, where polypeptide aggregates arranged in several droplet patterns can be seen. We think that the structures seen were not formed through an aggregation during drying but that this is material already associated in the solution. We further considered that the contribution to the aggregation of these particles by surface tension of the solvent on air drying might be eliminated by freeze drying. In the freeze-drying experiments, the electron microscope grids upon which the polypeptide solution was sprayed were cooled to −190° C. by liquid air and then warmed up slowly over a period of 3 to 4 hours at a pressure of 0.01 mm of mercury. The results are illustrated in Plate 24.1B. The polypeptide molecules were still aggregated, although into a somewhat less compact form than in the air-dried preparation, Plate 24.1A. These results strongly suggest that much of the molecular association was present in the solution before drying on the electron microscope grid.

When 1 per cent solutions of poly-γ-benzyl-L-glutamate in *m*-cresol or dimethylformamide were air dried, the particles were pulled together into the dried down droplets illustrated in Plate 24.2. A comparison of the morphology of the droplet as a whole between the chloroform or dioxane solutions (Plate 24.1) and the *m*-cresol (Plate 24.2A) or dimethylformamide (Plate 24.2B) solutions reveals an interesting difference. In both the *m*-cresol and dimethylformamide the droplet itself was made up of an aggregated mass of material in which many of the small units of the aggregate can be seen. This suggests to us that the polypeptide molecules were largely homogeneously dispersed in the solution but aggregated as the drop dried. In the dioxane and chloroform solutions the droplet itself was composed of groups of larger aggregates in which very few small units can be seen. This suggests to us that the polypeptide molecules were already aggregated in these solutions, indicating molecular association in chloroform or dioxane.

Doty, Bradbury, and Holtzer (*J. Am. Chem. Soc.* **78,** 947 [1956]) showed several years ago, from the dependence of reduced viscosity on concentration, that the molecular association of poly-γ-benzyl-L-glutamate in dioxane or chloroform is particularly great; in contrast, in *m*-cresol or dimethylformamide molecular association is absent. Since the molecular association of these polymers interpreted from electron microscopy correlates well with the molecular association inferred from the hydrodynamic properties in solution, these experiments may illustrate a case in which the conformation in the solid state was influenced by that in the solution from which the solid was deposited. In the electron micrographs of poly-γ-benzyl-L-glutamate air dried from a dimethylformamide solution, Plate 24.2A, a number of particles in a linear configuration can be seen near the edge of the droplet. Many of these appear to be composed of several strands, possibly three or four, which are entwined together (perhaps in a helical fashion) to form a larger linear rod. This is indicated with arrows in Plate 24.1B.

The X-ray diffraction results of Luzzati suggest that poly-γ-benzyl-L-glutamate may exist in two possible configurations, the paracrystalline or the complex, depending on the solution concentration. Luzzati further suggests that the complex phase is made up of particles in a compound helix of three α-helices. Because the concentration in the electron micrographs is unknown since the concentration is increasing as the droplet dries, and the exact width is uncertain because the shadow-to-height ratio for this electron micrograph is not known, it is difficult to evaluate the configuration seen in the electron micrograph with those obtained from X-ray diffraction data. However, one fact is clear: the width of particles which often appear to be made up of several strands is many times greater than that of a single α-helix and may therefore represent a coiled configuration of three or more strands.

PAPER 25

The Dependence of the Conformation of Polypeptides and Proteins upon Amino-Acid Composition

ELKAN R. BLOUT

During the last several years there have been many reports on the conformations in which synthetic polypeptides may exist in the solid state and in solution. The conclusions as to conformation have been obtained mainly from X-ray diffraction and infrared spectroscopy (for solid samples) and from optical rotatory dispersion and infrared flow dichroism measurements (for solutions). The recognition that helices are fundamental structures of polypeptides and proteins (*1*) of course does not preclude the possibility that these materials could exist in other conformations, such as extended (β) forms and random coils. However, with the explicit recognition of the α-helix as the fundamental structural unit, there has been some tendency to assume that synthetic polypeptides, and by inference proteins, exist only as helices, β-structures, and random chains, or mixtures of these structures.

In a recent communication (*2*) we have suggested that the conformations of synthetic homopolypeptides in the solid state depend on their amino-acid composition. In this paper I shall review the evidence for this hypothesis, indicate some preliminary data obtained from solution measurements, and show how this hypothesis has been used to predict the conformations of previously unknown synthetic polypeptides.

As a result of many measurements of the infrared dichroism of oriented solid samples of synthetic polypeptides, it has become apparent to us over the last few years that some of these materials always appear to prefer helical (α) conformations, whereas we were never able to obtain samples of other synthetic polypeptides showing the spectral properties of this conformation. Instead, they showed either β (extended) structures or no dichroism, characteristic of random chains. Some typical results for oriented solid samples are shown in Table 25.1.

The conformation hypothesis

On the basis of these (and other) measurements we proposed that there are two classes of α-amino acids: those which, when made into homopolypeptides, tend to form helical structures, and the second class, those which formed either random or β structures. We further proposed that the lack of helix formation of certain synthetic polypeptides may be due to at least two causes: (1) steric factors; (2) hydrogen-bonding factors. In the first class of non-helix-forming polypeptides we include those which are disubstituted by alkyl groups on the β-carbon atom—poly-L-valine is a typical member of this class. In the second class of non-helix-forming polypeptides are those which have a hetero atom, for example, oxygen or sulfur, attached to the β-carbon atom and thus may be capable of forming hydrogen bonds with the backbone amide group of the polypeptide chain or of forming side-chain–side-chain hydrogen bonds. The tendency to form this type of hydrogen bond may be sufficiently strong to overcome the *intra*chain backbone amide-amide hydrogen bonds of helical conformations. This is the hypothesis which we put forth, and for which we presented evidence in our preliminary communication.

If this is a useful hypothesis then it should

Table 25.1—Molecular and infrared spectral properties of various synthetic polypeptides (see Ref. *2*)

Polypeptide	$\eta_{sp/c}$*	Estimated† DP$_w$	Estimated† MW$_w$	Amide I‡ Frequency (cm^{-1})	Amide I‡ Dichroism	Amide II‡ Frequency (cm^{-1})	Amide II‡ Dichroism	Structure‡
Poly-L-valine	0.29	190	19,000	1638	⊥	1545	‖	β
Poly-L-leucine	0.11	35	3,900	1650	‖	1543	⊥	α
Poly-*S*-methyl-L-cysteine	0.24	140	16,000	1632	⊥	1540, 1525	‖	β
Poly-*S*-benzyl-L-cysteine	0.49	365	77,000	1632	⊥	1524	‖	β
Poly-L-methionine	0.22	135	18,000	1648	‖	1540	⊥	α
Poly-*O*-acetyl-L-serine	0.14	130	16,800	1637	⊥	1520	‖	β
Poly-L-serine	0.10§	130	11,300	1653	—	1525–35	—	random**
Poly-β-benzyl-L-aspartate	0.30	190	39,000	1668	‖	1563	⊥	α
Poly-γ-benzyl-L-glutamate	0.15	100	22,000	1645	‖	1542	⊥	α

* The reduced specific viscosities were determined at a concentration of 0.2 per cent in either dichloroacetic acid or trifluoroacetic acid to assure that the polypeptides were in the random form.

† The weight average degrees of polymerization, (DP$_w$), and the weight average molecular weights, (MW$_w$), were estimated from the correlation of the viscosities of the random form of poly-γ-benzyl-L-glutamate with light scattering and sedimentation measurements (see P. Doty, J. H. Bradbury, and A. M. Holtzer, *J. Am. Chem. Soc.* **78**, 947 [1956]; J. C. Mitchell, A. E. Woodward and P. Doty, *ibid.* **79**, 3955 [1957].)

‡ Of oriented films.

§ Intrinsic viscosity in water.

** This sample has been shown to contain ~5 per cent of the D-isomer. A more optically homogeneous preparation has recently been prepared and this material appears to be in the β-conformation (Z. Bohak and E. Katchalski, private communication).

have predictive value, and we should be able to predict the conformations of unknown synthetic polypeptides by the use of this hypothesis. We shall see that this is indeed the case, but first it is worthwhile to consider other factors which might affect the conformations of synthetic polypeptides.

Molecular weight

From the nature of the α-helical structure it is apparent that this structure will not be attained by materials having very low molecular weights, that is, low degrees of polymerization (DP's). In fact several years ago it was shown that the conformation of poly-γ-benzyl-L-glutamate depended directly on the molecular weight, that is, DP of the polypeptide chain (*3*). Also it was shown that, as the polymer chain grew in length during the polymerization of γ-benzyl-L-glutamate, the conformation changed from β to an α structure (*4*). In addition, there have been various estimates of the degree of polymerization necessary for the α-helical structure to be formed which ranged from about six units to twenty or more (*5, 6, 4, 7*). It was thus clear that in order to study the intrinsic conformations of synthetic homopolypeptides one should use materials having degrees of polymerization of over twenty and presumably of the order of one hundred so as to eliminate the effect of molecular weight on conformation. Thus one may eliminate the possibility that any lack of helix was due to an inadequate number of residues in the peptide chain.

As can be seen from the data presented in Table 25.1, this is what we have done in our work. The materials examined by us had degrees of polymerization (~100 or more) well above that necessary for the formation of helical structures. Using polypeptides with these DP's, we find that the conformation is independent of the molecular weight. It is also apparent from the foregoing brief summary and several reports in the literature that if one examined the conformations of polypeptides having DP's in the transition range before the helical structures were stable, that is, in the range 6 to ~30 residues, then one could be dealing with mixtures of helical and non-helical (or α and β) structures. In addition, since synthetic polypeptide preparations are often inhomogeneous in regard to molecular weight it is possible that physical or chemical treatments could transform some of these "transition range" polypeptides from one conformation to another.

Ionic side chains

In addition to the effect of molecular weight on conformation, it is, of course, well known that potentially ionic polypeptides such as poly-L-glutamic acid and poly-L-lysine can undergo conformational changes as a result of the ionization or deionization of the side chains. These conformational changes, which have been shown to exist both in the solid state and in solution (*8, 9*), appear to be independent of the degree of polymerization and can be attributed to the ionic repulsion of the side chains being sufficient to overcome the internal hydrogen bond stabilization of the helix. It also has been reported that polypeptides such as poly-L-lysine hydrohalides (*10, 11*) and poly-sodium glutamate (*12*) can undergo reversible $\alpha \rightarrow \beta$ conformational transitions. As far as we can determine at present, these transitions are a result of the effect of water molecules on the side-chain interactions and occur with high DP material as well as low DP material. This type of water-induced conformational change, however, appears to be limited to those polypeptides having ionic side chains.

Solvent interactions

Since, as indicated above, the stability of helical and other structures is dependent on hydrogen bonding, it is not surprising that synthetic polypeptides can have different conformations in the solid state than they do in solutions of strong hydrogen-bonding solvents. It is well known, for example, that strong acids such as dichloroacetic or trifluoroacetic acid tend to denature proteins and disrupt α-helical structures in polypeptides as evidenced by changes in rotatory dispersion data. On dissolution of a series of helix-forming poly-α-amino acids (e.g., poly-β-benzyl-L-aspartate and poly-γ-benzyl-L-glutamate) in strong hydrogen-bond-breaking solvents, and subsequent orientation during evaporation of solvent, dichroic infrared spectra indicative of α-helical structures are obtained (*13*). Thus, it seems that upon removal of the solvent in which the polypeptide is in the random form (in these cases trifluoroacetic acid), helix-forming synthetic polypeptides tend to return to the α-helical conformation. On the other hand, if polydisperse samples of synthetic polypeptides with lower molecular weight are used, then apparent changes in the conformation have been observed upon treatment with such solvents (*14*).

Classification of polypeptides

With these factors in mind we now can return to a consideration of the original hypothesis, namely that synthetic polypeptides of sufficiently high molecular weight tend to form helical structures or not, depending on the composition of their side chains. In Table 25.2 we list and classify polypeptides of high molecular weight whose solid state conformations have been determined (or predicted on the basis of the hypothesis). Using the postulate stated above we predicted that poly-L-isoleucine should not form helical structures, since isoleucine is a β-dialkyl-substituted amino acid. Recent experiments have confirmed this prediction (*15*). Similarly, it was predicted that poly-*O*-acetyl-L-threonine would also be non-helical on the basis of our hypothesis and this has been confirmed by experiment (*16*). Finally, one would also predict that poly-*O*-acetyl-L-homoserine should be helical, since it is a γ-carbon heteroatom-substituted amino acid, and experiments are now in progress to determine whether this is indeed the case.

In addition, we have recently been performing optical rotatory dispersion measurements on solutions of various synthetic polypeptides and have confirmed the fact that poly-L-methionine is helical in chloroform solution (*17*). No such data could be obtained with its lower homolog, poly-*S*-methyl-L-cysteine, since this material is insoluble in chloroform and all other weakly hydrogen-bonding solvents.

Finally, we have begun to examine the conformations of heteropolypeptides, that is, polypeptides composed of two or more amino acids. Although only relatively few experiments have been made so far with such materials, it appears that the general hypothesis of the dependence of conformation upon

Table 25.2—Conformational classification of α-amino acids

α-Helix forming	α-amino acid side chains (R) $-NH-CH(R)-C(=O)-$	Non α-helix forming	α-amino acid side chains (R) $-NH-CH(R)-C(=O)-$
Alanine	$-CH_3$	(A) *Steric Reasons*	
Aspartic acid esters	$-CH_2-C(=O)-$O-alkyl	Valine	$-CH(CH_3)-CH_3$
Glutamic acid	$-CH_2-CH_2-C(=O)-OH$	Isoleucine	$-CH(CH_3)-CH_2-CH_3$
Glutamic acid esters	$-CH_2-CH_2-C(=O)-$O-alkyl	(B) *Other Reasons*†	
Leucine	$-CH_2-CH(CH_3)-CH_3$	*O*-acetyl-serine	$-CH_2-O-C(=O)-CH_3$
Lysine	$-CH_2-CH_2-CH_2-CH_2-NH_2$	Serine	$-CH_2-OH$
Methionine	$-CH_2-CH_2-S-CH_2$	*S*-methyl cysteine	$-CH_2-S-CH_3$
Phenylalanine	$CH_2-C_6H_5$	*O*-acetyl threonine	$-CH(CH_3)-O-C(=O)-CH_3$
Tyrosine	$-CH_2-C_6H_4-OH$	*Threonine	$-CH(CH_3)-OH$
*Norleucine	$-CH_2-CH_2-CH_2-CH_3$		
*Norvaline	$-CH_2-CH_2-CH_3$		
**O*-acetyl homoserine	$-CH_2-CH_2-O-C(=O)-CH_3$		

* Predicted on basis of hypothesis, but not yet proved.
† Of course sarcosine, proline, and hydroxyproline must be included in this class.

amino-acid composition is tenable, provided one recognizes the limitations of molecular weight, ionic side chains, and solvent noted above.

Acknowledgment

Much of the work summarized here resulted from the collaborative efforts of my colleagues Drs. S. M. Bloom, C. de Lozé, and G. D. Fasman, to whom I am duly grateful, as is apparent from the references cited.

References

1 L. Pauling and R. B. Corey, *Proc. Nat. Acad. Sci.* **37,** 241 (1951).
2 E. R. Blout, C. de Lozé, S. M. Bloom, and G. D. Fasman, *J. Am. Chem. Soc.* **82,** 3787 (1960).
3 E. R. Blout and A. Asadourian, *ibid.* **78,** 955 (1956).
4 M. Idelson and E. R. Blout, *ibid.* **79,** 3948 (1957).
5 P. Doty and R. D. Lundberg, *ibid.* **78,** 4810 (1956).
6 J. C. Mitchell, A. E. Woodward, and P. Doty, *ibid.* **79,** 3955 (1957).

7 See, for example, M. Goodman and E. E. Schmitt, *Ann. N. Y. Acad. Sci.* **88,** 669 (1960).
8 E. R. Blout and M. Idelson, *J. Am. Chem. Soc.* **78,** 497 (1956); P. Doty, A. Wada, J. T. Yang, and E. R. Blout, *J. Polymer Sci.* **23,** 851 (1957).
9 J. Applequist, Ph.D. thesis, Harvard University (1959).
10 E. R. Blout and H. Lenormant, *Nature* **179,** 960 (1957).
11 A. Elliott, B. R. Malcolm, and W. E. Hanby, *ibid.* **179,** 963 (1957).
12 H. Lenormant, A. Baudras, and E. R. Blout, *J. Am. Chem. Soc.* **80,** 6191 (1958).
13 C. de Lozé and E. R. Blout, unpublished results.
14 C. H. Bamford, A. Elliott, and W. E. Hanby, "Synthetic Polypeptides," Academic Press, Inc., New York, 1956, pp. 302–10.
15 C. de Lozé, G. D. Fasman, and E. R. Blout, unpublished data.
16 M. Volz, G. D. Fasman, and E. R. Blout, to be published.
17 S. M. Bloom, G. D. Fasman, C. de Lozé, and E. R. Blout, *J. Am. Chem. Soc.* **84,** 458 (1962).

Discussion of Paper 25: Dependence of Conformation upon Amino-Acid Composition

C. H. BAMFORD

In the latter part of his paper Dr. Blout seems to imply that only polypeptides with molecular weights in the "transition range" may have their conformations changed by physical or chemical treatments. This is emphatically not so. In our early experiments we repeatedly demonstrated the transformations $\alpha \rightarrow \beta \rightarrow \alpha$ on high molecular weight polypeptides ($\bar{P}_n = 500$, approximately) effected by such treatments; indeed such methods provided (and still provide) the only route to the β-forms of many polypeptides. (A summary of the early work is given by Bamford, Elliott, and Hanby in "Synthetic Polypeptides," Academic Press, New York, 1956.) The transformations were demonstrated by X-ray, infrared and solubility properties, and by suitable choice of reagent could be confined to the amorphous regions of the polymers (Bamford, Hanby, and Happey, *Nature* **166,** 829 [1950]). To emphasize this point I should mention that when these experiments were done we were aware of the fact that a sufficiently low degree of polymerization may favor the adoption of the β-conformation (Bamford, Hanby, and Happey, *Proc. Roy. Soc.* **A206,** 407 [1951]). The existence of many high molecular weight polypeptides in both α and β forms is an undisputable fact. Naturally if one such conformation is sufficiently unstable for steric (or other) reasons, this form is unlikely to be attained by any treatment.

These early experiments also demonstrated that the conformation of the solid polymer obtained from solution depends critically on the precise method of drying (e.g., whether by precipitation or evaporation) and on the nature of the solvent. In many cases formic acid leads to a high content of the β form (of high polymers).

PART V

Biological Properties of Poly-α-Amino Acids

PAPER 26

Preparation, Properties, and Applications of Some Water-Insoluble Derivatives of Proteolytic Enzymes

EPHRAIM KATCHALSKI

The conversion of enzymes into water-insoluble products possessing specific catalytic activity is of interest since such "water-insoluble enzymes" can readily be removed from the reaction mixture, and can be used for the preparation of columns with enzymatic activity. If stable, they can be employed repeatedly to induce specific chemical changes in relatively large amounts of substrate. A short incubation of substrate with a water-insoluble enzyme may facilitate the detection and isolation of some important intermediates formed during an enzymatic reaction.

Adsorption techniques were applied to combine enzymes with insoluble carriers (*1, 2*). These procedures led, however, to partial denaturation and resulted in preparations from which the reversibly adsorbed enzyme was liberated while in contact with substrate. In order to strengthen the linkage between enzyme and carrier, attachment by chemical covalent bonds seemed preferable. Such links should obviously be carried out through functional groups nonessential for enzymatic activity.

Various attempts to attach antigens or antibodies by chemical bonds to water-insoluble high molecular weight compounds have been reported (*3, 4*). The immunologically active columns of this kind were used in the purification of both antibodies and antigens. Only few papers have been published until now on the chemical binding of enzymes to water-insoluble polymers. The coupling of pepsin and carboxypeptidase with a diazotized poly-*p*-aminostyrene was reported by N. Grubhofer and L. Schleith (*5*). The chemical properties and the enzymatic activity of the water-insoluble products were, however, poorly characterized. A. Bar-Eli and E. Katchalski (*6*) have recently described a water-insoluble trypsin derivative obtained by the coupling of polytyrosyl-trypsin with a diazotized copolymer of *p*-aminophenylalanine and leucine. The composition, stability, and mode of action of the water-insoluble enzyme on low and high molecular weight substrates have been tested. The synthesis of diazobenzylcellulose and carboxymethylcellulose derivatives of trypsin and chymotrypsin was given by Mitz and Summaria (*7*). Carboxymethylcellulose azide was used in the preparation of the carboxymethylcellulose derivatives. The water-insoluble enzymes described were found to be more stable than the corresponding water-soluble enzymes.

In the following a description will be given of the preparation, properties, and applications of some water-insoluble derivatives of proteolytic enzymes which have been prepared in our laboratory during the last two years. Wherever possible, their use in protein and polypeptide research will be illustrated. To insure high enzymatic activity different synthetic procedures were employed in the preparation of the various water-insoluble proteolytic enzymes. A direct coupling of the enzyme with the chemically reactive polymer was carried out in the case of papain. A water-insoluble pepsin was obtained by reacting pepsinogen with a water-insoluble polymer and subsequent acid activation of the insoluble pro-enzyme. The attachment of the enzyme to the carrier by polymeric side chains which ensure free movement of the catalyst molecules in the reaction mixture was used in the case of trypsin and chymotrypsin. In all the enzymatic preparations investigated a copolymer of *p*-amino-DL-phenylalanine and L-leucine was used as carrier.

Water-insoluble papain (8)

Since papain is known to have a relatively high content of tyrosine (13.7 per cent) (*9*) and approximately two-thirds of the molecule are nonessential for its activity (*10*), the enzyme was directly coupled with the diazotized copolymer of *p*-amino-DL-phenylalanine and L-leucine. The water-insoluble product obtained retained most of the activity of the original papain. It showed hydrolytic activity towards low and high molecular weight substrates parallel to that of the starting water-soluble papain. The insoluble papain derivative, which could easily be freed of reducing agents, was found to act on proteins in the presence of negligible amounts of cysteine. It was thus possible to carry out the digestion of proteins without reduction of substrate or products. Immune γ-globulin hydrolyzed briefly under such conditions, although susceptible to fragmentation by reduction, showed no alteration in its rate of sedimentation or in its ability to precipitate antigen. Antigen-antibody precipitates containing the papain-treated antibody dissolved on reduction to yield soluble complexes between antigen and fragments of antibody.

Preparation and properties of water-insoluble papain.—A colymer of L-leucine and *p*-amino-DL-phenylalanine hydrobromide (molar residue ratio 2.5:1) was prepared according to A. Bar-Eli and E. Katchalski (*6*). The copolymer was diazotized by nitrous acid and the water-soluble polydiazonium salt was precipitated by adding alkali to *p*H 8.0. Coupling with papain was effected at *p*H 7.6 and a temperature of 4°. The water-insoluble coupling product was stored at 4° under an aqueous solution of *p*H 6.0, 0.005 M in cysteine and 0.002 M in ethylene diamine tetra-acetic acid (EDTA).

The protein content (28 mg) of the water-insoluble papain (100 mg) was estimated by determining the amount of valine liberated on acid hydrolysis. The enzymatic activity of the insoluble papain was assayed by following the kinetics of the hydrolysis of benzoyl-L-arginine ethyl ester (BAEE) at *p*H 6.0 and 30°. By this assay, 100 mg of water-insoluble papain was found to equal in activity 20 mg of crystalline enzyme. Taking into consideration the protein content of the insoluble enzyme, an enzymatic activity of about 70 per cent was calculated for the bound protein.

Storage of the water-insoluble papain at 4° for a period of one month resulted in a 30 per cent loss in activity. When the enzyme preparation was heated for 30 seconds in a boiling water bath, the activity towards BAEE decreased to approximately a fifth of its original value.

A study of the activity of water-insoluble papain towards BAEE as a function of *p*H revealed a broad optimum in the region *p*H 6.0 to 7.5. A similar *p*H range for the optimal activity of water-soluble papain was observed by J. R. Kimmel and E. L. Smith (*11*), who used benzoyl-L-arginine amide as substrate.

Enzymatic hydrolysis of some native proteins.—The synthetic substrate BAEE could not be hydrolyzed by saline-washed, water-insoluble papain in the presence of EDTA unless cysteine was added. When gelatin, on the other hand, was used as substrate it was observed that the rate of hydrolysis was not markedly affected on decreasing the concentration of cysteine from 0.005 M to 0.0001 M. This finding suggested the possibility of hydrolyzing proteins by papain under conditions which do not cause protein reduction. Actually it was found that native serum albumin and hen ovalbumin, as well as gelatin, are digested appreciably on incubation with water-insoluble papain. The activity of the insoluble papain was similar to that of the soluble enzyme towards each of the substrates tested.

Hydrolysis of γ-globulin by water-insoluble papain.—Both titration and ninhydrin assay indicated that five-minute incubation of rabbit γ-globulin with insoluble papain at 37° suffices to hydrolyze approximately 3 to 5 peptide bond per molecule of globulin. Such short times of proteolysis did not result in any appreciable change in the original sedimentation constant (6.20 *S*). However, when the papain-free hydrolysate was treated with cysteine or thioglycolic acid the γ-globulin was completely transformed into fragments with a sedimentation constant of 3.3 *S* similar to those described by R. R. Porter (*12, 13*). Chromatography of the reduced 5-minute papain digest on carboxymethylcellulose resulted in 84 per cent recovery of the hydrolysate in three peaks in a weight ratio of 1:1:1.1 corresponding to fractions I, II, and III described by Porter (*13*).

The above results illustrate clearly that the cleavage of γ-globulin by cysteine-activated papain demonstrated by Porter (*12, 13*) consists of two separate stages: the first limited proteolysis, and the second reduction. The enzymatic hydrolysis of 3 to 5 peptide bonds per protein molecule suffices to render the γ-globulin susceptible to fragmentation by reduction. Such limited proteolysis does not result in any change in sedimentation, nor does it lead to the liberation of low molecular weight peptides. Recent data of Dr. Morawiecki (unpublished data from this laboratory) indicate that the optical rotatory properties of the intact γ-globulin closely resemble those of the papain-treated protein. Even after fragmentation by reduction the optical rotatory properties remained practically unchanged. The cleavage of rabbit γ-globulin into the above three fractions, under the conditions specified, does not seem therefore to affect markedly the secondary structure of the protein.

Immunological characteristics of enzymatically digested anti-ovalbumin.—The ability of rabbit anti-ovalbumin globulin to precipitate its specific antigen, after brief incubation with the water-insoluble papain in the absence of cysteine, was found to be similar to that of the intact antibody. The molar ratio of antibody to antigen in precipitates formed at the equivalence point is about 2.05:1. The number of combining sites that may be calculated for the ovalbumin antigen in this case is therefore 4.1.

In order to investigate the possibility of reducing the susceptible bonds of enzymatically digested antibody, even after it had been precipitated with antigen, the effect of reducing agents on such precipitates was tested. Complete dissolution of the precipitate occurred on incubation with thioglycolic acid. Precipitates formed by intact γ-globulin were unaffected under similar conditions.

The solublized precipitate gave two peaks in the ultracentrifuge with sedimentation coefficients of 9.4 *S* and 3.64 *S*. The ratio between the areas under these peaks was approximately 2:1. Since the number of combining sites calculated for hen ovalbumin is about 6.5, it was suggested that most of the material present in the solution (9.4 *S*) consists of molecular aggregates each composed of several antibody fragments bound to one antigen molecule. The slowly sedimenting material most likely represents the fragments of antibody which are devoid of any combining site.

Soluble molecular complexes of antigen and antibody fragments.—Dr. Cebra has now developed in our laboratory a procedure for the separation of antigen-antibody fragment complexes (9.4 *S*) from the immunologically inert part of the γ-globulin molecule. He thus obtained the "water-soluble complexes" in pure form and was able to determine their molecular weight and composition. For a soluble complex derived from hen ovalbumin and an enzymatically treated rabbit antibody, for example, a sedimentation constant of $s_{20,w} = 9.32$ *S* and a diffusion coefficient of $D_{20} = 3.5 \times 10^{-7}$ cm^2/sec was obtained. From these a molecular weight of 247,000 was calculated. The molecular weight of ovalbumin is 43,000, while that of the immunologically active antibody fragments is 50,000 (*13*). Since each of the soluble molecular complexes contains only one antigen molecule, it has been concluded that the complex consists on the average of 4.08 antibody fragment molecules attached to one antigen molecule. This picture is in good agreement with the number of combining sites of the ovalbumin, 4.1, derived from the molar ratio of antigen to antibody in precipitates formed at the equivalence point.

The soluble complexes discussed could be dissociated into their original components at neutral *p*H in the presence of detergents such as sodium dodecyl sulfate. Dissociation occurred also on lowering the *p*H to 4.0. These complexes thus provide a water-soluble reversible system most suitable for the evaluation of the nature of the binding forces between antigen and antibody.

Water-insoluble pepsin (14)

Pepsin is rapidly inactivated at neutral or alkaline *p*H values. The synthesis of the water-insoluble derivative of this enzyme was, therefore, carried out via the more stable zymogen.

Pepsinogen was coupled with the diazotized copolymer of *p*-amino-DL-phenylalanine and L-leucine (see ref. *6*) in 0.05 M phosphate buffer of *p*H 7.2 at 4° and the water-insoluble pepsinogen was isolated and stored in the cold. The insoluble zymogen contained approximately 10 mg of protein per 100 mg of dry solid as estimated by determining chromatographically the amount of valine liberated on acid hydrolysis. Activation to water-insoluble pepsin was effected with hydrochloric acid at room temperature. The peptic activity towards hemoglobin of the bound protein equalled, in the various water-insoluble pepsin preparations, 3 per cent to 5 per cent of that of an equivalent amount of intact pepsin.

The stability and mode of action of the water-insoluble pepsin towards low and high molecular weight substrates is now being investigated.

Water-insoluble trypsin (6)

Preliminary experiments have indicated that trypsin is readily inactivated by the diazonium salts of *p*-aminobenzoic acid or *p*-amino-DL-phenylalanine. Water-insoluble products of low enzymatic activity were also obtained on coupling the enzyme with diazotized poly-*p*-aminophenylalanine (*15*) or diazotized copolymer of *p*-aminophenylalanine and leucine. The direct binding of trypsin to a water-insoluble carrier via azo bonds was therefore abandoned and a different route for the preparation of the water-insoluble enzyme was worked out. A polytyrosyltrypsin containing tyrosyl peptide side-chains was prepared by initiating the polymerization of *N*-carboxy-L-tyrosine anhydride with trypsin. The trypsin derivative thus obtained was water-soluble and retained full enzymatic activity. On coupling with a diazotized copolymer of *p*-amino-phenylalanine and leucine it yielded the required water-insoluble trypsin.

Polytyrosyltrypsin.—In the preparation of this trypsin derivative the general procedure of Stahmann and Becker (*16, 17*) for the synthesis of polypeptidyl proteins was followed. Crystalline trypsin was dissolved in phosphate buffer, *p*H 7.2, and the solution was mixed with *N*-carboxy-L-tyrosine anhydride in dioxane. After 24 hours at 2°, the reaction mixture was dialyzed against a large volume of dilute hydrochloric acid, and then lyophilized. The water-soluble polytyrosyltrypsin obtained was free of monomeric tyrosine and of low molecular weight tyrosine peptides. A comparison of the ultraviolet absorption of polytyrosyltrypsin at 293.5 mμ in 0.1 N sodium hydroxide with that of trypsin revealed that the tyrosine content of the enzyme had increased from 2.8 per cent to 9.0 per cent. The average chain length ($\bar{n} \simeq 2$) of the attached tyrosine peptides was calculated after determining the number of free ϵ-amino lysine groups by dinitrophenylation. The high enzymatic activity of the polytyrosyltrypsin shows, in accord with previous findings of Sri Ram *et al.* (*18*), that acylation of the ϵ-amino groups of the enzyme does not affect its catalytic activity.

Preparation and properties of the water-insoluble enzyme.—The copolymer of *p*-amino-DL-phenylalanine and L-leucine (in molar residue ratio of 1:2.5) was prepared by copolymerization of the appropriate amounts of α,*N*-carboxy-*p*-amino-*N*-benzyloxycarbonyl-DL-phenylalanine anhydride, and *N*-carboxy-L-leucine anhydride in dioxane using triethylamine as initiator. The product formed after standing for three days at room temperature was precipitated with water, dried, and the benzyloxycarbonyl groups were removed with anhydrous hydrogen bromide in glacial acetic acid. The copolymer was diazotized with nitrous acid and the diazonium salt formed precipitated at *p*H 7.5 with a concentrated sodium acetate solution. Coupling with polytyrosyltrypsin was effected in phosphate buffer, *p*H 7.2, ionic strength 0.25, at 2° with magnetic stirring. The reaction mixture was kept in the cold for 20 hours and the light brown, water-insoluble trypsin was separated by centrifugation. The water-insoluble preparation was shown to contain 21 per cent protein as estimated from the amount of valine determined chromatographically after acid hydrolysis.

The rate of hydrolysis of benzoyl-L-arginine methyl ester by water-insoluble trypsin at *p*H 7.8 and 25° indicated that the catalytic activity of the bound protein amounts approximately to a fifth of that of the free enzyme.

The enzymatic activity of water-insoluble trypsin dropped slowly with time when kept under 0.0025 N hydrochloric acid at 2°. A decrease in activity of about 25 per cent was recorded after a period of two months. The relative stability of the water-insoluble trypsin is most likely due to the blocking of the ϵ-amino groups of the enzyme on the one hand, and to the fixation of the enzyme molecules to the water-insoluble carrier on the other. The former leads to a decrease in the number of enzyme peptide bonds susceptible to tryptic digestion, while the latter inhibits interaction between trypsin molecules, thus diminishing autodigestion. The proteolytic and esteratic activities of water-insoluble trypsin are inhibited by diisopropyl fluorophosphonate. The rate of inhibition is, however, considerably slower than that of trypsin. Addition of an excess of soybean or pancreatic trypsin inhibitor caused only partial inhibition of the enzymatic activity of the water-insoluble enzyme.

To test the action of the water-insoluble trypsin on a native protein, it was mixed with a solution of rabbit myosin in 0.5 M potassium chloride, 0.03 M phosphate, *p*H 7.0, and the reaction mixture was incubated at 22° C. After given time intervals, aliquots were withdrawn, the water-insoluble trypsin was removed by centrifugation, and the ultracentrifuge pattern of the supernatant determined. The pattern obtained showed clearly the presence of light and heavy meromyosin. When amounts of crystalline trypsin and water-insoluble trypsin possessing equal activities towards L-arginine methyl ester were reacted with myosin, it was found that the water-insoluble enzyme required 200 minutes of incubation to achieve the same extent of myosin cleavage as crystalline trypsin in 10 minutes (*19*).

A trypsin column.—A trypsin column of 0.8 cm diameter and 5.0 cm height was prepared by mixing water-insoluble trypsin (20 mg) with an inert polyvinyl resin ("Geon 426," B. F. Goodrich Co.) (2.0 gm) in 0.0025 N hydrochloric acid and by pouring the mixture into a glass column. Prior to use, the column was washed with a suitable buffer. The column retained full enzymatic activity, even after the hydrolysis of as much as 4.0 gm of L-arginine methyl ester.

On passing through the column a solution of poly-L-lysine hydrobromide in phosphate buffer *p*H 7.4, the effluent was found to contain dilysine, trilysine, tetralysine, and traces of lysine (*20*). Protamine and insulin oxidized by performic acid could be digested under similar conditions to yield products giving paper chromatographic patterns similar to those obtained by digestion with crystalline trypsin.

Column characteristics determining the extent of substrate degradation.—An equation describing the effect of enzyme concentration, (E_o), height of column, h, and rate of flow, V (given in cm/sec) of the substrate, S, on the extent of hydrolysis, can readily be derived from the well-known Michaelis-Mentcn relation:

$$-\frac{d(S)}{dt} = \frac{k(E_o)(S)}{K_M - (S)} \quad (1)$$

In equation 1, K_M is the Michaelis constant and k is the specific constant determining the rate of decomposition of the enzyme substrate complex. Equation 1 yields on integration:

$$(S_o) - (S_h) + K_M \ln \frac{(S_o)}{(S_h)} = k(E_o)t \quad (2)$$

where (S_o) is the initial concentration of substrate and (S_h) that of the effluent. Substituting the time of exposure of substrate to enzyme in the column by the ratio h/V, one obtains

$$(S_o) - (S_h) + K_M \ln \frac{(S_o)}{(S_h)} = k(E_o)\frac{h}{V} \quad (3)$$

Equation 3 shows that the extent of degradation of substrate is determined by the effective concentration of enzyme, by the height of the column, and the rate of flow of substrate.

When $(S) \gg K_M$, equation 3 reduces to equation 4:

$$\frac{(S_o) - (S_h)}{(S_o)} = \frac{k(E_o)h}{(S_o) \cdot V} \quad (4)$$

The last equation could be verified experimentally for the case of L-arginine methyl ester. In the presence of a relatively large excess of substrate the amount of ester hydrolyzed was inversely proportional to the rate of flow and directly proportional to the height of the column.

Water-insoluble chymotrypsin (*21*)

The water-insoluble chymotrypsin was prepared by a procedure analogous to that given for the water-insoluble trypsin (*6*). A polytyrosylchymotrypsin was prepared by initiating the polymerization of *N*-carboxy-DL-tyrosine anhydride in aqueous dioxane with chymotrypsin. The polytyrosylchymotrypsin obtained contained a 7 per cent excess by weight of DL-tyrosine and showed an esteratic activity towards L-tyrosine ethyl ester corresponding to approximately two-thirds of that of crystalline chymotrypsin. On coupling polytyrosylchymotrypsin with the polydiazonium salt derived from the copolymer of *p*-amino-DL-phenylalanine and L-leucine (molar residue ratio of 1:1) the water-insoluble chymotrypsin was obtained. It contained 22 per cent protein by weight and showed towards L-tyrosine ethyl ester approximately one tenth of the esteratic activity of crystalline chymotrypsin.

The water-insoluble chymotrypsin lost about half of its activity when stored for seven months under distilled water at −10°. A *p*H optimum of 6.25 to 6.5 for the hydrolysis of tyrosine ethyl ester was found for water-insoluble polytyrosylchymotrypsin, water-insoluble chymotrypsin, and crystalline chymotrypsin.

The examples given above illustrate some of the possible methods for the synthesis of water-insoluble enzymes. The principles used, particularly that of attachment of the enzyme to the water-insoluble carrier via a linear polymeric chain, might be used for the preparation of other water-insoluble proteins with biological activity. In such cases it will be of primary importance to determine the functional groups which are nonessential for the activity, and to find the suitable chemical reactions for forming the desired biologically active water-insoluble products.

The water-insoluble proteolytic enzymes prepared so far should be of particular use, as demonstrated above, in the degradation studies of peptides and proteins.

The columns with enzymatic and other biological activities open up new possibilities for basic and applied research. They may lead to new models for the study of specific surface reactions, and to the development of novel laboratory and industrial techniques for the isolation and modification of biologically important compounds.

Acknowledgment

The investigations described in this article were supported by grant A-3083 of the National Institutes of Health, United States Public Health Service, and grant AF-61(052)-391 of the U.S. Air Force.

References

1 I. Langmuir and V. J. Schaefer, *J. Am. Chem. Soc.* **60,** 1351 (1938).
2 L. B. Barnett and H. B. Bull, *Biochim. et Biophys. Acta* **36,** 244 (1959).
3 D. H. Campbell, E. Luescher, and L. S. Lerman, *Proc. Nat. Acad. Sci,* **37,** 575 (1951).
4 H. C. Isliker, *Ann. N. Y. Acad. Sci.* **57,** 225 (1953).
5 N. Grubhofer and L. Schleith, *Hoppe-Seyler's Z. physiol. Chem.* **297,** 108 (1954).
6 A. Bar-Eli and E. Katchalski, *Nature* **188,** 856 (1960).
7 M. A. Mitz and L. J. Summaria, *ibid.* **189,** 576 (1961).
8 J. J. Cebra, D. Givol, H. I. Silman, and E. Katchalski, *J. Biol. Chem.* **236,** 1720 (1961).
9 E. L. Smith, A. Stockell and J. R. Kimmel, *ibid.,* **207,** 551 (1954).
10 R. L. Hill and E. L. Smith, *ibid.* **235,** 2332 (1960).
11 J. R. Kimmel and E. L. Smith, *ibid.* **207,** 515 (1954).
12 R. R. Porter, *Nature* **182,** 670 (1958).
13 R. R. Porter, *Biochem. J.* **73,** 119 (1959).
14 H. Neumann and E. Katchalski, unpublished results.
15 M. Sela and E. Katchalski, *J. Am. Chem. Soc.* **76,** 129 (1954).
16 M. A. Stahmann and R. R. Becker, *ibid.* **74,** 2695 (1952).
17 R. R. Becker and M. A. Stahmann, *J. Biol. Chem.* **204,** 745 (1953).
18 J. Sri Ram, L. Terminiello, M. Bier, and F. F. Nord, *Arch. Biochem. Biophys.* **52,** 464 (1954).
19 S. Lowey and A. Holtzer, *Biochim. et Biophys. Acta* **34,** 470 (1959).
20 S. G. Waley and J. Watson, *Biochem. J.* **55,** 328 (1953).
21 H. I. Silman and E. Katchalski, unpublished results.

Discussion of Paper 26: Water-Insoluble Derivatives of Proteolytic Enzymes

C. H. BAMFORD

Professor Katchalski's work on "tethered" enzymes is of great interest and potential importance. However, it would seem that the chains anchoring the enzymes are not necessarily limited to polypeptides and that other types of polymer chains which can be prepared by sufficiently gentle means could be used. The properties of the resulting material might depend on the character (e.g., flexibility) of the link. Could Professor Katchalski indicate whether any observations on non-polypeptide links have been made, other than those in the two references in his paper?

EPHRAIM KATCHALSKI

Dr. Bamford's suggestion is a good one. I am certain that it will be possible, in the future, to prepare water-insoluble enzymes by the attachment of the catalyst to insoluble carriers via polymers of different type. Preliminary experiments, carried out recently in our laboratory, indicate that some of the properties of the water-insoluble enzymes, such as their affinity to low and high molecular weight substrates, vary with the length, flexibility, and affinity to solvent of the "anchoring" side chains.

PAUL MAURER

I would like to add to Dr. Katchalski's remarks that Nisonoff has shown that pepsin can also degrade 7 *S* γ-globulin to a 5 *S* fragment analogous to Porter's combined fraction I and II. When this altered γ-globulin is treated with cysteine and an -*S*-*S* linkage is split, 3.5 *S* fragments like those obtained by Porter with papain are produced. The results of Porter are attributed to the combined action of papain and cysteine. It is also of interest that the specificities of pepsin and papain towards splitting of γ-globulin appear to be the same.

PAPER 27

Enzymic Degradation of Poly-α-L-Glutamic Acid

WILMER G. MILLER

Enzymic degradation of proteins has been studied extensively even though interpretation of the data is generally difficult due to lack of knowledge of the structure of the protein. To avoid the structural unknowns, dipeptides and small peptide derivatives are frequently used as model compounds, although they lack several features common to proteins. Synthetic polypeptides of known primary and secondary structure are much more realistic models; however, few extensive studies have been made (*1, 2*).

Evidence that the relatively nonspecific enzyme papain hydrolyzes poly-α-L-glutamic acid (PGA) was first reported by Green and Stahmann (*3*). We have extended their work considerably, and it is with the system PGA-papain that this report is concerned. In contrast to the unique molecular weight of native proteins, synthetic polypeptides have a distribution of molecular weights, thus turning the enzyme-catalyzed reaction into a problem primarily in polymer degradation.

The PGA used in this study has a weight average molecular weight (MW_w) of 50,000–60,000, a random molecular weight distribution, and undergoes a sharp helix–random coil transition when approximately half of the side-chain carboxyls are ionized. The transition occurs at approximately *p*H 5 in the solvent-buffer-activator system consisting of 0.2 M NaCl, 5 mM cysteine, 1 mM EDTA, and 10 mM citrate. To characterize the degradation, the molecular weight distribution as a function of degree of degradation must be determined in addition to the rate of hydrolysis. The rate of hydrolysis was determined quantitatively by viscosity measurements, and the molecular weight distribution by quantitative ion exchange chromatography using diethylaminoethyl cellulose (*4*).

Knowledge of the concentration dependence of the viscosity, the molecular weight dependence of the concentration dependence, and the intrinsic viscosity–molecular weight calibration curve permit the viscosity data to be converted to MW_w. Graphs of $1/MW_w$ as a function of degree of degradation are linear and this linear relation is to be expected only when the initially random distribution is degraded randomly (*5*). Furthermore, the theory of random degradation predicts that an average of one bond per molecule has been hydrolyzed

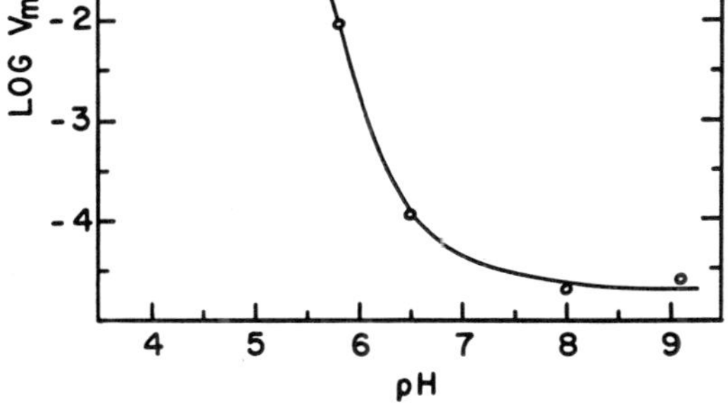

Fig. 27.1. Plot of log V_m as a function of *p*H. For PGA, values above *p*H 6 are *estimated maximum values.* Curves for α-benzoyl-L-benzoyl-L-argininamide (BAA) and carbobenzoxyglycylglycine (CGG) were obtained from data of Smith *et al.* (*6, 7*).

when the molecular weight has dropped to one-half of its initial value, thus permitting quantitative calculation of rate of bond breakage from viscosity data. Variation of PGA concentration at a fixed pH showed the rate to follow the Michaelis-Menton rate law typical of enzyme-catalyzed reactions. The rate data were compiled within the framework of this rate law and the profile of log of maximum rate (V_m) against pH is shown in Fig. 27.1, together with the pH profile for small synthetic substrates studied extensively by Smith (*6, 7*). There is a marked difference in the behavior of the small substrates and the more protein-like polymer.

To further elucidate the important factors in the enzymic degradation and also to study the degradation in regions inaccessible to viscosity measurements, the molecular weight distribution was determined by column chromatography. Chain lengths up to approximately ten can be resolved and this is sufficient to characterize the weight distribution. Resolution has been achieved with PGA samples ranging from 2 mg to 3 gm. A particularly good separation is shown in Fig. 27.2. Extensive analysis at pH 4.85 and 25.0° C. (see reference *4* for details) indicates that all peptide bonds are equally susceptible to enzymic hydrolysis except the two bonds nearest either end of the chain, which are much less susceptible; that chain length is unimportant above a chain length of about six residues; that the enzyme hydrolyzes dimer and trimer at a rate several orders of magnitude slower than the longer chain peptides, if at all; that rates determined by viscosity measurements quantitatively predict weight distribution patterns for much larger extents of degradation; and that the ratio of enzyme to substrate, when varied by a factor of 10^4, produces no difference in the distribution. At pH 4.85 PGA is near the midpoint of the helix-coil transition. Less extensive analysis of the weight distribution for reactions carried out with the PGA predominantly in the helical conformation or in the random-coil conformation shows no change in weight distribution from that observed at pH 4.85.

Chromatographic analysis together with the use of a polypeptide containing a single type of residue thus allows definite statements to be made concerning the effect of chain length as well as the relative probability of hydrolysis as a function of position in a given chain. Questions of this nature are of interest in protein degradation, but can rarely be answered due to the complexity of a protein. The large difference in rate of hydrolysis of dipeptide and tripeptide in comparison with longer chain peptides also points out a danger in using small substrates as models for protein degradation. Smith in his extensive studies using small substrates has found that positively charged substrates are hydrolyzed at a much faster rate than negatively charged substrates. However, Fig. 27.1 shows that long-chain, negatively charged polyglutamic acid is an excellent substrate for papain, being virtually as good a substrate as the best of the positively charged small substrates.

The most interesting aspect of the hydrolysis is the pH dependence of the maximum rate (V_m). Although on the logarithmic scale used in Fig. 27.1 the two small substrates appear to have a negligible pH dependence, on a linear scale V_m shows a pH dependence often found in enzyme kinetics and generally interpreted in terms of the active site of the enzyme having two ionizable groups. In the case of polyglu-

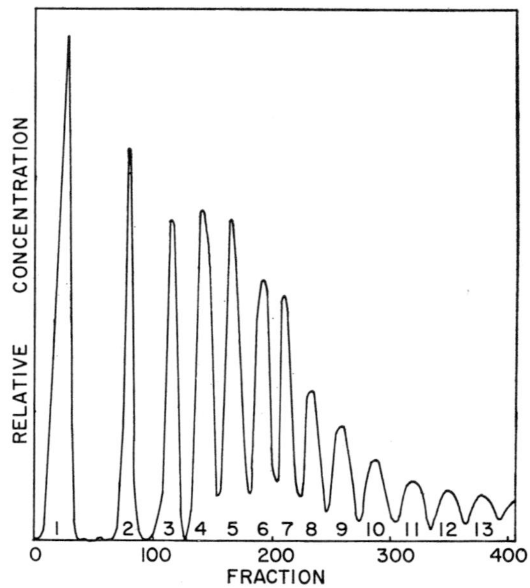

Fig. 27.2. Chromatograph of 3 gm. of low molecular weight PGA (not degraded with enzymes) on DEAE cellulose using a linear NaCl gradient for elution. Numbers under peaks indicate chain length.

tamic acid this factor is completely overshadowed by some property of the polypeptide, probably charge or conformation.

In Fig. 27.3 the maximum rate, taken from the smooth curve through the data in Fig. 27.1, is plotted along with optical rotation and titration curves. It has been suggested earlier (*4*) that the most reasonable interpretation is that the enzyme attacks preferentially at helix-coil junctions, as this is the only single parameter that would show a *p*H dependence similar to that exhibited by V_m. It is important to know whether the *p*H dependence in the region 4.5 to 5.5 is really as sharp as shown in Fig. 27.3; the data in Fig. 27.1 are not precise enough to answer this question. Considerably better precision can be obtained by systematically measuring the rate of hydrolysis at a single polypeptide concentration through this *p*H region. The *p*H-rate data at 25.0 and at 5.7° C. for a 7.5 gm/l. polypeptide solution are shown in Fig. 27.4. The polypeptide concentration is at least as large as the Michaelis constant in this *p*H

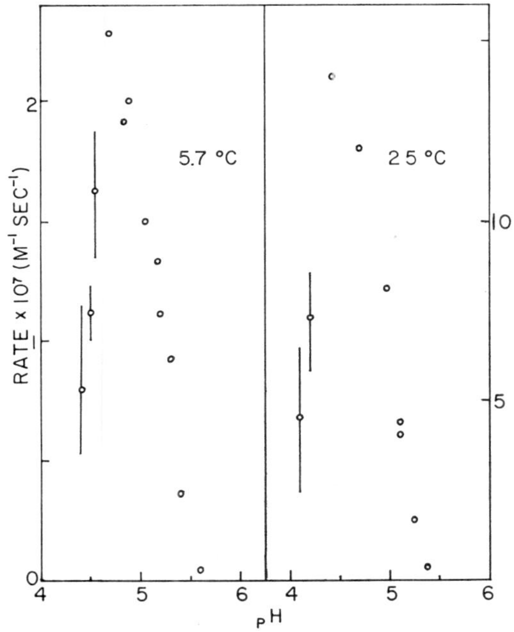

Fig. 27.4. Initial rate of hydrolysis of 7.5 gm/l. PGA solution as a function of *p*H at 5.7 and at 25.0° C.

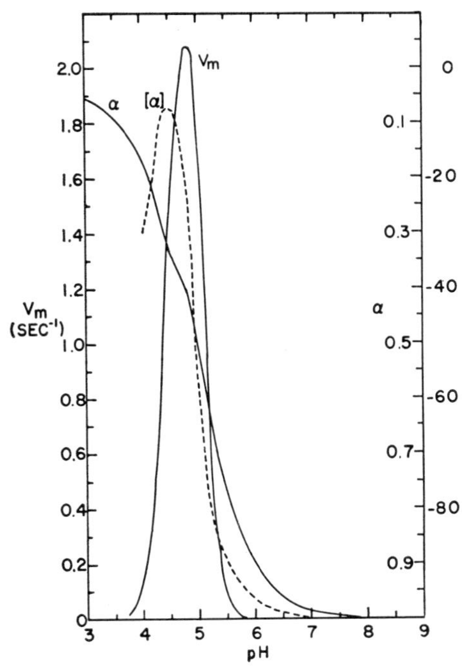

Fig. 27.3. Plots of V_m, specific rotation ($[\alpha]_{546}^{25}$), and degree of dissociation (α) in same activator-buffer-solvent system as function of *p*H. V_m was obtained from the smooth curve through the data in Fig. 27.1.

region. A sharp maximum is observed which occurs at a *p*H slightly lower than the midpoint of the titration or optical rotation curves. The curve shifts slightly with temperature and in the same direction as that of titration, and optical rotation shifts with temperature change. It seems unlikely that charge or charge together with another factor can be used to explain the rate data. Attack at helix-coil junctions is still consistent with the *p*H-rate data and is also consistent with the effect of chain length and position in the chain and with the random hydrolysis of bonds. The data could also be interpreted as attack of the helical region with falling off of the rate at low *p*H due to tightening of the helix. Either viewpoint runs counter to the commonly held view that proteolytic enzymes attack preferentially at unfolded regions in proteins.

In summary, the most interesting result obtained from a study of the hydrolysis of poly-α-L-glutamic acid by papain is that the enzyme preferentially attacks at a region other than unfolded or random-coil regions. All bonds are shown to be equally susceptible to attack, except for bonds near the end of a chain or in chains containing five or six residues or less. Polyglutamate is shown to be an excellent substrate for papain even when carrying a relatively large negative charge.

References

1 S. G. Waley and J. Watson, *Biochem. J.* **55,** 328 (1953).
2 M. Sela and E. Katchalski, *Advances in Protein Chem.* **14,** 391 (1959).
3 M. Green and M. A. Stahmann, *J. Biol. Chem.* **197,** 771 (1952).
4 W. G. Miller, *J. Am. Chem. Soc.* **83,** 259 (1961).
5 R. Simha, *J. Appl. Phys.* **12,** 569 (1941).
6 E. L. Smith and M. J. Parker, *J. Biol. Chem.* **233,** 1387 (1958).
7 E. L. Smith, V. J. Chavre, and M. J. Parker, *ibid.* **230,** 283 (1958).

PAPER 28

Electron Transfer Mediated by Polypeptides

KENNETH D. KOPPLE, ROY R. MILLER, and THOMAS C. MULLER

In several of the steps in the oxidation-reduction chain from which living organisms derive their energy, electrons, rather than atoms, are transferred. The role of protein in electron-transfer reactions is therefore of interest. Enzyme-catalyzed electron transfer may well occur directly between prosthetic groups, but hypotheses involving transfer through the peptide backbone must be considered. Electron transfer mediated by bridging groups in general has been discussed in detail (*1*, *2*). We focus attention on the polypeptide system as a bridging group.

Evans and Gergely have made molecular orbital calculations for an infinite repeating system of peptide bonds with pi orbital overlap between spatially adjacent (hydrogen-bridged) nitrogen and oxygen atoms. These calculations indicate that the system described possesses three narrow electronic energy bands (*3*). The lower two are fully occupied and the upper one is empty. Although excitation of an electron in the upper filled band to the empty one is in principle possible, the gap between these bands is over three electron volts, so that thermally excited protein semiconduction, once suggested by Szent-Györgyi (*4*), is highly improbable. Of course, holes or electrons introduced from an outside source would be mobile in this system. It might be possible to demonstrate the delocalization of a charge carrier introduced to a polyamide array; observable spectral and magnetic consequences should result. These have not been reported.

On the other hand, the ordered system of peptide bonds in a conformation like the α-helix is a one-dimensional analog of a crystal composed of conjugated unsaturated molecules such as anthracene. If sufficient pi orbital overlap occurs between individual units—amide groups in the polypeptide case—an electron or hole introduced to one unit may move to a neighbor. The extensive delocalization of charge carriers implied by the band system of Evans and Gergely is not involved. In an α-helix the NH···O distance is about 2.8 Å (*5*); since carriers exchange between molecules in the anthracene crystal (*6*), where the nearest approach of carbon nuclei is almost 4 Å (*7*), a similar process cannot be excluded for the peptide.

We may represent mediation of electron transfer by a polypeptide array by steps 1–3. (A similar set of steps could be written for positive charge transfer.)

$$\text{Donor} + \text{Amide} \rightleftarrows \text{Donor}^{+}\ \text{Amide}^{-} \quad (1)$$

$$\text{Amide}_i^{-} \cdots \text{Amide}_j \rightleftarrows \text{Amide}_i \cdots \text{Amide}_j^{-} \quad (2)$$

$$\text{Amide}^{-} + \text{Acceptor} \rightleftarrows \text{Amide} + \text{Acceptor}^{-} \quad (3)$$

Step 1 requires a reducing agent capable of transferring a single electron to an amide function, a requirement which may not be easy to meet. Step 2, electron exchange between amides, requires the suitable spatial arrangement that will be provided in a polypeptide by the hydrogen bonds, side-chain interactions, etc., which influence polymer conformation. The over-all results of these steps will be electron transfer between a donor and an acceptor separated by unusual distances. A demonstration of this process will necessarily be circumstantial, for in general there will always be a direct path for transfer between oxidant and reductant, although in properly chosen systems this may be made unlikely. Mediation by a single amide group will be another source of ambiguity. A dependence of those effects suggesting electron transfer at a distance on the conformation of the polypeptide mediator must be part of any evidence for step 2.

The reduction of carboxylatopentammine

cobalt(III) by aquo chromium(II) appeared to us to be a tool with which to test the possi-

$$Cr^{++} + RCOOCo(NH_3)_5^{++} \rightarrow RCOOCr^{++} + Co^{++} + 5HN_3 \quad (4)$$

bility of electron transfer through polypeptides. This system is an appropriate, if far from biological, one. Detailed studies of this reaction by Taube and co-workers have elicited kinetic criteria for mediation of electron transfer by conjugated ligands (*8, 9, 10*). Further, aquo chromium(II) ion, though it does not appear to reduce amide bonds, does transfer an electron to, and thus effect chemical reduction of, many conjugated unsaturated organic molecules (*11*).

Taube's studies of reaction 4 have shown that two reaction paths, depending on the ligand RCOO——, are possible. The first of these is direct reaction at the carboxyl group, involving transfer of an electron to cobalt and transfer of the carboxyl group to the chromium(III) so produced (reaction 5). The second route, represented in reaction 6 for terephthalic acid as ligand, involves attack of chromium at a site removed from the cobalt-bearing carboxyl, electron transfer through an extended pi orbital system, and transfer of the ligand to chromium at a site different from that at which the cobalt was bound. The existence of this second path has been demonstrated by product identification and the restrictions on the nature of carboxy ligands with which it occurs (*10*). Because of lessened repulsion of the like-charged ions in the activated complex, remote attack of chromium is associated with rate constants 100–1000 times greater than those for direct attack.

By the reaction of aquopentammine cobalt(III) perchlorate with poly-DL- and poly-L-glutamic acid, we have prepared the corresponding carboxylatopentammine complexes, (I). For comparison there were also prepared a series of copolymers of acrylamide and acrylatopentammine cobalt(III) perchlorate(II).

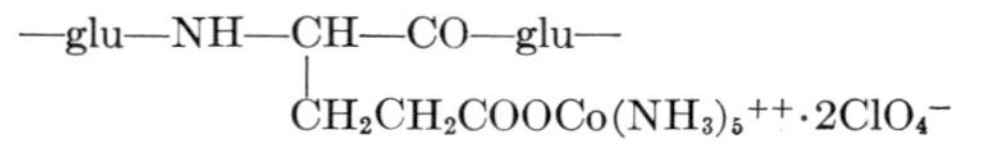

I

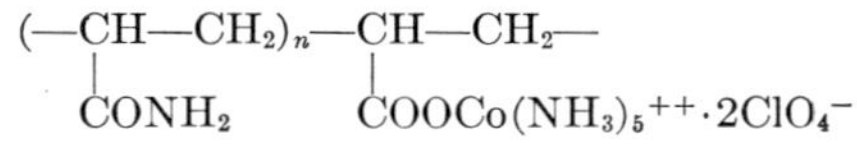

II

The reduction of these polymeric complexes by chromium(II) was studied spectrophotometrically, using techniques similar to those developed by Taube's group (*9*). The reactivities of the two groups of complexes differed strikingly, although in both the product chromium(III) was tightly bound to the polymeric ligand, as was observed in Taube's studies.

The acrylic polymers were reduced at a relatively slow rate by chromium (II), and the reaction exhibited partial inhibition by products. Table 28.1 gives the results of these experiments. Two rate constants are given, the initial rate, characteristic of polymeric complex bearing only cobalt, and the final rate, characteristic of carboxylatopentammine cobalt(III) groups with only chromium(III) neighbors. These constants are substantially independent of the acidity of the medium. It will be observed that the extent of inhibition is dependent on the spacing of the complex ions along the polymer chain. The nature of this inhibition is still under investigation. The initial reaction rate is about that to be expected for the cobalt(III) pentammine complex of a secondary carboxylic acid, and only weakly de-

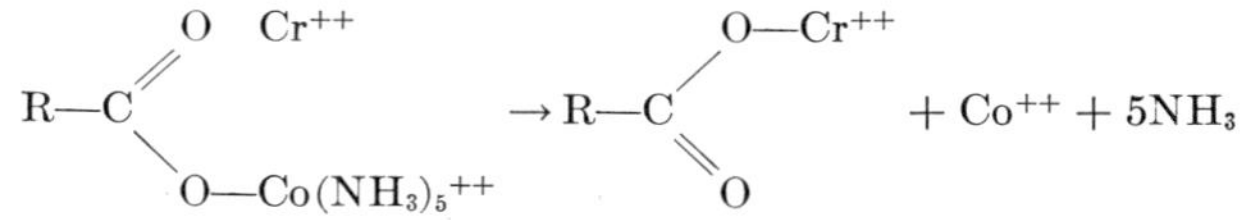

(5)

$k_{25} = 0.18\ M^{-1}\ sec^{-1}$

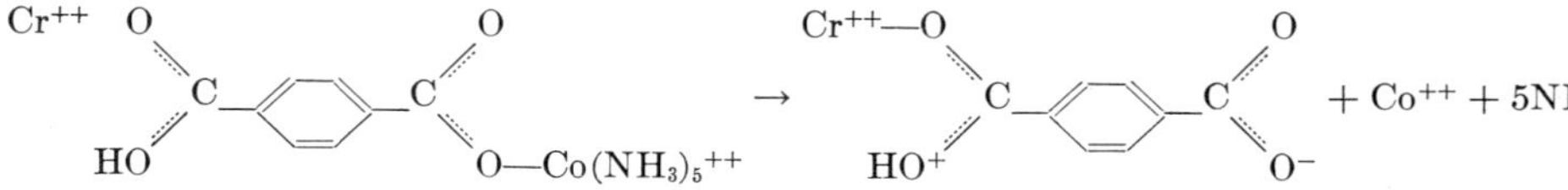

(6)

$k_{15.6^0} = 36\ M^{-1}\ sec^{-1}$

pendent on the frequency of complex ions along the chain. Since polyacrylamide exists as a random coil in aqueous solution (*12*), no remote attack would be expected since the lack of ordering of amide groups precludes process 2; the results cited in Table 28.1 are consistent with this hypothesis.

Table 28.1—Cr(II) reduction of acrylamide-acrylatopentammine cobalt(III) copolymers*

CONH/Co	k_{initial}	k_{final}
0	.037	.004
4		.011
13		.016
25	.044	.028

* 25°, ionic strength 0.25, $[H^+]=0.25$ M; rate constants in M^{-1} sec^{-1}.

For the poly-L-glutamic acid complexes, the rate of reduction is much higher and strongly dependent on the acidity of the reaction medium. These reactions also exhibit inhibition effects, which are a function of a number of variables, including polymer composition, chromium(II) concentration, and acidity; they are undoubtedly related to changes in polymer conformation occurring during reaction as well as to whatever effects are operative in inhibition of reaction of the acrylic polymers. Figure 28.1 shows initial rate constants obtained for one of these polymeric complexes. The second-order rate constant for reduction of a number of poly-L-glutamic acid–cobalt(III)pentammine complexes, of varying glu/Co ratios, at 0.1 N hydrogen ion is 24 M^{-1} sec^{-1}, 300 times that determined by Taube for the corresponding complex of butyric acid (*8;* see Table 28.2). For the one cobalt(III) pentammine-poly-DL-glutamic acid complex prepared to date (glu/Co 12), the measured constant is 0.85 M^{-1}sec^{-1} at 0.1 N hydrogen ion, one-thirtieth that for the L polymer complex. The acid dependence of reduction of the DL complex has not yet been determined.

It is apparent that the presence of a helical structure in the polymeric-cobalt complex facilitates its reduction. Two interpretations of these observations are possible. We currently favor that involving mediation of electron transfer between separated oxidant and reductant by the ordered peptide structure. On this hypothesis, acid dependence is most likely due

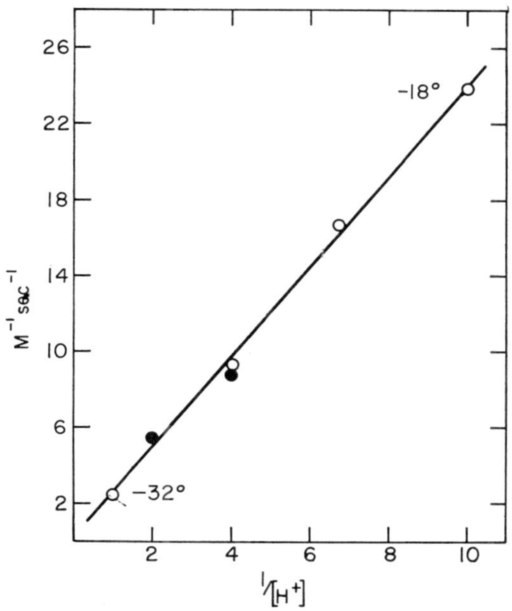

Fig. 28.1. Acid dependence of initial rate of reduction of polypeptide-pentamminecobalt(III) complexes by Cr^{++}, 25°, ionic strength 1.0. Open circles are for a poly-L-glutamic acid complex containing 13.2 glutamic acid residues per cobalt, and represent the rate constant for the first 80 per cent of reaction. The solid circles are data from a copolymer of L-glutamic acid and DL-alanine (7.4/1) containing 4.2 glutamic acid residues per cobalt and represent the rate constants obtained for the initial 20 per cent to 30 per cent of reaction. At two points the molecular rotation of the sodium *D* line of the homopolymer under the reaction conditions is given.

to a requirement for carboxylate ion to facilitate the initial transfer of an electron from chromium(II) to a peptide bond (reaction 7). Experiments are underway to determine whether chemical reduction of simple amides

$$\text{—COO}^- + \text{Cr}^{++} + \text{—CONH—} \rightarrow \text{—COOCr}^{++}\text{—CONH}^-\text{—} \quad (7)$$

by chromium(II) may be made to proceed at appreciable rates in the presence of high concentrations of carboxylate ions or other ligands.

Another interpretation of the inverse acid dependence shown in Figure 28.1 is that the fast reduction is the consequence of the presence of an ionized carboxyl group that stabilizes the activated complex for direct attack by coördination with chromium(II). It is instructive to compare, therefore, the acid dependence determined by Taube and Sebera (*9*) for reduction of the succinate and *o*-phthalate complexes (pK_A of adjacent carboxyl 3.8 and 2.6 respectively), with that measured for the poly-L-glutamic acid complex. This comparison, in the region of interest, is made in Figure 28.2. The variation in succinate complex reduction rate is negligible on the scale used, about one-tenth that of the *o*-phthalate complex, but recent results (*13*) suggest that a carboxylate ion path may be extremely important for the corresponding complex of malonic acid even at these acidities. Assuming a rigid helical peptide backbone and free carboxyethyl side chains, it would not be possible to account for the rapid reaction observed for the poly-L-glutamic acid complex on the basis of results observed in these models. If, however, the glutamic-acid side chains are highly ordered themselves, and evidence for some degree of order does exist (*14*), a direct reaction of the type shown in reaction 5, facilitated in some manner by carboxylate ion, may account for the observed phenomena. Although further model experiments are under way to determine the structural requirements for carboxylate catalysis in nonpolymeric systems, it seems unlikely that in aqueous solutions the side chains of the polymer are sufficiently constrained to provide the necessary effect.

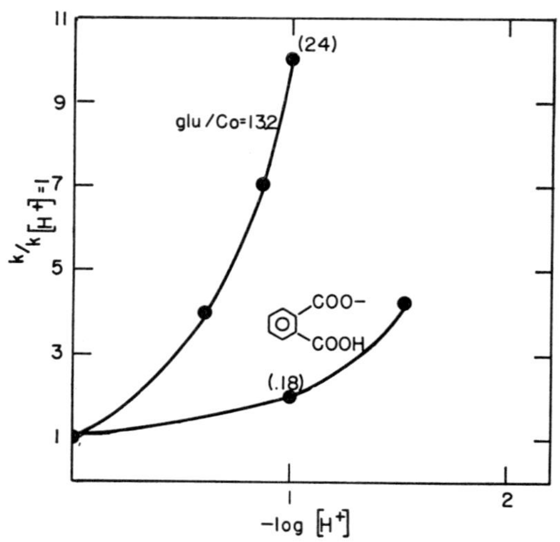

Fig. 28.2. Rate constants relative to rate in 1 N acid for *o*-phthalatopentamminecobalt(III) perchlorate (reference *9*) and a poly-L-glutamic acid–cobaltpentammine complex. Values in parentheses are second-order rate constants in M^{-1} sec^{-1}.

Although the behavior of the acrylic polymers indicates that the simple presence of many amide groups in the carboxylate ligand of carboxylatopentammine cobalt(III) complex has little effect on their reduction by chromium(II), we have looked at some simple peptide complexes as well. Table 28.2 summarizes the results obtained. It is apparent upon comparison with the result for the acetato complex that the presence of amide groups does facilitate reaction, probably by coördination of the amide link with chromium, but the effects are small.

Table 28.2—Cr^{++} reduction of carboxylatopentammine cobalt(III) complexes

R	*k*, M^{-1} sec^{-1}*
CH_3—	.18†
$CH_3(CH_2)_2$—	.08†
$HOOC(CH_2)_2$—	.32†‡
$CH_3CONHCH_2$—	.31
$CH_3CONHCH_2CH_2$—	.24
$CH_3CONHCH_2CONHCH_2$—	.40
$CH_3CONHCH(CH_2CONH_2)$—	.48

* 25°, ionic strength 1.0.
† Reference *9*.
‡ For the unionized species.

In summary, then, there has been observed a facilitation of reduction of carboxylatopentammine cobalt(III) complexes by chromium(II) in aqueous solution when the carboxyl ligand is on a side chain of a helical polypeptide. A similar phenomenon has not been noted in nonpolymeric model systems containing similar functional groups or in nonhelical polymeric systems. The observations may be explained by mediation of electron transfer by an ordered system of amide links or by the assumption that the polypeptide side chains are restricted to conformations particularly favorable for direct electron transfer from carboxyl-coördinated chromium(II) to another carboxyl bearing the cobalt(III) species. Work is under way to distinguish between these possibilities.

Acknowledgments

This work has been supported by National Science Foundation Grants G-5717 and

G-14324. We are grateful to Professors James Franck and Henry Taube for helpful discussions.

References

1 P. George and J. S. Griffith, "The Enzymes," Academic Press, Inc., New York, 1959, Vol. I., p. 347.
2 J. Halpern and L. E. Orgel, *Trans. Faraday Soc.* **29,** 32 (1960).
3 M. G. Evans and S. Gergely, *Biochim. et Biophys. Acta* **3,** 188 (1949).
4 A. Szent-Györgyi, *Science* **93,** 609 (1941).
5 L. Pauling and R. B. Corey, *Proc. Nat. Acad. Sci.* **37,** 235 (1951).
6 R. G. Kepler, *Phys. Rev.* **119,** 1226 (1960).
7 J. M. Robertson, *Proc. Roy. Soc.* **A140,** 79 (1933).
8 H. Taube, *Can. J. Chem.* **37,** 129 (1959).
9 D. K. Sebera and H. Taube, *J. Am. Chem. Soc.* **83,** 1785 (1961).
10 R. T. M. Fraser and H. Taube, *ibid.* **81,** 5000, 5514 (1959); R. T. M. Fraser, *ibid.* **83,** 564 (1961).
11 J. B. Conant and H. B. Cutter, *ibid.* **48,** 1016 (1926).
12 W. Scholtan, *Makromol. Chem.* **14,** 169 (1954).
13 H. Taube, private communication.
14 See the Paper 12 by Dr. A. Wada in this book.

Discussion of Paper 28: Electron Transfer by Polypeptides

C. H. BAMFORD

It is very desirable that the "mediated electron-transfer" mechanism favored by Dr. Kopple and his colleagues should be tested as completely as possible, since its establishment could be very important. A decision as to whether the reducing agent (Cr^{++}) reacts before or after complex formation with poly-L-glutamic acid would be helpful. Evidence on this point might be obtained kinetically if it is possible to work under conditions such that the Cr^{++} concentration is greater than that of the glutamyl residues. If at high [Cr^{++}] the order in [Cr^{++}] falls below unity, initial complex formation is indicated. On the other hand, if complexing can be excluded, several of the mechanisms alternative to Dr. Kopple's may be eliminated.

PAPER 29

Properties of Polypeptidyl Derivatives of Chymotrypsin and Ribonuclease

R. R. BECKER

Chemically modified proteins have long been used in studies directed toward determining the structures responsible for the specific biological action and function of proteins (*1*). Interest in both synthetic polypeptides and polypeptidyl proteins as model systems for such studies has increased sharply in recent years (*2*, *3*).

The work reported here represents an attempt to gain further information on the utility of polypeptidyl proteins as one approach to the more general problem of relating structure and function. Enzymes have been chosen for study since their readily measured activities afford a useful indicator of the integrity of the protein molecule after the modification reaction. However, it is clear that retention or loss of enzymatic activity alone provides little useful information with regard to the mechanism of the enzymatic reaction or to structural changes produced. It has therefore been necessary to devote considerable effort to the problem of determining the general course and the specificity of the reaction of the *N*-carboxy amino acid anhydrides (NCA's) with proteins. Accordingly, in this paper we will be concerned with the characterization, physical properties, and enzymatic activity of polypeptidyl chymotrypsins. In addition, preliminary results on polypeptidyl ribonuclease will be discussed.

Polypeptidyl chymotrypsins

STRUCTURE AND END-GROUP STUDIES

Our initial interest in preparing polypeptidyl proteins containing amino acid residues with bulky *R* groups was based upon the fact that the differences noted in the biological properties of several polyglycyl proteins, compared to the native proteins, appeared to be due to simple steric interference in the interactions involved (*4*, *5*, *6*). Konigsberg noted that in the hydrolysis of polyglycyl bovine plasma albumin by chymotrypsin the rate of hydrolysis of the modified protein was slower than that of the native, but the *p*H optima of both reactions and the extent of hydrolysis were the same (*7*). It was shown that tryptic activation of polyglycyl chymotrypsinogen produced an enzyme with the esterolytic activity of native chymotrypsin, although its rate of activation was slower than that of the native zymogen (*7*). Since these results suggested that steric effects were operating, we decided to work with valine, an amino acid expected to possess the desired properties because of the unusual stability of valine peptides toward acid hydrolysis (*8*, *9*). In the course of studying the effects of valine modification upon the chemical and biological properties of proteins, several particularly interesting results were noted (*10*, *11*). DNP-peptides isolated from DNP-polyvalyl proteins contained, in addition to the valine added in the modification reaction, amino acids from the protein chain (*10*). The introduction of valine peptides into chymotrypsin brought about a decrease in the esterolytic activity compared to either native chymotrypsin or the polyglycyl derivative (*11*). With these observations at hand, we decided to prepare a series of polypeptidyl derivatives of chymotrypsin, so as to determine the effect of variation of the *R* group on both peptide isolation and enzymatic activity.

The series of derivatives prepared (*12*) was chosen on the basis of the relative size of the α-substitutent of the amino acid used in the

Table 29.1—Analytical data for polypeptidyl chymotrypsins

	Moles added/ mole*	Sites acylated/ mole†	Average chain length	DNP-ileu/ mole‡	DNP-ala/ mole‡	DNP-peptides/ mole‡
α-chymotrypsin	0	0	0	.5	.7	0
polyglycyl chymotrypsin §	55	10	5	—	—	—
polyglycyl chymotrypsin	5.6	3	1.7	.7	.4	0
poly-DL-ala chymotrypsin	22.8	6	3.6	.5	6.0	0
poly-D-val chymotrypsin	34.1	10	3.4	.5	0	.6
poly-DL-val chymotrypsin	31.7	9	3.5	.5	0	.7
poly-L-val chymotrypsin	33.0	9	3.7	.5	0	.5
poly-DL-ileu chymotrypsin	28.0	8.6	3.3	4.3	0	.5
poly-DL-t-leu chymotrypsin	17.3	9.8	1.8	1.5‖	0	2.1

* Determined by method of D. H. Spackman, W. H. Stein, and S. Moore, *Anal. Chem.* **30,** 1190 (1958).
† Determined by method of Sanger.
‡ 16-hour hydrolysis.
§W. H. Konigsberg and R. R. Becker, *J. Am. Chem. Soc.* **81,** 1428 (1959).
‖ DNP-*t*-leu and DNP-ileu combined.

modification.* Thus the series included derivatives where *R* is hydrogen, methyl, isopropyl, sec-butyl, and *t*-butyl. The isomers of valine were used in order to determine whether the extent of acylation or the properties of the derivative might be stereochemically dependent.

The modified proteins were analyzed for total amino-acid composition (Table 29.1) after hydrolytic periods of 24, 48, 72, and 120 hours in constant boiling HCl, since the valyl, isoleucyl, and *t*-leucyl derivatives are very resistant to acid hydrolysis (*10, 11, 12*). Even 120 hours appears to be an insufficient hydrolytic period for the *t*-leucyl derivative in that only 10 of the 13 lysines present were recovered. The value given for the number of moles of *t*-leucine added is therefore somewhat low. Since the polyglycyl derivative prepared in this series was unaccountably low in extent of modification, an earlier preparation which also showed complete activity but which was heavily modified is included. The sites acylated were calculated by analyzing the DNP-protein derivatives after complete hydrolysis, for both lysine and ε-DNP-lysine, and by assuming that two α-amino groups have been acylated (*11*). The data recorded under DNP-ileu, DNP-ala, and DNP-peptides were obtained from 16-hour hydrolyses. It is apparent that the N-terminal isoleucine was not acylated in the modification reaction. The N-terminal isoleucine is also unavailable for reaction with FDNB under conditions where the enzyme is not denatured (*13*). The data show that the N-terminal alanine has been acylated (except in the case of the very poorly modified polyglycyl derivative). The α-amino groups were acylated also in the reaction of the NCA's of glycine, phenylalanine, and valine with BPA, chymotrypsin, insulin, and lysozyme (*11, 14*). Thus for the calculation of the average chain length of the added peptides, we assumed that two α-amino groups (alanine and 1/2 cystine) have been acylated in the reaction in addition to the lysines which were determined. The average length of the attached peptides is relatively low for most of the derivatives. No significant differences were found among the L-, D- and DL-valyl derivatives in terms of extent of acylation, average chain length, or in peptide isolation.

DNP-peptides were isolated only from those proteins modified with amino-acid anhydrides containing rather bulky *R* groups. In addition to the amino acid added in the acylation reaction, all of these peptides contained the same amino acids: asp, glu, lys, gly, ala, ser, val, ileu. It is clear that all of the sites acylated do not give rise to peptides, and the particular peptide isolated from the chymotrypsin derivatives probably arises from a definite portion of the protein chain. The stabilization of only some portion of the protein to acid hydrolysis

* The preparation, end-group analysis, and enzymatic activity studies of these modified proteins is largely the work of Dr. John H. Hash and a preliminary account of part of it has been presented (*12*). The details of this study will be reported in a later paper.

by adding nonpolar peptides would indicate a particularly strong interaction with the chain that depends on a distinct property of this area of the protein backbone. For example, the interaction could depend on composition and conformation of the areas surrounding lysine. The result of the interactions could produce a nonpolar area which prevents the approach of the hydronium ions for the hydrolytic reaction. Support for this suggestion comes from experiments with poly-DL-valyl insulin (*10*). Both N-terminal groups and the lysine in position 29 of the *B* chain were acylated. A DNP-decapeptide containing about 7 moles of valine attached to the ε-amino group of lysine was isolated from DNP-poly-DL-valyl insulin. Analysis showed the peptide to be from the C-terminal end of the *B* chain. In this instance, the interaction must be unusually strong, since the DNP-peptide was isolated in almost quantitative yields from a 16-hour hydrolysate. Apparently the peptide also prevented the reaction of the tyrosine in the 26 position of the *B* chain with FDNB, since tyrosine was liberated from the peptide upon further hydrolysis, and not *O*-DNP-tyrosine. Consistent with these results are the preliminary experiments of Karliner and Hochstein (*15*), who found that poly-DL-valyl insulin is not hydrolyzed by trypsin under conditions where the native hormone is cleaved at positions 22 (arginine) and 29 (lysine) to yield alanine and a heptapeptide. The lack of cleavage at the arginine position is best explained as interference by the added peptide.

It has been found that poly-DL-valyl chymotrypsin (*16*) and poly-DL-*t*-leucyl chymotrypsin (*17*) react mole for mole with diisopropylfluorophosphate and are completely inhibited. Attempts to isolate pure phosphate-containing peptides from these derivatives after partial acid hydrolysis or enzymatic hydrolyses have been unsuccessful.

Physical properties

For most of those polypeptidyl proteins studied by electrophoresis and ultracentrifugation, the results have indicated reasonable purity (*5, 18, 19, 20*). The series of derivatives under discussion have not been studied by these techniques. Some information on the physical properties of the chymotrypsin derivatives is available through the experiments carried out by Dr. Alfred Stracher while at the Rockefeller Institute. He studied their rotatory dispersion properties and their escape rates through cellophane membranes according to the method of Craig, King, and Stracher (*21*). The optical rotatory dispersions were determined using the Model 610 photoelectric spectropolarimeter manufactured by O. C. Rudolph & Sons. All of the measurements were made on approximately 1 per cent protein solutions in 0.1 M KCl and the dispersion data plotted according to the method of Doty and Yang (*22*). The data plotted in this manner gave straight lines and followed a one-term Drude equation. It is apparent from the data of Table 29.2 that only very minor changes in the over-all configuration, by this criterion, are brought about by the modification reaction. The only significant changes in molar rotation noted occur in the L- and D-valyl derivatives where changes would be expected. The slightly lower values of λ_c for the poly-DL-valyl and poly-DL-*t*-leucyl derivatives indicate no major change in configuration (*20*). On the other hand, when the rate of change of optical rotation of poly-DL-valyl chymotrypsin in 6 M urea

Table 29.2—Optical rotation data for polypeptidyl chymotrypsins

	Molecular weight*	$-[\alpha]_D^{20}$	$-[M]_D^{20}\times10^{-4}$	λ_c
chymotrypsin†	24,500	58.1	1.42	227
polyglycyl chymotrypsin	27,600	50.5	1.42	228
poly-DL-ala chymotrypsin	26,120	57.6	1.50	228
poly-D-val chymotrypsin	27,880	42.9	1.20	222
poly-DL-val chymotrypsin	27,640	51.4	1.42	224
poly-L-val chymotrypsin	27,770	62.9	1.74	230
poly-DL-*t*-leu chymotrypsin	26,450	53.5	1.42	222

* Based upon amino-acid analysis (Table 29.1).
† Crystalline chymotrypsin gave a λ_c of 231, a value in good agreement with that of Jirgensons (B. Jirgensons, *Arch. Biochem. Biophys.* **85**, 532 [1959]), and a b_o of −110. The sample for which the data are recorded here is a control sample that had been lyophilized. From this sample and the others listed, nonlinear plots were obtained when attempts were made to evaluate b_o and a_o.

at 25° C. was measured and compared to the unmodified enzyme, it was found that chymotrypsin reached its maximum value in about one hour, but the modified enzyme showed a slower rate and did not reach its maximum value until 24 hours. Other evidence for changes in behavior of the modified enzymes from the native was found in dialysis experiments. All of the modified enzymes with the exception of polyglycyl chymotrypin (55 moles of glycine/mole) failed to pass through the membranes whereas the native protein had a 50 per cent escape time of 5 hours (*21*). The 50 per cent escape time of polyglycyl chymotrypsin was increased to 16 hours. Since the polyglycyl derivative is relatively highly substituted compared to the others (Table 29.1), it is reasonable to assume that something other than the increased size of the molecules from added chains prevents the other substituted molecules from passing through the membranes. It appears quite probable that an intermolecular reaction of the modified proteins due to hydrophobic forces (*23*) might well account for some of the properties noted. Recently we have examined a preparation of poly-L-valyl chymotrypsin using the Archibald method for determining molecular weights under the same conditions used by Rao and Kegeles (*24*) in studying the polymerization of the native enzyme. That the poly-L-valyl chymotrypsin is a highly polymerizing system is shown by the following data: In approximately 1 per cent solution, α-chymotrypsin gave a molecular weight of 28,000; the value for poly-L-valyl chymotrypsin was approximately 120,000 (*25*). These data support the suggestion that those polypeptidyl proteins modified with nonpolar amino acids aggregate, probably through the formation of hydrophobic bonds. Hydrophobic bonding is also suggested by our observations that several poly-*t*-leucyl and polyvalyl chymotrypsin derivatives form turbid solutions at room temperature that clear upon cooling to 0°.

Enzymatic properties

Of the proteolytic enzymes, α-chymotrypsin has been studied quite extensively from the point of view of its formation from chymotrypsinogen (*26*), its structure (*27, 28, 29*), and its mechanism of action (*30, 31, 32*). For our purposes in comparing the activities of various polypeptidyl derivatives, the large number of well-characterized α-chymotrypsin-substrate systems available owing to the extensive work of Niemann and his co-workers was particularly valuable.

The enzymatic activities of the series of polypeptidyl derivatives toward an ester substrate and an amide-type substrate are shown in Table 29.3. The inhibition noted for the ester substrates was not unexpected (*11*). However, the increased activities noted for the amide-type substrates were quite surprising. Using *N*-nicotinyl-L-tryptophanamide (*33*) as substrate, similar but smaller increases over the control activity were noted. Since the evidence is very strong that one active center is responsible for both the esterolytic and proteolytic activities of chymotrypsin, it was felt that the differences found would not be due to mechanistic differences, but could probably be understood in kinetic terms. It was therefore decided to study one of the derivatives in some detail, including the evaluation of the kinetic constants for several substrates.

Poly-DL-*t*-leucyl chymotrypsin was chosen for this purpose (*34*). The particular preparation used in these studies contained 10.5 moles of *t*-leucine per mole corresponding to a molecular weight of 25,700. End-group studies indicated 7.5 moles/mole of N-terminal *t*-leucine and that 6 ε-amino and 2 α-amino groups had been acylated. Thus the *t*-leucine is attached largely as one unit per site acylated. To test the enzymatic properties of poly-DL-*t*-leucyl

Table 29.3—Inverse relationship between esterase and amidase activities in polypeptidyl chymotrypsins

	Percentage of control activity*	
	Esterase†	Amidase‡
control chymotrypsin	100	100
polyglycyl chymotrypsin	100	105
polyalanyl chymotrypsin	93	134
poly-DL-valyl chymotrypsin	76	193
poly-L-valyl chymotrypsin	69	167
poly-D-valyl chymotrypsin	80	172
poly-DL-isoleucyl chymotrypsin	62	130
poly-DL-*t*-leucyl chymotrypsin	64	190

* Comparisons to control based upon initial velocities.

† Assay of R. E. Parks and G. W. E. Plaut, *J. Biol. Chem.* **203,** 775 (1953), using L-phenylalanine ethyl ester as substrate.

‡ Assay of B. M. Iselin, H. T. Huang, and C. Niemann, *J. Biol. Chem.* **183,** 403 (1950), using benzoyl-L-phenylalanyl hydroxamide as substrate.

chymotrypsin, a series of L-tyrosine derivatives was synthesized for use as substrates. *N*-acetyl-L-tyrosine ethyl ester (ATEE) (*35, 36*) and L-tyrosine ethyl ester hydrochloride (TEE.HCl) were used for esterase measurements, and *N*-acetyl-L-tyrosinhydrazide (ATH) (*37*) and *N*-acetyl-L-tyrosinhydroxamide (ATHx) (*38*) were used for the amidase measurements. In addition, benzoyl-L-valine methyl ester (BVME) (*39*), an ester substrate which is hydrolyzed at a rate comparable to that of ATH, was prepared.

The results of *p*H-optimum studies using both α-chymotrypsin and poly-DL-*t*-leucyl-α-chymotrypsin indicated a shift to a higher value for the modified enzyme when ATH was used as substrate. The value of 7.8 ± 0.1 obtained for α-chymotrypsin–ATH system at $25.0 \pm 0.1°$ C. using 0.02 M THAM-HCl buffer is in reasonable agreement with the reported value for 7.95 ± 0.20 for the same system (*40*). In the case of the modified chymotrypsin-ATH system a *p*H-optimum of 8.3 ± 0.1 was observed. A comparison of the two *p*H-activity curves (Fig. 29.1) reveals that, at *p*H values below 7.9, poly-DL-*t*-leucyl chymotrypsin showed a lower rate of hydrolysis, whereas above this value its activity continued to increase until it reached an optimum *p*H of 8.3 ± 0.1. Thus the shift in the *p*H-optimum value was accompanied by an increase in proteolytic activity. To show that this observation was due to the alteration of the chemical structure of the enzyme and not due to the process of modification, the *p*H-optimum of a sample of crystalline α-chymotrypsin which had undergone all the processes of modification such as dissolving in aqueous phosphate buffer, dialysis against distilled water, and lyophilization was determined. As seen in Fig. 29.1, the two *p*H-activity curves of α-chymotrypsin and control-α-chymotrypsin were indeed identical within the limits of experimental error.

Other studies using ATHx, ATEE, and TEE.HCl as substrates showed no significant changes in the *p*H optimum for both α-chymotrypsin and poly-DL-*t*-leucyl chymotrypsin. However, a decrease in the activity of the modified enzyme was observed for these substrates, the loss in esterolytic activity being significantly greater than the loss observed with ATHx. The observed loss in amidase activity, in the case of the ATHx-poly-DL-*t*-leucyl chymotrypsin system, can be accounted for at

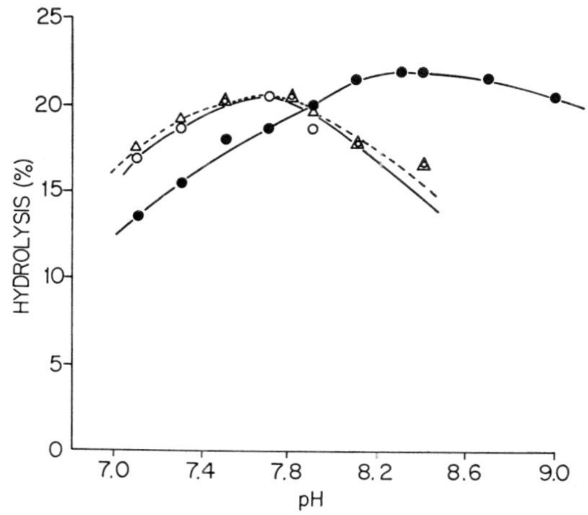

Fig. 29.1. *p*H Optima of modified and native chymotrypsin with *N*-acetyl-L-tyrosinhydrazide as substrate. $[S]_o = 4.6 \times 10^{-4}$ M, $[E]_o = 2.6 \times 10^{-5}$ M, $t = 60$ min; buffer, 0.02 M THAM-HCl at 25.0 $\pm 0.1°$ C.
● = poly-DL-*t*-leucyl-α-chymotrypsin
○ = crystalline α-chymotrypsin
△ = control-α-chymotrypsin

least in part by assuming the substrate was ionized. This substrate exists in both the uncharged and negatively charged forms at the *p*H values involved (Fig. 29.2). Hogness and Niemann (*41*) determined the pK_a' of this hydroxamide to be 9.0, and concluded from their study of the ATHx-α-chymotrypsin system that the active specific substrate is ATHx and not the *N*-acetyl-L-tyrosinhydroxamate ion. If one calculates the results on the basis of the anion substrate, then a shift to higher *p*H optimum would be expected.

It should be mentioned that shifts in *p*H optimum have been reported both with cases of modified enzymes and with variations in the structure of substrates. Jansen, Curl, and Balls (*42*) in their work with acetylated and oxidized chymotrypsin noted a shift of about 0.3 *p*H unit in the *p*H optimum using TEE as substrate, whereas no change was observed using either ATEE or hemoglobin as substrates. From these results, they concluded that the shift in the *p*H optimum is not a general case but dependent upon the nature of the substrate. Their results differed from those reported here in that changes were noted only on the acid side of the *p*H optimum; it is apparent from Fig. 29.1 that the entire curve was shifted to higher *p*H values for the poly-DL-*t*-leucyl chymotrypsin-ATH system. Manning

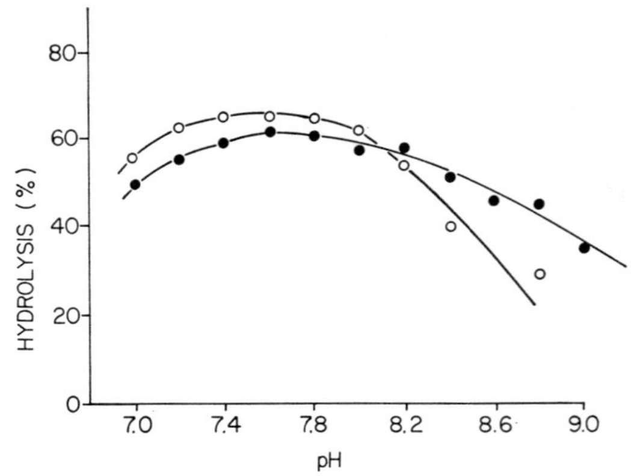

Fig. 29.2. pH Optima of modified and native chymotrypsin with N-acetyl-L-tyrosinhydroxamide as substrate. $[S]_o = 10 \times 10^{-3}$ M, $[E]_o = 7.8 \times 10^{-6}$ M, $t = 60$ min; buffer, 0.3 M THAM-HCl 25.0±0.1° C.
● = poly-DL-t-leucyl-α-chymotrypsin
○ = crystalline α-chymotrypsin

and Niemann (*43*) have reported that the pH-optimum for the α-chymotrypsin catalyzed hydrolysis of α-N-carboethoxy-L-tyrosinamide is 8.2 ±0.1; the corresponding value for α-N-acetyl-L-tyrosinamide is 7.9 ±0.1. Another example of a pH-optimum shift to higher values is found in experiments of McLaren and Estermann (*44*), who studied the hydrolysis of denatured lysozyme adsorbed on kaolinite particles as catalyzed by α-chymotrypsin. They indicated that the H^+ concentration at the interface where the reaction took place is actually higher than that in the medium.

After determining the pH optima for these systems, it was possible to evaluate the kinetic constants K_s and k_3. The classical intermediate enzyme-substrate complex theory was applied using Huang and Niemann's (*45*) forms of the equations in the evaluation of K_s and k_3. A summary of the kinetic constants for the various enzymatic systems is given in Table 29.4. All determinations were made under essentially the same experimental conditions as reported in the literature and at the corresponding pH optimum for each system.

The results for the various enzymatic systems studied, with the exception of the BVME substrate, indicate that the K_s values obtained for both chymotrypsin and the t-leucyl derivative are in reasonable agreement and the differences noted are considered not to be significant (Table 29.4). However, a significant change in the k_3 values was observed. A decrease in the k_3 values resulting from the chemical modification of chymotrypsin was noted using ATEE, TEE, and ATHx as substrates. The percentage of activities retained by the modified enzyme for the substrates mentioned are 74, 61, and 86 per cent respectively. On the other hand, an increase in proteolytic activity was obtained using ATH as substrate. For purposes of comparison, a "slow" ester substrate, benzoyl-L-valine methyl ester (BVME), was tested, with the expectation that it might yield results similar to those with ATH. An increased rate of hydrolysis similar to that in the ATH system was found, but in this case the change was accompanied by a decrease in K_s.

The differences noted in the amidase and esterase activities cannot be explained by the steric shielding of the active center by the bulky group of the amino acid residues added during modification since both inhibition and stimulation have been observed. It would be reasonable to assume that, if steric hindrance alone were involved, the effect on the enzymatic activities using the various specific substrates should be a decrease.

A possible explanation for these differences

Table 29.4—Kinetic constants as a function of type of substrate

	α-chymotrypsin			poly-DL-t-leucyl-α-chymotrypsin		
Substrate*	K_s†	k_3(min^{-1})	pH	K_s†	k_3(min^{-1})	pH
ATH	30.0 ±3	4.1± 0.4	7.8	36.5 ±5	6.3± 0.4	8.3
ATHx	36.4 ±3	99.3± 5	7.6	33.2 ±2	84.9± 4	7.6
ATEE	1.18±0.2	11,300 ±100	7.9	1.25±0.2	8,400 ±20	7.9
TEE	11.5 ±2	2,300 ± 20	6.3	10.9 ±2	1,400 ±20	6.3
BVME	3.1	1.88	7.9	1.8	1.85	7.9

* ATH = Acetyl-L-tyrosinhydrazide; ATHx = Acetyl-L-tyrosinhydroxamide; ATEE = Acetyl-L-tyrosine ethyl ester; TEE = L-tyrosine ethyl ester; BVME = Benzoyl-L-valine methyl ester.
† In units of 10^{-3} M.

in enzymatic behavior may be based upon a change in rate of one or more of the individual steps involved in the enzyme-catalyzed reaction. It has been suggested that the reaction of α-chymotrypsin with specific substrates involves three kinetically distinct steps.* The three steps include the rapid formation of a complex, acylation of the enzyme, and finally deacylation. The rate of formation of the acyl-enzyme intermediate and the rate of deacylation presumably are of the same order of magnitude and thus both contribute to the overall rate of hydrolysis (*32*, *46*). On this basis, it would appear that in the hydrolysis of ATEE and TEE by the modified enzyme, deacylation may indeed become rate limiting, as indicated by the decrease in k_3 observed. However, since the acyl-enzyme intermediate would be the same with both ATEE ($k_3=11{,}000$) and ATH ($k_3=4.1$) as substrates, it follows that the deacylation step cannot be rate limiting for the amide hydrolysis. Gutfreund and Sturtevant (*46*) proposed that the slow step in amide hydrolysis is acylation of the enzyme. Could it then be that the increase in rate noted in the ATH-poly-*t*-leucyl enzyme system is due to an increase in rate of one of the earlier steps? If this were the case, the reasons for this increase are not clear, since no significant change in K_s was noted. Although attempts have been made to "explain" results of this sort on the basis of formation of a "better" complex, the fact remains that the reaction cannot be described in detail, and possible conformational changes which may be brought about by the modification reaction cannot be described. The increased activity for the BVME-modified chymotrypsin system is accompanied by a decrease in K_s, suggesting enhancement of ES complex formation.

The three-step mechanism used in the above discussion is not universally accepted, and Bernhard, Coles, and Nowell (*47*) have pointed out that acyl-enzyme intermediates have been clearly demonstrated only with labile acyl anhydrides and nitrophenyl esters as substrates. The recently reported experiments of Bender and Zerner (*48*), however, have demonstrated that the hydrolysis of nonlabile trans-cinnamic esters proceeds through an acyl-enzyme intermediate. Obviously the kinetic experiments carried out here do not provide information regarding detailed mechanisms. However, one can speculate that the decrease in k_3 noted for the fast ester substrates may be brought about by a subtle change in the water site (*47*, *49*, *50*) caused by the introduction of nonpolar groups into the protein. It will be recalled that polyglycyl chymotrypsin shows enzymatic activities not significantly different from the native enzyme. On the basis of the analytical data and properties cited earlier, the significant difference between polyglycyl chymotrypsin and, for example, poly-DL-valyl chymotrypsin is not in the number of sites acylated nor in the average chain length of the added peptides, but in the nature of the *R* group of the added amino acid. Indeed, in looking at the data of Table 29.3, we can see that the decrease noted in esterolytic activity becomes larger roughly as the size of the *R* group increases in going from the polyalanyl to the poly-*t*-leucyl derivative. In the same series, increases in the hydrolysis of amide-type substrates are noted. A comparison of the change in *p*H optimum and relative rates of hydrolysis of ATH by poly-DL-valyl chymotrypsin substituted to the same extent as those listed in Table 29.3 gave results almost identical to those for the lightly substituted *t*-leucyl derivative (*17*). Hence it is probable that the data obtained with this *t*-leucyl derivative are typical of those modified with the nonpolar amino acids studied, and the rather striking difference in effect on the two activities can be explained at least partially by a kinetic and not a mechanistic change. In addition, this system is an example of a situation in which the necessity for considering the role of water in enzymatic hydrolysis is demonstrated (*47*).

* The experimentally determined kinetic constants, K_s and k_3, represent only apparent values and the manner in which they are related to the true constants, according to the three-step mechanism, is included in reference *32*.

Polypeptidyl ribonucleases

Interest in the study of ribonuclease by chemical modification (*51*) has increased with the knowledge of the amino-acid sequence (*52*) and the location of the disulfide bonds in the molecule (*53*). Most of these studies have been primarily concerned with determining which functional groups are necessary for enzymatic activity by the use of specific reagents. However, experiments providing information regarding some of the secondary forces responsible for the native structure of the protein have been carried out by Richards and Vithayathil (*54*) and White and Anfinsen (*55, 56*). Our initial interest in preparing polypeptidyl derivatives of ribonuclease was in connection with studies on the specificity of the reaction with *N*-carboxy amino acid anhydrides and on peptide isolation from the DNP-derivatives. In addition, ribonuclease offers an excellent framework for the attachment of a variety of peptides for studies of both biological and physical properties. Nonpolar peptides are of particular interest in this regard, since the addition of hydrophobic areas on the surface of a protein of known covalent structure provides a useful tool for the study of the problem of hydrophobic bonding.

In preliminary experiments, several poly-DL-valyl and poly-DL-isoleucyl ribonuclease derivatives were prepared, and end-group studies carried out (*57*). The results agreed with those reported for other proteins, in that only amino groups were acylated (including the α-amino group), and DNP-peptides were isolated in comparable yields.

The availability of solvent systems for countercurrent distribution of ribonuclease offered a technique for studying several important properties of these modified proteins. One should be able to obtain information regarding homogeneity additional to that available by electrophoretic and ultracentifugal studies. In addition, changes in solubility properties as well as in molecular weight that should be detectable by countercurrent distribution would be expected by the introduction of nonpolar groups into the molecule. Dr. A. Stracher carried out the countercurrent distribution of one of our preparations of poly-DL-valyl ribonuclease as well as experiments on optical rotatory dispersion and enzymatic activity. Approximately eight amino groups were acylated and a total of 20 moles of valine added per mole in the preparation examined by these techniques. Figure 29.3 shows the distribution pattern obtained in a 100-tube distribution of poly-DL-valyl ribonuclease. Comparison of the curve for modified ribonuclease with that calculated indicates reasonable homogeneity of this preparation. The absence of a peak at tube 44, the position of native ribonuclease in this distribution, indicates that no native enzyme remains in the preparation. Of particular interest is the very large change from the native enzyme noted in $K_{\text{traveling}}$. This value of 5.5 is to be compared with that of 0.78 for ribonuclease (*58*). The change noted represents a sevenfold increase in the solubility of the modified enzyme in the organic phase. Whether this peak represents monomeric poly-DL-valyl ribonuclease or some aggregate is not known. Poly-DL-valyl ribonuclease (22 moles of valine added per mole) in aqueous solution does aggregate on the basis of ultracentrifugal experiments (*25*).

The results of the optical rotatory dispersion experiments gave the following values: ribonuclease, $\lambda_c = 232$, $-[M]_D^{20} = 1.04 \times 10^4$; poly-DL-valyl ribonuclease, $\lambda_c = 229$, $-[M]_D^{20} = 0.95 \times 10^4$. The results indicate no major structural change has occurred in the modification reaction. When the data were plotted according to the equation of Moffitt and Yang (*59*), a value for b_o of -118 was obtained for ribonuclease and a value of 0 for ribonuclease in 9 molar urea. For the poly-DL-valyl ribonuclease, these plots were nonlinear. The rotation data obtained on the modified ribonuclease after isolation from the distribution were not significantly different from those indicated above.

The enzymatic activity of the modified protein after isolation from the distribution toward yeast ribonucleic acid was found to be 46 Kunitz units per mg, compared to 41 units

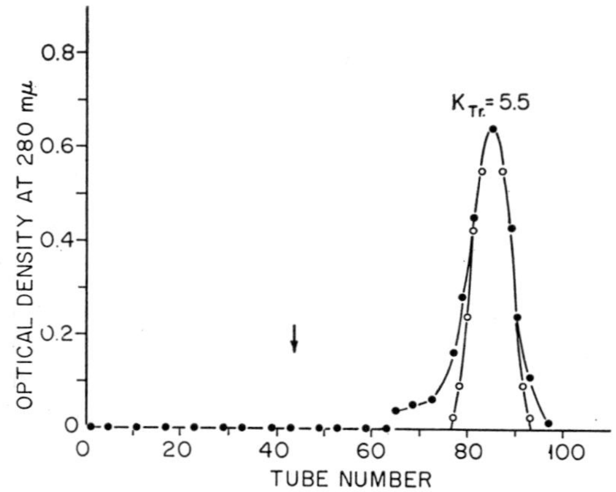

Fig. 29.3. Counter current distribution of poly-DL-valyl ribonuclease. ● Upper Phase. ○ Calculated. Solvent System: 2 parts EtOH (abs.); 1 part H_2O; 3 parts $(NH_4)_2SO_4$ (2.53 M). Arrow indicates peak tube for native enzyme.

per mg for the unmodified enzyme. The difference in activities noted may not be significant and more experiments are obviously required. Increased activities of deaminated ribonuclease derivatives toward yeast ribonucleic acid have been reported at *p*H 5.0 and low ionic strength, but no change was noted at higher ionic strengths, more alkaline *p*H's, or toward 2′, 3′-cyclic phosphate substrates (*60*). White (*61*) has reported increased activities toward yeast ribonucleic acid by thiolated ribonuclease derivatives. Under the same conditions, he found that modified enzyme showed decreased specific activity toward the cyclic phosphate substrates. From the preliminary data shown here, it can be said that it is remarkable that the large change in hydrophobic character of the protein, indicated by countercurrent distribution, is not reflected in a greater change in enzymatic activity. Obviously, a detailed study of these derivatives should provide useful information with regard to ribonuclease activity.

Recently, we have examined the NMR spectra of poly-L-valyl ribonuclease (22 moles of valine added per mole), to obtain information on the extent to which the added residues are restricted in their motion (*25*). These experiments were suggested by the work of Kowalsky and Boyer (*62*), who demonstrated that a marked sharpening of the spectra of several proteins, including ribonuclease, could be brought about by 8 M urea, or by oxidizing the disulfide bridges. The sharpening effect was explained largely as an increase in configurational freedom of the various residues brought about by the unfolding of the molecules. In the case of poly-L-valyl ribonuclease, one would then expect a rather broad spectrum if strong interaction is occurring, and a sharpening of this spectrum by the addition of urea. In the presence of urea, the spectrum is sharpened considerably, although it is still more broad than that of a physical mixture of ribonuclease and either L-valine or L-valyl-L-valine. Of special interest is the fact that solutions of poly-L-valyl ribonuclease in D_2O at the concentrations used (10 per cent to 20 per cent) show a turbidity that can be removed by cooling or by the addition of urea. In addition, the NMR spectrum of this derivative is sharpened on cooling, and broadens as the sample is permitted to warm to room temperature. This behavior is consistent with other observations mentioned in regard to hydrophobic interactions.

Conclusions

That changes in enzymatic activity, stabilization of portions of the protein chain toward acid hydrolysis, and aggregation reactions all occur in polypeptidyl proteins modified with nonpolar amino acids is apparent from the results of these experiments. All of these phenomena, therefore, are related in some way to the hydrophobic properties of the added peptides. The attached peptides can behave in several ways: they can interact strongly with the adjacent protein backbone, as evidenced by the isolation of DNP-peptides from certain portions of the protein chain; or they can interact with other hydrophobic areas in the same molecule and/or in other molecules. The latter, intermolecular hydrophobic bonding, is illustrated by the aggregation phenomena shown. Clearly, these results demonstrate that polypeptidyl proteins provide an excellent model for the study of hydrophobic interactions.

Acknowledgment

I would like to express my indebtedness to the following: Drs. J. H. Hash, R. G. Ona, A. Stracher, W. H. Konigsberg, Mrs. Y. Tsuji, Mrs. O. A. Darling, Dr. J. Kirby-Smith, and H. G. Jones. A major portion of this work was carried out in the Department of Chemistry, Columbia University, New York, and was supported by grants from the United States Public Health Service (A-1816) and the National Science Foundation (NSF-G-2132).

References

1 F. W. Putnam, in H. Neurath and K. Bailey (eds.), "The Proteins," Academic Press, Inc., New York, N. Y., 1953, Vol. I, Part B, p. 893.
2 E. Katchalski and M. Sela, *Advances in Protein Chem.* **13**, 244 (1958).
3 M. Sela and E. Katchalski, *ibid.* **14**, 391 (1959).
4 M. A. Stahmann and R. R. Becker, *J. Am. Chem. Soc.* **74**, 2695 (1952).
5 R. R. Becker and M. A. Stahmann, *J. Biol. Chem.* **204**, 745 (1953).
6 M. Green and M. A. Stahmann, *ibid.* **213**, 259 (1955).

7 W. H. Konigsberg, Ph.D. thesis, Columbia University (1956).
8 R. L. M. Synge, *Biochem. J.* **39,** 351 (1945).
9 H. S. Rhinesmith, W. A. Schroeder, and L. Pauling, *J. Am. Chem. Soc.* **79,** 609 (1957).
10 A. Stracher, W. H. Konigsberg, and R. R. Becker, *Biochim. et Biophys. Acta* **20,** 595 (1956).
11 A. Stracher and R. R. Becker, *J. Am. Chem. Soc.* **81,** 1432 (1959).
12 J. H. Hash, R. G. Ona, and R. R. Becker, *Federation Proc.* **17,** 238 (1958).
13 V. Massey, *Discussions Faraday Soc.* **20,** 66 (1955).
14 W. H. Konigsberg and R. R. Becker, *J. Am. Chem. Soc.* **81,** 1428 (1959).
15 J. Karliner and P. Hochstein, private communication.
16 A. Stracher, Ph.D. thesis, Columbia University (1956).
17 Y. Tsuji and R. R. Becker, unpublished experiments.
18 H. Tsuyuki, H. Van Kley, and M. A. Stahmann, *J. Am. Chem. Soc.* **78,** 764 (1956).
19 H. Van Kley and M. A. Stahmann, *J. Phys. Chem.* **60,** 1200 (1956).
20 H. Van Kley and M. A. Stahmann, *J. Am. Chem. Soc.* **81,** 4374 (1959).
21 L. C. Craig, T. P. King, and A. Stracher, *ibid.* **79,** 3729 (1957).
22 J. T. Yang and P. Doty, *ibid.* **79,** 761 (1957).
23 W. Kauzmann, *Advances in Protein Chem.* **14,** 1 (1959).
24 N. Rao and G. Kegeles, *J. Am. Chem. Soc.* **80,** 5724 (1958).
25 L. M. Krausz, F. C. Scaduto, and R. R. Becker, unpublished results.
26 H. Neurath, *Advances in Protein Chem.* **12,** 320 (1957).
27 B. Meedom, *Acta Chem. Scand.* **10,** 150, 881 (1956).
28 B. Meedom, *Biochim. et Biophys. Acta* **30,** 429 (1958).
29 B. S. Hartley, *J. Cellular Comp. Physiol.* **54,** Suppl. 1, 203 (1959).
30 H. Neurath and B. S. Hartley, *ibid.* p. 179 (1959).
31 M. L. Bender, *Chem. Revs.* **60,** 53 (1960).
32 J. M. Sturtevant, *Brookhaven Symposia in Biol.,* Number **13,** p. 151 (1960).
33 B. M. Iselin and C. Niemann, *J. Biol. Chem.* **182,** 821 (1950).
34 R. G. Ona, Ph.D. thesis, Columbia University (1960); R. R. Becker, O. A. Darling, and R. G. Ona, *Federation Proc.* **20,** 380 (1961). Details of this work will appear in a later paper.
35 L. Cunningham and C. Brown, *J. Biol. Chem.* **221,** 287 (1956).
36 R. Bixler and C. Niemann, *J. Am. Chem. Soc.* **81,** 1412 (1959).
37 J. Braunholtz, R. Kerr, and C. Niemann, *ibid.* **81,** 2825 (1959).
38 R. Foster and C. Niemann, *ibid.* **77,** 1886 (1955).
39 T. Applewhite, H. Waite, and C. Niemann, *ibid.* **80,** 1465 (1958).
40 R. Lutwack, H. Mower, and C. Niemann, *ibid.* **79,** 2179 (1957).
41 D. Hogness and C. Niemann, *ibid.* **75,** 884 (1953).
42 E. Jansen, A. Curl, and A. Balls, *J. Biol. Chem.* **189,** 671 (1951).
43 D. Manning and C. Niemann, *J. Am. Chem. Soc.* **80,** 1478 (1958).
44 A. McLaren and A. Estermann, *Arch. Biochem. Biophys.* **68,** 157 (1957).
45 H. Huang and C. Niemann, *J. Am. Chem. Soc.* **73,** 1541 (1951).
46 H. Gutfreund and J. Sturtevant, *Proc. Nat. Acad. Sci.* **42,** 719 (1956).
47 S. A. Bernhard, W. C. Coles, and J. F. Nowell, *J. Am. Chem. Soc.* **82,** 3043 (1960).
48 M. L. Bender and B. Zerner, *ibid.* **83,** 2391 (1961).
49 D. B. Sprinson and D. Rittenberg, *Nature* **167,** 484 (1951).
50 D. G. Doherty and F. Vaslow, *J. Am. Chem. Soc.* **74,** 931 (1952).
51 W. H. Stein, *Brookhaven Symposia in Biol.,* Number **13,** p. 104 (1960).
52 C. H. W. Hirs, S. Moore, and W. H. Stein, *J. Biol. Chem.* **235,** 633 (1960).
53 D. H. Spackman, W. H. Stein, and S. Moore, *ibid.* **235,** 648 (1960).
54 F. M. Richards and P. J. Vithayathil, *Brookhaven Symposia in Biol.,* Number **13,** p. 115 (1960).
55 F. H. White, Jr., and C. B. Anfinsen, *Ann. N. Y. Acad. Sci.* **81,** 515 (1959).
56 F. H. White, Jr., *J. Biol. Chem.* **235,** 383 (1960).
57 R. G. Ona, Y. Tsuji, and R. R. Becker, unpublished results.
58 T. P. King and L. C. Craig, *J. Am. Chem. Soc.* **80,** 3366 (1958).
59 W. Moffitt and J. T. Yang, *Proc. Nat. Acad. Sci.* **42,** 596 (1956).
60 H. Van Vunakis, E. Leikhim, R. Delaney, L. Levine, and R. K. Brown, *J. Biol. Chem.* **235,** 3430 (1960).
61 F. H. White, Jr., *Federation Proc.,* **20,** 221 (1961).
62 P. D. Boyer, *Brookhaven Symposia in Biol.,* Number **13,** p. 1 (1960).

Discussion of Paper 29: Polypeptidyl Chymotrypsin and Ribonuclease

MICHAEL SELA and
CHRISTIAN B. ANFINSEN

We have recently investigated some polypeptidyl derivatives of ribonuclease and their usefulness in studies of the role of side-chain interactions in the conversion of reduced ribonuclease to the native enzyme as well as in studies of the chemical topography of protein molecules in solution.

Ribonuclease was reacted with *N*-carboxy-DL-alanine anhydride at *p*H 6.8 in a phosphate buffer-dioxane mixture. The polyalanyl ribonuclease obtained appeared homogeneous when chromatographed on phosphorylated cellulose. No native ribonuclease was present at the end of the alanination reaction. An artificial mixture of poly-DL-alanine, poly-DL-alanyl ribonuclease, and native ribonuclease was easily separated chromatographically. On the average 40 alanine residues were bound to each ribonuclease molecule, and dinitrophenylation and desamination experiments showed that they were attached to 8 amino groups per molecule (100% of the one α-NH_2 and 70% of the 10 ϵ-NH_2 groups).

The alaninated enzyme was fully active towards ribonucleic acid, but it had only 60 per cent of the activity of the unmodified ribonuclease toward 2′,3′-cyclic uridine phosphate. Similarly to the native enzyme, activity was completely restored after oxidation of the fully reduced and inactive polyalanyl ribonuclease.

Re-alanination of the polyalanyl ribonuclease sample yielded, after column chromatography, a ribonuclease with 72 alanine residues attached per molecule. While the enrichment in alanine content was highly significant, there was no change in the number of unreacted amino groups of lysine. The biological activity towards ribonucleic acid and cyclic substrate was in this case 70 per cent and 15 per cent respectively, compared to the native enzyme. Upon re-oxidation of the reduced material, most of the activity was recovered. The ultraviolet absorption of the richly alaninated sample had a peak at 277.5 mμ, like the native enzyme. The optical rotation recalculated for its ribonuclease content was $[\alpha]_D = -77°$ (to be compared with $[\alpha]_D = -74.4°$ for ribonuclease). It sedimented with $S_{20,w} = 3.5$.

It seems that the enzyme may be peptidylated without loss of activity, and that additional enrichment resulted in a general decrease in turnover rate rather than in total inactivation of part of the molecule.

Preliminary experiments with polytyrosyl ribonucleases show that extensive tyrosination results in the insolubilization of the enzyme, while a limited tyrosination (6 residues per molecule) yielded a fully active material. The activity could be completely restored upon oxidation of the reduced polytyrosyl ribonuclease.

Previous studies have suggested that the entire chemical information required for the correct pairing of the half-cystine residues of reduced ribonuclease in disulfide bonds, and for the subsequent internal coiling and tertiary folding of the regenerated native enzyme, is contained in the coded message represented by the amino-acid sequence of the extended polypeptide chain. From experiments reported here it may be concluded that the open chain obtained upon the reduction of polyalanyl ribonuclease is still able to refold correctly. Thus it seems that as long as the half-cystines are fixed in the main chain, the electrostatic equilibrium is not disturbed (an α-NH_2 group of alanine appears for each ϵ-NH_2 group of lysine that had reacted), and the side chain of the residue attached does not contain very reactive groups, the peptidylation does not interfere with the correct translation of the chemical information contained in the open-chain sequence.

It is an inevitable corollary of the present work that the ribonuclease molecule in solution is arranged in such a way as to permit some amino groups to attach long alanine peptides without distorting the molecule to the extent of significant losses in activity, while other seem unreactive. In view of the knowledge of the complete chemical formula of ribonuclease, it should be possible to elucidate the location of the three unavailable lysines along the chain. This could lead to conclusions concerning the three-dimensional structure of the enzyme and might be of use in mapping the topography of other proteins whose primary structure is known.

C. H. BAMFORD

In connection with Dr. Sela's remarks on the relative reactivities of the different amino groups in ribonuclease, I would like to suggest that it might be of interest to react those groups readily accessible to NCA's with *N*-phenyl glycine NCA. This compound, while reacting with unhindered amino groups, does not polymerize on account of the steric obstruction introduced by the phenyl group (cf. Paper 7 by Dr. Block and myself in this book). Thus, after the first addition the accessible groups will be blocked, and it may be possible to examine the behavior of the other base group in effective isolation (e.g., towards higher concentrations of different NCA's).

PAPER 30

Studies on the Catalytic Activity of Synthetic Polyamino Acids Having an Imidazole Group in the Active Site

JUNZO NOGUCHI and TOMOO SAITO

It has been recently assumed that one of the active sites of some hydrolytic enzymes is the imidazole group of histidine (*1*). Various investigations of hydrolytic action of histidine, imidazole, or their derivatives (*2–6*) have been made for studying the enzymatic mechanism. Lukton and Olcott (*3*) found that carnosine, anserine, 1-methylhistidine, histidylhistidine, and β-asparatylhistidine hydrolyzed *p*-nitrophenylacetate at *p*H 6.2 in phosphate buffer. The rates of catalytic hydrolysis were 2.7, 3.6, 2.0, 2.1, and 1.0, respectively, against 1.0 of histidine. Merrifield and Woolley (*4*) have recently reported that chymotrypsin, poly-L-histidine, (poly-L-histidyl)-L-glutamic acid, and (poly-L-histidyl)-L-seryl-L-glutamic acid hydrolyzed ethyl-*p*-nitrophenyl carbonate at *p*H 7.6 in phosphate buffer and at 25° C. The numbers turned over, defined as moles of substrate per milligram catalyst per minute, were 4.4, 70, 80, and 40 against 7.7 of L-histidine and the catalysts were similarly active for *p*-nitrophenylacetate. E. Katchalski and M. Sela (*5*) suggested that one or more histidyl residues in the above polymers might be more active than those of the polymers having only one imidazole group. They also recommended that the catalytic activity should be calculated for unionized imidazole. E. Katchalski *et al.* (*6*) measured the catalytic hhdrolysis for *p*-nitrophenylacetate with poly-L-histidine and the histidine copolymers having serine, glutamic acid, aspartic acid, or lysine prepared by Leuchs' method and showed that the catalytic activity was proportional to the quantity of the unionized imidazole groups of the polymers.

The objective of this study was to confirm how the neighboring groups affected a histidyl residue of polymer in its hydrolysis of *p*-nitrophenylacetate, to study the nature of enzymatic activity using synthetic polyamino acids, and to obtain some perceptions of the synthesis of artificial enzymes in the future.

Experimental

Copoly (asp, his·HCl, 1:1) and copoly (glu, his·HCl, 1:1) by the Fox-Harada method (*7*): The mixture of powdered L-histidine monohydrochloride and L-aspartic acid or L-glutamic acid in equimolar portions in liquid paraffin covered with carbon dioxide was heated at 180° C. for two hours, then at 200° C. for two hours, and then for two hours more at water pump pressure. The reaction mixture was washed with ligroin, with acetone, and then dried. The product dissolved in water was reprecipitated with alcohol or acetone, and the precipitate was washed with alcohol (or acetone) and ether. The samples for analysis were dried over phosphorus pentoxide at 110° C. for two hours in vacuo.

Copoly (L-his·HCl, L-glu), copoly (L-his·HCl, L-asp), copoly (L-his·HCl, L-lys·HCl), copoly (L-his, L-ser), and copoly (L-his·HCl, L-cysteine) by Leuchs' method: Hydrogen chloride was removed from 1-*N*-benzyl-L-histidine *N*-carboxylic acid anhydride hydrochloride (1-*N*-$PhCH_2$-L-his NCA·HCl) suspended in dioxane by adding an equimolar amount of triethylamine. The filtered solution was immediately mixed with each dioxane solution of γ-$PhCH_2$-L-glu NCA (γ-benzyl-L-glutamic acid *N*-carboxylic acid anhydride), β-$PhCH_2$-L-asp

Table 30.1—Thermal copolymers of aspartic acid: histidine HCl, 1:1 and glutamic acid: histidine HCl, 1:1

Amino acids used (gm)	Product (gm)	Amino acid residue ratio in copolymer: Paper chromatography	Amino acid residue ratio in copolymer: Colorimetry of diazo reaction	Molecular weight by: Van Slyke method	Molecular weight by: End group titration with CH_3ONa	Elementary analyses Cl (%): found	Elementary analyses Cl (%): calcd.	Elementary analyses N (%): found	Elementary analyses N (%): calcd.
Asp 8.9; His·HCl·H_2O 14	18.7	1 : 1	1 : 1	425	420	11.61	12.30	19.25	19.41
Glu 9.8; His·HCl·H_2O 14	18.6	1 : 1	1 : 1	—	579	11.43	11.73	18.20	18.51

NCA, ε-Cbzo-L-lys NCA, O-$PhCH_2$-L-ser NCA, and *S*-Cbzo-L-cysteine NCA, respectively.

The mixture was polymerized at room temperature for seven days by stirring and then heated for one hour in a boiling water bath. The product was taken up by evaporating dioxane and washing the residue with water.

Each was reduced with an equimolar amount of sodium in liquid ammonia, and the residue obtained by the evaporation of ammonia was dissolved in a small quantity of water. The solution was adjusted to *p*H 4.5 with Congo Red and in HCl, and evaporated to dryness. The residues were separated from sodium chloride with glacial acetic acid and the extracts were precipitated with ether, centrifuged, and dried. Copoly (L-his·HCl, L-glu NH_2), copoly (L-his, L-asp NH_2), copoly (L-his·HCl, L-lys·HCl), copoly (L-his·HCl, L-ser), and copoly (L-his ·HCl, L-cysteine) were prepared. Copoly (L-his·HCl, L-asp) and copoly (L-his·HCl, L-glu) were obtained from copoly (L-his·HCl, L-asp NH_2) and copoly (L-his·HCl, L-glu NH_2), respectively, by the following method.

Each reaction mixture was dissolved in an equimolar amount of 1 N sodium hydroxide, kept overnight at room temperature, and then neutralized with 1 N hydrochloric acid to *p*H 4.5. The solutions were evaporated to dryness and the acetic acid extracts separated from sodium chloride were precipitated with ether.

The molecular weights of these copolymers (Table 30.2) were determined by Van Slyke's amino N method before removal of protecting groups.

After removal of the protecting groups by

Table 30.2—Intermediate copolymers having protecting groups

Components of copolymer	NCA used (gm)	Products (gm)	Van Slyke's method: Mol. wt.	Van Slyke's method: Total amino acid residues
$PhCH_2$-L-his γ-$PhCH_2$-L-glu	1 0.865	1.16	494	2.2
Bz-L-his β-$PhCH_2$-L-asp	1 0.811	1.21	1330	6.4
$PhCH_2$-L-his ε-Cbzo-L-lys	1 0.996	1.50	1670	6.8
$PhCH_2$-L-his O-Bz-L-ser	1 0.719	1.06	1350	7.0
$PhCH_2$-L-his S-Cbzo-L-cys	1 0.912	1.02	800	3.4

Table 30.3—Copolymers with protecting groups removed

Components of copolymer	Intermediate copolymers used (gm)	Products (gm)	Amino-acid residue ratio by diazo reaction	Total amino acid residue in copolymer
L-His HCl L-Glu NH_2	1	0.687	10 : 26	2.2
L-His HCl L-Glu	0.520	0.520	10 : 26	2.2
L-His HCl L-Asp NH_2	1	0.615	10 : 22	6.4
L-His HCl L-Asp	0.380	0.380	10 : 22	6.4
L-His HCl L-Lys HCl	1	0.194	10 : 16	6.8
L-His HCl L-Ser	1	0.481	10 : 18	7.0
L-His HCl L-Cysteine	0.900	0.493	10 : 37	3.4

reduction with liquid ammonia and sodium, the polymers (Table 30.3) were hydrolyzed and assayed for histidine content by the diazo color reaction. The other amino acids were assayed by quantitative paper chromatography (Table 30.4).

Copoly (ϵ-aminocaproic acid, his·HCl) by Ham's method (*8*): L-Histidine HCl·H_2O and powdered ϵ-caprolactam were mixed in various molar ratios and heated at 190°–20° C. for 17–22 hours in an atmosphere of nitrogen. The copolymers produced were removed by rubbing with water, and dissolved in water or acetic acid.

The solutions were dialyzed with a cellophane membrane and brought to *p*H 7.5–8 with aqueous ammonia.

The precipitates separated out were centrifuged and lyopholized. Among these copolymers, copoly (ϵ-amino-caproic acid, his·HCl 3:2) was soluble in water and was used as a sample in the test of catalytic hydrolysis.

Poly (ϵ-aminocaproyl-L-histidine HCl) by carbothiophenyl method (*9*): ϵ-Aminocaproic acid ethyl ester hydrochloride (20 gm) dissolved in ethanol (60 cc) was mixed with sodium methylate (Na 5 gm) and carbothiophenyl chloride (17.6 gm) by adding each to the other in small portions and the reaction mixture was stirred for thirty minutes at −5° C. and for one hour at room temperature.

The filtered solution was acidified with acetic acid and concentrated at below 35° C. Syrupy product (30.5 gm) was obtained. The syrup was dissolved in acetone (175 cc) and heated with 10 N hydrochloric acid (90 cc) for thirty minutes at 100° C. The solution was immediately cooled under the tap and a small amount of water was added to it.

The solution was then concentrated in vacuo at below 50° C. A yellow precipitate was obtained. It was washed with water, dissolved in ethyl acetate, concentrated, and washed with ether. *N*-carbothiophenyl-ϵ-aminocaproic acid was obtained. Yield 18.5 gm (67.5%); m.p. 97° C.

The product (10 gm) suspended in chloroform (30 cc) was mixed with thionyl chloride (5 gm) and heated for thirty minutes at 40°–50° C.

The clear solution was evaporated, washed with ether, and redissolved in chloroform (20 cc).

Table 30.4—Copoly (ε-aminocaproic acid, histidine) by thermal condensation of caprolactam and histidine · HCl

No.	Components of copolymers used for polymerization: Substance	(gm)	Molar ratio	Product (gm)	Mol. wt. by Van Slyke method	Amino-acid residue ratio by: Paper chromatography	Amino-acid residue ratio by: Diazo reaction	Total amino-acid residue in copolymer	N analysis: found %	N analysis: calcd. %
1	ε-Caprolactam	4.95	1 : 1	1.5	1650	3 : 2	3 : 2	13.5	19.72	20.55
	L-His HCl · H_2O	10.5								
2	ε-Caprolactam	15	3.3 : 1	6	1620	3.3 : 1	*	13.0	17.11	17.29
	L-His HCl · H_2O	15								
3	ε-Caprolactam	12	7.4 : 1	4.5	1792	5.7 : 1	*	14.8	15.63	15.63
	L-His HCl · H_2O	3								
4	ε-Caprolactam	13.5	15.5 : 1	5.5	1720	5.7 : 1	*	14.0	15.47	15.63
	L-His HCl · H_2O	1.5								

* The colorimetric determination of diazo reaction of the samples 2, 3, and 4 was impossible to measure due to their insolubility in water.

Half of the chloroform solution (10 cc) was added dropwise to L-histidine methyl ester prepared from its 2 HCl salt (10 gm) and triethylamine (10.4 cc) in chloroform (20 cc). This was followed by the addition of triethylamine (5.2 cc) and another half portion of chloroform solution.

The reaction mixture was stirred for thirty minutes at −5° C. and kept overnight at room temperature. The reaction mixture was washed with water, dried with sodium sulphate, and evaporated.

Syrup (15 gm) was obtained. The *N*-carbothiophenyl-ε-aminocaproyl-L-histidine methylester (14 gm) dissolved in acetone (10 cc) was heated with 8 N hydrochloric acid (45 cc) for thirty minutes at 100° C. After cooling, the reaction mixture was decolorized with charcoal, acetone was distilled off, water was added, and the mixture was concentrated. The residue was washed with ether, petroleum ether, and dried. *N*-carbothiophenyl-ε-aminocaproyl-L-histidine hydrochloride was obtained. Yield 8.5 gm; m.p. 93°–95° C.

Analysis: Found, C 51.0%, H 5.80%, N 12.6%, S 7.31%; equiv. wt. by CH_3ONa titration 448/3. Calcd. for $C_{19}H_{25}O_4N_4SCl$, C 51.8%, H 5.66%, N 12.7%, S 7.26%; mol. wt. 440.5.

N-carbothiophenyl-ε-aminocaproyl-L-histidine HCl (5 gm) was heated with dioxane (1.5 cc) at 120° C. for thirty-six hours. The product was removed with acetone, crushed to a powder, and washed with acetone and ether. The polymer was obtained in the yield of 3 gm.

Analysis: Found, C 49.2%, H 6.74%, N 19.54%, Cl 12.72%; mol. wt. by Van Slyke's method 983. Calcd. for $(C_{12}H_{19}O_2N_4Cl)_n$, C 50.0%, H 6.60%, N 19.55%, Cl 12.38%.

The molar ratio of ε-amino-caproic acid : histidine HCl measured by the colorimetric determination of paper chromatography with ninhydrin developer was 1 : 1.

Poly (β-alanyl-L-histidine HCl) by carbothiophenyl method (9): The sodium ethylate solution in alcohol (60 cc, Na 2.8 gm) was dropped into a solution of β-alanyl-L-histidine ethyl ester 2 HCl (20 gm) in ethanol (60 cc) and then carbothiophenyl chloride (5.25 gm) was added to it at below −5° C. by stirring. After standing for ten minutes, sodium ethylate in ethanol (30 cc, Na 1.4 gm) and carbothiophenyl chloride (5.25 gm) were mixed portionwise with each other.

This reaction mixture was allowed to stand for thirty minutes at −5° C. and was stirred constantly for one hour at room temperature.

After separation of the crystals of sodium

chloride, the ethanol solution was concentrated in vacuo. The syrupy residue was dissolved in ethyl acetate, the solution was washed with water, dried with sodium sulphate, and evaporated to dryness. The residue was washed with ether and concentrated in a vacuum desiccator. Syrup (18 gm, 75%) was obtained.

N-carbothiophenyl-β-alanyl-L-histidine ethyl ester (18 gm) in acetone (110 cc) was hydrolyzed with 8 N hydrochloric acid (55 cc) by refluxing for thirty minutes at 100° C. The reaction mixture was evaporated to remove acetone and then water (50 cc) was added. The aqueous solution was evaporated at 35°–40° C. under reduced pressure and the syrup obtained was washed with ether and dried. It was reprecipitated with ether from the ethanol solution as crystals. The yield of *N*-carbothiophenyl-β-alanyl-L-histidine HCl was 15.5 gm (84.2%); m.p. 105° C. (dec.).

Analysis: Found, C 47.8%, H 4.59%, N 14.0%, Cl 9.25%; equiv. wt. by CH_3ONa titration 136.5. Calcd. for ($C_{16}H_{18}O_4N_4S \cdot HCl$), C 48.1%, H 4.52%, N 14.1%, Cl 8.90%; mol. wt./3 133.

N-carbothiophenyl-β-alanyl-L-histidine HCl (3 gm) was heated with dioxane (1.5 cc) at 120° C. for thirty hours and then for five hours in an atmosphere of nitrogen.

The product was washed with acetone, boiled with alcohol, and washed again with acetone and ether. Yield 1.4 gm (76%). The polymer is hygroscopic and gives the biuret reaction. It is soluble in water, formic acid, a mixture of acetic acid and monochloracetic acid (1:1), and dichlor acetic acid, sparingly soluble in acetic acid, and insoluble in acetone, benzene, ether, and petroleum ether. Analysis: Found, C 43.5%, H 5.61%, N 22.72%, Cl 14.50%; equiv. wt. by $HClO_4$ titration 6,100, $n \doteqdot 25$. Calcd. for $(C_9H_{12}O_2N_4 \cdot HCl)_n$, C 44.1%, H 5.30%, N 22.90%, Cl 14.52%.

L-Histidyl-L-glutamic acid hydrobromide: α-Ethyl, γ-benzyl-L-glutamic acid diester was prepared from its hydrochloride salt (3.42 gm) and triethylamine (1.6 cc) in tetrahydrofurane (10 cc) and the solution filtered from triethylamine hydrochloride was mixed with dicarbobenzyloxy-L-histidine hydrate (*10*) (5.0 gm) in tetrahydrofurane (10 cc). Dicyclohexylcarbodiimide (*11*) (4.7 gm) in tetrahydrofurane (10 cc) was added to the mixture and stirred at room temperature for four hours and then left standing overnight. The crystals which separated out were filtered and tetrahydrofurane was evaporated. The residue was dissolved in ethyl acetate and washed with dilute acid, dilute alkali, and water. The ethyl acetate layer was dried with sodium sulphate and was evaporated (6.0 gm, 79%).

α-Ethyl, γ-benzyl-dicarbobenzyloxy-L-histidyl-L-glutamate (6 gm) in dioxane (9 cc) was hydrolyzed with 1 N sodium hydroxide (9 cc) at room temperature for fifty minutes. The reaction mixture was brought to pH 2 with 1 N hydrochloric acid and was washed with ethyl acetate. The aqueous layer brought to pH 4.5 was concentrated. γ-Benzyl-dicarbobenzyloxy-L-histidyl-L-glutamate was obtained as crystals (4.5 gm, 61.8%); m.p. 83°–86° C. γ-Benzyl-dicarbobenzyloxy-L-histidyl-L-glutamate (4.5 gm) was dissolved in glacial acetic acid (20 cc) and hydrogen bromide was passed in at 70° C. for one hour. After standing overnight, the acetic acid was evaporated and the residue was washed with ether and centrifuged. Yield, 2.12 gm (67.8%); m.p. 92°–95° C. (dec.). $[\alpha]_D^{22} = +9.9°$ (C 2.623, H_2O).

Analysis: Found, C 29.1%, H 4.52%, N 12.2%, Br 36.0%; equiv. wt. by CH_3ONa titration 115. Calcd. for ($C_{11}H_{18}N_4O_5Br_2$), C 29.6%, H 4.40%, N 12.6%, Br 35 9%; mol. wt./4 111.5.

L-Histidyl-L-lysine·3HBr: ϵ-Carbobenzyloxy-L-lysine ethyl ester in tetrahydrofurane (10 cc) prepared from its hydrochloride (3.9 gm) and triethylamine (1.6 cc) was mixed with dicarbobenzyloxy-L-histidine (5 gm) in tetrahydrofurane (10 cc) and dicyclohexylcarbodiimide (4.7 gm) in tetrahydrofurane (10 cc). The mixture was stirred for four hours and kept overnight at room temperature. The precipitate was filtered off and the filtrate was concentrated to dryness. The ethyl acetate solution of the residue was washed with dilute acid, dilute alkali, and water, and evaporated after drying with sodium sulphate.

Dicarbobenzyloxy-L-histidyl-ϵ-carbobenzyloxy-L-lysine ethyl ester was obtained as a syrup. Yield, 6.5 gm (80%). It was dissolved in dioxane (9.1 cc) and mixed with 1 N sodium hydroxide (9.1 cc). After stirring at room temperature for fifty minutes, the reaction mixture was acidified to pH 2 with 1 N hydrochloric acid and washed with ethyl acetate.

The aqueous layer was brought to pH 4.5

and evaporated. Dicarbobenzyloxy-L-histidyl-ε-carbobenzyloxy-L-lysine was obtained as a syrup. Yield, 4 gm (67%).

Dried hydrogen bromide was passed through the solution of the syrup (4 gm) in acetic acid (20 cc) at 70° C. for one hour. The solution was kept overnight at room temperature, and evaporated to dryness in vacuo.

The residue was washed with ether and centrifuged. Yield 1.92 gm (62.6%); m.p. 97°–99° C. (dec.). $[\alpha]_D^{22} = +17.2°$ (C 1.980, H_2O).

Analysis: Found, C 27.3%, H 4.65%, N 13.1%, Br 45.0%; equiv. wt. by CH_3ONa titration 135. Calcd. for ($C_{12}H_{24}N_5O_3Br_3$), C 27.4%, H 4.57%, N 13.3%, Br 45.6%; mol. wt./4 131.5.

L-Glutamyl-L-histidine hydrobromide: γ-Methyl-*N*-carbobenzyloxy-L-glutamate (9 gm), which was prepared from the mixture of γ-methyl L-glutamate (5 gm) in aqueous solution (10 cc) of sodium bicarbonate (2.61 gm), carbobenzyloxy chloride (5.85 gm), and sodium carbonate (1.81 gm) in water (10 cc), was dissolved in tetrahydrofurane. The solution was mixed with L-histidine methyl ester in tetrahydrofurane (20 cc), which was prepared from L-histidine methyl ester·2HCl (7.4 gm) in methanol (100 cc) and the methanol solution (30.5 cc) of 2 N sodium methylate.

Dicyclohexylcarbodiimide (6.3 gm) in tetrahydrofurane (10 cc) was added to the mixture and stirred at room temperature. The precipitates which separated were filtered. The filtrate was concentrated and the residue dissolved in ethyl acetate was washed with dilute acid, dilute alkali, and water, and the ethyl acetate layer dried with sodium sulphate was concentrated to dryness and crystallized with petroleum ether. *N*-Carbobenzyloxy-γ-methyl glutamyl-L-histidine methyl ester (9.5 gm, 70%) was obtained. The diester (7 gm) dissolved in dioxane (31.4 cc) was mixed with 1 N sodium hydroxide (31.4 cc) and stirred for one hour and then acidified to *p*H 2 with 2 N hydrochloric acid. The solution washed with ethyl acetate was brought to *p*H 4 and evaporated to dryness. The residue extracted with alcohol to remove sodium chloride was concentrated and taken out with ether. *N*-Carbobenzyloxy-L-glutamyl-L-histidine was obtained.

Yield 6.5 gm (99.2%); m.p. 76°–77° C.

The product (6 gm) was dissolved in glacial acetic acid (40 cc), hydrogen bromide passed in for one hour, and then the mixture stood overnight.

The filtered precipitate and the precipitate separated from the concentrated filtrate were combined and washed with ether. L-Glutamyl-L-histidine 2 HBr was obtained. Yield 5.7 gm (89%).

Analysis: Found, C 29.4%, H 4.62%, N 12.4%, Br 35.2%; equiv. wt. with CH_3ONa 109. Calcd. for ($C_{11}H_{18}N_4O_5Br_2$), C 29.6%, H 4.40%, N 12.6%, Br 35.9%; mol. wt./4 111.5.

Pyridine was added to the solution of L-glutamyl-L-histidine·2 HBr (3.5 gm) in alcohol (10 cc) until crystals separated and the reaction mixture was kept in an ice box. The crystals were filtered and washed with alcohol. L-Glutamyl-L-histidine hydrobromide was obtained. Yield 2 gm (70%); m.p. 137°–139° C. (dec.). $[\alpha]_D^{22} = +20.4°$ (C 0.687, H_2O).

Analysis: Found, C 36.6%, H 4.76%, N 15.30%, Br 21.4%; equiv. wt. 124.5. Calcd. for ($C_{11}H_{17}N_4O_5Br$), C 37.1%, H 4.65%, N 15.3%, Br 22.0%; mol. wt./3 122.

L-Lysyl-L-histidine hydrobromide: L-Histidine methyl ester was prepared from L-histidine methyl ester 2 HCl (4.09 gm) in methanol (50 cc) and 2 N sodium methylate (16.9 cc). To the solution of tetrahydrofurane (10 cc), dicarbobenzyloxylysine (*12*) (7 gm) in the mixed solvent of dioxane and tetrahydrofurane (1:1) (30 cc) was added and mixed with dicyclohexylcarbodiimide (3.5 gm) in tetrahydrofurane (10 cc). This was stirred and kept overnight at room temperature, filtered, and concentrated to dryness.

The residue was dissolved in ethyl acetate and this solution was washed with dilute acid, dilute alkali, and water, dried with sodium sulphate, and concentrated. α-ε-Dicarbobenzyloxy-L-lysyl-L-histidine methyl ester was obtained as crystals by treating with petroleum ether. Yield 2.5 gm; m.p. 84°–86° C.

The methyl ester (2 gm) dissolved in dioxane (3.54 cc) was mixed with 1 N sodium hydroxide (3.54 cc) and stirred for one hour at room temperature. The reaction mixture acidified to *p*H 2 was washed with ethyl acetate, brought to *p*H 4, and evaporated to dryness.

The residue dissolved in alcohol, after separation of sodium chloride, was crystallized with ether. α-ε-Dicarbobenzyloxy-L-lysyl-L-histidine was obtained. Yield 1.7 gm (87.2%); m.p. 85°–86° C.

Analysis: equiv. wt. by CH_3ONa titration 550; calcd. 551.

The carbobenzyloxy derivative (1 gm) dissolved in glacial acetic acid (40 cc) was treated with hydrogen bromide at 50° C., and was kept overnight at room temperature and concentrated in vacuo.

The crystals which separated out were washed with ether. L-Lysyl-L-histidine 3 HBr was obtained. Yield 0.94 gm (98.5%); m.p. 98°–100° C.

Analysis: Found, C 27.1%, H 4.63%, N 13.0%, Br 45.1%; equiv. wt. by CH_3ONa titration 130. Calcd. for ($C_{12}H_{14}N_5O_3Br_3$), C 27.4%, H 4.57%, N 13.3%, Br 45.7%; mol. wt./4 131.5.

L-Lysyl-L-histidine·2 HBr was prepared by the treatment of L-lysyl-L-histidine 3 HBr (0.5 gm) with pyridine in alcohol (5 cc). Yield 0.35 gm (83%); m.p. 118°–120° C. $[\alpha]_D^{22} = +11.7°$ (C 0.513, H_2O).

Analysis: Found, C 33.0%, H 5.22%, N 15.7%, Br 35.7%; equiv. wt. by CH_3ONa titration 145. Calcd. for ($C_{12}H_{23}N_5O_3Br_2$), C 32.3%, H 5.16%, N 15.72%, Br 36.0%; mol. wt./3 148.5.

2-C-Acetyl-L-histidine HBr·1/2 H_2O: The solution of 1-N-benzyl-L-histidine (10 gm) in glacial acetic acid (200 cc) was saturated with hydrogen bromide for two hours at 70° C. The solvent was evaporated in vacuo and the residue was washed with ether, dissolved in alcohol, precipitated with pyridine, and the precipitate was recrystallized from water and alcohol. Yield 6.5 gm; m.p. 219°–220° C.

Analysis: Found, C 33.3%, H 3.92%, N 14.2%, Br 28.8%, amino N (Van Slyke) 4.86%, H_2O 3.37%. Calcd. for ($C_8H_{11}O_3N_3Br$ ·1/2 H_2O), C 33.6%, H 3.85%, N 14.7%, Br 27.9%, amino N 4.88%, H_2O 3.14%. Equiv. wt. by $HClO_4$ titration 291 (calcd. 286) and by CH_3ONa titration 146 (calcd. M/2 = 143).

L-Tyrosyl-L-histidine hydrochloride: *N*-Carbobenzyloxy-*O*-benzyl-L-tyrosine (8 gm) in tetrahydrofurane (20 cc) and L-histidine methyl ester in tetrahydrofurane (20 cc) prepared from L-histidine methy ester 2 HCl (4.8 gm) and 2 N sodium methylate (19.8 cc) were mixed. To the mixture, the tetrahydrofurane solution (10 cc) of dicyclohexylcarbodiimide (4.1 gm) was added with stirring at room temperature. The precipitate which separated was filtered off and the filtrate was concentrated to dryness. The residue dissolved in ethyl acetate was washed with dil. acid, dil. alkali, and water. The ethyl acetate layer was dried with sodium sulfate and evaporated. *N*-Carbobenzyloxy-*O*-benzyl-L-tyrosyl-L-histidine methyl ester was obtained as crystals by treating the residue with ether and petroleum ether. Yield 8.0 gm (72.5%); m.p. 71°–74° C. (dec.).

The ester (7 gm) in dioxane (12.6 cc) was hydrolyzed with 1 N sodium hydroxide (12.6 cc) under cooling. After one hour, the reaction mixture was neutralized with 2 N hydrochloric acid. *N*-Carbobenzyloxy-*O*-benzyl-L-tyrosyl-L-histidine separated out as crystals. These were filtered and washed with water. Yield 6.7 gm (98.5%); m.p. 171°–172° C.

Analysis: Found, equiv. wt. by CH_3ONa titration 544; calcd. for ($C_{30}H_{30}N_4O_6$), 542.

N-Carbobenzyloxy-*O*-benzyl-L-tyrosyl-L-histidine (6 gm) was dissolved in liquid ammonia (100 cc) and reduced with sodium (1.1 gm). The blue color of the solution due to excess sodium was decolorized with ammonium chloride and then ammonia was evaporated at room temperature and the residue was dried over sulphuric acid in a desiccator. It was dissolved in 2 N hydrochloric acid and brought to *p*H 4.5. The solution was washed with ether and the water layer was evaporated to dryness in vacuo. The residue extracted with alcohol was separated from sodium chloride and pyridine was added to the solution until crystals separated out. These were kept in an ice box and filtered. The crystals were recrystallized with a mixture of methanol and alcohol. L-Tyrosyl-L-histidine HCl was obtained in a yield of 1.14 gm (29.0%); m.p. 207°–208° C. (dec.). $[\alpha]_D^{22} = +36.0$ (C 0.473, H_2O). Millon's reaction: positive.

Analysis: Found, C 51.2%, H 5.40%, N 15.7%, Cl 10.1%, equiv. wt. by CH_3ONa titration 177. Calcd., C 50.8%, H 5.35%, N 15.8%, Cl 10.0%; mol. wt./2 177.3.

Anhydro-L-aspartyl-L-histidine·H_2O·HBr: N-Carbobenzyloxy-β-benzyl-L-aspartate (10 gm) and L-histidine methyl ester prepared from L-histidine methyl ester 2 HCl (6.78 gm) were dissolved in tetrahydrofurane (40 cc) and mixed with dicyclohexylcarbodiimide (5.8 gm) in tetrahydrofurane (10 cc) at room temperature. The precipitate which separated out was filtered, the tetrahydrofurane solution was evaporated to dryness, and the residue dissolved in ethyl acetate was washed with dil. acid, dil. alkali, and water. The ethyl acetate layer was dried with sodium sulfate and evaporated to dryness in vacuo. *N*-Carbobenzyloxy-β-benzyl-L-aspartyl-L-histidine methyl ester was obtained as crystals by treating the residue with ether. Yield 12 gm (84.5%); m.p. 100°–103° C.

The above ester (6 gm) in dioxane (5.9 cc) was hydrolyzed with 1 N sodium hydroxide (5.9 cc) for one hour at room temperature and the reaction mixture was brought to *p*H 2 with 2 N hydrochloric acid. It was washed with ethyl acetate and then the water layer was adjusted to *p*H 4.5. It was evaporated to dryness and the residue dissolved in alcohol was separated from sodium chloride and the alcohol was evaporated. The syrupy residue was crystallized with ether. *N*-Carbobenzyloxy-β-benzyl-L-aspartyl-L-histidine was obtained. Yield 5.5 gm (94%); m.p. 108°–111° C.

Analysis: Found, equiv. wt. by CH_3ONa titration 451; calcd. for ($C_{25}H_{26}O_7N_4$), 494.

N-Carbobenzyloxy-β-benzyl-L-aspartyl-L-histidine (5 gm) was dissolved in glacial acetic acid (30 cc) and saturated with hydrogen bromide at 50° C. The reaction mixture was kept at room temperature overnight and was evaporated in vacuo. The residue was washed with ether, dried, and redissolved in alcohol. Pyridine (0.9 cc) was added to the alcohol solution and it was kept in an ice box. The precipitate was filtered, washed with alcohol, and dried. Yield 2.0 gm (54.7%); m.p. 154°–155° C. $[\alpha]_D^{22} = -3.62°$ (C 0.831, H_2O).

Analysis: Found, C 33.7%, H 4.35%, N 15.9%, Br 23.1%; equiv. wt. by CH_3ONa titration 182. Calcd. for ($C_{10}H_{15}O_5N_4Br$), C 34.2%, H 4.28%, N 16.0%, Br 22.8%; mol. wt./2 175.5.

L-Cysteinyl-L-histidine hydrochloride: L-histidine methyl ester in tetrahydrofurane (20 cc) prepared from L-histidine methyl ester 2 HCl (6.22 gm) and *N,S*-dicarbobenzyloxy-L-cysteine (10 gm) in tetrahydrofurane (20 cc) were mixed and dicyclohexylcarbodiimide (5.3 gm) in tetrahydrofurane (10 cc) was added to the mixture by stirring at room temperature. The precipitate separated out overnight, was filtered, and the filtrate was evaporated. The residue dissolved in ethyl acetate was washed with dil. acid, dil. alkali, and then water. The ethyl acetate layer was dried with sodium sulfate and evaporated to dryness. The residue was crystallized with ether. *N,S*-Dicarbobenzyloxy-L-cysteinyl-L-histidine methyl ester was obtained. Yield 9 gm (65%); m.p. 112°–114° C.

The ester (8 gm) was dissolved in dioxane (14.8 cc) and hydrolyzed with 1 N sodium hydroxide (14.8 cc) at room temperature for one hour. The reaction mixture was brought to *p*H 2 with 2 N hydrochloric acid, washed with ethyl acetate, and the aqueous solution was adjusted to *p*H 4.5. It was evaporated to dryness and extracted with alchol to separate sodium chloride. The alcohol solution was evaporated and the residue was crystallized with ether. *N,S*-Dicarbobenzyloxy-L-cysteinyl-L-histidine was obtained in a yield of 6 gm (77%); m.p. 66°–69° C.

Analysis: Found, equiv. wt. by CH_3ONa titration 531; calcd. for ($C_{25}H_{26}O_7N_4S$), 526.

N,S-Dicarbobenzyloxy-L-cysteinyl-L-histidine (5.7 gm) was reduced with sodium (1.5 gm) in liquid ammonia. The solvent was evaporated at room temperature and the residue was dried over sulfuric acid in a desiccator in a hydrogen atmosphere. It was dissolved in 2 N hydrochloric acid and the solution was brought to *p*H 4.5 and then washed with ether. The aqueous solution was evaporated to dryness and the residue was extracted with methanol to separate sodium chloride. The methanol solution was evaporated and the residue was crystallized with acetone. Yield 2.2 gm (69%); m.p. 96°–98° C. (dec.). $[\alpha]_D^{22} = +7.30°$ (C 1.370, H_2O). Nitroprusside reaction: positive.

Analysis: Found, C 36.5%, H 4.92%, N 18.9%, Cl 12.4%, equiv. wt. by CH_3ONa titration 148. Calcd. for ($C_9H_{15}O_3N_4SCl$), C 36.7%, H 5.09%, N 19.0%, Cl 12.1%; mol. wt./2 147.3.

L-Seryl-L-histidine hydrochloride: *N*-Carbobenzyloxy-*O*-benzyl-L-seryl-L-histidine *(12)* (0.2 gm) was reduced with sodium (0.5 gm) in liquid ammonia (20 cc). After evaporation of the ammonia at room temperature, the residue was dried in a desiccator. It was dissolved in 2 N hydrochloric acid and the solution was brought to *p*H 4.5 and then washed with ether. The aqueous solution was evaporated to dryness in vacuo and the residue was extracted with alcohol. Pyridine was added to the alcohol solution and the precipitate which separated out was filtered. It was recrystallized with methanol-alcohol. Yield 40 mg (33.9%); m.p. 188°–191° C. (dec.).

Analysis: Found, C 38.5%, H 5.42%, N 19.8%, Cl 12.5%, equiv. wt. by CH_3ONa titration 142. Calcd. for ($C_9H_{15}N_4O_4Cl$), C 38.8%, H 5.39%, N 20.1%, Cl 12.8%; mol.wt./2 139.3.

Results and discussion

Some imidazole derivatives, histidine derivatives, dipeptides of L-histidyl-L-glutamate, L-glutamyl-L-histidine, L-histidyl-L-lysine, L-lysyl-L-histidine, β-alanyl-L-histidine, and ε-amino-caproyl-L-histidine, and many copolymers containing histidyl residue were prepared for study of catalytic activity. The velocity constants of the hydrolysis for *p*-nitrophenylacetate by the imidazole groups of histidyl residues in these compounds were measured and the rate of activity of each imidazole group was shown against a standard of 1.0 of L-histidine in order to compare the effect of the neighboring groups to the histidyl residue. The results are listed in Table 30.5.

The activity of imidazoylmethyl alcohol is greater than that of methylimidazole and less than that of histidine. This difference might be due to the activation of the OH group adjacent to the imidazole group. The activity of benzyl-L-histidine having the $C_6H_5CH_2$ group on the *N* of imidazole is completely absent, in contrast to the activity in the case of methyl-L-histidine with a CH_3 group in the same position. When the acetyl group is on the *C* atom of —N=CH—N— of imidazole, the activity disappears too. The structure of the acetyl derivative was inferred from the following data.

The color of the diazo reaction is orange-yellow. After treatment with 28 per cent aqueous ammonia or 20 per cent hydrochloric acid the acetyl derivative was recovered without any change. It was converted into L-histidine by reduction with liquid ammonia and sodium.

The acetyl group should accordingly be in a position other than the *N*-1 of imidazole because the *N*-acetyl derivative is thought to be easily hydrolyzed. 1-*N*-Benzyl-L-histidine shows a strong absorption band at 13.8 μ due to the benzyl group, but it disappears by treating with hydrogen bromide in acetic acid.

The four strong bands of L-histidine hydrochloride, perhaps due to imidazole, appear in the range of 11.5–12.5 μ, and they are weak and deformed in the case of 1-*N*-benzyl-L-histidine. The absorption of the acetyl derivative obtained is different from that of L-histidine and is intermediate between those of L-histidine and *N*-benzyl-L-histidine in intensity. These facts suggest some displacement of the imidazole of histidine. The stretching vibration band of the carbonyl near 6.2 μ of L-histidine is similar to that of benzyl-L-histidine but that of the acetyl derivative seems to be composed of two kinds of carbonyl groups.

From these data, the acetylhistidine is 2-*C*-acetyl-histidine. It has no catalytic activity for *p*-nitrophenylacetate. All of the histidine copolymers debenzylated by treatment with hydrogen bromide in acetic acid were completely inactive in contrast to those debenzylated by liquid ammonia and sodium. The activation or inactivation may depend on the nucleophilic or nucleophobic replacement on the imidazole group, as this may in turn affect the resonance state. L-Glutamyl-L-histidine and L-lysyl-L-histidine were more active than L-histidyl-L-glutamate and L-histidyl-L-lysine and the activity of L-lysyl-L-histidine was greater than that of L-glutamyl-L-histidine. The activities of the mixture of dipeptides having the histidyl group showed the additive activities of the respective components without additional activation.

Carbobenzyloxy derivatives showed generally strong activity. The benzyl group might allow *p*-nitrophenylacetate to come close to the catalyst and to keep the substrate in high concentration near the active site of the catalyst. The activities of copoly (L-his, L-glu), copoly (L-his, L-asp), copoly (L-his, L-lys), and copoly (L-his, L-ser) of low molecular weights prepared by Leuchs' method were a little greater than that of histidine, but the activities of copoly (his, glu 1:1) and copoly (his, asp 1:1) prepared by the Fox-Harada method were much greater than those of Leuchs' polymers despite their low molecular weights. The activity of β-alanyl-L-histidine was 2.40 against 1.00 of histidine and that of ε-aminocaproylhistidine was the same as that of histidine.

Table 30.5—Catalytic activities of compounds containing imidazole groups

Catalyst	$k \times 10^{4}$* moles/l. min.	Activity of 3.311 m moles imidazole derivative relative to histidine	Total amino-acid residues
Imidazoyl alcohol[13]	17.0	2.52	—
Methylimidazole[14]	4.05	0.60	—
His-HCl	6.75	1.00	—
Cbzo-his[15]	23.70	3.49	1
His methyl ester·2HCl[16]	10.04	1.51	1
N-(1)-Methyl-his	7.23	1.07	1
N-(1)-Benzyl-his	0.00	0.00	1
C-(2)-Acetyl-his	0.00	0.00	1
β-Ala-his(carnosine)·HCl[17]	16.30	2.40	2
ε-Aminocaproyl-his	6.75	1.00	2
Cbzo-carnosine[17]	18.53	2.73	2
Carnosine methyl ester·HCl	9.11	1.34	2
His-glu·2HBr	7.56	1.12	2
Glu-his·HBr	11.06	1.64	2
Cbzo-glu-his	16.50	2.45	2
His-lys·3HBr	9.76	1.44	2
Lys-his·2HBr	18.30	2.72	2
Anhydroasp-his·HBr	12.17	1.81	2
Ser-his·HCl	9.36	1.39	2
Tyr-his·HCl	21.40	3.18	2
Cys-his·HCl	14.70	2.18	2
Poly-L-his·HCl†	9.46	1.40	3.2
Copoly(L-asp(NH_2), L-his·HCl 22:10)†	8.67	1.28	6.4
Copoly(L-asp, L-his·HCl 22:10)†	8.61	1.27	6.4
Copoly(L-glu(NH_2), L-his·HCl 26:10)†	7.08	1.05	2.2
Copoly(L-glu, L-his·HCl 26:10)†	7.62	1.13	2.2
Copoly(L-lys·HCl, L-his·HCl 16:10)†	8.82	1.30	6.8
Copoly(L-ser, L-his·HCl 18:10)†	8.23	1.22	7.0
Copoly(L-cys, his·HCl 37:10)†	9.94	1.47	3.4
Copoly(glu, his·HCl 1:1)‡	17.30	2.56	4.0
Copoly(asp, his·HCl 1:1)‡	13.50	2.00	3.0
Poly(β-ala-his·HCl)§	16.65	2.44	50.0
Copoly(ε-aminocaproic acid, his·HCl 3:2)‖	6.76	1.00	13.5
Poly(ε-aminocaproyl-his·HCl)§	6.71	1.00	6.6
Tyr-his·HCl + Cys-his·HCl	18.08	2.68(2.66)	
Tyr-his·HCl + Ser-his·HCl	15.31	2.27(2.29)	
Tyr-his·HCl + Carnosine	18.20	2.70(2.79)	
Glu-his·HBr + Carnosine	13.20	1.96(2.02)	
Poly-L-his·HCl + Tyr-his·HCl	15.52	2.30(2.29)	
Poly-L-his·HCl + Lys-his·2HBr	13.50	2.00(2.06)	
Tyr-his·HCl + Lys-his·2HBr	19.56	2.90(2.95)	

* k is the reaction velocity constant of hydrolysis for p-nitrophenylacetate at pH 6.4 with phosphate buffer and at 25° C. The quantity of the substrate used was about 7–8 times that of catalyst. The reaction time was measured in the early stage at 30 min.

† By Leuchs method.

‡ By Fox-Harada method.

§ By N-carbothiophenyl method.

‖ By block polymerization of caproamide and his·HCl.

Abbreviations—Cbzo: Carbobenzyloxy; β-ala:β-alanine; glu: glutamic acid; asp: aspartic acid; lys: lysine; glu(NH_2): glutamine; asp(NH_2): asparagine.

Poly (β-ala-his) and poly (ε-aminocaproyl-his) prepared by polymerization of their monomeric dipeptides had the same activities as their monomers respectively; no significant activation by polymerization was observed. If we neglect the influence of the secondary structure of polypeptide in placing both types of amino-acid residue in three-dimensional proximity, the activity of the histidine residue in polypeptides might be decided mainly by the identity and arrangement of amino-acid residues adjacent to histidyl residues. The meaning of the polymeric structure in the enzyme might then be to keep the substrate in a high concentration near to the active site, or to have a part in maintaining the special shape of the polypeptide in the region of the active site, but these questions require further study in the future. Related to such studies, we had reported (*18*) the phenomenon of heat-reversible coagulation of poly (ε-aminocaproyl-ala) or poly (ε-aminocaproyl-pro) in water and had suggested that it would be due to the change of structure involving the alanyl or the prolyl residue.

We have recently found by six kinds of physical measurement that polyalanine or copolymers containing alanine undergo change in conformation at about 60° C. (*19*). E. Katchalski *et al.* (*20*) reported also a change in the secondary structure in poly-L-proline by heating. The alanyl or prolyly residue contributes so significantly to the secondary structure of the polymers that it is interesting to learn whether these groups play a role in the flexible part near the active site of enzymes or not.

Copoly (his, glu) and copoly (his, asp) prepared by the Fox-Harada method (*7*) were more active than the Leuchs copolymers, as mentioned above. The structures and catalytic activities of these copolymers were therefore studied in more detail. L-Aspartic acid and L-histidine hydrochloride or L-glutamic acid and L-histidine hydrochloride were mixed in various proportions and copolymerized thermally. The activities of these copolymers were measured qualitatively by comparing the yellow color of *p*-nitrophenol produced by the hydrolysis of *p*-nitrophenylacetate. The results are shown in Table 30.6. In any case, the copolymers from a 1:1 molar ratio had the greatest activity compared to other molar ratios. The copolymers from a 1:1 molar ratio at 180° were not so active, but heating to 200° C. gave more active polymers.

The amino-acid residues of the components in the copolymers were completely racemized in the thermal polymerization and the copolymers were not obtained in high molecular weight. The active copolymers from a 1:1 molar ratio of amino acid were assumed on the basis of their analyses to be the following structures.

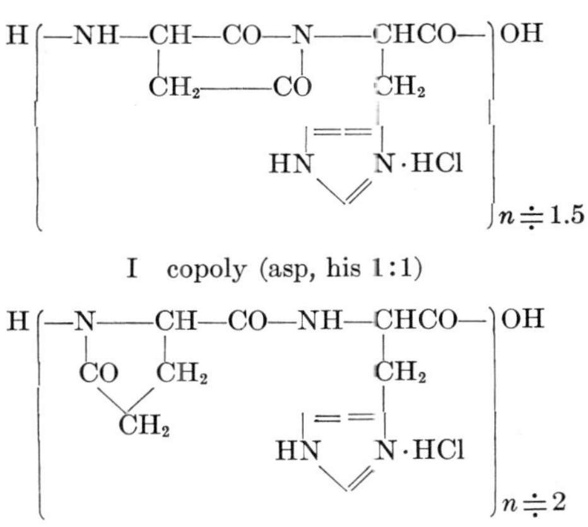

I copoly (asp, his 1:1)

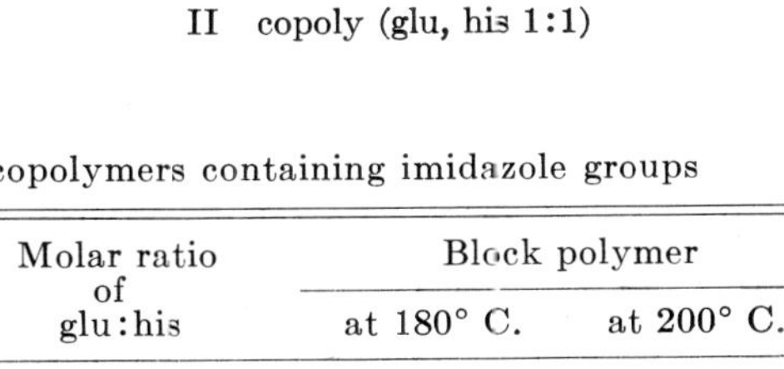

II copoly (glu, his 1:1)

Table 30.6—Composition and catalytic activities of copolymers containing imidazole groups

Molar ratio of asp:his	Block polymer at 180° C.	Block polymer at 200° C.	Molar ratio of glu:his	Block polymer at 180° C.	Block polymer at 200° C.
2:1	+	++	2:1	+	++
1:1	+	+++	1:1	+	+++
3:1	−	−	3:1	−	+
5:1	−	−	5:1	−	−
8:1	−	−	8:1	−	−
10:1	−	−	10:1	−	−
0:1	++	++	0:1	++	++

+ weakly active ++ active +++strongly active − inactive

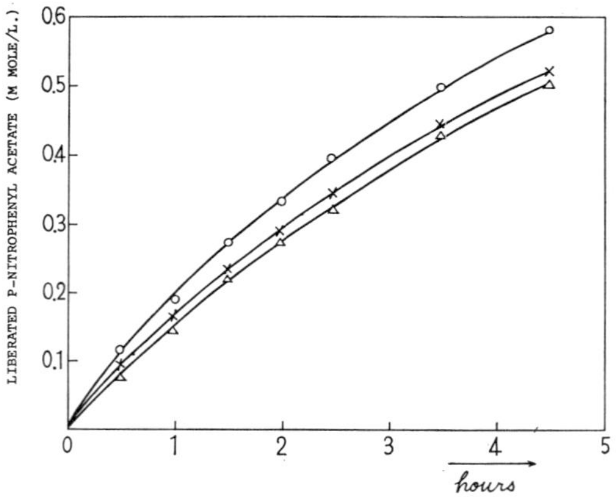

Fig. 30.1. The relationship between the reaction time and the catalytic hydrolysis of *p*-nitrophenylacetate.
○—○ copoly(his, glu)
×—× copoly(his, asp)
△—△ histidine·HCl

Structure I has both free amino and carboxyl groups and the β-carboxyl groups of the aspartyl residues are cyclized in the succinimide type as in the case of polyaspartate (*21*). Structure II has a carboxyl end but no free amino end. The γ-carboxyl groups of glutamyl residues are cyclized in the pyrrolidone type (*22*). When the mixture of L-glutamic acid and L-histidine is heated at 180° C. for six hours the product is no more than a dimer of amino acids (dipeptide). The tetramer II, however, is obtained by heating the dimer at 200° C. for six hours. Higher polymers were not obtained by longer heating. It seems that this polymerization proceeds in steps through (*a*) the condensation of acidic amino acid and histidine and (*b*) the coupling of the two dipeptides. Structure I is perhaps the mixture of dimer and tetramer. Structures I and II are soluble in water, formic acid, and acetic acid, sparingly soluble in pyridine, and insoluble in ether, benzyl ether, benzene, acetone, dioxane, and tetrahydrofurane. Structure II is soluble in methanol, ethanol, and benzyl alcohol, but Structure I is sparingly soluble in methanol and insoluble in ethanol and benzyl alcohol.

The catalytic activities of I and II for the hydrolysis of *p*-nitrophenylacetate were determined against L-histidine by measuring the yellow color of *p*-nitrophenol liberated from *p*-nitrophenylacetate in phosphate or veronal buffer with Hitachi E.P.U. Type 2 spectrophotometer. All of the data were collected under the same conditions of 100 mg/l. of catalyst (*c*), 3.311 millimoles/l. of substrate (*S*), reaction time (*t*) of 30 min., *M*/30 phosphate buffer, and temperature of 25° C. The reaction mixture contained 10% (*v*/*v*) acetone.

The relationship between the reaction time and the *p*-nitrophenol produced was shown by a straight line for thirty minutes, as in the early stage of Fig. 30.1. The reaction time for the determination of activity was thirty minutes, in all data of this paper.

The optimal *p*H of the catalytic hydrolysis was 6.9 in phosphate buffer and 7.4 in veronal buffer, as shown in Fig. 30.2. This suggests that the hydrolysis involves three components: copolymer, *p*-nitrophenylacetate, and buffer. The succinimide ring of Structure I was opened to aspartyl residues by treatment with sodium methylate in methanol, but the pyrrolidone ring of II was not attacked by this reagent. The sample of Structure I in which the succinimide

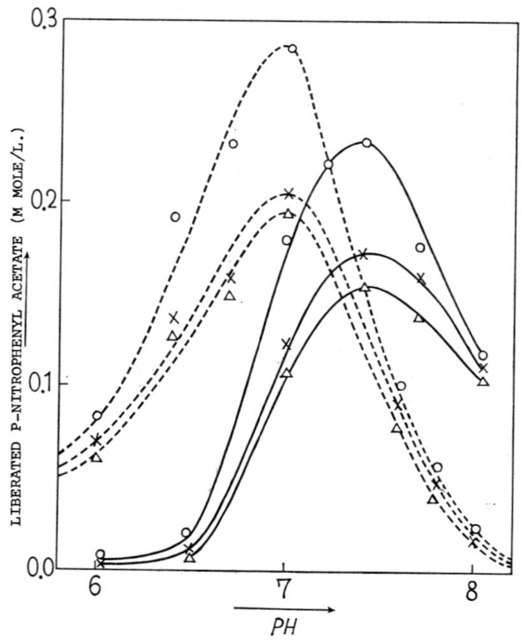

Fig. 30.2. The effect of *p*H on the catalytic hydrolysis of *p*-nitrophenylacetate.
○—○ copoly(his, glu)
×—× copoly(his, asp)
△—△ histidine·HCl
——— phosphate buffer
--------- veronal buffer

ring had been opened had the same activity as the cyclic type. It seems to make no difference in the activity whether the amino-acid residue is ringed or open. The copoly (glu, his 1:1) dimer was not so active but the tetramer II was more active (Table 30.5). In the catalytic hydrolyzate, acetic acid was detected. This suggests that the copolymer behaves as a catalyst hydrolyzing *p*-nitrophenylacetate into *p*-nitrophenol without being acetylated. The reaction velocities of hydrolysis were measured at 25° C. in various concentrations of *p*-nitrophenylacetate at a constant concentration of copolymer. The velocities were proportional to the concentrations of the substrate and led to a first-order dependence of reaction rate as shown in Fig. 30.3. From the gradient of each straight line of Fig. 30.3, the velocity constants were calculated. They are listed in Table 30.5. The relationship between the reaction temperature and the reaction velocity at a constant ratio of catalyst to substrate is shown in Fig. 30.4. The reaction velocity increases with increase in temperature and the curve obtained is not similar to that found for natural enzymes, which shows a temperature optimum. This might be due to the fact that the copolymers used as catalysts are of such low molecular weight as to lack a secondary structure subject to thermal change recognizable as "denaturation" in aqueous solution. When Fig. 30.4 was redrawn to show the relationship between the logarithm of the velocity constant k and the reciprocal of the temperature T, the straight lines of Fig. 30.5 were obtained. The activation energies ΔE calculated from the gradient of each line in Fig. 30.5 were the following; ΔE copoly (his, asp 1:1)$_{1.5}$ 10.63 Kcal, ΔE copoly (his, glu 1:1)$_2$ 9.91 Kcal, and ΔE L-his 11.37 Kcal, respectively. The inhibitory effects on the catalytic action of these copolymers were tested with various metallic ions in amounts equimolar to the histidyl residues in the copolymers. The inhibition was in the following order,

$$Cu^{2+} > Co^{2+} > Cr^{3+} > ICH_2\text{-}COOH.$$

The cupric ion, Cu^{2+}, was the most effective,

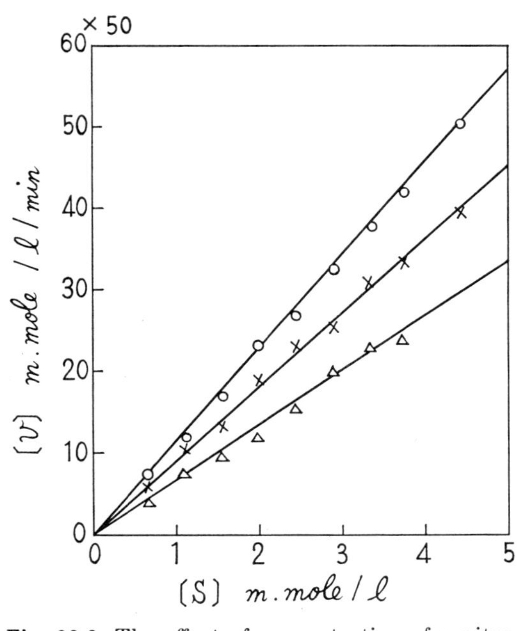

Fig. 30.3. The effect of concentration of *p*-nitrophenylacetate on the velocity of the catalytic hydrolysis.
○—○ copoly(his, glu)
×—× copoly(his, asp)
△—△ histidine·HCl

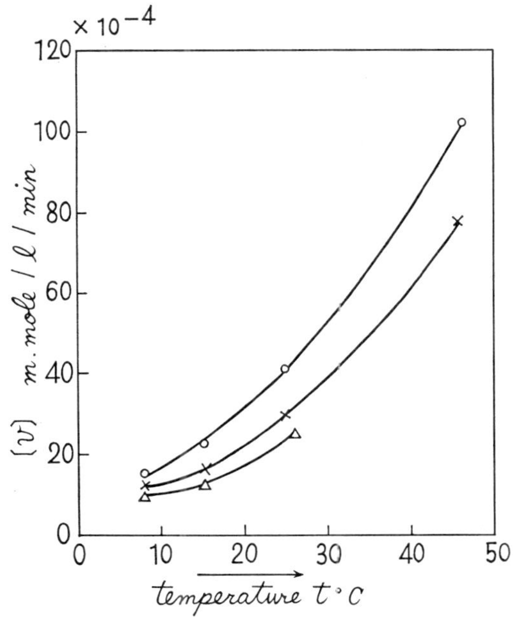

Fig. 30.4. The relationship between reaction temperature and the velocity of the catalytic hydrolysis of *p*-nitrophenylacetate.
○—○ copoly(his, glu)
×—× copoly(his, asp)
△—△ histidine·HCl

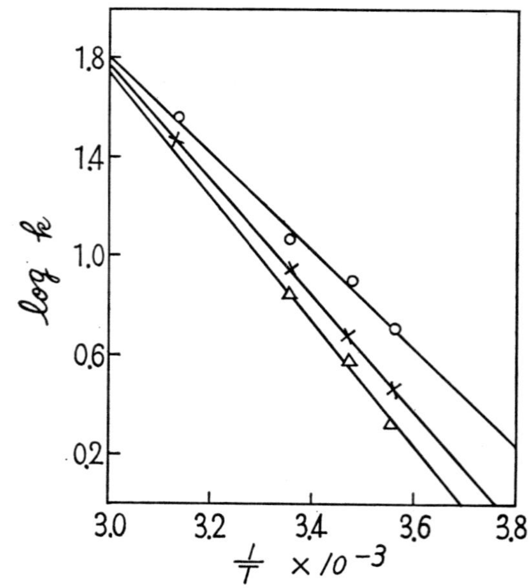

Fig. 30.5. The relationship between the velocity constant of the hydrolysis of *p*-nitrophenylacetate and the reaction temperature.
○—○ copoly(his, glu)
×—× copoly(his, asp)
△—△ histidine·HCl

inhibiting about 80 per cent of the hydrolytic activity of the catalyst. Ca^{2+}, Mg^{2+}, and Mn^{2+} ions were almost uninhibitory (*23*).

It is planned to continue this study in the following directions in the future: preparation of oligopeptides having more active combinations with a definite arrangement of amino-acid residues including histidine; preparation of polymers containing the active oligopeptide and a study of their activities; preparation of polymers having alanyl or prolyl residues adjacent to the active oligopeptide as a model of the flexible part of the active center; and preparation of compounds of an active polymer and another polymer which easily adsorbs the substrate.

Summary

1. The catalytic activity of imidazoyl alcohol for the hydrolysis of *p*-nitrophenylacetate was a relatively high value of 2.52 compared to 1.00 of L-histidine. This might be due to the activation of the OH group adjacent to the imidazole group but copoly (L-ser, L-his 18:10) had an unexpectedly low value of 1.22.

2. The hydrolytic activity is lost when an electrophilic group such as benzyl or acetyl is substituted on the imidazole of histidine.

3. In the case of dipeptides such as L-his-L-glu, L-glu-L-his, L-his-L-lys, and L-lys-L-his the compounds with histidine in the C-terminal position were more active than those with histidine N-terminal. The activity of L-lysyl-L-histidine was the highest, at 2.72.

4. Carbobenzyloxypeptides were more active, 2.5–3.5, than the corresponding peptides. Carbobenzyloxyhistidine was more active than histidine and had unusually high activity.

5. The copolymers of L-histidine and L-glutamic acid, L-aspartic acid, L-lysine, or L-serine were unexpectedly not so active. This low activity might be due to the difficulty of homogeneous copolymerization between L-histidine Leuchs anhydride and another Leuchs anhydride or the difficulty of a high degree of polymerization. Further study on such copolymerization should be worth while.

6. The active rates of β-alanyl-L-histidine and ε-aminocaproyl-L-histidine were 2.40 and 1.00, and poly (β-ala-L-his) and poly (ε-aminocaproyl-L-his) had also the same activities as the monomers. No activation was inferred for the polymerization.

7. Copoly (asp, his $1:1)_{1.5}$ I and copoly (glu, his $1:1)_2$ II prepared by polymerization at 200° C. showed the high activities of 2.00 and 2.56 and other proportions of these amino-acid residues in the reaction mixtures were almost inactive. Structures were postulated for I and II and the mode of catalytic action of hydrolysis for *p*-nitrophenylacetate was studied in detail. The catalytic reaction seems to proceed with the mutual involvement of the three components: catalyst, substrate, and buffer. The inhibition of the catalytic activity was the order of $Cu^{2+} > Co^{2+} > Cr^{3+} > ICH_2COOH$ and was virtually unaffected by Ca^{2+}, Mg^{2+}, and Mn^{2+}.

Acknowledgement

The authors express their sincere thanks to Dr. E. Katchalski and Dr. G. D. Fasman for their kind present of L-histidine, to the Ajinomoto Co., Ltd., and the Ministry of Education, Japan, for their support, and to Dr. Sidney W. Fox for discussions.

References

1 T. Spencer, *J. Am. Chem. Soc.* **81,** 1874 (1959).

2 B. S. Hartley and V. Massey, *Biochim. et Biophys. Acta* **21,** 58 (1956); A. S. Brecher and A. K. Balls, *J. Biol. Chem.* **227,** 845 (1957); M. L. Bender and B. W. Turnquest, *J. Am. Chem. Soc.* **79,** 1663 (1957); **80,** 148, 1173 (1958); T. G. Bruice, *ibid.,* **81,** 2860 (1959).

3 A. Lukton and H. S. Olcott, *Biochim. et Biophys. Acta* **32,** 267 (1959).

4 R. B. Merrifield and D. W. Woolley, *Federation Proc.* **17,** 275 (1958).

5 M. Sela and E. Katchalski, *Advances in Protein Chemistry* **14,** 420 (1959).

6 E. Katchalski, G. D. Fasman, E. Simons, E. R. Blout, F. R. N. Gurd, and W. L. Koltun, *Arch. Biochem. Biophys.* **88,** 361 (1960).

7 A. Vegotsky, K. Harada, and S. W. Fox, *J. Am. Chem. Soc.* **80** 3361 (1958); K. Harada and S. W. Fox, *ibid.* **80,** 2694 (1958); K. Harada and S. W. Fox, *Arch. Biochem. Biophys.,* **86,** 274 (1960).

8 G. E. Ham, U. S. Pat. 2,892,817 (1959).

9 J. Noguchi, *J. Chem. Soc. Japan* **74,** 961 (1953); J. Noguchi and T. Hayakawa, *J. Am. Chem. Soc.* **76,** 2846 (1954).

10 S. Akabori, K. Okawa, and F. Sakiyama, *Nature* **181,** 772 (1958).

11 J. C. Sheehan and G. P. Hess, *J. Am. Chem. Soc.* **77,** 1067 (1955).

12 K. Okawa, *Bull. Chem. Soc. Japan* **29,** 488 (1956); *ibid.* **30,** 976 (1957).

13 R. Weidenhagen, R. Herrmann, and H. Wegner, *Ber., deut. chem. Ges.* **70,** 570 (1937).

14 A. Windaus and F. Knoop, *ibid.* **38,** 1166 (1905).

15 A. Patchornik, A. Berger, and E. Katchalski, *J. Am. Chem. Soc.* **79,** 6416 (1957).

16 E. Fischer and L. H. Cone, *Ann. Chem. Liebigs* **363,** 1070 (1908).

17 R. H. Siefferd and V. du Vigneaud, *J. Biol. Chem.* **108,** 753 (1935).

18 J. Noguchi, T. Hayakawa, J. Suzuki, and M. Ebata, *J. Chem. Soc. Japan* **76,** 548 (1955); *J. Polymer Sci.* **23,** 843 (1957); M. Ebata, *Bull. Chem. Soc. Japan* **33,** 899 (1960).

19 J. Noguchi *et al.,* unpublished.

20 J. Kurtz, A. Berger, and E. Katchalski, "Recent Advances in Gelatin and Glue Research," Pergamon, London, 1958, p. 131; J. Kurtz, G. D. Fasman, A. Berger, and E. Katchalski, *J. Am. Chem. Soc.* **80,** 393 (1958).

21 J. Noguchi, T. Saito, and M. Asai, *J. Chem. Soc. Japan* **81,** 620 (1960).

22 M. Asai, K. Imai, and J. Noguchi, *ibid.* **81,** 624 (1960).

23 A part of this was reported by J. Noguchi *et al., Ann. Rept. Fibre Research Inst., Osaka Univ.* **13,** 139 (1960).

Discussion of Paper 30: Catalytic Activity of Synthetic Polyamino Acids

S. W. FOX

The process referred to by Dr. Noguchi as the Fox-Harada method differs from our thermal process by the use of paraffin and by deletion of the dialysis step which we include. One or both of these differences may explain why the molecular weights by Dr. Noguchi's process are so very much lower than those recorded by us or others and may also explain his higher yields.

At this time, due to the work of Luke Rohlfing and Peter Hoagland, we can generalize that some of our proteinoids, which contain histidine, have much more catalytic activity than the histidine-containing simple polymers so far tested. In tests on thirty preparations of proteinoid, we found nine to be at least three times as active as histidine when compared on a basis of histidine equivalence, and two of these more than ten times as active. One of the latter is shown in Fig. 5.1 in Paper 5.

The fact that our catalytically active proteinoids can be thermally inactivated in aqueous solution indicates subtle intramolecular structures. One would not expect such behavior from simple derivatives of histidine such as lysylhistidine, but the experiment has not yet been performed.

PAPER 31

Chemotherapeutic Possibilities of Polyamino Acids

MARK A. STAHMANN

In consideration of the chemotherapeutic possibilities of polyamino acids, it should first be pointed out that a major defense of higher animals including man against bacteria or virus infections rests upon the biological properties of proteins or polypeptides. These include proteins such as antibodies with a very specific action against a given infective agent, other proteins with a more general action against many pathogens such as the recently described interferon (*1*) which inhibits virus multiplication, properdin (*2*) which destroys bacteria and neutralizes viruses in the presence of complement, and some nonspecific proteins which cause agglutination or lysis of bacteria. In addition, there are many very active antibiotics that can be considered to be polypeptides or derivatives of polypeptides. Secondly, the regulation and control of many biochemical reactions or physiological processes rests upon the biological properties of polypeptide or protein hormones. Since the polyamino acids have been shown to share many of the chemical and physical, and some biological properties of proteins and certain polypeptides, it would seem that we should consider whether or not it may be possible to prepare polyamino acids whose biological properties overlap or modify those of the proteins or polypeptides that are involved in the defense or regulatory reactions.

Possibilities against viruses

The first chemotherapeutic possibility to be considered may be stated as a question: Is it possible to prepare polyamino acids whose biological activities will overlap those of the proteins or of polypeptides that inhibit the multiplication of infectious agents? I believe that this is a very real possibility and that already some progress has been made in this direction.

Our own interests in this possibility developed shortly after the war when we were studying a virus inhibitor protein from spinach leaf tissue (*3*) that rapidly and reversibly inhibited or reduced the infectivity of tobacco mosaic virus. We obtained evidence that this inhibition involved an interaction between the virus and the spinach protein. Such an interaction seemed to involve the acidic and basic groups of the virus and spinach proteins. My interest in polyamino acids crystallized in 1947 when Dr. Katchalski published his communication to the editor describing polylysine (*4*). Within a few weeks, we had synthesized polylysine with the hope that it might serve as a model of the spinach protein and tested it for virus inhibitory activity. The data obtained clearly indicated that the synthetic lysine polypeptide did indeed produce a reversible inhibition or reduction in local lesions formed by tobacco mosaic virus (*5*). This inhibition was closely similar to that produced by the spinach protein; thus polylysine did serve as a biologically active model of a plant protein and the virus-inhibitory activity of polylysine did overlap that of the spinach protein.

An ionic combination formed between many epsilon amino groups of polylysine and acidic groups of the virus nucleoprotein (*6*). This combination was rapid and reversible. It produced an aggregation and precipitation of virus particles and seemed to block essential groups at the surface of the virus. We soon showed that other plant viruses and a bacterial virus were also inhibited by polylysine.

The results suggested that the basic polylysine interfered with cell-virus interaction or virus multiplication by forming an ionic combination with essential acidic groups of the virus or host cell. This concept was supported by other work which demonstrated an ionic interaction in the attachment of bacterial viruses to host cells. Consequently, it seemed likely that polylysine might inhibit other viruses, including those attacking animals, and thus provide a basis for a new approach into virus chemotherapy. Experiments which showed an inhibition of influenza, infectious bronchitis virus, Newcastle disease virus, and mumps virus by polylysine supports this view. These viruses were cultured in chick embryos and the extent of virus inhibition was determined by the reduction in the infectivity or the hemagglutination titers, or by prevention of death of the infected embryos.

When 0.2 mg of polylysine was injected into chick embryos one hour prior to infection with 1000 infective doses of bronchitis virus, 80 per cent of the embryos survived while all of the control embryos died (*8*). With influenza virus it was possible to measure the reduction in the virus multiplication by the hemagglutination technique. Ninety-four to 96 per cent inhibition of virus multiplication was obtained by treatment with polylysine (*7*). Polylysine was more active against the virus causing mumps (*9*). Some typical results on the inhibition of mumps virus by polylysine are presented in Table 31.1. The inhibition was first measured by hemagglutination, then confirmed by infectivity titrations. When chick embryos were treated with 0.04 mg of polylysine followed by 30 infectious virus doses, there was 99.9 per cent inhibition of growth of the mumps virus as measured by infectivity. At 30,000 infectious doses, 98 per cent inhibition was obtained. As little as 1 μg per embryo of polylysine caused an 85 per cent inhibition of mumps virus multiplication when 30 infectious doses were administered. If this polylysine were distributed throughout the incubating egg, then mumps virus inhibition was manifest at a concentration of only one part in 50 million. Hence, polylysine is active in inhibiting mumps virus multiplication at very low concentrations. From experiments where the time interval between inoculation of the mumps virus and administration of the polyamino acid was varied, it was shown that the polylysine could be injected into the embryo as late as 36 hours after infection and still significantly inhibit virus multiplication.

It appeared that polylysine inhibited tobacco mosaic virus by combining with virus particles. The combination could cause a precipitation which showed an equivalence zone like that with antisera. Reduction in infectivity occurred only when the virus and the polypeptide were mixed; very little inhibition was obtained when the polypeptide was applied to the leaf after the virus. The inhibition of animal virus is more complex. The results cannot be explained adequately by a simple combination with virus or host cells. It appeared that polylysine exerted its inhibitory action by an intracellular mechanism, possibly through an effect on intermediates in virus synthesis, perhaps in protein synthesis. The basic polylysine readily combines with virus nucleic acids and hence may compete with the virus protein for viral nucleic acid. Polylysine inhibits some enzymes and may block essential enzymes.

We did not have the facilities in our department to work with infective animals. However, Dr. Cochran will tell you of the inhibition of poliomyelitis in mice by polylysine.

In summary, there is now good evidence that

Table 31.1—Effect of the amount of polylysine and the time of inoculation upon the inhibition of mumps virus multiplication in the chick embryo

1st injection	2nd injection	Interval (hours)	Per cent inhibition
0.04 mg PL	30 E.I.D.*	1	99.9
0.01 mg PL	30 E.I.D.	1	98
0.001 mg PL	30 E.I.D.	1	85
30 E.I.D.	0.4 mg PL	8	99.9
30 E.I.D.	0.4 mg PL	36	99.9
30 E.I.D.	0.4 mg PL	56	0

* E.I.D. = embryo infectious doses

some polyamino acids will inhibit the multiplication of viruses that infect bacteria, plants, and experimental animals. More research is needed to further evaluate a wider variety of polyamino acids against other viruses particularly in experimental animals.

Possibilities against bacteria and fungi

Following the observations that polylysine would inhibit virus multiplication and agglutinate virus particles, we turned to a study of its effect on bacteria. We observed that polylysine would cause agglutination of bacterial cells and inhibition of respiration and growth *(10)*. Independently Katchalski *et al.* *(11, 12)* observed the inhibition of bacterial growth by polylysine and studied the antibacterial action of some basic amino acid copolymers related to gramicidin. These studies revealed that amino-acid polymers containing the basic amino acid, lysine, or ornithine were antibacterial at levels comparable to that of some antibiotics. Growth inhibition was observed at 1 to 5 μg per ml. The inhibition could be reversed by acidic polymers as nucleic acid or polyglutamic acid. Resistance toward some of the copolymers could be developed; however, attempts to obtain resistant strains against polyornithine or polylysine were unsuccessful. De Vries *et al.* *(13)* reported that the basic polyamino acids promote phagocytosis of starch particles by white blood cells.

We have studied the action of polylysine on the growth and phagocytosis of some pathogenic bacteria in vitro *(14)* and found that the basic polypeptide markedly increases the phagocytosis of pneumococcus and streptococcus which have an acidic capsule. On the other hand in studies with staphylococcus, no opsonizing effect was seen even though the polylysine inhibited growth of this pathogen. In a study of the action of polylysine on experimental infection in animals *(15)*, it was observed that polylysine significantly protected mice against infection with types I, II, and III pneumococcus or with spores of *Bacillus anthracis* when the animals were inoculated intraperitoneally and treated intraperitoneally. When inoculation was at a site other than that of the treatment, there was little protection.

It appears that the basic polyamino acids are effective antibacterial agents in vitro but much less effective in vivo. This reduction in their action in vivo may be caused by the great tendency of the basic polymers to combine with acidic substances of the host and to be removed from circulation.

Preliminary experiments *(16)* have shown that polylysine has an anti-fungal activity against some plant pathogens and also against two human pathogens in vitro. It inhibited penetration of *Ceratocystis fimbriata* into sweet potato slices and the infection of tomato cuttings by Fusarium.

With Dr. Don Slagel we have made an electron microscope study of the effect of polylysine on bacterial cells *(17)*. In these experiments, a culture of *Escherichia coli* was incubated overnight; one portion of the culture was then suspended for 10 minutes in a poly-L-lysine solution containing 1 mg per ml and another portion in buffer alone. The bacterial cells were fixed in buffered 1 per cent osmium tetraoxide, suspended in agar, dehydrated, and embedded in an epoxy resin. Thin sections were cut and observed in the electron microscope. Pronounced morphological changes, produced by the polylysine were seen in the electron micrographs.

The electron micrographs of untreated cells (Plate 31.1) showed a cell wall, cytoplasmic membrane, an osmophilic cytoplasm which contains ribonucleoprotein particles, and a nuclear vacuole along the longitudinal axis of the cell. This nuclear vacuole seemed to be filled with a nucleic-acid network which did not stain with the acidic osmium tetraoxide. The most prominent morphological change produced by exposure to polylysine was the appearance of a marginal vacuole just inside the cytoplasmic membrane (Plates 31.2 and 31.3). Ferric-chloride staining revealed that this marginal vacuole also contained nuclear material, probably nucleic acids, which under the influence of the polylysine had migrated to the outer portions of the cell. These results suggest that exposure to polylysine induced profound changes in the internal morphology of the bacterial cell. Following polylysine treatment, the internal morphology resembled that of untreated cells in the senescent phase of growth. Both had their nuclear material dispersed throughout the cytoplasm (or concentrated along the cell wall).

Amino-acid analyses of the supernate above

treated and untreated cells revealed that polylysine caused an increased leakage of amino acids from the bacterial cells. Thus the amount of valine in the supernatant above-treated cells was increased 200 per cent; leucine, 100 per cent; aspartic acid, 80 per cent. Two unknown peaks were increased 320 per cent and 60 per cent respectively. We conclude from this that polylysine changed the character of the cell wall to allow the leakage of amino acid from the bacterial cell. It seems likely that the transfer of other compounds through the cell might also be effected.

I suggest that further study be made of the action of other polyamino acids on bacteria and fungi and that consideration be given to the topical application of basic polyamino acids in superficial bacterial infections. The evidence on which this suggestion is based may be summarized as follows: (*a*) Basic polyamino acids have high antibacterial action in vivo. (*b*) They enhanced phagocytosis of certain bacteria; other unpublished data suggested that basic polyamino acids stimulate phagocyte infiltration and increased capillary permeability. (*c*) Resistant strains to some basic polyamino acids could not be developed. Thus, although present evidence suggests that polyamino acids may not have a high systemic antibacterial effect, it seems likely that they may be of value for local topical applications.

Possibilities against tumors

The action of polylysine in suppressing the development of ascites tumors in mice was reported by Richardson *et al.* (*18*). These workers observed that all untreated mice transplanted with Erhlich ascites tumor cells developed large tumors after twelve days and died between 12 and 25 days following transplantation. In contrast, 60 per cent of the mice treated with 24 mg per kg of polylysine daily for six days were alive when all of the control mice had died and 28 per cent survived over 100 days without any sign of tumor. The development of the ascites tumors appeared to be largely suppressed; most of the mice subsequently died from solid tumors which developed often at the site of transplantation.

When polylysine contacts the tumor cells within the peritoneal cavity it seems capable of inhibiting their development. It does not act against solid tumors outside of the peritoneal cavity; apparently polylysine may not reach solid tumors in sufficient concentration to suppress their development. Polyglutamic acid, copolymers of leucine and lysine, glutamic acid and lysine, or valine and lysine, did not inhibit Erhlich ascites tumor. Study of the tumor cells from mice given polylysine showed that the mitotic index decreased from 3.6 to 1.6, which suggested an inhibition at prophase. This observation suggests that polylysine interferes with cell division, particularly with the latter stages of the mitotic cycle.

Subsequent studies have shown that polylysine was effective in inhibiting ascites tumors if injected immediately or within 24 hours after the tumor cells were transplanted but was not effective if administered much later. Subcutaneous or intravenous injection had no effect.

We have recently examined the effect of polylysine on the morphology of Erhlich ascites tumor cells (*19*). Washed tumor cells were suspended for ten minutes in dilute solutions of a fluorescent polylysine derivative and then viewed under an ultraviolet microscope. The fluorescence was seen bound to the surface of the cell. It did not bind uniformly but was concentrated at certain spots on the cell wall. When the cells were agglutinated, the fluorescence could be seen in the narrow space between cells. Fractionation of the cells revealed that the binding was most strong to the lipoprotein portion of the cell wall.

Following polylysine treatment the mitochondria and nuclei were seen near the cytoplasmic membrane. This disorganization of the internal structure of the ascites cells may have been the cause of the inhibition of mitosis. Cytoplasmic agglutination of nucleic acid and proteins was probably due to the basic polyelectrolyte character of the polycation. The polylysine-treated cells showed cytoplasmic aggregates that were sulfhydryl positive. An interaction between polylysine and sulfhydryl groups was suggested by a shift of the absorption maximum of thioglycolic acid when polylysine was allowed to react with it under alkaline conditions. The morphological changes observed following the interaction of polylysine with Erhlich ascites tumor cells closely resembled those noted by Lindner (*20*) following the reaction of these cells with specific antibody and complement.

By observing in the electron microscope thin sections cut through Ehrlich ascites tumor cells that had been exposed to polylysine, Dr. Don Slagel has recently obtained further information relating to the morphological changes induced by the polypeptide and the mechanism whereby a polyamino acid might penetrate into such cells. In these studies, tumor cells were suspended in a solution containing 1 mg of polylysine per ml for a period of twelve minutes. They were then collected by centrifugation, resuspended in a buffered osmium fixative, embedded, and the thin sections cut. In untreated cells (Plate 31.4) these sections revealed a granular cytoplasm, within which could be seen a nuclear area of marked osmophilic nature. The cell membrane was made up of numerous microvilli which may function in the pinocytotic intake of nutrient materials. The section through cells exposed to polylysine appeared to have less microvilli along the cell membrane and many of these seem to be surrounding particles of dense osmophilic material (Plate 32.5). We feel that these particles may be polylysine or aggregates of polylysine with material from the surface of the cells. It may be significant that some of the osmophilic particles seemed to be in a process of being phagocytized and that a few sections contained what appeared to be particles that had been taken into the cytoplasm. Perhaps this is the way such large molecules as the polyamino acids penetrate into cells. In the sections through cells exposed to polylysine, the mitochondria appeared greatly swollen and the nucleoplasma had a somewhat more highly aggregate appearance. Polylysine caused visible agglutination of the ascites cells. In the electron micrograph, agglutinated cells often appeared to share the same osmophilic clumps. Thus the agglutination process may involve the neutralization of the negative charge on the cell surface by the polylysine and the binding together of cells through aggregates containing polylysine. Thus polylysine may have a marked effect on cells by virtue of its combination at the cells' surface and perhaps by being taken into the cells where it seems to produce an aggregation of both cytoplasmic and nuclear material.

I believe that the search for antitumor agents among the polyamino acids should be continued. Ambrose *et al.* (*21*) have reported that tumor cells develop an abnormally high negative surface charge. They suggest that this surface charge may be one factor which allows metastasis of tumor cells, facilitates their migration away from the site where they initially form, and enables them to pass into surrounding tissues. The effectiveness of the basic polyamino acids in inhibiting a tumor development may be associated with their high positive charge, which would favor a selective electrostatic interaction with such negatively charged tumor cells. The report by Regelson *et al.* (*22*) that other synthetic polyelectrolytes act as tumor inhibitors would support the view that cancer cells may have an altered surface charge.

The antitumor action of polylysine apparently requires an intimate contact with the tumor cells which has thus far been shown to exist only within the peritoneal cavity and not in other tissues where solid tumors may develop. Nevertheless, it may be possible to develop polyamino acids which would be effective against other tumors, or to couple other antitumor agents to basic polypeptides and thereby hope to enhance their activity. I think more serious study and search for anti-cancer agents in the polyamino acids should be made.

Possibilities from their antigenic activity

The close structural relationship between the proteins and the polyamino acids suggested that the latter would be of great value in immunochemical studies of the chemical basis of antigenicity. The first published experimental work in this direction appears to be that of Makinodon *et al.* (*23*), who studied the immunochemical properties of polyglycyl bovine albumin and showed that the polypeptidyl albumin could elucidate as good an antibody response as the unmodified protein. Because these results were encouraging, we decided to further study the immunochemical properties of polyamino acids and to conduct additional immunochemical experiments at the Institute Pasteur where immunochemistry had its beginning and where our results could be evaluated by experienced immunologists. With Drs. P. Garbar and C. Lapresle we studied the antigenicity of synthetic polypeptides and polypeptidyl proteins. Preparations were selected that would allow us to answer the following questions: (*a*) Are antibodies formed to the polypeptidyl portion of polypeptidyl heterologous proteins (polypeptidyl bovine albumins) when injected into rabbits? (*b*) Are antibodies formed to the polypeptidyl portion of polypeptidyl homologous non-antigenic proteins (polypeptidyl rabbit albumins) when injected into rabbits? (*c*) Can antibodies be formed to multichain, branched co-polypeptides whose properties suggest that they resembled globular proteins? (*d*) Can antibodies be formed against linear polar polyamino acids? The results obtained in Paris in 1955 (*24*) clearly indicated that an affirmative answer could be given to all of these questions.

Preparations of bovine serum albumin modified by the addition of peptides of glutamic acid, lysine, leucine, or phenylalanine were strongly antigenic in rabbits (*24, 25, 26*). Precipitin reactions were obtained when antisera to each polypeptidyl bovine albumin were reacted with similarly modified rabbit albumin or unmodified bovine albumin indicating the presence of antibodies that were specific for each of the added peptides and for the carrier protein. Antisera to the polyleucine and polyphenylalanine bovine albumins were able to precipitate the homologous antigen after absorption with a similarly modified rabbit albumin and with unmodified bovine albumin, which suggested that a third type of antibody had been formed that required the added peptide and a part of the carrier protein.

Similarly modified rabbit albumins were also antigenic in rabbits (*24, 25, 26*). Antisera to polypeptidyl rabbit albumins formed precipitates with the correspondingly modified bovine albumin as well as with the polypeptidyl rabbit albumin, indicating that coupling polyamino acids to the non-antigenic rabbit protein had converted it into a strong antigen.

Although the antisera to the polar polyglutamic acid, polylysine, or the multichain polypeptide did not form precipitates with the homologous synthetic polypeptides, their antisera did cross react and precipitate with related polypeptidyl albumins and with certain purified protein preparations (*24, 27*). These results clearly indicated that these synthetic polyamino acids were antigenic. This conclusion was supported by studies in guinea pigs in which it was shown, as illustrated in Table 31.2, that the antisera prepared against a multichain synthetic polyamino acid produced fatal anaphylactic shock in guinea pigs (*28*).

Because antisera against many synthetic polyamino-acid preparations may not form precipitates with the homologous polyamino acid, we should not conclude that the preparations are not antigenic. Other tests such as the ability of the preparations to induce anaphylactoid symptoms or skin reactions in sensitized animals are equally important in deciding antigenicity. I suggest as evidence of antigenicity the ability of the antisera against synthetic polypeptides to cross react with and precipitate purified protein preparations when the control sera taken from the same animal just before immunization by the polyamino acid show no precipitation. It is possible that non-precipitating antibodies may be formed. Sera from patients with such hypersensitive diseases

Table 31.2—Passive anaphylaxis in guinea pigs sensitized with anti-multichain-polypeptide-17 rabbit serum

1st challenge with	No. of pigs and reaction	2nd challenge with	No. of pigs and reaction
MCP	2-+++	BA PG-BA	1-++++* 1-++++*
PG-BA	1-++++*		
BA	1-++++		

MCP Multichain-polypeptide-17, mol. wt. 90,000
BA Bovine albumin
PG-BA Polyglutamyl bovine albumin

\+ Slight shivering
++ Marked shivering with 1° to 2° temperature drop
+++ Respiratory distress and shivering
++++ Convulsions and collapse

* Died from anaphylactic shock.

Plate 31.1. Thin section of normal *E. coli* cells grown in inorganic salt media and fixed with a tryptone fixative. The section shows the cell wall and the cytoplasmic membrane enclosing the cytoplasm. The cytoplasm appears to be made up almost entirely of microsomal particles with no evidence of membrane structures as endoplasmic reticulum or mitochondria which are found in higher plant and animal cells. The nuclear vacuole is filled with a fine network of fibrils of approximately 30 Ångstroms in diameter which we believe to be nucleic acid molecules. The tryptone has increased the contrast by binding with these structures and with the osmium tetraoxide. Magnification 24,000.

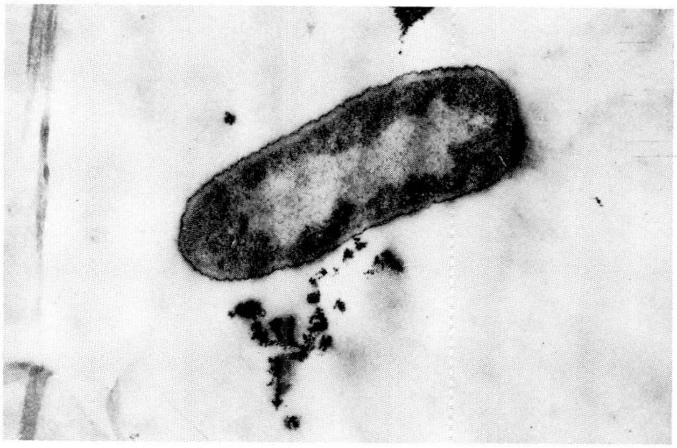

Plate 31.2. Thin section of polylysine treated *E. coli* cells. The cell walls and the adjacent cytoplasmic membrane are separated from the cytoplasm by a marginal vacuole. In other experiments the marginal vacuole, which was particularly large at the poles of the cells, was stained with ferric chloride suggesting that it is actually filled with nuclear material that had migrated toward the surface of the cell as a result of the exposure to polylysine. Magnification 5,500.

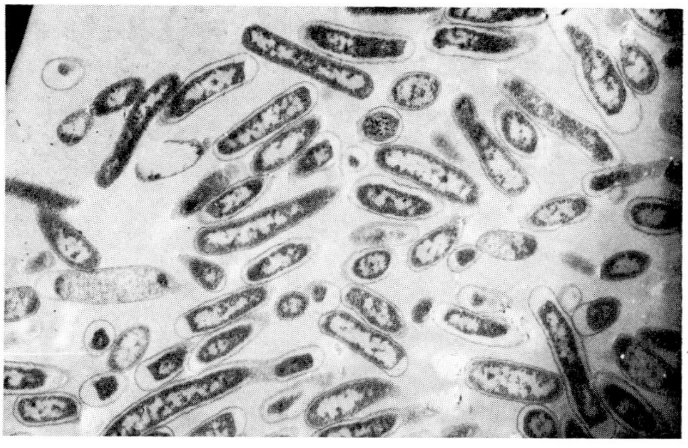

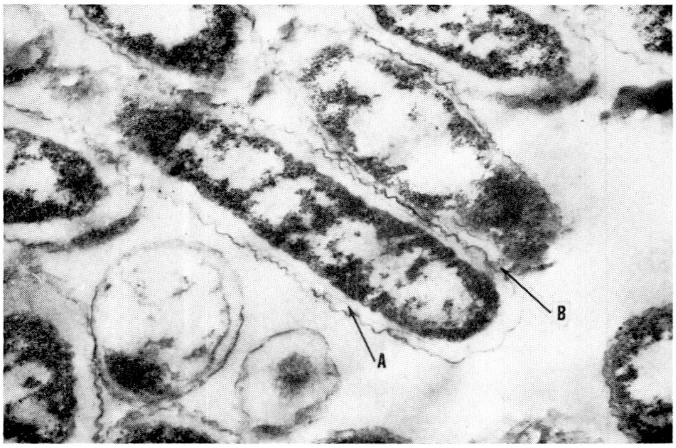

Plate 31.3. Thin section of polylysine treated *E. coli* cells exposed to polylysine. This section shows the cytoplasmic membrane (*A*) with the marginal vacuole lying within it. The cell walls of adjacent cells are closely enmeshed (*B*) illustrating the agglutination of cells caused by the polylysine. Magnification 24,500.

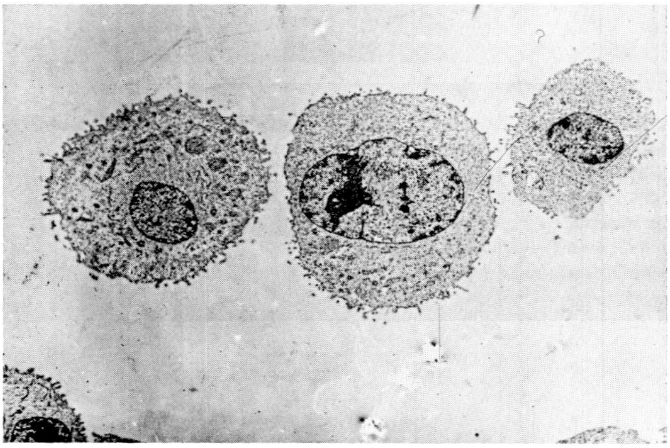

Plate 31.4. Thin section of normal Erhlich ascites cells which had been grown seven days in Swiss mice. The cell membrane shows many microvilli that may function in the pinocytotic intake of large soluble compounds. The cytoplasm appears granular and shows some endoplasmic reticulum characteristic of these cells. The mitochondria, although badly broken, can be seen and contain few cristae. Within the nucleus are areas which have a marked osmophillic nature. Magnification 2,150.

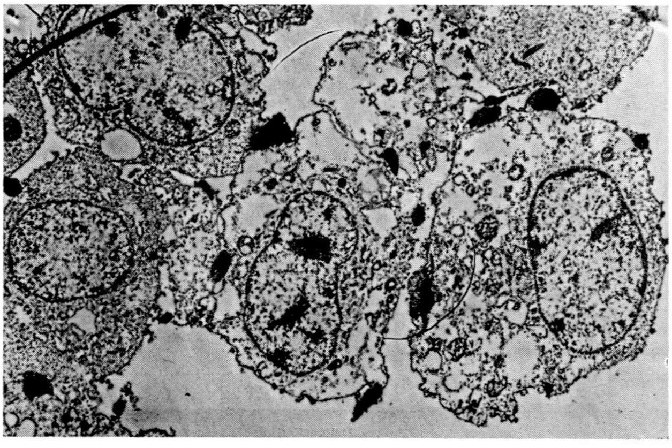

Plate 31.5. Thin section of Erhlich ascites cells which had been exposed in vitro to polylysine. Attached to the cell wall are several dense osomophillic particles which are thought to be aggregrates of polylysine. Less microvilli are seen on the cell membrane and many appear to be surrounding a clump of the dense osmophillic material as though it were in the process of being phagocytized. The mitochondria appear swollen. The nucleoplasm has a more aggregated appearance than normal. The agglutinated cells can be seen sharing the same osmophillic clumps which may bind them together. Magnification 2,750.

Table 31.3—Cross reactions of antisera from synthetic polypeptides and polypeptidyl bovine albumins with crystalline proteins

	Rabbit antisera against						
Test proteins	Poly-lysine	Poly-glu-tamic acid	Multi-chain peptide	Poly-leucyl albumin	Poly-phenyl-alanyl albumin	Poly-glutamyl albumin	Poly-lysyl albumin
Bovine fibrinogen	+++	++	+++	++++	+++	++	+++
Bovine γ globulin	+	−	+	++	−	−	+
β-Lactoglobulin	−	−	+	+	+	−	−
Ovalbumin	+	−	−	+++	+	−	++++
Catalase	+	++	++	−	−	++	+++
Lysozyme	−	−	−	+	+	+	−
Egg albumin	−	−	+	−	−	−	+
Trypsinogen	+	−	−	−	+++	−	++++
Ovomucoid	+	−	−	−	+	−	+
Tobacco mosaic virus	+	++	+++	+++	++++	++	−

++++ heavy positive ring test.
− negative ring test.

as rheumatic fever, asthma, or allergic encephalomyelitis seem not to contain precipitating antibodies although such diseases are generally thought to arise from an antigenic response. Thus our early results as summarized in 1955 in a communication entitled "L'antigenicité des polypeptides synthetiques" (*24*) clearly indicated that some synthetic polyamino acids were indeed antigenic (*29*). As will be discussed later in this symposium, Drs. Sela (Paper 33) and Gill (Paper 35) have extended these studies and have shown that many other copolypeptides and multichain polyamino acids are also antigens, much stronger than those we studied, and that antisera can be obtained which precipitate some synthetic polypeptides. From their results it appears that when the polar nature of the polyamino acid is sufficiently reduced, the antisera may then combine with and precipitate the homologous polyamino acids. Such studies may be expected to add much to our knowledge of the structural requirements for antigenicity and for antibody-antigen reactions.

Our initial observations (*24, 27*) that antisera to synthetic polypeptides and polypeptidyl proteins will cross react or form precipitates with a number of purified protein preparations are summarized in Table 31.3. These results suggest that it may be possible to prepare synthetic polyamino acids whose antigenic properties are such that antibodies prepared against them will cross react with some foreign proteins or perhaps even infectious agents whose presence in the body is undesirable. I recommend that in studies of the immunochemical properties of polyamino acids we give more consideration to this possibility and that we examine antisera for cross reactions with purified protein preparations, with viruses, and bacteria. If cross reactions whose strength is sufficiently high can be found, it may make possible a new method of immunization with synthetic antigens.

Kornguth *et al.* (*30*) have recently studied the effect of polylysine on experimental allergic encephalomyelitis. These workers found that polylysine injected into rabbits enhances the paralysis and increases the frequency and severity of lesions in the central nervous system that were induced by brain extracts. A basic protein extracted from brain tissue has been shown to have encephalitogenic activity (*31*). When a group of 19 rabbits were injected with a brain homogenate alone, only two animals developed partial paralysis of the hind limb. Ten of 19 animals developed paralysis when a single injection of polylysine was administered eight days after the brain extract. Polylysine alone showed no paralysis. Among other possibilities, the authors suggested that the polylysine may enhance the production of antibodies which react with the central nervous system by virtue of its chemical similarity to histones found in this tissue.

Thus, in view of the known information on the antigenicity of polyamino acids, it would seem that further immunochemical studies should be encouraged. Such studies may significantly increase our knowledge of the chemical basis for antigenicity and for antibody-antigen reactions. A study of the cross reaction of antisera to synthetic polypeptides with purified proteins, viruses, or bacteria could lead to a new process of immunization using such synthetic antigens. Studies of the effects of selected polyamino acids on animals with experimental allergic diseases may contribute to our understanding of the cause of such diseases and may suggest a new treatment. For example, the results with experimental allergic encephalitomyelitis would suggest that the administration of an acidic polyelectrolyte such as polyglutamic acid may reduce the disease by combining with the basic proteins that seem to be its cause.

Possibilities in complexing or releasing hormones

Polylysine-insulin.—Several years ago we carried out experiments which showed that polylysine would readily complex with insulin to form a polylysine-insulin precipitate whose appearance resembled protamine-insulin. Because of the similarities between the polylysine-insulin and protamine-insulin, we decided to study the action of polylysine-insulin in lowering blood sugar. Results of preliminary rabbit assays showed that this action of polylysine-insulin is considerably different from that of commercial protamine-zinc-insulin or of insulin alone. These differences seemed to us to be sufficiently great to suggest that polylysine-insulin may have some definite advantages over protamine-insulin and to justify reporting our results to the Second International Congress of the International Diabetics Federation at its meeting in Cambridge, England, in 1955 (*32*). There we learned of an independent study on the design of insulin trials that included a polylysine-insulin (*33*).

Our studies showed that the action of insulin in lowering blood sugar can be markedly altered by complexing with polylysine, and that the type of insulin response depends upon the kind and ratio of the polylysine to insulin in the complex. Some polylysine-insulin preparations showed a quick onset of action; others with a high ratio of polylysine to insulin showed a slow onset with a blood-sugar action like protamine-insulin. One preparation produced a reduction in blood sugar which was much greater than that from the same amount of either insulin alone or of protamine-insulin, and the blood-sugar response was depressed for a much longer period.

Figure 31.1 summarizes the results obtained when one unit of insulin was administered to rabbits as zinc-insulin or as polylysine-insulin complexes with different ratios of polylysine to insulin. The insulin alone and the polylysine-insulin complexes containing either a low (curve *B*) or a high (curve *D*) ratio of polylysine to insulin produce a very rapid fall in blood sugar. In these three cases, the level dropped to about 30 per cent of normal within an hour after the injection.

The polylysine-insulin preparation with an intermediate ratio (curve *C*) was slow in its action, for there was no significant drop in the blood sugar 45 minutes after injection. In the case of the polylysine-insulin complex with a high level of polylysine, both the extent and duration of the fall in blood sugar was greater than with the same amount of insulin alone or as protamine-insulin. We believe that this enhanced action of the insulin when complexed with polylysine may be due to both a change in the rate at which the insulin is absorbed from the site of injection and a decrease in the rate at which the insulin is hydrolyzed by proteolytic enzymes.

Polylysine-insulin complexes dissolve in the presence of an excess of either insulin or polylysine. If only the soluble portion of a complex can be absorbed from the site of injection and exert its hormone action, then it would follow that the absorption should be most rapid with those polylysine-insulin complexes that contain either a low or a high ratio of polylysine to insulin. Indeed, these preparations had the most rapid action. The enhanced blood-sugar lowering activity shown by the polylysine-insulin with a high ratio of polylysine to insulin may be due in part to an inhibition of the enzymatic destruction of the insulin. Polylysine

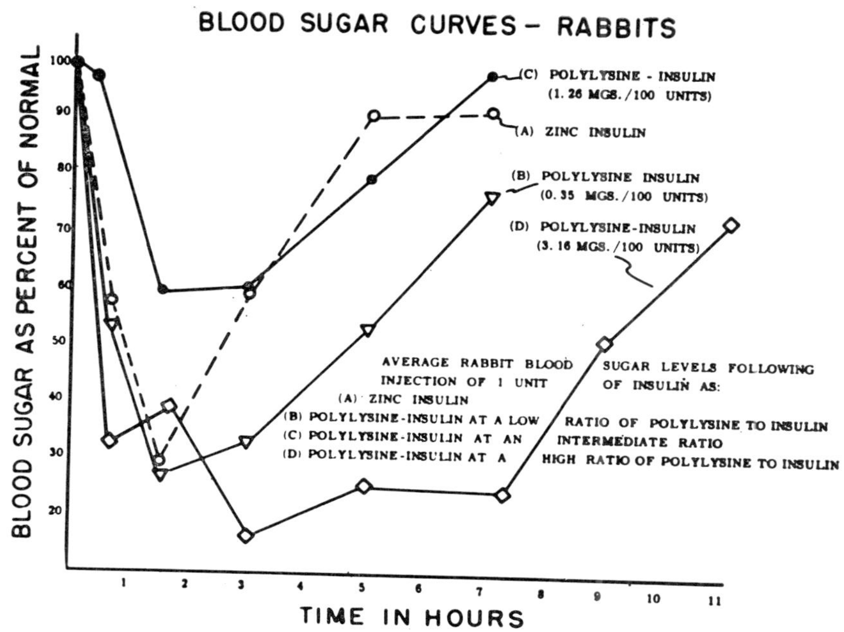

Fig. 31.1. Change in the average blood-sugar levels of rabbits following injection of one unit of insulin as: (*A*) zinc insulin, (*B*) a polylysine-insulin complex with a low ratio of polylysine to insulin, (*C*) a polylysine-insulin complex with an intermediate ratio of polylysine to insulin, and (*D*) a polylysine-insulin complex at a high ratio of polylysine to insulin.

inhibits the activity of pepsin; we have shown that a pancreas extract which clarifies a suspension of zinc-insulin in about 30 minutes requires 180 minutes to clarify a similar suspension of polylysine-insulin. The polylysine-insulin appears to be less susceptible to enzymatic breakdown. We suggest that this protection from enzymatic destruction may explain part of the greatly enhanced and prolonged action shown by the polylysine-insulin.

We believe that our results clearly indicate that it may be possible to prepare polylysine-insulin complexes with many desirable properties for the control of blood-sugar level. They suggest that it may be possible to prepare complexes of polyamino acids and other protein or polypeptide hormones that will alter, prolong, or otherwise influence their hormonal activity.

Histamine release.—With Dr. T. Richardson, we have studied the action of synthetic lysine polypeptides on isolated guinea pig ileum and shown that polylysine will displace or release histamine from this tissue (*34*). In these experiments, the polyamino acid was tested on the isolated guinea pig ileum using the technique described by Code and McIntire (*35*). Sections of ileum were mounted in Tyrode's solution and the contraction induced by the test material added to the bath was recorded. Fig. 31.2 shows that the tissue responded with the characteristic contraction when histamine was added, that there was no response to the free amino acid lysine, and that there was a strong contraction when polylysine was added. This contraction was closely similar to that induced by histamine. The muscles were generally refractory to a second addition of polylysine, but were always sensitive to subsequent additions of histamine. In exceptional cases it was possible to demonstrate as in Figure 31.2 very weak response to a second challenge with polylysine.

These data indicate that polylysine was not mediating the contraction but stimulated the release of an active material. The first contact with polylysine deleted the available contractant and subsequent additions were ineffective.

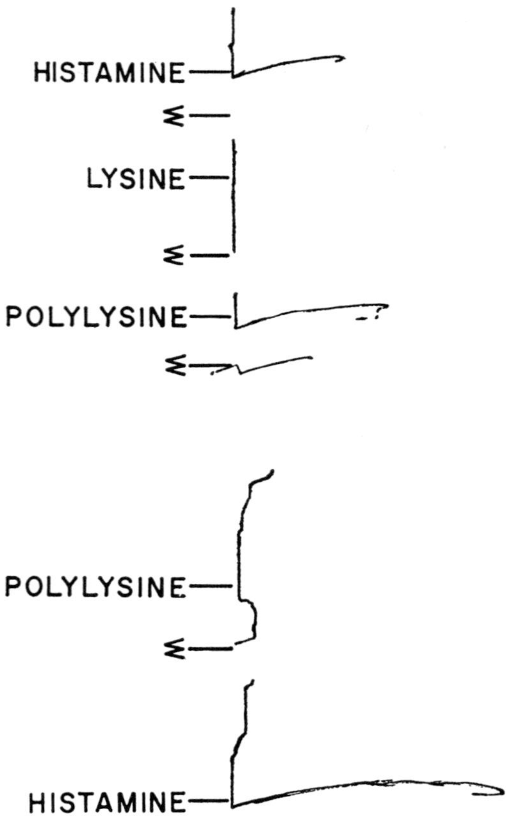

Fig. 31.2. The effect of polylysine on the isolated guinea pig illeum.

Benadryl at a concentration of 1 in 10,000,000 completely prevented the intestinal response to polylysine, thus implicating endogenous histamine or a histaminelike compound as the contractive agent.

These results clearly indicate that the basic polyamino acids of lysine are capable of displacing histamine from the isolated guinea pig ileum. It suggests the possibility that some polyamino acids may displace other hormones from an inactive form in which they are bound to tissue. I suggest that the effect of polyamino acids upon the activity and release of other hormones would merit further study.

Possibilities in haeomatology

Because of their high molecular weight and their chemical or structural similarities to plasma proteins, it is only natural that consideration should be given to polyamino acids as plasma expanders. It would seem that such polypeptide expanders would have a definite advantage over other types of expanders such as the polydextrans or polyvinylpyrollidone carboxylic acids because they would be metabolized to amino acids. Thus a polyamino-acid plasma expander may not only maintain the desired blood volume, but also might contribute to the amino-acid nutrition of the patient. However, before synthetic polyamino acids can be seriously considered as plasma expanders, it is necessary that experiments be carried out to show that it is possible to synthesize soluble polypeptides of a size and shape comparable to and compatible with the plasma proteins, and that such polypeptides do not possess antigenic or other toxic actions toward the patient.

Inasmuch as Drs. Fasman and Klein (Paper 36) will discuss their biological studies with copolymers of glutamic acid and lysine that has been related to the development of polyamino acids as plasma expanders, I will limit my discussion of this possibility. However, I would like to point out that in order to make polyamino acids soluble under physiological conditions, it is necessary to build into the molecule considerable amounts of polar amino acids such as glutamic acid or lysine for the polyamino acids of the neutral amino acids are insoluble in blood. These same polar amino acids may have a great tendency to interact or complex with plasma proteins, tissue proteins, or essential enzymes with which the plasma expander must contact. Care should be taken to ascertain that the extent of such interactions from polyamino acids being considered as plasma expanders are not so great as to impair vital functions. Inasmuch as it is now clear that some polyamino acids may have antigenic properties, we must also ascertain that any polyamino-acid preparations being considered for use as plasma expanders must not induce antibody formation that would lead to undesirable immunochemical reactions or would sensitize the individual to proteins that he may contact. These are serious limitations to the use of polyamino acids as plasma expanders, but they should not preclude the continued search for polyamino-acid structures that would circumvent these difficulties and be suitable for use as plasma expanders.

The marked effect of the polyamino acids, particularly the basic polyamino acids, upon the blood-clotting process has been reviewed by Sela and Katchalski (*36*). The basic polyamino acids such as polylysine cause agglutination of blood cells (*37*), inhibit blood coagulation (*38, 39*), accelerate the conversion of fibrinogen to fibrin by thrombin (*40*), and retard the lysis of fibrin (*41*). An interesting observation is the recent report that polylysine will activate prothrombin (*42*) and may be a model for the biological non-enzymatic reactions providing the unknown mechanism by which blood coagulation is initiated.

Hahn (*43*) observed that when dogs were injected with heparin there was a decrease in the lipemic turbitidy of their blood and that such blood contained a clearing factor that would hydrolyze the fat in the chylomicrons or in an artificial fat emulsion. Stein *et al.* (*44*) observed that polylysine caused an increase in the turbidity and fatty-acid content of plasma and reversed the lipemia-clearing action of heparin.

We have recently carried out experiments which show that a number of polyglutamic-acid preparations cause the release of an enzyme, a lipoprotein lipase (clearing factor), into the blood when injected intravenously into rabbits. The clearing factor activity was estimated by determining the glycerol liberated from a fat emulsion by serum taken from rabbits at intervals from ten minutes to three hours after the animals were injected with either 24 mg of heparin or 200 mg of polyglutamic acid synthesized with a high anhydride to initiator ratio. The 200 mg of polyglutamic acid caused the release of lipoprotein lipase to about the same extent as did 24 mg of heparin. The injection of lower amounts of the polyglutamic acid released less clearing factor; polymers of lower molecular weight also showed a reduced activity. We conclude that some anionic polyamino acids have the ability to release a lipoprotein lipase (clearing factor) in experimental animals.

The chemical, physical, and biological properties of polyamino acids would suggest that we continue the search for amino-acid polymers that might be suitable as plasma expanders. However, such preparations must be carefully examined for possible antigenic activity or undesirable interaction with blood or tissue components in their evaluation as plasma expanders. The lipemic clearing action of polyglutamic acid would suggest that other anionic amino-acid polymers be examined for such biological activity. Continued studies of the effect of polyamino acids on the clotting processes may be expected to reveal more basic information about this very complex but important process.

Acknowledgment

These studies were aided by research grants (No. E-101) from the National Microbiological Institute of the National Institutes of Health, U.S. Public Health Service.

References

1 A. Isaacs, *Scientific American* **204,** 5 (1961).
2 L. Pillemer, L. Blum, L. H. Lepow, O. A. Ross, W. E. Todd, and A. C. Wardlaw, *Science* **120,** 279 (1954).
3 J. E. Kuntz and J. C. Walker, *Phytopathology* **37,** 561 (1947).
4 E. Katchalski, I. Grossfeld, and M. Frankel, *J. Am. Chem. Soc.* **69,** 2564 (1947).
5 M. A. Stahmann, L. H. Graf, E. L. Patterson, J. C. Walker, and D. W. Watson, *J. Biol. Chem.* **189,** 45 (1951).
6 W. C. Burger and M. A. Stahmann, *ibid.* **193,** 13 (1951).
7 J. R. Rubini, A. F. Rasmussen, Jr., and M. A. Stahmann, *Proc. Soc. Exp. Biol. Med.* **76,** 662 (1951).
8 M. Green, M. A. Stahmann, and A. F. Rasmussen, Jr., *ibid.* **83,** 641 (1953).
9 M. Green and M. A. Stahmann, *ibid.* **83,** 852 (1953).
10 W. C. Burger and M. A. Stahmann, *Arch. Biochem. Biophys.* **39,** 27 (1952).
11 E. Katchalski, L. Bichowski-Slomnicki, and B. E. Volcani, *Bull. Research Council Israel* **1,** 153 (1951).
12 L. Bichowski-Slomnicki, A. Berger, J. Kurtz, and E. Katchalski, *Arch. Biochem. Biophys.* **65,** 400 (1956).
13 A. De Vries, J. Salgo, Y. Matoth, A. Nevo, and E. Katchalski, *Arch. intern. pharmacodynamie* **104,** 1 (1955).
14 D. J. Buchanan-Davidson, E. V. Seastone, and M. A. Stahmann, *J. Bact.* **80,** 590 (1960).
15 D. J. Buchanan-Davidson, M. A. Stahmann, C. A. Neeper, C. V. Seastone, and J. B. Wilson, *ibid.* **80,** 595 (1960).
16 D. J. Buchanan-Davidson, D. C. Deese, I. Uritani, and M. A. Stahmann, *Science* **132,** 1664 (1960).
17 D. E. Slagel, Ph.D. thesis, University of Wisconsin, 1961.
18 T. Richardson, J. Hodgett, A. Lindner, and M. A. Stahmann, *Proc. Soc. Exp. Biol. Med.* **101,** 382 (1959).

19 S. E. Kornguth, M. A. Stahmann, and J. W. Anderson, *Expt. Cell Research* **24,** 484 (1961).
20 A. Lindner, *Am. J. Clin. Pathol.* **34,** 426 (1960).
21 E. J. Ambrose, A. M. James, and J. Lowick, *Nature* **177,** 576 (1956).
22 W. Regelson, S. Kuhar, and M. Tunis, *Nature* **186,** 778 (1960).
23 T. Makinodan, R. R. Becker, H. R. Wolfe, and M. A. Stahmann, *J. Immunol.* **73,** 159 (1954).
24 M. A. Stahmann, H. Tsuyuki, K. Weinke, C. Lapresle, and P. Grabar, *Compt. rend. acad. sci.* **241,** 1528 (1955).
25 M. A. Stahmann, C. Lapresle, D. J. Buchanan-Davidson, and P. Grabar, *J. Immunol.* **83,** 534 (1959).
26 D. J. Buchanan-Davidson, E. E. Dellert, S. E. Kornguth, and M. A. Stahmann, *ibid.* **83,** 543 (1959).
27 D. J. Buchanan-Davidson, M. A. Stahmann, C. Lapresle, and P. Grabar, *ibid.* **83,** 552 (1959).
28 D. J. Buchanan-Davidson, M. A. Stahmann, and E. E. Dellert, *ibid.* **83,** 561 (1960).
29 M. A. Stahmann, D. J. Buchanan-Davidson, C. Lapresle, and P. Grabar, *Nature* **184,** 549 (1959).
30 S. E. Kornguth, D. R. Bennett, H. G. Thompson, Jr., G. M. ZuRhein, and M. A. Stahmann, *Nature*, in press.
31 M. W. Kies and E. C. Alvord, Jr., in M. W. Kies and E. C. Alvord, Jr., Editors, "Allergic Encephalomyelitis," Charles C Thomas Publishers, Springfield, Ill., 1959, Chapter 12.
32 M. A. Stahmann, A. W. Chow, A. Kammeraad, C. O. Miller, and R. F. Hanzel, Intern. Congr. Intern. Diabetics Federation, 2nd Congr., Cambridge, England (1955).
33 G. R. Fryers, R. G. Paley, and R. E. Tunbridge, *Lancet* **16,** 782 (1954).
34 T. Richardson and M. A. Stahmann, *Proc. Soc. Exp. Biol. Med.* **107,** 366 (1961).
35 C. F. Code and F. C. McIntire, *Methods of Biochem. Anal.* **8,** 62 (1956).
36 M. Sela and E. Katchalski, *Advances in Protein Chem.* **14,** 391 (1959).
37 J. R. Rubini, M. A. Stahmann, and A. F. Rasmussen, Jr., *Proc. Soc. Exp. Biol. Med.* **76,** 659 (1951).
38 A. De Vries, A. Schwager, and E. Katchalski, *Biochem. J.* **49,** 10 (1951).
39 J. R. Rubini, R. R. Becker, and M. A. Stahmann, *Proc. Soc. Exp. Biol. Med.* **82,** 231 (1953).
40 E. Shafrir, A. De Vries, and E. Katchalski, *ibid.* **87,** 63, (1954).
41 I. Ginsburg, A. De Vries, and E. Katchalski, *Science* **116,** 15 (1952).
42 K. D. Miller, *J. Biol. Chem.* **235,** PC63 (1960).
43 P. F. Hahn, *Science* **98,** 19 (1943).
44 Y. Stein and A. DeVries, *Arch. intern. pharmacodynamie* **102,** 274 (1955).

Discussion of Paper 31: Chemotherapeutic Possibilities of Polyamino Acids

PAUL MAURER

I was wondering whether the polylysine could be active as an adjuvant in the production of encephalomyelitis in a way similar to the action of lipo-polysaccharides from gram negative bacteria or actually the purified lipid A of Westphal. The inflammatory reaction which polylysine produces may be sufficient to act as an adjuvant. I don't know.

DOROTHY J. BUCHANAN-DAVIDSON

In discussing the chemotherapeutic actions of polylysine and other synthetic polymers, it would be wise to consider whether the action is directly on the infecting organism or on the defense mechanisms of the animal. Polylysine was effective against some bacteria in vitro but had no detectable effect in vivo. Despite fears that polylysine would be rapidly digested in several experiments, it had a protective effect even if injected 24 or more hours before infection but had no protective effect if injected after the infective agent. Either it remained in the animal to kill the infecting organism as soon as it was injected or it stimulated some of the animal's defense mechanisms so that they could combat the infection. Such a possible mechanism would be that of phagocytosis. Of the few bacteria studied in vivo, those which were not phagocytozed were infectious.

PAPER 32

Chemoprophylaxis of Poliomyelitis Using Polylysine and Other Polymers

KENNETH W. COCHRAN and MARK A. STAHMANN

The general biological activities of synthetic poly-α-amino acids have been reviewed by Sela and Katchalski (*1*), their broader chemotherapeutic potentials have been discussed in the preceding paper (*2*), and their antiviral activities in vitro and in ovo have been summarized (*3*). Certain other polymers have been reported to have action against influenza and vaccinia (*4*, *5*). The present study was made to determine if polylysines and other selected polymers possessed antiviral activity in vivo, particularly against poliomyelitis.

Materials and methods

Webster-strain mice, weighing 10 grams or less, except as noted, were inoculated intraperitoneally with approximately ten infectious doses (ID_{50}'s) of the MEF_1 strain of type 2 polio virus. For initial studies of toxicity and antiviral activity, aqueous solutions or suspensions of the test compounds were given daily for four days, beginning the day before virus inoculation, and using the intraperitoneal route except for the second dose, which was given subcutaneously. This scheme of treatment was designed to avoid nonspecific effects from the injection of virus and drug by the same route and at nearly the same time. Control animals were similarly treated with equal volumes of buffered saline, and all mice were checked regularly for paralysis throughout 21 days of observation. Each compound was tested at least twice using groups of 20 mice. In one experiment young cynomolgus monkeys were inoculated orally with an estimated 100 ID_{50}'s of the Mahoney strain of type 1 polio virus, and observation extended to 30 days. In other studies, the Holt strain of ECHO-9 virus was given to young mice in the same manner as polio virus, and the PR8 strain of type A influenza virus was given intranasally.

Effectiveness against polio virus and ECHO-9 virus was assessed on the basis of relative morbidity in treated and control groups, and by a survival index, the ratio of the harmonic mean time to the onset of paralysis in the treated groups to that in control groups. Survival indices greater than one indicate relative effectiveness in delaying or preventing paralysis or death. Representative statistical analyses by Drs. Wilcox and Remington, School of Public Health, indicated that with an index of 1.4 protection was usually significant at or beyond the 95 per cent level. Effectiveness against influenza was adjudged by comparing lung lesion scores, as described (*6*), in treated and control animals killed at designated intervals.

Polylysines were prepared by polymerization of *E*-carboxybenzoxy-α-carboxyl-L-lysine anhydride (*7*). Other polymers included 20-unit polylysine, supplied by Kremers Urban Company; polyphloroglucinol and polyphloretin phosphates, Aktiebologet Leo and National Drug Company; and phosphorylated hesperidin, National Drug Company.

Results

Toxicity.—The supply of most of these compounds was so limited that precise determination of lethal dose (LD_{50}) was omitted. Nevertheless, approximations were possible from data secured in establishing maximum tolerated doses. These estimates of toxicity to mice are included in Tables 32.1 and 32.2. In the first series of polylysines, where the degree of

Table 32.1—Effect of various polymers on poliomyelitis in mice

Compound	Est. LD_{50}* mg/kg/day	Dose mg/kg/day	Morbidity Treated %	Morbidity Control %	Survival index†
Lysine		250	88	96	1.2
Polylysine, av. DP=8	250	150	91	96	1.1
Polylysine, av. DP=9	100	50	91	96	1.2
Polylysine, av. DP=20	50–100	25	61	90	1.9
Polylysine, av. DP=58	>100	100	87	90	1.1
Polylysine, av. DP=83	>100	50	82	90	1.1
Polylysine, av. DP=83		25	53	85	2.4
Polylysine, av. DP=240	25–50	25	68	92	1.7
Polyglutamic acid, av. DP=20	>250	250	95	98	1.2
Polyglutamic acid, av. DP=20		25	75	90	1.0
Polyhomoarginine	75	25	73	99	1.9
Phloroglucinol	>500	250	92	86	1.1
Polyphloroglucinol phosphate, National	>500	250	95	86	.9
Polyphloroglucinol phosphate, Leo	250	150	98	98	.9
Polyphloretin phosphate, National	>250	250	92	87	.9
Polyphloretin phosphate, Leo	>500	250	98	100	1.0
Phosphorylated hesperidin	1000	250	75	89	1.5
Heparin		10	90	89	1.1

* Four doses starting one day before virus inoculation and given intraperitoneally, except second dose given subcutaneously.
† Ratio of harmonic mean survival time of treated group to that of control group.

polymerization was of greater certainty, it may be seen that as chain length increased from 8 through 20 units, toxicity also increased. Twenty- and 240-unit polylysines did not differ markedly in toxicity, but certain polymers of intermediate size appeared to be less toxic. Whereas all mice died from four daily doses of 100 mg/kg of polylysine-20, none died from this amount of polylysine-58. Higher doses of polylysine-58 could not be studied. The LD_{50} for polylysine-83 was estimated to be less than that for polylysine-58 since four daily doses of polylysine-83 were fatal to 3 of 10 mice. There appeared to be some cumulative toxicity since the approximate LD_{50} of single intraperitoneal injections was 250–500 mg/kg for polylysine-8; 100–250, polylysine-20; 100–200, polylysine-240; *ca.* 100, polyhomoarginine. Toxicity estimates for the second series of polylysines are included in Table 31.2. All of these preparations were tolerated at daily doses of at least 50 mg/kg given as described above. All, with the possible exception of preparation 1N9, were uniformly lethal when the rate was increased to 150 mg/kg/day. Polylysine-20, 50 mg/kg/day intraperitoneally for five days,

Table 32.2—Effect of selected polylysines on poliomyelitis in mice*

	A/I	Est. LD_{50}†	% Morbidity Treated	% Morbidity Control	Survival index
1N7	10	125	63	91	2.0
−8	40	75	52	91	2.4
−9	160	50	90	96	1.1
−10	320	75	68	91	2.0
−12	50	75	75	91	1.6
−14	20	125	66	91	1.8
−16	50	75	80	91	1.4

* 25 mg/kg/day; IP×4 given as described elsewhere.
† mg/kg/day; IP×4 given as described elsewhere.
Note.—The data of Paper 10 indicate that these samples are mixed population of polymers. This does not alter the finding that the degree of polymerization influences toxicity and antiviral activity.

killed 1 of 2 cynomolgus monkeys. Estimated LD_{50}'s appeared to differ for two samples each of polyphloretin and polyphloroglucinol phosphates. In view of the results with polylysines such differences in toxicity may reflect differences in molecular size.

Antiviral activity.—Results of testing the first series of polylysines and other selected polymers for activity against poliomyelitis in mice are presented in Table 31.1. Approximately maximum tolerated doses were employed except for monomers and polylysine-58. One hundred mg/kg was the largest daily dose that would allow a minimum of two experiments with the amount of polylysine-58 available. In this series 20-, 83-, and 240-unit polylysines and polyhomoarginine, all at 25 mg/kg/day, provided significant protection to mice. Polylysine-83 at a daily rate of 50 mg/kg/day was not active, however. Other inactive polypeptides did not become active when doses were reduced. Among the other polymers tested, only phosphorylated hesperidin appeared to exert any effect and its action was a delay in onset rather than a reduction in incidence of paralysis.

The apparent variation of chemoprophylactic activity with molecular size led to study of another series of polylysines. These results are summarized in Table 32.2, with compounds identified by sample code and ratio of anhydride to initiator. The samples in this series appeared to have the same range of toxicity as did the first series. With the exception of preparation 1N9, all were active against poliomyelitis in mice.

Effective polylysines also were active when given in single doses, depending on time and route of administration, as shown in Table 32.3. Single doses of polylysine-20 or polylysine-240, given intraperitoneally the day before virus, provided as much protection as did four doses. Protection was even greater if the single dose of polylysine-20 was given three or four days before virus inoculation, although under these conditions the virus challenge was less severe. Reducing the dose of polylysine-20 to 10 mg/kg/day reduced but did not eliminate its activity; but if administration of a single dose of polylysine-20 was delayed until the day after virus inoculation, there was no protection. Neither were single doses effective if given subcutaneously one day or two hours before virus.

Table 32.3—Effect of single doses of polylysine on poliomyelitis in mice

Polylysine av. DP	Treatment*	Survival index
20	25, D−1, IP	2.1
20	25, D−3 or 4, IP	3.6
240	25, D−1, IP	2.0
20	10, D−1, IP	1.4
20	25, D+1, IP	.9
20	25, H−2, SC	.9
20	25, D−1, SC	1.3
8	25, D−1, IP	.9
9	25, D−1, IP	.9

* mg/kg given I(ntra) P(eritoneally) or S(ub) C(utaneously) either 3–4 days before virus, D−3 or 4; 1 day before, D−1; 2 hours before, H−2, or 1 day after, D+1.

Other tests were made to determine the effect of varying the host or route of infection. To ascertain if its protective effect in young mice represented some idiosyncracy of immaturity, polylysine-20 was also tested in 15-gram mice. Older, heavier mice are less susceptible to polio virus inoculated intraperitoneally so that the virus challenge was less potent. Nevertheless polylysine-20 was active as shown by a survival index of 1.8 in older mice compared to 1.9 in younger. Polylysine-20 was also active when polio virus was inoculated subcutaneously but not when virus was given intracerebrally.

Results with polylysines against poliomyelitis in mice seemed to justify testing in monkeys. Two groups of six monkeys each were inoculated orally with Mahoney polio virus. One group was treated with five daily doses of 25

Table 32.4—Effect of varying host and route of infection on the antipoliomyelitis activity of polylysine

Host	Virus route	No. of doses*	Survival index
10-gram mice	IP	4	1.9
10-gram mice	IP	1	2.1
10-gram mice	IP	4	1.8
15–20-gram mice	1C	1	.8
10-gram mice	SC	1	1.8
monkey	Oral	5	.7

* 25 mg/kg/day, intraperitoneally starting one day before virus.

mg/kg of polylysine-20, injected intraperitoneally beginning one day before virus administration, while the other group was similarly treated with equal volumes of sterile buffered saline. No protection was observed. Phosphorylated hesperidin, 250 mg/kg/day intraperitoneally for six days, was also ineffective in monkeys. The regimen of treatment with polylysine-20, 25 mg/kg/day for four days, which protected mice from poliomyelitis, failed to protect young mice from ECHO-9 virus, inoculated intraperitoneally, or from PR8 influenza virus given intranasally.

Discussion

Certain lysine and homoarginine polypeptides were partially effective against poliomyelitis in mice. In this investigation general toxicity to the host may be eliminated as a factor in the antiviral activity of polylysines. While it is true that the most effective polylysines, 20 and 240 units, were the most toxic, their protective ability would not seem to be the result of nonspecific toxicity since most of these compounds were administered in doses representing approximately the same relative toxicity. Single doses, even further below toxic levels, were as effective as more prolonged treatment. Furthermore, lowering the dose of polylysine-83 increased its effectiveness, leading one to consider that surreptitious toxicity of polylysines might tend to decrease rather than to increase apparent chemoprophylactic efficiency. Variation in chain length which appeared, at least in the first series of polylysines, to modify antiviral effectiveness has also been reported to alter antitumor activity (*8*). Polylysine-20, one of the polypeptides partially effective against poliomyelitis in mice, failed to protect monkeys, but variations in dosage and molecular size sufficiently changed chemoprophylactic activity in mice to allow the possibility that other combinations of these factors might be effective in higher species.

Among the other compounds tested only phosphorylated hesperidin seemed to exhibit activity, and its major effect was only to delay paralysis. Although phloroglucinol had been found to have some effect against polio virus in vitro (*9*), the investigators did not classify it with their active compounds, and the current study did not reveal any activity in vivo. Polyphloretin and polyphloroglucinol phosphates, which inhibit alkaline phosphatase and hyaluronidase (*10, 11*) and influenza virus in ovo (*4*), as well as heparin, inhibiting ribonuclease (*12*), failed to inhibit poliomyelitis in mice. Gebhardt and Bachtold (*13*) noted a possible correlation between antihyaluronidase and antipolio virus activities, but in the present study inhibitors of these enzymes failed to provide protection. Heymann *et al.* (*5*) found antiribonuclease activity "a necessary but not a sufficient prerequisite for activity" against influenza and vaccinia viruses in ovo.

The potentiality of this general category of polymers for providing antiviral agents has been cited (*14*). Possible mechanisms of the antiviral action of polylysines were considered by Green and Stahmann (*15*), who favored the intracellular combination of polypeptide with intact virus, virus fragments, or intermediates. More recently, treatment with polylysines was found to have no effect on the cellular concentration of nucleic acids, although others report that polylysine interferes with certain nucleic-acid functions (*16*). Although it remains to be demonstrated that this is the mechanism of the antiviral action of polylysines, such a biochemical function would be compatible with antiviral activity. Consideration must be given the specificity of antiviral activity of differing polymers against several viruses in various hosts.

Summary

Lysine polypeptides with average chain lengths of 20, 83, and 240 residues, as well as polyhomoarginine, partially protected mice infected intraperitoneally with approximately ten ID_{50}'s of polio virus. Polylysines with average sizes of 8, 9, and 58 residues were inactive. Polylysine-20, although active in mice, failed to protect monkeys infected orally with polio virus.

Acknowledgment

These studies were aided by a grant from the National Foundation.

References

1 M. Sela and E. Katchalski, *Advances in Protein Chem.* **14,** 391 (1959).
2 M. A. Stahmann, this book, Paper 31.
3 F. L. Horsfall, Jr., and I. Tamm, *Ann. Rev. Microbiol.* **11,** 339 (1957).
4 K. Penttinen, *Ann. Med. Exp. et Biol. Fenniae (Helsinki)* **34,** 88 (1956).
5 H. Heymann *et al., Arch. Biochem. Biophys.* **73,** 366 (1958).
6 F. M. Schabel, Jr., and H. E. Skipper, *Cancer Research Suppl.* **3,** 52 (1955).
7 R. R. Becker and M. A. Stahmann, *J. Am. Chem. Soc.* **74,** 38 (1952).
8 T. Richardson *et al., Proc. Soc. Exp. Biol. Med.* **101,** 382 (1959).
9 P. E. Kramer *et al., J. Pharmacol. Exp. Therap.* **113,** 262 (1955).
10 E. Diczfalusy *et al., Acta Chem. Scand.* **7,** 913 (1953).
11 O. Ferno *et al., ibid.* **7,** 921 (1953).
12 L. Vandendriessche *Arch. Biochem. Biophys.* **65,** 347 (1956).
13 L. P. Gebhardt and J. G. Bachtold, *Proc. Soc. Exp. Biol. Med.* **88,** 103 (1955).
14 A. Todd, *Brit. Med. J.* **2,** 517 (1959).
15 M. Green and M. A. Stahmann, *Proc. Soc. Exp. Biol. Med.* **83,** 852 (1953).
16 V. Allfrey and A. E. Mirsky, *Trans. N. Y. Acad. Sci.*, Series II, **21,** 3 (1958).

PAPER 33

Some Contributions of the Study of Synthetic Polypeptides to the Understanding of the Chemical Basis of Antigenicity

MICHAEL SELA

The considerable progress in our understanding of the chemistry and biology of proteins in the last decade permitted an attack on a molecular level, aiming towards the elucidation of the correlation between the biological function and the chemical structure of natural polypeptides and of proteins. Significant information has been obtained while studying various enzymes and hormones as well as protein antigens.

When discussing the immunologic activity of a protein, I would like to distinguish between its antigenicity and its antigenic specificity. We may draw a parallel with enzymes, where a certain area may be assumed to possess the catalytic activity, while a different area of the molecule (even though sometimes it may overlap or be identical with the first one) is responsible for the enzymic specificity. In the case of antigens it seems desirable to discuss, separately, the moiety that must be present in the molecule to render it antigenic, and a different moiety responsible for the serological specificity. We might ask ourselves, therefore: (*a*) What chemical structure must be present in the protein molecule to endow it with antigenic properties? (*b*) Can the antigenicity be enhanced through a chemical modification? (*c*) What is the chemical nature of the site responsible for the antigenic specificity? By enhancement or increase in antigenicity I mean an increase in the ability of a chemically modified antigen to elicit antibodies, relative to the unmodified substance, even if they are serologically different.

A vast immunochemical literature exists, concerned with the definition of the serological specificity of various antigens, as well as with the changes in specificity induced through chemical modifications. The effort to answer the first two questions mentioned above has been much more limited. These problems, like many others discussed in this book, can obviously be approached from two directions, degradative or synthetic.

The degradative approach, first used by Landsteiner and Chase in 1933 (*1*), consists of an immunologic study of the partial degradation products of proteins. Such products may be investigated either for their ability to react with antibodies to the undegraded protein (looking for the smallest fragment carrying the antigenic specificity of the protein) or for their capacity to elicit antibodies when injected into an animal (looking for the smallest immunogenic unit). The studies of human (*2*, *3*, *4*) and bovine (*5*) serum albumin, eye lens crystallin (*6*), as well as silk fibroin (*7*, *8*) may serve as examples of this direction of research.

An alternative approach consists of the study of the serological activity and specificity of either completely synthetic polypeptides or of proteins modified by attachment of peptides.

Polypeptidyl proteins

The polymerization of *N*-carboxy-α-amino acid anhydrides in aqueous media using proteins as homogeneous multifunctional initiators was first described by Becker and Stahmann (*9*). The synthesis of polypeptidyl proteins thus formed was found to proceed under mild conditions (aqueous media, low temperature, neutral *p*H) in which most proteins remain native. In this way peptides of different sizes can be built on the free amino groups of the protein (see also *10, 11, 12*).

Definite changes in the immunologic specificity have been observed when polypeptidyl derivatives of strongly antigenic proteins have been compared serologically with the unmodified proteins (*13, 14, 15, 16*). Moreover, the attachment of polypeptide chains to rabbit albumin converted it to an antigen in rabbit (*15, 16*).

We have asked ourselves whether peptidylation of a very poor antigen can result in the enhancement of its antigenicity, and we have chosen gelatin for this study. Gelatin is such a poor antigen that for many years its antigenicity escaped notice (*17, 18*) and could be demonstrated only with refined precipitin techniques (*19*). Among a number of possible reasons for its low antigenicity (*20*), the deficiency of aromatic groups in gelatin has been mentioned. As a matter of fact, the attachment of *O*-β-glucosido-*N*-benzyloxycarbonyltyrosine (*21*) or *N*-benzyloxycarbonyltyrosine (*22, 23*) to gelatin gave substances which elicited the production of antibodies. These antibodies could not be precipitated with homologous antigens, but could be detected by precipitation with a similarly modified globulin preparation.

In our own experiments we have sensitized guinea pigs with a polytyrosyl gelatin containing 16 per cent tyrosine (*24*). Typical anaphylactic shocks were obtained after intracardial injection of the same compound (*25*). On the other hand, no sensitization was observed by a similar treatment with unmodified gelatin, or with a copolymer of tyrosine and aspartic acid. (Polytyrosine is insoluble in water in the neutral *p*H range.) Strong anaphylactic shocks were also observed in guinea pigs sensitized with poly-*p*-arsanilazotyrosyl gelatin (*11*). Antibodies to poly-*p*-arsanilazotyrosyl gelatin could also be detected in rabbits by means of the precipitin test (*26*).

The above investigations prompted us to undertake a systematic survey of the possible enhancement of the antigenicity of gelatin by the attachment of various polypeptidyl chains (*20*). The gelatin derivatives given in Table 33.1 were prepared and injected into rabbits in a complete Freund adjuvant (*20, 27, 28*). The antisera were investigated by means of the precipitin reaction and, whenever negative, the

Table 33.1—Polypeptidyl gelatins

Gelatin derivative (enriched with amino acid *X*)	Percentage of the *X* amino acid in the gelatin derivative	Enrichment (calculated assuming gelatin as 100%)	Ref.
*p*TyrGel A	10.3	11.1	*20, 27*
*p*TyrGel B	5.7	5.7	*27*
*p*TyrGel C	2.7	2.4	*27*
*p*TryGel	25.4	34.0	*20*
*p*PheGel	8.9	7.4	*20*
*p*CysGel	7.1	7.6	*20*
*p*Cha*Gel A	4.4	4.6	*28*
*p*Cha*Gel B	10.9	12.2	*28*
*p*AlaGel	18.4	11.8	*20*
*p*GluGel	26.6	22.5	*20*
*p*LysGel	22.6	24.1	*20*
*p*SerGel	10.2	7.9	*28*
*p*ProGel	25.3	16.3	*28*
*p*GluTyrGel Glu	13.0	3.5	*27*
*p*GluTyrGel Tyr	5.2	6.0	*27*

* Cha = cyclohexylalanine.

modification used by Maurer to prove the antigenicity of gelatin was employed. Enrichment with tyrosine, tryptophan (Fig. 33.1), or phenylalanine (Fig. 33.2) converted gelatin into relatively powerful antigens, as followed by the precipitin reaction. The attachment of cysteine (Fig. 33.2) to gelatin caused a limited enhancement of antigenicity (*20*). The attachment of alanine, glutamic acid, lysine (*20*), serine (*28*), or proline (R. Arnon and D. Givol, unpublished data) did not convert gelatin into a more powerful antigen.

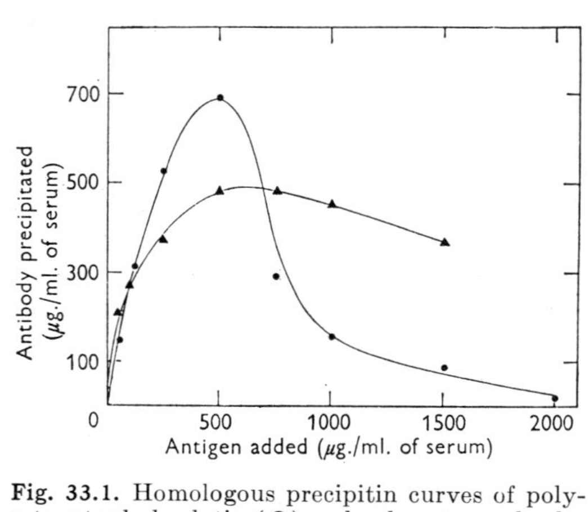

Fig. 33.1. Homologous precipitin curves of poly-L-tryptophyl gelatin (●) and poly-L-tyrosyl gelatin (▲). From Sela and Arnon (*20*).

We have assumed that the enhancement of the antigenicity of gelatin by the attachment of the aromatic α-amino acids and of cysteine was due to an increase in the rigidity of the active site in the molecule. According to Haurowitz (*29*), a rigid structure of the determinant groups is a prerequisite for antigenicity. Even though the content of sulfhydryl groups in polycysteinyl gelatin did not change significantly during a period of several months, it is possible that disulfide bonds, which would increase the rigidity in limited areas of the macromolecule, are formed in vivo. The lack of effect of the attachment of serine to gelatin supports this suggestion.

In order to elucidate the role in the enhancement of antigenicity of the aromatic character of the aromatic amino acids attached, we have also investigated the immunologic activity of polycyclohexylalanyl gelatins (*28, 30*). The attachment of amino acids containing the fully saturated cyclohexane rings caused a strong increase in antigenicity, as compared with that of the original gelatin. Although the cyclohexane ring is not as rigid as the aromatic benzene ring, it is known that certain energetically favored configurations may become the stable ones. The chair conformation of cyclohexane is favored energetically over that of the boat form by approximately 6 kcal. For methylcyclohexane an equatorial conformation is more stable than the axial one by 1.8 kcal (*31, 32*). Thus, it may be concluded that the nonflexibility of a determinant part in the antigenically active site of the molecule may serve as the common denominator for all the cases where peptidylation resulted in the enhancement of the antigenicity of gelatin. This does not necessarily mean the converse, namely that the presence of such a rigid site must confer antigenic properties on a molecule.*

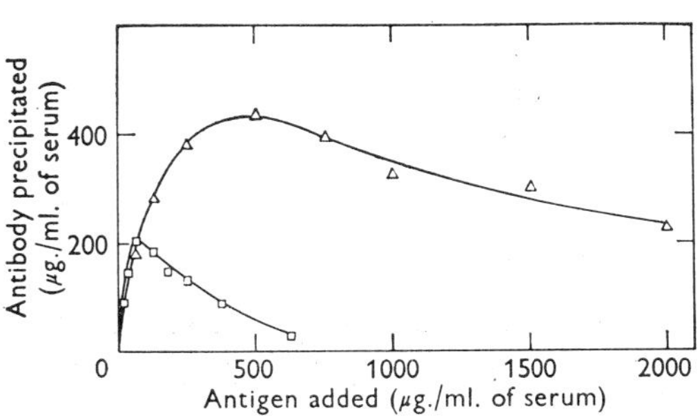

Fig. 33.2. Homologous precipitin curves of poly-L-phenylalanyl gelatin (△) and poly-L-cysteinyl gelatin (□). From Sela and Arnon (*20*).

The antibodies to the various gelatin derivatives are very specific and may be precipitated almost exclusively by the homologous antigens. However, it was possible to interfere with the precipitin reactions by means of specific inhibitors such as peptidyl proteins and copolymers of amino acids (*20*). The serological specificity of the gelatin derivatives varied according to the nature and the length of the peptides attached. Thus the specificities of a polytyrosyl gelatin containing 10 per cent of tyrosine residues and of a polytryptophyl gelatin containing 25 per cent of tryptophan residues were due, respectively, to tyrosine peptides and tryptophan peptides, but not to gelatin. On the other hand, with polyphenylalanyl gelatin (9 per cent of phenylalanine residues), it could be shown that the specificity was associated not only with the phenylalanine chains but also with gelatin itself (*20*). The same is true for polycyclohexylalanyl gelatin (*28*).

A subsequent study of three polytyrosyl gelatins differing in their tyrosine content (see Table 33.1) showed that the extent of the antigenicity depends on the amount of tyrosine attached (*27*). As little as 2 per cent of tyrosine sufficed to enhance, strongly, the antigenicity of gelatin (Fig. 33.3). At the same time the serological specificity changed markedly with variations in the tyrosine content. Cross-precipitation and inhibition experiments demonstrated that the specificity of the antigens is determined by the length of the peptide chains attached to gelatin. Thus the antigenic specificity of *p*TyrGel A resides exclusively in the tyrosine peptide chains. With *p*TyrGel B both gelatin and compounds containing tyrosine have an effect on the specific precipitin reaction, whereas with *p*TyrGel C almost all the serological specificity is due to the gelatin moiety (*27*). In the last case the attachment of monomeric units of tyrosine to half the amino groups of gelatin caused a strong enhancement in antigenicity essentially without changing

* In polysaccharide antigens the pyranose or furanose rings might serve as the rigid sites. In this connection, it would be of interest to know whether they may still be immunogenic after reacting with periodate.

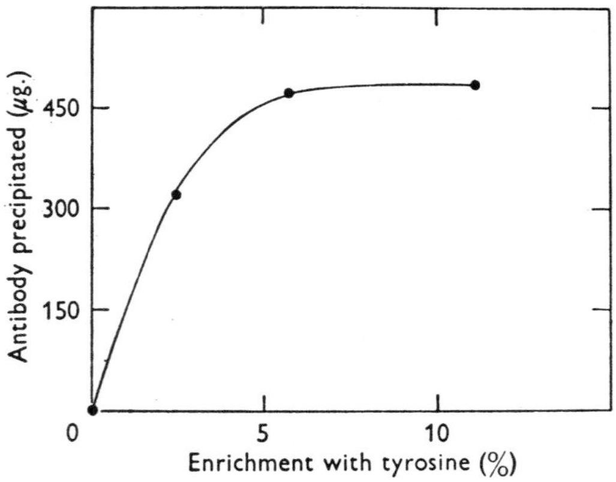

Fig. 33.3. Maximal amounts of antibodies precipitated in 1 ml of serum from the antisera against the polytyrosyl gelatins included in Table 33.1 and against gelatin, as a function of the extent of the enrichment of gelatin with tyrosine residues. From Arnon and Sela (*27*).

the specificity. Experiments on the cross reaction of the various polytyrosyl gelatins with the antiserum to gelatin showed, again, that the increase in the tyrosine content diminished considerably the extent of precipitation (*28*).

Whereas the attachment of polar amino-acid residues to gelatin caused almost no change in the extent of antigenicity (*20*), the enrichment of gelatin with both tyrosine and glutamic acid (Table 33.1) yielded a powerful antigen of narrow specificity. The only substance that interacted with the antibodies was a copolymer of glutamic acid and tyrosine (*27*). This is in agreement with the common assumption that the polar groups strongly influence the antigenic specificity of proteins.

With none of the gelatin derivatives did the precipitates in the equivalence zone contain all the antigen added. Nevertheless, no precipitate occurred upon addition of either antigen or antibody to the supernatant fluid. The polydispersity of gelatin could not explain this phenomenon since at high antibody-antigen ratios all the antigen is found in the precipitate. This implied the presence of soluble antigen-antibody complexes in the equivalence zone. The existence of such a soluble complex was shown unequivocally in the case of *p*GluTyrGel (*27*).

Increasing amounts of I^{131}-labelled *p*GluTyrGel were added to the antiserum. The precipitates formed after 48 hours were centrifuged off and the supernatant fluids were subjected to electrophoresis on paper strips. No radioactivity was found in the supernatant fluid from the antibody-excess zone. In the supernatant fluid from the equivalence zone all the radioactivity was found in the γ-globulin region (Fig. 33.4). In the antigen-excess region one large black area appeared on the radioautogram, covering the zones both of the γ-globulin and of the free antigen. In a control experiment (Fig. 33.4) it was shown that the antigen investigated does not react with normal serum to give nonspecific complexes that would migrate with γ-globulin. It may thus be inferred that the soluble complex formed results from the interaction of the antigen with specific antibodies.

Summarizing the experiments with polypeptidyl gelatins, the following conclusions may be drawn: (*a*) The attachment of aromatic amino acids enhances the antigenicity, while the attachment of polar amino acids contributes to narrowing the antigenic specificity. (*b*) A limited tyrosination resulted in enhancement accompanied by very little change in specificity. The potentiation of antigenicity by tyrosination may be of more general character.

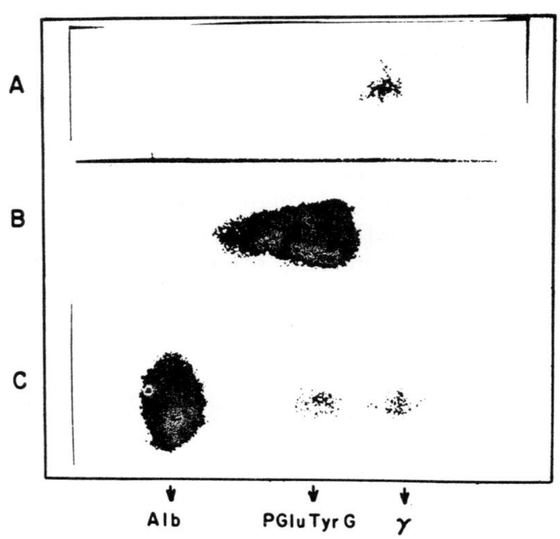

Fig. 33.4. Paper strip electrophoresis of I^{131}-labeled *p*-GluTyrGel. *A:* Radioautogram of a mixture of rabbit normal serum and the antigen. *B:* Radioautogram of the supernatant fluid from the equivalence zone. *C:* Paper strip electrophoresis of a mixture of rabbit normal serum and the antigen, detected with naphthalene black.

(*c*) A more extensive increase in the amount of tyrosine or tryptophan, or of tyrosine and glutamic acid, together bound to gelatin results in drastic changes not only in the extent of antigenicity but also in the serological specificity of the gelatin derivatives obtained. As the specificity was due in these cases entirely to the polypeptides attached, the question arose whether gelatin still contributes in any way toward the immunologic properties of the new antigens, or whether it is simply an inert carrier, in which case it could be replaced by a synthetic material, thus yielding completely synthetic antigens.

Synthetic polyamino acids

The investigations of the immunological properties of polyamino acids, up to 1959, have been summarized recently (*11*). The first studies on the possible antigenicity of polyamino acids were reported by Stahmann and his colleagues (*14, 33, 34, 35, 36*). Of a large series of synthetic polypeptides investigated only one particular sample of polyglutamic acid and a multichain copolymer of glutamic acid, leucine, glycine, and lysine were reported to be antigenic. The antibodies formed did not precipitate with the homologous synthetic polypeptides, but cross-reacted with related polypeptidyl albumins as well as with various unrelated proteins. The synthetic polypeptides did not inhibit precipitin reactions between their antisera and proteins. The two antisera caused only a weak anaphylactic reaction in guinea pigs. The authors concluded that the two polymers had the capacity to induce antibody formation but lacked features necessary to permit the reaction between the homologous antigen and antibody (*36*).

A detailed serological investigation of polyglutamic acid has been reported by Maurer (*37*). No antibodies could be detected in man, rabbit, or guinea pig. Similarly, attempts to produce antibodies against various synthetic polymers of a single amino acid by methods both in vitro and in vivo were unsuccessful (*38, 39*). Antibodies to a linear copolymer of glutamic acid and lysine could be detected by passive cutaneous anaphylaxis, but not by the precipitin reaction (*38*). Only tobacco-mosaic–virus protein inhibited the anaphylaxis reaction. The large number of reported failures to detect antibodies against various homopolymers of α-amino acids suggests that the homopolymers studied are at best very weak antigens (*11*).

In connection with our studies on the sensitization of guinea pigs with polytyrosyl gelatin, we have observed that the animals were not sensitized by a low molecular weight copolymer of tyrosine and aspartic acid (*25*). Similarly, we failed to detect, in rabbits, antibodies to a poly-*p*-arsanilazo-L-tyrosine of low molecular weight (*26*).

As the previously described conversion of gelatin into strong antigens resulted, in some cases, in materials whose serological specificity was due entirely to the attached polypeptides, it seemed worth while to investigate the immunologic properties of synthetic macromolecules in which peptides of tyrosine, seemingly important for immunogenicity, and glutamic acid, playing a role in defining the role of specificity, were built on "inert carriers" of polyamino acid nature. We have chosen as a potential "inert carrier" a multichain polymer of DL-alanine (*24, 40*).

Multichain polyamino acids are branched polymers whose molecules are composed of linear polymeric chains attached to a polyfunctional core. They may be obtained by the initiation of *N*-carboxy-α-amino acid anhydrides with multifunctional initiators (*11, 24, 40, 41, 42*). Multichain poly-DL-alanine was prepared using poly-L-lysine as initiator, and tested in rabbits for antigenicity. No antibodies could be detected either by the usual homologous precipitin test or by Maurer's modification, nor could any cross precipitation be observed on reaction with related polypeptidyl derivatives (*43*). The immunologic activity of a multichain poly-L-glutamic acid (*44*) of an average molecular weight 160,000 was similarly completely negative as followed both by precipitin and hemagglutinin reactions.

In contrast to the above multichain polyamino acids we could show that a multichain copolymer, in which chains of polypeptides containing L-tyrosine and L-glutamic acid were built on a multichain poly-DL-alanine,* is a powerful and sharply specific antigen in rabbits (*43*). This material contained 10 per cent

* We abbreviate the name of this multichain copolymer to: (*T, G*)-*A*-*L*.

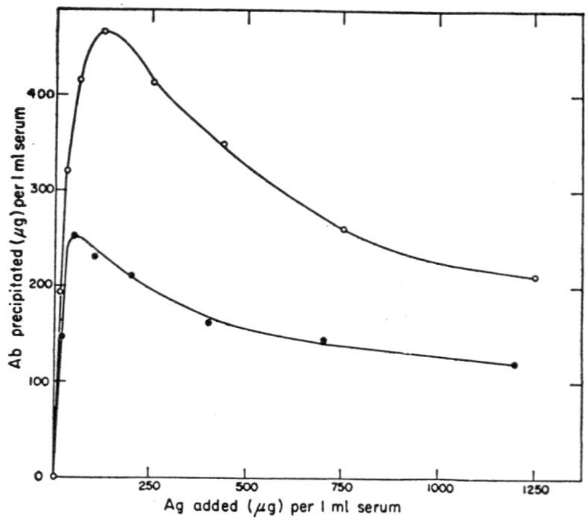

Fig. 33.5. Precipitin curves of (T,G)-A-L (○) and pTyrGluGel (●) with the antiserum to (T,G)-A-L. The amounts of antibodies were obtained from extinction at 280 mμ after deducting the calculated extinction of the antigens. The amount of antigen in the precipitate was obtained for the synthetic material from the radioactivity data after tagging it with I^{131}, and for the gelatin derivative from hydroxyproline data. From Sela and Arnon (*43*).

tyrosine residues and 15 per cent glutamic-acid residues, and it had an average molecular weight of 70,000, calculated from sedimentation and diffusion data. The antibodies formed could be demonstrated by the usual precipitin reaction (Fig. 33.5). Except for the homologous antigen, the antibodies could be cross-precipitated only with the related *p*GluTyrGel (see Table 33.1). The homologous reaction could be inhibited completely by a copolymer of L-tyrosine and L-glutamic acid in a residue molar ratio of 1:1, only partially by a copolymer of the same amino acids in a ratio of 1:4, and not at all by a copolymer with a ratio of 1:9. Neither cross reaction nor inhibition was obtained with many polymers lacking tyrosine (such as poly-DL-alanine or multichain polyglutamic acid) or with unrelated proteins (e.g., egg albumin and human γ-globulin).

We have thus shown that a completely synthetic material composed of α-amino acids bound through peptide linkages can be a powerful and specific antigen. At the same time Gill and Doty (*45*) have reported the antigenicity of a synthetic linear polypeptide containing tyrosine. I shall not discuss this study here since Dr. Gill will consider their recent work in detail in this book.

The synthetic approach offers the advantage that, once a biological property has been unequivocally demonstrated, tens of analogs may be prepared and tested. Knowing the chemistry of the synthetic materials it seems possible, through a study of copolymers showing only limited variations in their chemical formulae, to arrive at conclusions concerning the role of various structural features of the molecules in their biological function.

In the case of antigenicity, the problems to be considered should include the role of size, shape, composition, as well as the locus of specific amino acids, in conferring immunogenic properties on a molecule. We have begun, in collaboration with Dr. Ruth Arnon and Mrs. Sara Fuchs, a comprehensive study attempting to elucidate the points mentioned above, and I would now like to summarize some of the results obtained.

We have mainly used multichain copolymers in our studies. In order to avoid a more cumbersome nomenclature, we have decided to adopt the following conventions: homopolymers are denoted by the first letter of the amino acid alone (thus L stands for polylysine); a random copolymer, be it in the backbone of the multichain polymer or in its polymeric side chains, is described by letters of the amino acids composing it, separated by commas and enclosed in parentheses, [thus (L,A) stands for a copolymer of lysine and alanine, while (T,G) for a copolymer of tyrosine and glutamic acid]; the link between the side-chain polymer and the backbone polymer, as well as the link between the side-chain polymer and a polypeptide attached to elongate it, are indicated by hyphens. According to this simplified nomenclature, our first synthetic antigen is abbreviated to (T,G)-A-L.

Table 33.2 lists a series of synthetic materials shown to be antigenic as well as a series of materials which have not elicited the formation of precipitable antibodies in rabbits. Before deciding that a substance is nonantigenic by the criterion of precipitability, we have in all cases tried cross precipitation in addition to the homologous reaction. We have not injected polymers with a strong over-all positive charge since these often tend to give precipitates with normal serum. A specific antigen was obtained

Table 33.2—Immunogenicity of some polyamino acids investigated

Ag	Non-Ag*
(*T*, *G*)-*A*-*L* *T*-*A*-*L* (*T*, *G*)-*L* (*T*, *G*)-(*L*, *A*) (*T*, *G*)-*A*-(*L*, *A*) *A*-(*T*, *G*)-(*L*, *A*) (where the residue molar ratio of L-lysine to DL-alanine is 1:6) (*G*, *T*, *A*) (4:4:3) (Average molecular weight 5,000)	*G*-*A*-*L* *A*-(*T*, *G*)-*L* *A*-*T*-(*L*, *A*) (where the residue molar ratio of L-lysine to DL-alanine is 1:3) *A*-*T*-*L* (*G*, *T*) (1:1) (average molecular weight 4,500)

* As followed by the precipitin test.

upon attachment of chains containing L-tyrosine alone to the multichain poly-DL-alanine. This material, *T*-*A*-*L*, had a molecular weight of 40,000 and contained 12.5 per cent tyrosine residues. On the other hand, the attachment of L-glutamic acid to yield *G*-*A*-*L* (with a molecular weight of 43,500 and 15 per cent glutamic acid residues) did not convert multichain poly-DL-alanine into an antigenic substance.

The multichain polymers present a unique opportunity to investigate the role of the locus in the molecule of the amino acid important for immunogenicity. This critical amino acid was tyrosine in the case of (*T*,*G*)-*A*-*L*. We have, therefore, prepared a material with a similar shape, size, and composition, but with the tyrosine-glutamic acid peptide linked directly to the backbone polylysine, and further extended with poly-DL-alanine chains on the (*T*,*G*) side chains (Fig. 33.6). This material, *A*-(*T*,*G*)-*L* (10 per cent tyrosine residues), is nonantigenic, presumably because of the inaccessibility of the tyrosine to the site of biosynthesis of the antibody. A similar result was obtained with *A*-*T*-*L* (molecular weight 32,000, 8.5 per cent tyrosine residues), in contrast to the strongly antigenic *T*-*A*-*L*.

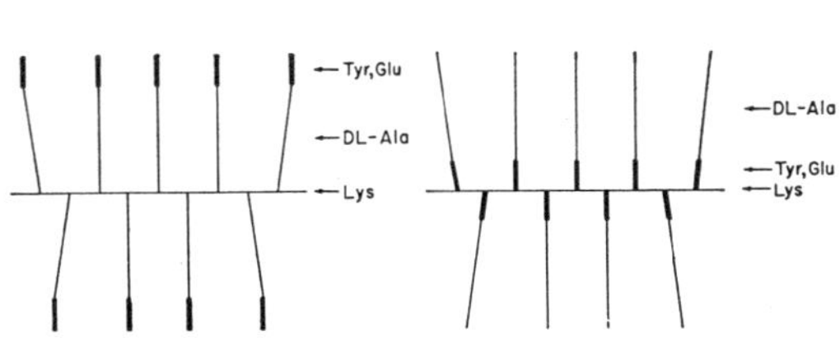

Fig. 33.6. Schematic diagram of (*T*,*G*)-*A*-*L* and *A*-(*T*,*G*)-*L*.

To further check this hypothesis, we have prepared materials in which we have used linear copolymers of L-lysine and DL-alanine, rather than polylysine, as the backbones of the multichain polymers. In the macromolecules thus obtained the average distance between polymeric side chains may thus be increased at will. When side chains of L-tyrosine were attached to a copolymer of lysine-alanine with a ratio of 1:3, and then covered with poly-DL-alanine [*A*-*T*-(*L*,*A*; 1:3)], no antigenicity was observed. On the other hand, when tyrosine and glutamic acid were attached as peptides to a copolymer of lysine and alanine with a ratio of 1:6 and then covered with polyalanine, the resulting *A*-(*T*,*G*)-(*L*,*A*; 1:6) (12 per cent tyrosine residues) is a good antigen as seen in Fig. 33.7. It seems that in the last case the average distance between the polymeric side chains is sufficient to permit a high degree of flexibility in the molecule and, therefore, to render the "internal" tyrosine-rich region accessible, as in a linear copolymer.

By diluting, in the above manner, the density of the side chains, it is possible to obtain multichain polymers of different shapes. Several polymers of the general formula (*T*,*G*)-*A*-(*L*,*A*) were shown to be antigenic. We have also

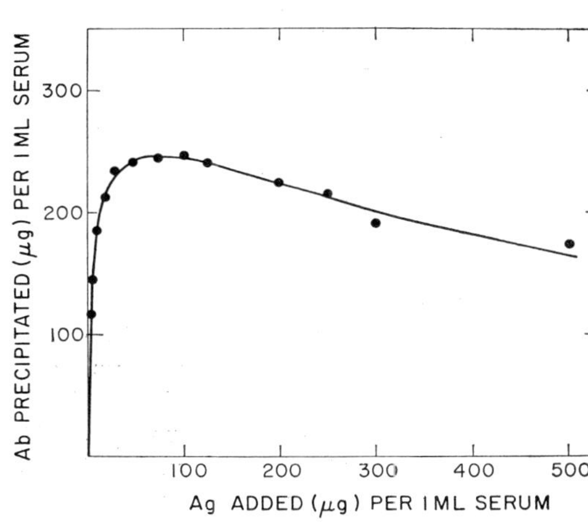

Fig. 33.7. Homologous precipitin curve of *A*-(*T*,*G*)-(*L*,*A*; 1:6). The amount of antigen was obtained from the radioactivity data, after tagging it with I^{131}.

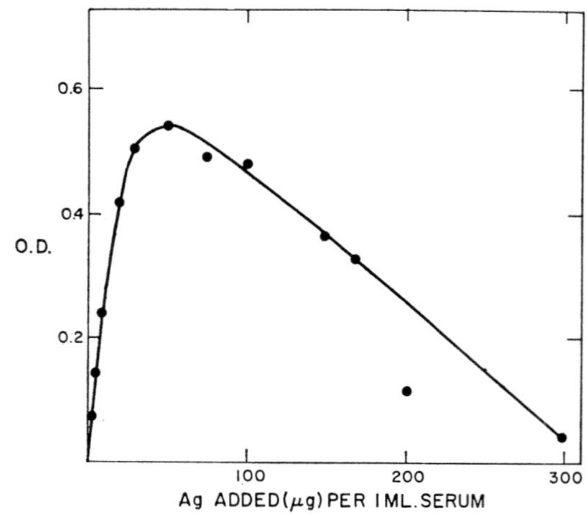

Fig. 33.8. Optical density, at 280 mμ of neutral solutions of precipitates (dissolved in 0.1 N sodium hydroxide and neutralized with N hydrochloric acid) obtained by the addition of *(G,T,A)* (4:4:3) to the homologous antiserum, as a function of the amount of the antigen.

checked a linear copolymer of L-glutamic acid, L-tyrosine, and L-alanine, in a residue molar ratio of 4:4:3. Fig. 33.8 shows the homologous precipitin curve obtained with this copolymer. Gill and Doty have previously reported the antigenicity of a linear copolymer (*45*), and it thus appears that the over-all shape of the molecule is not an important prerequisite for immunogenicity.

The question as to the lowest molecular weight at which a substance may still be antigenic has been of great concern to the immunologist. We do not yet know the lower limit, but some synthetic materials of an average molecular weight of around 10,000 are good antigens. Thus both *(T,G)-L*, a material with a molecular weight of 10,200 (the residue molar ratio of Tyr:Glu:Lys was 1.3:2.2:1), and a *(T-G)-(L,A)* with a molecular weight of 10,100 elicited antibodies in rabbits (Fig. 33.9). The molecular weights were calculated from sedimentation and diffusion data. The molecular weight of the linear copolymer *(G,T,A)* (4:4:3) was even lower, 5000, but as it was obtained from end-group determination (Van Slyke), this value should be considered only a rough approximation. Nevertheless, it is interesting to compare the positive result obtained with this compound, with the complete lack of antigenicity of a linear copolymer of L-tyrosine and L-glutamic acid in a ratio of 1:1 which had a molecular weight of 4500, determined by the same method.

The polymer *(T,G)-L* was an intermediate in the synthesis of *A-(T,G)-L*. The striking difference in the serological behavior of these two materials leads to the conclusion that the attachment of long polyalanine chains to an antigenic molecule has completely repressed its immunogenic capacity.

Until now, we have discussed the extent of antigenicity of the various polymers investigated, but not their antigenic specificity. In all cases, the antisera were cross-reacted with related and unrelated copolymers and proteins. The reaction is always quite specific. *(T,G)-L* may serve as an example. Fig. 33.9 and Fig. 33.10 show the quantitative results of cross precipitation with substances containing both tyrosine and glutamic acid on the outside. The homologous reaction is inhibited by linear copolymers containing the same two amino acids. The copolymeric inhibitors are quite effective. Thus 50 per cent inhibition at the equivalence

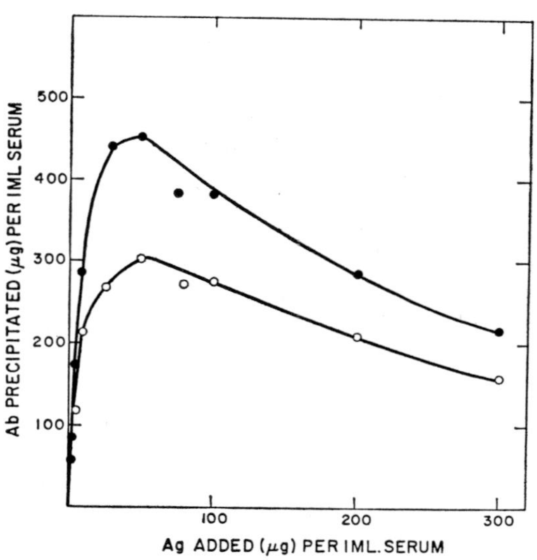

Fig. 33.9. Precipitin curves of *(T,G)-L* (●) and *(T,G)-(L,A*; 1:6) (○) with the antiserum to *(T,G)-L*. The amounts of antibodies were obtained from extinction at 280 mμ after deducting the calculated extinction of the antigens. The amount of antigen in the precipitate was obtained from the radioactivity data, after tagging it with I^{131}.

zone is obtained at a ratio of 2 μg copolymer per μg precipitable antibody. Neither cross reaction nor inhibition was observed when the antiserum was exposed to materials lacking in either tyrosine or glutamic acid or both.

Isolation of antibodies

Any progress in our understanding of the chemical nature of the immune reactions will be a function of our knowledge of the chemistry of the antibody as well as that of the antigen. The successful splitting of antibodies into serologically active fragments by means of proteolytic enzymes (*46, 47*) raises the hope that it will be possible, in the not too distant future, to obtain low molecular weight antibody fragments containing the active site of the molecule, and to define their chemical formula. This in turn will yield an answer to the question of whether all antibodies are chemically identical. If differences, chemical or physicochemical, between antibodies to different antigens can be proved unequivocally, their meaning will depend on the extent to which it will be possible to correlate them with differences in the chemistry of the active site on the antigen molecule. Thus a comparison of antibodies to various polypeptidyl gelatins, where groupings of different antigenic specificity are attached to the same "inert carrier" or of antibodies to various tyrosinated "carriers," where the antigens possess the same active site, could be of appreciable interest.

Based on the above arguments we have decided, in collaboration with Dr. Ruth Arnon and Mr. David Givol, to develop methods for the immunospecific purification of antibodies to some polypeptidyl gelatins as well as some synthetic multichain polymers. Antibodies to gelatin could be isolated, at *p*H 7.4, by digestion of the gelatin-antigelatin complex with collagenase and removal of the gelatin fragments by dialysis (*48*). The pure antibody could be precipitated anew by the antigen. Antibodies to tyrosylgelatin were isolated in a similar way, but in this case peptides containing the active antigenic sites remained bound to the antibody. In order to remove them, the solution must be acidified to *p*H 3 and either dialyzed or passed through a column of Sephadex G50.

Preliminary experiments suggest that antibodies to polytyrosyl gelatin may also be purified on a cross-linked polytyrosine column. The reaction of the diazonium salt of poly-*p*-aminophenylalanine (*49*) with polytyrosine (*50*) yields an insoluble polypeptidic azo dye (*51*). A column with better properties is obtained when a copolymer of leucine and *p*-aminophenylalanine (*52*) is used for the diazotization. The cross-linked polytyrosine removes all the antibodies from the antiserum. After repeated washing with saline the antibodies may be eluted from the column upon acidification to *p*H 3. The neutralized antibody solution gives anew an immunospecific precipitate with the antigen. The use of columns containing the active antigenic site, rather than the whole antigen, increases their efficiency and seems of considerable promise.

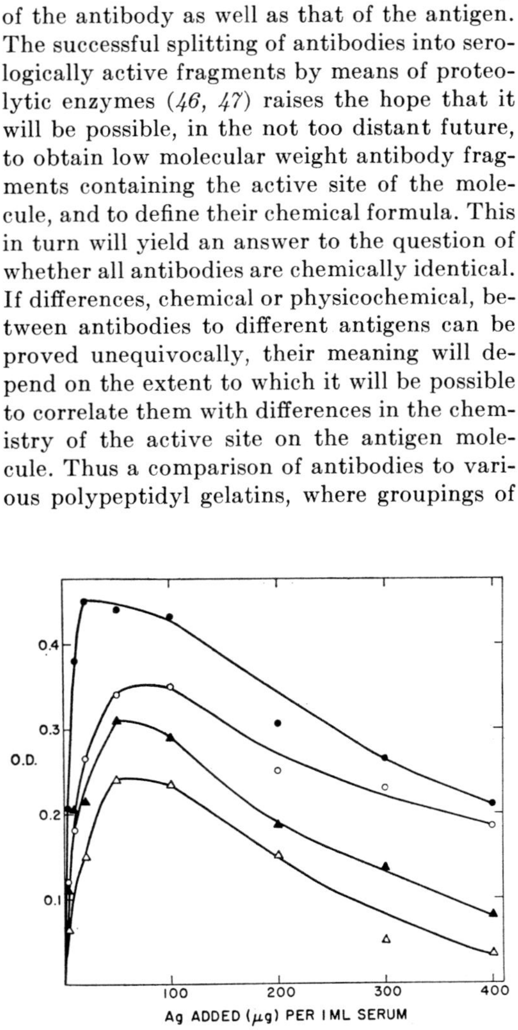

Fig. 33.10. Optical density, at 280 mμ, of neutral solutions of precipitates obtained by the addition of *(T,G)-L* (●), *(T,G)-(L,A;* 1:6) (○), *(T,G)-A-L* (▲) and pGluTyrGel (△) to the antiserum to *(T,G)-L*, as a function of the amount of the precipitant.

Conclusions

The results with multichain polyamino acids in which tyrosine is present in different parts of the molecule have shown that the antigenically important amino acid (or antigenically important area in the active site) cannot be hidden in the interior. The immunogenicity of molecules containing tyrosine close to the backbone polymer, but with an increased distance from one side-chain polymer to another, permits the conclusion that the important factor is *accessibility* of the particular amino acid (or rigid grouping) to the biosynthetic site rather than its presence at the end of a peptide chain. Especially in globular proteins, the amino and carboxyl terminal ends of polypeptide chains are not necessarily expected to be on the surface of the molecule. In the case of linear polypeptides the antigenicity may, therefore, be attributed to *any* segment in the chain, rather than to the terminal portions.

The above results also suggest that the synthetic polypeptide antigens investigated are not split in vivo to a significant extent before reaching the site of the biosynthesis of the antibody.

Multichain polymers of different side-chain densities as well as linear polymers may be antigenic. Thus the over-all shape of the molecule does not seem to be an important factor in determining the immunogenicity.

Synthetic materials with molecular weights around 10,000 may be good antigens. We do not yet know the lowest molecular weight limit for antigenicity, but it should be possible to guess at this value by studying synthetic substances possessing the required immunogenic features.

In our recent studies we have concentrated mainly on molecules containing the same amino acids that were present in our first synthetic antigen. While these always contained tyrosine as the main cause for antigenicity, it is to be expected that tryptophan, phenylalanine, and possibly also cysteine, which were capable of enhancing the immunogenicity of gelatin, will also be able to induce antigenicity in synthetic polypeptides. We do not know yet about histidine, methionine, or leucine.

If I may extrapolate from the results obtained with modified gelatins to synthetic antigens, I would expect that as little as 2 per cent tyrosine (or any other rigid immunogenic grouping) would be enough to make it a good antigen. In such a case the specificity would certainly be controlled by the neighboring amino acids.

Although tyrosine is certainly not unique in its serological role, two recent reports concerning proteins confirm the importance of this amino acid in immune reactions. Cebra (*8*) has shown that removal of tyrosine from peptide fragments of silk fibroin decreases dramatically their cross reaction with anti-silk fibroin. Richmond (*53*) has reported that a penicillinase containing *p*-fluorophenylalanine instead of tyrosine and phenylalanine is enzymatically as active as penicillinase, but it reacts only in a limited way with antipenicillinase. Both these reports deal with the role of tyrosine in conferring the serological specificity on the protein molecules investigated rather than their immunogenic capacity.

I want to avoid the dangerous tendency of excessive generalization. It seems that tyrosine, and some other amino acids discussed, are good enough reasons for antigenicity when present in relatively simple polymeric models. They might have a similar role in some proteins. I would rather believe, nevertheless, that proteins also possess other rigid "immunogenic" sites which result from their intricate macromolecular configuration, and which may be completely destroyed upon transconformation. The very recent results of Maurer (*54*) and of T. J. Gill, III, and P. Doty (private communication) showing that some high molecular weight copolymers of glutamic acid and lysine are good immunogens may be pertinent in this connection. The attachment of peptides of either glutamic acid alone or lysine alone to gelatin did not enhance its antigenicity, nor are polyglutamic acid or polylysine good antigens. It would seem thus that the antigenicity of the polyampholytic molecule is due to small rigid sites formed because of macromolecular interactions in the polymeric chain.

The manner in which tyrosine (or any other immunogenically critical structural feature of antigens) is involved in the biosynthesis of antibodies is still in the realm of complete speculation. It might be that such groupings are necessary to give the imprint on the biosynthetic "template." It is also possible, however, that the main role of groups such as tyrosine is to give strong short-range bonds with the area responsible for the biosynthesis and thus permit the synthesis of the antibodies to occur.

One other area of investigation should be mentioned, in which polymers and copolymers of amino acids might be of considerable help in increasing our knowledge of the serological phenomena. I refer to their use as immunospecific inhibitors of protein antigen-antibody reactions. We have been unsuccessful in experiments with the gelatin-antigelatin system (*28*), but this was due probably to the imperfection of our synthetic models, rather than to the inadequacy of the approach, which might eventually yield the chemical structure of the active center of specificity in the antigen molecule.

I would like to conclude by expressing the hope that, as a result of the synthetic approach to the chemical basis of antigenicity, we shall be able in the not too distant future to return to the immunological investigation of proteins and to study them with a deeper understanding.

Acknowledgment

This paper was written while the author was a visiting scientist at the National Institutes of Health. The warm hospitality of Dr. Christian B. Anfinsen is gratefully acknowledged.

References

1 K. Landsteiner and M. W. Chase, *Proc. Soc. Exp. Biol. Med.* **30,** 1413 (1933).
2 C. Lapresle, *Ann. inst. Pasteur* **89,** 654 (1955).
3 C. Lapresle and J. Durieux, *ibid.* **92,** 62 (1957).
4 C. Lapresle, M. Kaminski, and C. E. Tanner, *J. Immunol.* **82,** 94 (1959).
5 R. R. Porter, *Biochem. J.* **66,** 677 (1957).
6 T. Hara, *J. Biochem.* (*Japan*) **43,** 263 (1956).
7 K. Landsteiner, *J. Exp. Med.* **75,** 269 (1942).
8 J. J. Cebra, *J. Immunol.* **86,** 190, 197, 205 (1961).
9 R. R. Becker and M. A. Stahmann, *J. Biol. Chem.* **204,** 745 (1953).
10 E. Katchalski and M. Sela, *Advances in Protein Chem.* **13,** 243 (1958).
11 M. Sela and E. Katchalski, *ibid.* **14,** 391 (1959).
12 E. Katchalski and I. Z. Steinberg, *Ann. Rev. Phys. Chem.* **12,** 433 (1961).
13 T. Makinodan, R. R. Becker, H. R. Wolfe, and M. A. Stahmann, *J. Immunol.* **73,** 159 (1954).
14 M. A. Stahmann, H. Tsuyuki, K. Weinke, C. Lapresle, and P. Grabar, *Compt. rend.* **241,** 1528 (1955).
15 M. A. Stahmann, C. Lapresle, D. J. Buchanan-Davidson, and P. Grabar, *J. Immunol.* **83,** 534 (1959).
16 D. J. Buchanan-Davidson, E. E. Dellert, S. E. Kornguth, and M. A. Stahmann, *J. Immunol.* **83,** 543 (1959).
17 F. Haurowitz, "Chemistry and Biology of Proteins," Academic Press, Inc., New York, 1950.
18 W. C. Boyd, "Fundamentals of Immunology," Interscience Publishers, Inc., New York, 1956.
19 P. H. Maurer, *J. Exp. Med.* **100,** 497, 515 (1954).
20 M. Sela and R. Arnon, *Biochem. J.* **75,** 91 (1960).
21 R. F. Clutton, C. R. Harington, and M. E. Yuill, *Biochem. J.* **32,** 1111 (1938).
22 J. H. Humphrey and M. E. Yuill, *ibid.* **33,** 1826 (1939).
23 F. Micheel and E. Schallenberg, *Z. Physiol. Chem.* **291,** 87 (1952).
24 M. Sela, *Bull. Research Council Israel* **4,** 109 (1954).
25 M. Sela, E. Katchalski, and A. L. Olitzki, *Science* **123,** 1129 (1956).
26 M. Sela and F. Haurowitz, *Experientia* **14,** 91 (1958).
27 R. Arnon and M. Sela, *Biochem. J.* **75,** 103 (1960).
28 M. Sela and R. Arnon, *ibid.*, **77,** 394 (1960).
29 F. Haurowitz, *Biol. Rev.* **27,** 247 (1952).
30 M. Sela and R. Arnon, *J. Am. Chem. Soc.* **82,** 2625 (1960).
31 C. W. Beckett, K. S. Pitzer, and R. Spitzer, *J. Am. Chem. Soc.* **69,** 2488 (1947).
32 M. S. Newman, "Steric Effects in Organic Chemistry," John Wiley and Sons, Inc., New York, 1956.
33 M. A. Stahmann, H. Tsuyuki, K. Weinke, C. Lapresle, and P. Grabar, *Federation Proc.* **15,** 361 (1956).
34 E. Dellert, D. J. Buchanan, and M. A. Stahmann, *ibid.* **16,** 171 (1957).
35 D. J. Buchanan-Davidson, M. A. Stahmann, C. Lapresle, and P. Grabar, *J. Immunol.* **83,** 552 (1959).
36 D. J. Buchanan-Davidson, M. A. Stahmann, and E. E. Dellert, *J. Immunol.* **83,** 561 (1959).
37 P. H. Maurer, *Proc. Soc. Exp. Biol. Med.* **96,** 394 (1957).
38 P. H. Maurer, D. Subrahmanyam, E. Katchalski, and E. R. Blout, *J. Immunol.* **83,** 193 (1959).
39 D. Subrahmanyam and P. H. Maurer, *Federation Proc.* **18,** 600 (1959).
40 M. Sela, E. Katchalski, and M. Gehatia, *J. Am. Chem. Soc.* **78,** 746 (1956).
41 M. Sela and E. Katchalski, *Experientia* **11,** 62 (1955).
42 E. Katchalski, M. Gehatia, and M. Sela, *J. Am. Chem. Soc.* **77,** 6175 (1955).
43 M. Sela and R. Arnon, *Biochim. et Biophys. Acta* **40,** 382 (1960).
44 A. Yaron and A. Berger, *Bull. Research Council Israel* **7A,** 96 (1958).
45 T. J. Gill, III, and P. Doty, *J. Mol. Biol.* **2,** 65 (1960).
46 R. R. Porter, *Biochem. J.* **73,** 119 (1959).

47 A. Nisonoff, F. C. Wissler, L. N. Lipman, and D. L. Woernley, *Arch. Biochem. Biophys.* **89,** 230 (1960).
48 R. Arnon and M. Sela, *Science* **132,** 86 (1960).
49 M. Sela and E. Katchalski, *J. Am. Chem. Soc.* **76,** 129 (1954).
50 E. Katchalski and M. Sela, *ibid.* **75,** 5284 (1953).
51 M. Sela and E. Katchalski, *ibid.* **77,** 3662 (1955).
52 A. Bar-Eli and E. Katchalski, *Nature* **188,** 856 (1960).
53 M. H. Richmond, *Biochem. J.* **77,** 112, 121 (1960).
54 P. H. Maurer, *Federation Proc.* **20,** 21 (1961).

Discussion of Paper 33: Synthetic Polypeptide Antigens

DOROTHY J. BUCHANAN-DAVIDSON

Before definite evaluation can be made of the chemotherapeutic action or the antigenicity of synthetic polymers of certain molecular weights, more information should be known about their exact structure—exact size, cyclic forms, terminal end, and so on. From the work of Mr. Stewart and Dr. Sober, it would appear that preparations with an A/I ratio of a certain value contain a range of molecular species and perhaps several different polymers with the same molecular weight. It is possible that only one of these polymers is responsible for the biological activity. Attempts were made to synthesize more of the polyglutamic acid which was found to be antigenic, but they were not successful.

PAPER 34

Immunological Studies with Synthetic Polymers

PAUL H. MAURER

One of the most intriguing problems facing immunologists in the past and through the present has been the inability to define completely the criteria which a compound must fulfill in order to be antigenic. Although we can list many of the factors (*1*) that determine the extent of an immune response in a given host such as the dose injected, the route of injection, the incorporation of the antigen in an adjuvant vehicle, the experimental animal under investigation, and the "foreignness" of the antigen, we really know little about the physical, chemical, and structural requisites a material must have in order to be antigenic. However, by using tailor-made synthetic antigens consisting of polymers of single amino acids, or of random copolymers of two amino acids or terpolymers containing three amino acids or more, we can begin to gain some insight into the requirements for antigenicity. In addition to the possibility of solving the above question, the availability of synthetic antigens can also aid considerably in arriving at the size of an antigenic unit against which antibody is produced. In the polysaccharide anti-polysaccharide system Kabat has shown that antibodies are produced against 4–6 glucose units having the proper configuration (*2*). The use of inhibition techniques in the immunochemical studies with oligosaccharides has allowed a calculation of the approximate size of the antibody site which is $32 \times 12 \times 7$ Å. As for the protein antigens, for years it was thought that the presence of aromatic residues is an absolute requisite for antigenicity (*3*). However, when it was shown (*4*) and confirmed (*5*) that gelatin can be antigenic by itself, this requisite no longer appeared necessary.

The several approaches employed during the past few years to study the chemical basis for antigenicity have been: (*a*) the introduction of polypeptides into several proteins such as serum albumin (*6*) or gelatin (*7*); (*b*) the introduction of tailor-made polypeptides into a non-antigenic linear polymer (*8*); and (*c*) the use of the polypeptides themselves as antigen (*9, 10, 11*). The first two approaches have been used by Stahmann, and Sela, Arnon, Katchalski, and co-workers and are discussed elsewhere in this book.

It is about the last method that I should like to present data which have been accumulated in my laboratory during the past few years. Although the research was initiated because of the possible utility of the synthetic poly-amino acids and copolymers of glutamic acid and lysine as plasma expanders and the need for testing their antigenicity, the problem has developed into a search for the chemical basis of the antigenicity of protein-like substances. When one talks about the antigenicity of a substance, one usually refers to the ability of the material to cause the production of circulating antibody in the injected host. Because one is usually looking for minimal responses, one should use the most sensitive method for analysis of the response and also use enough animals to make the data statistically significant.

If a sensitized guinea pig having circulating antibody is given an intradermal injection of the antigen, an erythematous lesion is produced in about 4–6 hours. This is referred to as an immediate reaction. If upon intradermal injection of the antigen into a sensitized guinea pig, one does not observe an immediate reaction but rather an erythematous reaction which reaches a maximum size in about 24–48 hours, it is said that the animal is exhibiting a "delayed cellular" reaction. It is not the intent of this presentation to compare the immunology and histology of the immediate and delayed reactions. Suffice to say that there are several real differences between these two reactions. Among immunologists there is interest not only in the chemical basis of antigenicity for the production of circulating antibody, but also for the production of delayed sensitivity.

In previous publications we reported that in our hands, employing fairly sensitive methods of analysis, such as the passive cutaneous anaphylaxis (P.C.A.) test (*12*) which can measure a few thousandths of a microgram of antibody, and adjuvant immunization, no antibody could be detected in rabbits injected with the many polymers of single amino acids (*9*). The method and amounts of materials used in the immunization are listed in Table 34.1. The same sort of negative results have since been obtained in guinea pigs. Also, a single preparation of polyglutamic acid having a molecular weight of 80,000 was found not to be antigenic in man. These results are somewhat different from the initial observations of Stahmann with polyglutamic acid and polylysine. The differences in responses may be due to some of the factors discussed later on in the text.

The first synthetic random copolymer of glutamic acid and lysine against which antibody was produced in the rabbit was a 6:4 preparation having a molecular weight of about 38,000. Initially, the antibody could be detected only by the P.C.A. reaction. When more copolymers containing various ratios of glutamic acid and lysine were made available by Drs. Blout and Fasman, rabbits were injected with 50 mg or more of the copolymer in complete adjuvants. Table 34.2 shows the copolymers studied so far. None of the twelve rabbits injected with the 5:5 or 7:3 copolymers reacted. The reactions against the 6:4 copolymers as measured by P.C.A. reactions are summarized in Table 34.3. It is apparent that the response was not uniform and was not related to the molecular weight of the antigen. There was no response in six rabbits to the copolymer 6:4 C-2 which has a molecular weight of 64,000 whereas 3/6 rabbits responded to

Table 34.2—Copolymers of glutamic acid and lysine studied for antigenicity

Copolymer Ratio glu/lys	Lot number	Molecular weight
6:4	I (450–200B)	38,000
6:4	(II) C–1	36,000
6:4	C–2	64,000
6:4	C–3	70,000
6:4	C–4	24,000
7:3	I 407–200B	110,000
7:3	II 437–200B	75,000
5:5	I 368–200B	32,000
5:5	II 430–200B	42,000

Table 34.1—Synthetic polymer studies for antigenicity

Polymer	Average molecular weight	Method of immunization (mgm injected): Adjuvant course 1	Adjuvant course 2	Adjuvant course 3	Alum. course 1	Alum. course 2
Poly DL-alanine	3,500	25	12			
Polysarcosine	3,500	25	22			
Poly L-glutamic acid	6,400	25	25	25	50	
Poly L-glutamic acid	80,000	28	22		50	50
Poly L-aspartic acid	4,000	25	25	25	50	50
Poly L-lysine HBr	18,000	25	30		50	50
Poly L-proline II:						
a	50,000	18	25		50	50
b	50,000	8	5		50	
Poly L-hydroxyproline	7,000	30	30			
Copoly L-glutamic acid: L-lysine HBr 450–200B (6:4)	38,000	7	12	7		

Table 34.3 Response in rabbits to 6:4 copolymers

Polymer injected	Reactors (P.C.A.)	% Reactors
6:4 I	4/6	66.7
6:4 C-1	1/5	20.0
6:4 C-2	0/6	0.0
6:4 C-3	3/6	50.0
6:4 C-4	3/6	50.0

preparation 6:4 C-3 having a similar molecular weight. The over-all response among the rabbits was about 38 per cent.

In several of these sera the antibody concentration was sufficient so that positive reactions were also obtained by agar diffusion and quantitative precipitin techniques. Plate 34.1 and Fig. 34.1 show the agar diffusion patterns in Preer tubes (*13*) and the quantitative precipitin reaction respectively. From the single band in agar diffusion and the smooth precipitin curve it appeared that the 6:4 I copolymer was homogeneous. When two different sera prepared against this same copolymer were studied by P.C.A. and the quantitative complement fixation reaction (*14*), it was obvious that the random synthetic copolymers were different and not at all homogeneous. Plate 34.2 shows the positive P.C.A. reactions of two antisera with the homologous antigen and the reaction of copolymer 6:4 C-1 with only one of the sera, which was the stronger of the two.

By the quantitative complement fixation reactions with antiserum 6842, the same sort of

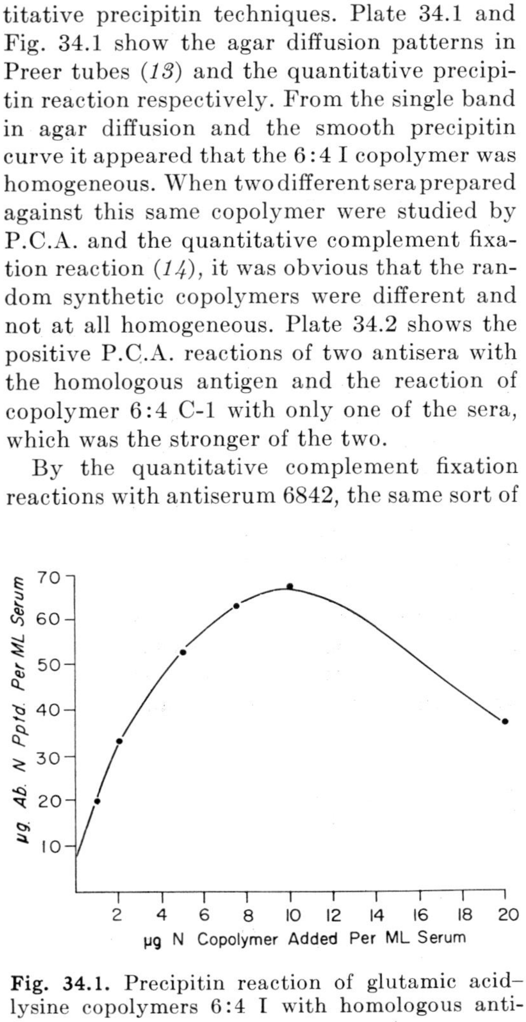

Fig. 34.1. Precipitin reaction of glutamic acid–lysine copolymers 6:4 I with homologous antiserum.

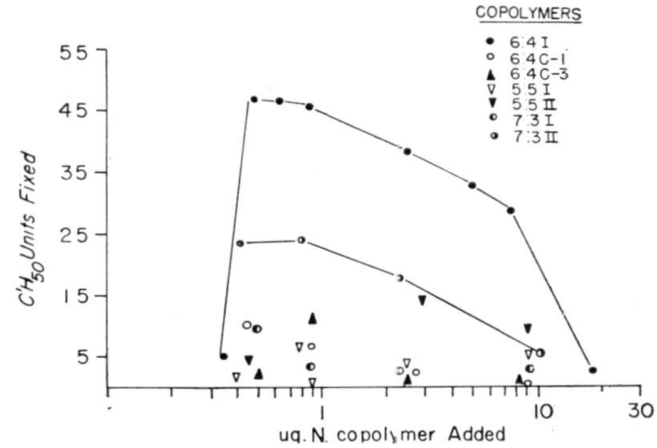

Fig. 34.2. Fixation of C′ by varying quantities of glutamic acid–lysine copolymers and anti 6:4 I serum (*R* 6842 dil. 1→3).

differences were noted (Fig. 34.2). Fortuitously, the two rabbits produced antibodies against different configurations in the same antigen. Moreover, the configurations were not present in all of the random copolymers prepared by the same synthetic method. These data show the importance of not pooling antisera when one is to undertake immunochemical studies.

The heterogeneity in the copolymers was also shown with rabbit antiserum prepared against the 6:4 C-1 copolymer. The reaction in agar diffusion in Preer tubes (Plate 34.3) shows two definite bands that with increasing antigen concentrations migrate to the antiserum area. When the reaction was studied by the Öuchterlony method of double diffusion in agar (*15*), two bands were also obtained. However, as shown in Plate 34.4 not all of the other copolymers gave two bands. One of the 5:5 preparations gave only a single band, whereas the other preparations produced two continuous bands indicating a good deal of similarity among the products. The differences between the different preparations were more evident when they were studied by quantitative immunochemical methods. The results of the quantitative complement fixation deserve comment (Fig. 34.3). First of all, the far from smooth curve obtained with the homologous antigen indicates heterogeneity in the copolymer. Moreover, it is evident that the other 6:4, 5:5, and 7:3 preparations were all different in their reactivities with this antiserum. These data point up a major problem in work-

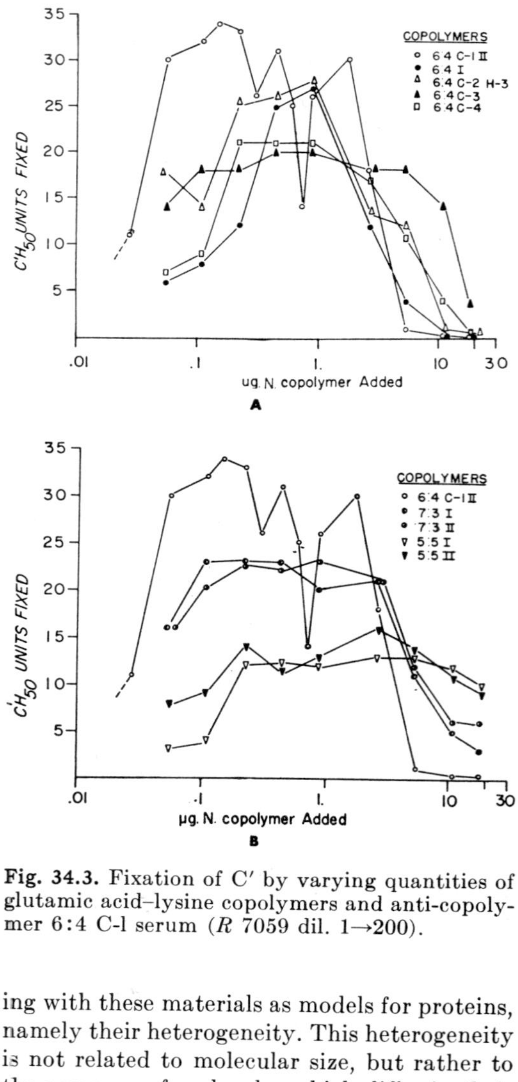

Fig. 34.3. Fixation of C′ by varying quantities of glutamic acid–lysine copolymers and anti-copolymer 6:4 C-l serum (*R* 7059 dil. 1→200).

ing with these materials as models for proteins, namely their heterogeneity. This heterogeneity is not related to molecular size, but rather to the presence of molecules which differ in their amino acid sequences.

As was to be expected, it was possible to inhibit the in vitro antigen-antibody reaction with glutamic acid or lysine. The size of the reactive antigenic units will not be known until inhibition studies are completed with lysine and glutamic acid peptides of varying sizes and composition.

The responses of guinea pigs to immunization with different copolymers incorporated in complete Freund's adjuvant were greater than in rabbits. Positive reactions as measured by skin reaction and P.C.A. reactions with the sera from the guinea pigs showed that all of the copolymers including the 5:5 and 7:3 preparation could be antigenic in a certain percentage of the animals (Table 34.4).

Table 34.4—Response of guinea pigs to injection with various copolymers*

Copolymer	Amt. injected† (μgm)	Reactors	% Reactors
6:4 C-1	10 (50)	1/3	33.3
	50 (50)	4/14	28.6
	250 (50)	1/5	20.0
	1000 (50)	1/6	16.7
	2000 (100)	2/11	18.2
	6000 (100)	3/10	30.0
6:4 C-3	25	2/6	33.3
	1000	4/6	66.7
6:4 C-4	25	2/5	40.0
	1000	4/5	80.0
7:3 I	25	2/5	40.0
	1000	3/5	60.0
5.5 II	25	4/5	80.0
	1000	3/5	60.0

* All injections were with complete Freund's adjuvant.
† Values in parentheses refer to amount injected one week after first injection.

From a large-scale experiment dealing with the effect of the dose injected on the immediate and delayed reactions, it became obvious that the immune response depended more on the guinea pigs' make-up rather than on the dose employed. Some of the results of the experiment shown in Table 34.5 show that varying the dose over a 1000-fold range had no significant effect on the percentage of reactors observed over a three-week period. Reimmunization of the nonreactors with 200 γ of copolypeptide consisting of glutamic acid, lysine, and alanine or glutamic acid, lysine, and tyrosine, which will be shown to be good antigens in the guinea pigs, led to no further response to these

Table 34.5—Immune response of guinea pigs to 6:4 C-3 copolymer

Amt. inj. (μgm)	Total reactors (3-week period)	% Reactors
1000	8/23	34.8
200	5/23	21.7
50	14/24	58.3
5	11/20	55.0
1	6/22	27.3

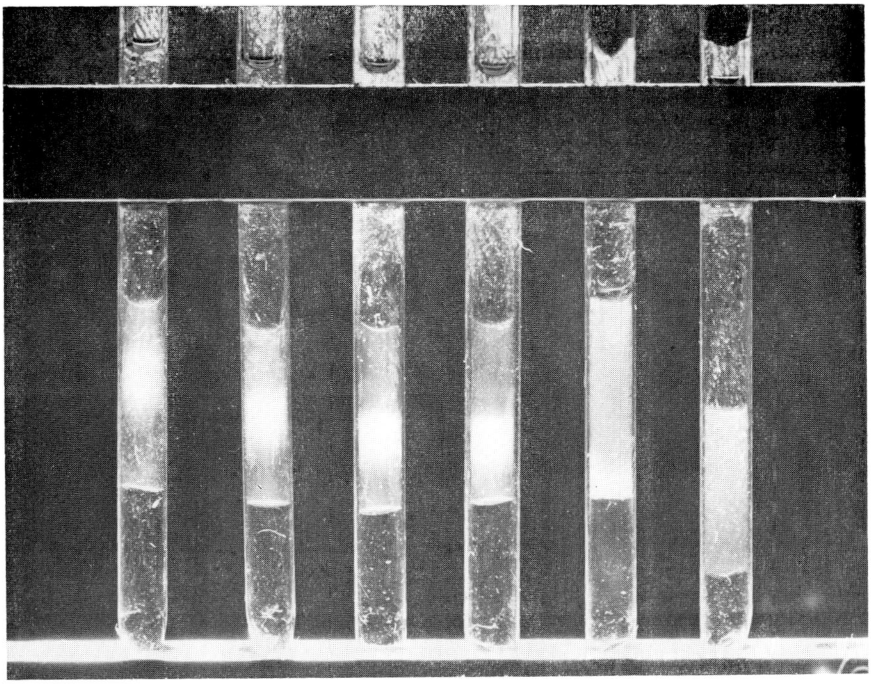

Plate 34.1. Agar diffusion in Preer tubes of reactions between rabbit anti 6:4 I serum and 6:4 I copolymer (concentrations of antigen, *left to right:* 2.5, 5, 10, 30, 40 μgm N/ml and 0.15 M NaCl.

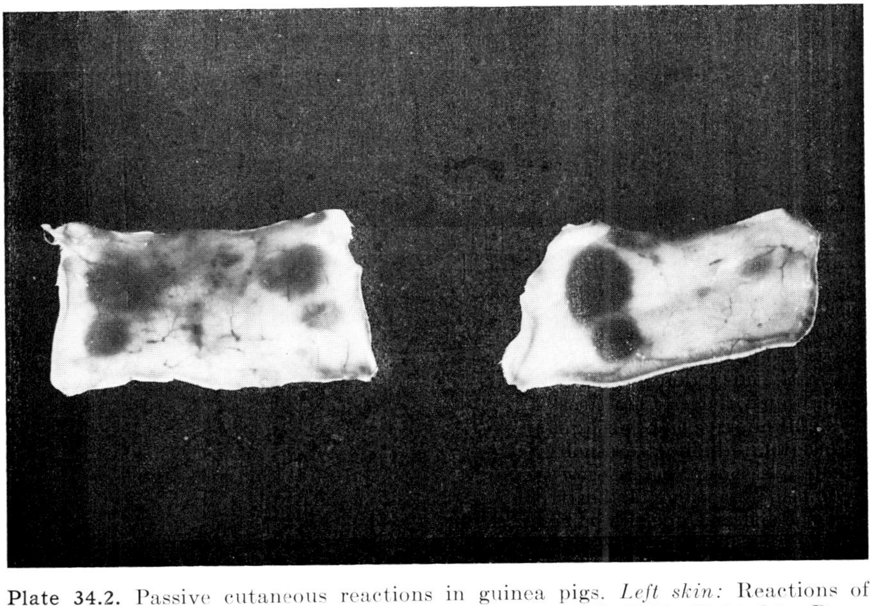

Plate 34.2. Passive cutaneous reactions in guinea pigs. *Left skin:* Reactions of copolymer 6:4 I with homologous antisera R 6845 and R 6842. *Right skin:* Reactions of copolymer 6:4 C-1 with same antisera.

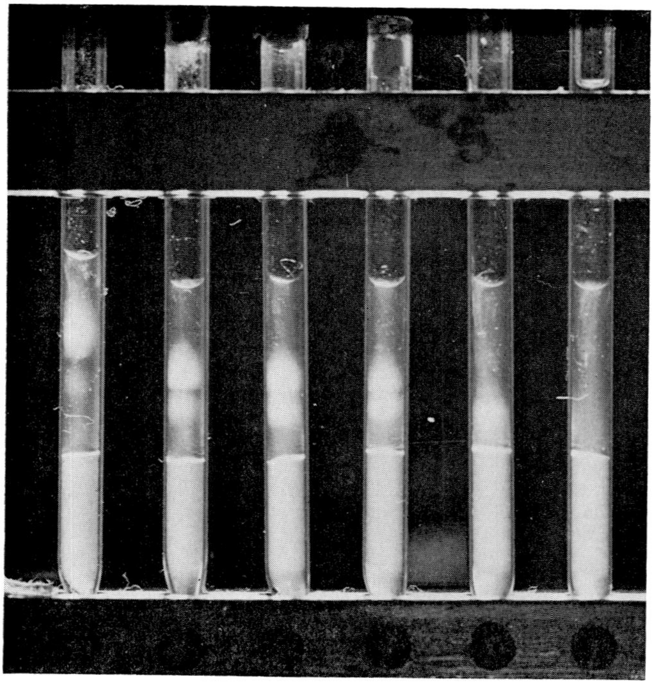

Plate 34.3. Agar diffusion in Preer tubes of reactions between rabbit anti C-1 serum and 6:4 C-1 copolymer. (Note two bands in all tubes except control.)

Plate 34.4. Agar diffusion patterns of reaction of various glutamic acid–lysine copolymers with antiserum against copolymer 6:4 C-1 (R7059).

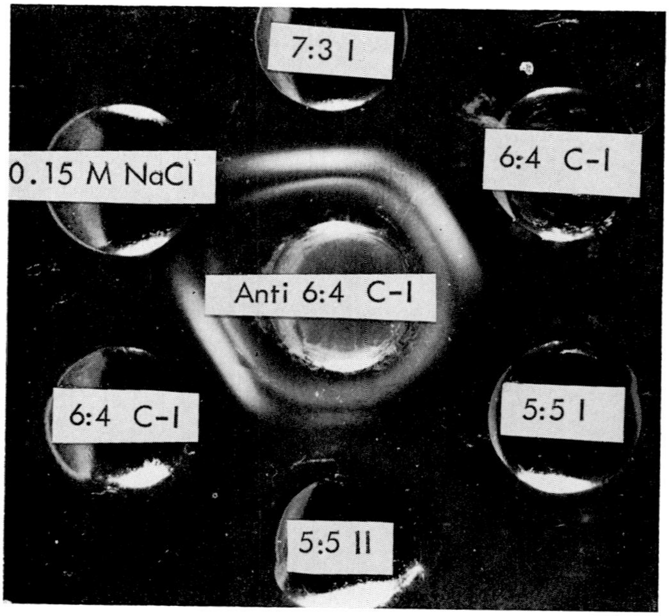

materials either. It would therefore appear that the genetic make-up of the animal is a very important factor in determining the extent of the immune response. This may account for some of the differences in responses by the rabbits against polyglutamic acid and polylysine observed by Dr. Stahmann and ourselves and the inability of Dr. Stahmann to obtain an immune response against rabbits in the United States, whereas he did get antisera in Paris.

Experiments are under way in our laboratory to breed positive reactors and to immunize their offspring to determine if we can obtain a stock which will show a 100 per cent reaction to the synthetic poly-α-amino acids.

As for the response in humans, it can be mentioned in passing that there was no detectable response against three different 6:4 copolymers injected intramuscularly in a saline solution (10–200 mg injected).

When we turned to a study of delayed hypersensitivity, it was possible to produce it in guinea pigs against the 6:4, 7:3, and 5:5 copolymers. The guinea pigs were injected with the antigen in complete adjuvants and were then skin tested on days 3, 7, 14, and 21. The doses injected were the same as shown in Table 34.5. None of the guinea pigs reacted when tested on day 3. However, on day 7 there were positive delayed reactions in all of the groups except the 1 γ group. Table 34.6 shows some of the data from this experiment.

From the studies reported here and from some of our previous studies on the antigenicity of gelatin (*4*) and polyvinylpyrrolidone in man (*16*) and the extremely poor antigenicity of polypeptides consisting of single amino acids, it would appear that we are getting information as to when antigenicity can begin in proteins. It is fairly obvious that some complexity and rigidity is necessary, and two amino acids arranged in a random sequence and having a reasonable molecular weight, possibly of the order of 10,000, can be antigenic.

The question of the role of tyrosine versus another amino acid such as alanine in enhancing antigenicity was studied with the preparations shown in Table 34.7. The two different preparations of glu-lys-ala polymers were employed (*a*) to compare the introduction of 5 tyrosine residues versus 5 alanine residues and (*b*) to learn whether a theoretically more complex molecule having 30 residues of alanine would be more antigenic than one having 5 residues. Although we have much immunochemical data with the antisera against these preparations which show them to be quite heterogeneous (i.e., 2–4 bands have been observed

Table 34.6—Immediate and delayed reactions in guinea pigs injected with 6:4 C-3 copolymer*

		Response to intradermal injection of copolymer					
		6-hr. reactions			24-hr. reactions		
Amt. inj. μg	G.P. No.	100γ	10γ	1γ	100γ	10γ	1γ
1000	61	8	3	—	15	12	3
	69	6	3	—	18	8	—
	71	8	3	—	18	5	—
200	86	6	4	3	25	15	3
	91	4	3	—	18	12	3
	95	3	4	3	22	15	—
50	110	—	—	—	20	10	—
	113	3	—	—	20	16	3
	114	8	5	3	22	15	8
	116	3	3	—	18	15	—
	117	3	2	—	16	10	3
	118	4	3	—	22	15	3
	119	8	8	—	22	15	3
5	133	6	3	—	16	12	3
	135	8	—	—	12	3	—
	140	4	4	—	20	18	—

* Reactions reported as diameter of erythema.

Table 34.7—Immune response of rabbits and guinea pigs to terpolymers*

Polymer injected	Ratio of amino acids	Molecular weight	Animal injected	No. of reactors	% Reactors
Glu-Lys-Tyr	57:38:5	91,000	Rabbits	7/12†	58.3
			Guinea pigs (5 μg)‡	2/12	16.7
			Guinea pigs (100 μg)§	4/12	33.3
Glu-Lys-Ala	57:38:5	50,000	Rabbits	4/12‖	33.3
			Guinea pigs (5 μg)‡	4/10	40.0
			Guinea pigs (100 μg)§	7/12	58.3
Glu-Lys-Ala	42:28:30	62,000	Rabbits	8/12†	66.7
			Guinea pigs (5 μg)‡	4/12	33.3
			Guinea pigs (100 μg)§	9/12	75.0

* Measured by P.C.A. Reaction.
† Based on 2 injections of antigen in complete adjuvants.
‡ G.P. injected with 5 μgm in complete adjuvants.
§ G.P. injected with 100 μgm in complete adjuvants.
‖ Based on 1 injection of antigen in complete adjuvants.

in agar diffusion studies) the data presented here refer only to the number of positive reactors in rabbits and guinea pigs as judged by the P.C.A. reaction. As shown in Table 34.7 there were real differences in the number of positive reactors in rabbits and guinea pigs as judged by the P.C.A. reaction.

As shown in Table 34.7 there were no real differences in the number of positive reactors observed with the polymers of three amino acids versus the copolymer. It is true that the number of sera containing precipitating antibody was greater with the polymers of three amino acids than with two. These data would indicate that with this system the introduction of tyrosine does not enhance the antigenicity of copolymer any more than the introduction of alanine. However, there is a definite specificity against tyrosine produced by the G-L-T polymer. Although the GLT and GL polymers reacted well with the antisera against G-L-A, the GL and GLA preparations did not react as well with the anti-GLT sera.

As for the amino acids which can contribute to building the requisite structure, I agree with Sela that they may be the ones which can form areas of fixed rigidity, which may be some of the antigenic sites. For this purpose, the amino acids having oppositely charged groups should contribute more than those without polar groups unless, as is now believed, areas of rigidity can also form between hydrocarbon residues.* In this regard we are testing copolymers of neutral amino acids to see whether these can be antigenic. Also the random sequences produced must be foregin to the host. An animal doesn't ordinarily produce antibody against configurations which are native to his own individual make-up. As one introduces more amino acids into a synthetic polymer, the chances of its becoming a good antigen increase. The importance of molecular size and perhaps rigidity is shown by the non-antigenicity of some proteinoid preparations obtained from Dr. Sidney Fox. The preparation and properties of these materials are discussed elsewhere in this book.

In summary, it has been shown that the copolymers of glutamic acid and lysine are definitely antigenic in rabbits and guinea pigs but not in humans under the conditions employed. They can also induce delayed sensitivity in guinea pigs. Polymers consisting of glutamic acid, lysine, and alanine have also been shown to be good antigens in rabbits and guinea pigs. Although the introduction of tyrosine into copolymers of glutamic acid and lysine appears to be important in determining the specificity of the antibody produced, it does not appear to enhance the antigenicity as measured by the number of positive reactors. The variation in

* *Added in proof:* Because of the recent observations on the exceptionally good antigenicity in rabbits of glu-ala copolymers, it may be necessary to revise our ideas on the role of polar versus nonpolar amino-acid residues in enhancing antigenicity.

response among rabbits and guinea pigs may be due to the genetic make-up of the animal which may govern its ability to recognize a configuration as foreign. Therefore, in establishing whether or not any material is antigenic, the material should be tested in large numbers of animals under extreme conditions of immunization.

Immunochemical studies have shown that the synthetic copolymers are definitely heterogeneous and this accounts for the production of antibody against different copolymer molecules.

Acknowledgment

Supported by Grant E3514 from the U.S· Public Health Service and by contract with the Office of the Surgeon General No. DA-49-193-MD-2113.

References

1 M. Cohn in "Methods in Medical Research," Yearbook Publishers, Inc., Chicago, 1952, Vol. V, p. 271.
2 E. A. Kabat, *J. Am. Chem. Soc.* **76,** 3709 (1954); *J. Immunol.* **77,** 377 (1956); **84,** 82 (1960); *J. Cellular Comp. Physiol.* **50,** Suppl., 79 (1957).
3 F. Haurowitz, "Chemistry and Biology of Proteins," Academic Press, Inc., N. Y., 1950; W. C. Boyd, "Fundamentals of Immunology, "Interscience Publishers, Inc., N. Y., 1956.
4 P. H. Maurer, *J. Exp. Med.* **100,** 497, 515 (1954).
5 P. Grabar, *Ann. inst. Pasteur* **88,** 11 (1955).
6 R. R. Becker and M. A. Stahmann, *J. Biol. Chem.* **204,** 745 (1953); T. Makinodan, R. R. Becker, H. R. Wolfe, and M. A. Stahmann, *J. Immunol.* **73,** 159 (1954).
7 M. Sela and R. Arnon, *Biochem. J.* **75,** 91 (1960); R. Arnon and M. Sela, *ibid.* **75,** 103 (1960).
8 M. Sela, summarized in Paper 33, this book.
9 D. J. Buchanan-Davidson, M. A. Stahmann, C. Lapresle, and P. Grabar, *J. Immunol.* **83,** 552 (1959).
10 P. H. Maurer, D. Subrahmanyam, E. Katchalski, and E. R. Blout, *ibid.* **83,** 193 (1959).
11 T. J. Gill III and P. Doty, *J. Mol. Biol.* **2,** 65 (1960).
12 Z. Ovary, *Progress in Allergy* **5,** 459 (1958).
13 J. R. Preer, Jr., *J. Immunol.* **77,** 52 (1956).
14 A. G. Osler, J. H. Strauss, and M. M. Mayer, *Am. J. Syphilis, Gonorrhea, Venereal Diseases,* **36,** 140 (1952).
15 O. Öuchterlony, *Arkiv. Kemi* B26 (1949), **1,** 43 (1949).
16 P. H. Maurer, *J. Immunol.* **77,** 105 (1956).

Discussion of Paper 34: Immunological Studies with Synthetic Polymers

PAUL MAURER

To answer Dr. Sela's comment concerning the data I presented with GLA and GLT, I think the antigenicity of the glutamic-lysine copolymers is of the same order as gelatin. Therefore, the effect of the introduction of tyrosine into glutamic-lysine certainly has to be reckoned with in discussing criteria for immunogenicity.

In agreement with the comments of Dr. Sela I, too, have tried the reaction of bovine albumin, bovine γ-globulin, hemoglobin, fibrinogen, catalase, and ovalbumin, and have observed no reaction with antisera against the glu-lys, glu-lys-ala, or glu-lys-tyr polymers.

Concerning the role of charges in the antigen-antibody reaction, I do not think they are as important for the interaction as for the precipitation. Several years ago we showed that if the —COOH groups of bovine serum albumin were esterified the derivative would not precipitate with the homologous antibody, but would interact as shown by complement fixation and passive cutaneous anaphylaxis tests. If the derivative were de-esterified, then precipitation of the antigen-antibody complexes could be observed.

To answer Dr. Katchalski's question on how guinea pigs are tested, if I had a choice of how to test the immunogenicity of a new polymer of amino acids I would test it in guinea pigs because we have observed a more uniform response. The method we use is to inject 5–100 μg of material in complete adjuvants. Two to three or four weeks later the guinea pigs are bled and then skin tested with different amounts of the polymer as described in the text. The serum obtained is tested for antibody by P.C.A. tests in other guinea pigs. As for the correlation of antibody with skin reaction, there are instances when there are no skin reactions observed after 6–8 hours, but a definite one appears after 24 hours and there is no detectable antibody. This type of behavior is associated with delayed sensitivity. If there are positive skin reactions in 6–8 hours, then the intensity of these reactions do correlate well with the amount of antibody present in the guinea pig serum.

T. J. GILL, III

Our experience with different species of animals has been somewhat different in that we have found synthetic polypeptides to be good antigens in rabbits and poorer, erratic antigens in guinea pigs (Gill and Doty, *J. Biol. Chem.* **236**, 2677 [1961]).

PAUL MAURER

In response to Dr. Katchalski's question on the homogeneity of antibody, I believe that Dr. Katchalski is aware of the work of Lapresle, Kaminski, and co-workers in Paris with antigen fragments of human serum albumin obtained by enzymatic digestion of the antigen. In immunoelectrophoresis studies, whereas the whole antigen after electrophoresis in agar gives a single band in its reaction with antibody, the enzyme digest which can be obtained by treatment with pepsin or chymotrypsin can with many rabbit antisera give three distinct bands. This shows, as we know from other studies, too, that the antibodies produced against a protein antigen are not homogenous in reactivity with antigen. I would agree that the heterogeneity which was shown to be present in the copolymers of α-amino acids also exits with the antibodies.

In response to Dr. Wetlaufer's comment on whether Lapresle and co-worker have produced univalent nonprecipitating antigenic fragments, the answer is no. These workers have cleaved the albumin molecule into three large fragments, each of which precipitates with some of the antibody molecules. The precipitation observed indicates that each of these fragments is at least bivalent.

I would agree with Dr. Doty that we do not have to be completely pessimistic about the situation. The gross heterogeneity of the immune response with these polymers as judged by the presence of two or more bands in agar diffusion doesn't occur with all rabbit sera. We, and I'm sure Dr. Gill, also, get antisera which give a single band in agar. With such antisera one can perform immunochemical studies to determine the nature of the antigenic units against which antibody has been produced.

I would agree with Dr. Sela that other materials besides tyrosine can enhance the antigenicity of gelatin. Dr. Sela has shown that cyclohexyl alanyl gelatin has enhanced the antigenicity of gelatin. Several years ago we showed that succinylated gelatin had enhanced antigenicity for rabbits. Some precipitating antibody was produced. I think that almost anything done to gelatin which increases its structural rigidity will also enhance the antigenicity.

As for Dr. Sela's discussion on the antigenicity of polysaccharides, his ideas are confirmed by the known enhanced antigenicity of branched polysaccharides compared to the straight-chain polysaccharides.

In the gelatin-antigelatin system we were as unsuccessful as Dr. Sela in obtaining inhibition with polymers of glycine, proline, and hydroxy proline. However, we did get inhibition with digests of gelatin formed by pepsin or chymotrypsin.

PAPER 35

The Immunological and Physico-Chemical Properties of a Group of Linear-Chain Synthetic Polypeptides

THOMAS J. GILL, III, and PAUL DOTY

The ubiquitous structural and functional roles of proteins make them of paramount interest in the study of biological systems. Despite the superficial similarity among proteins from the same source and performing the same function, there is a high degree of biological individuality associated with each protein that extends not only across species lines, but is sometimes manifest among different individuals of the same species. The problem of the basis of this individualtiy has long been a central and vexing one in immunology and one which the great diversity and complexity of native proteins has rendered difficult to study.

Landsteiner's classical studies on the specificity of serological reactions (*1*) did much to elucidate the qualitative nature of the antibody-antigen reaction. Subsequent quantitative studies, which Heidelberger and Kendall pioneered (*2, 3, 4*), further clarified the relationship between the antibody and antigen. Most of these studies, however, were carried out on modified native proteins in which specific chemical groups were added to a protein that was already antigenic. Hence these systems were admirably suited for studying the specificity of the antibody-antigen reaction, but not for studying the minimum structural requirements for antigenicity and the size and composition of the antigenic site on the unmodified protein.

With the development of practical methods for producing synthetic poly-α-amino acids and the elucidation of their unique chain configurations, it was natural to see whether these materials were antigenic. If they were antigenic, they would offer a unique means of exploring the minimum structural requirements for antigenicity.

Several early attempts to find antigenic behavior upon the injection of synthetic polypeptides either failed or produced unpromisingly low levels of antibody production (e.g., Stahmann [*5*] and Maurer [*6*]). Following these attempts with the polypeptides alone, Stahmann and his collaborators turned to the use of polypeptidyl proteins to study the antigenic function of synthetic polypeptides (*7–10*). They found that the antibody response was directed toward the polypeptide side chain in large measure as Landsteiner had found earlier (*1*) when he attached short peptides to protein carriers. They also found that the antisera to synthetic polypeptides alone would occasionally yield small precipitates with the polypeptidyl proteins.

During the latter stages of these studies, Sela and Katchalski and their collaborators were exploring the function of tyrosine in antigenicity. Following the earlier suggestions in the literature about the importance of tyrosine (*11, 12, 13*), these workers studied the effects of coupling poly-L-tyrosine to gelatin (*14, 15*), a protein of quite poor antigenicity. They found that this would induce the formation of antibodies detectable by the in vivo anaphylactic shock method in sensitized guinea pigs. Further studies with polypeptidyl gelatins using rabbits showed that they induced the formation of precipitating antibodies with a fairly sharp antigenic specificity (*16, 17*).

At the same time a reinvestigation of the antigenicity of polypeptides by Maurer *et al.* (*18*) did reveal the occasional immunologic response to a copolypeptide of L-glutamic acid and L-lysine. However, this could only be detected by the extremely sensitive passive cutaneous anaphylaxis test. Clearly, a much stronger response was needed to justify further work.

Such a strong immunologic response to synthetic polypeptides was achieved last year when it was shown by ourselves (*19*) and by Sela and Arnon (*20, 21*) that tyrosine-containing polypeptides could elicit fairly high levels of precipitating antibodies. We employed linear polypeptides containing L-lysine, L-glutamic acid, and L-tyrosine while Sela and Arnon used a highly branched multichain polymer.

With the further investigation of this display of antigenicity, it has been found that the simple copolymer of only L-glutamic acid and L-lysine is indeed antigenic (*22, 23*) and this has been confirmed by Maurer (*24*). This report aims at presenting and discussing our recent work in this area.

All of the polypeptides used in our studies contained residues linked only by peptide bonds involving their α-functional groups and did not contain any groups or linkages foreign to native proteins. The estimate of the amount of antibody formation induced by these synthetic antigens will depend, of course, on the method of antibody detection and the species used for immunization. Since the antibody detected by the precipitin reaction is the most amenable to quantitative chemical study, we decided to restrict our efforts to the study of precipitating antibody production. Thus rabbits were the animals of choice since they are good producers of this type of antibody. The immunization schedule was designed to give only moderate antibody titers and to avoid hyperimmunization which would bring out minor antigenic determinants thus further complicating the issue. In addition to measuring the amount of antibody produced, the immunized animals were skin-tested with the immunizing antigen and the resulting reactions interpreted by their gross and histological appearance. Thus we have two criteria to assess the immune response, one giving the humoral antibody production and the other giving one aspect of the cellular response.

We have devoted our recent studies on synthetic antigens to the examination of three basis questions: (*a*) the minimum structural requirements for polypeptide antigenicity, (*b*) the role of the aromatic amino acids in antigenicity, and (*c*) the interrelationships among the various synthetic polypeptides and their antisera. Two other specific problems were suggested by these general questions: first, whether rigidity, particularly that conferred by the aromatic amino acids, is necessary for antigenicity; and, second, what the size and composition of the antigenic site is and how it varies among the different polypeptides. This latter study can eventually be expanded into investigating the relationship between the site against which the antibody is formed and the minimum structural requirements for the elicitation of antibody production.

The synthetic polypeptides

Nomenclature.—As the number of polypeptides studied for their antigenic activity increases, referring to them only by number becomes quite confusing. Therefore we have decided to assign symbols to each polypeptide which will reflect its amino-acid composition as determined by amino-acid analysis. The numbers will be retained for secondary identification and for correlation with previous work. Also, a polypeptide will be referred to by its number until its amino-acid composition has been determined exactly. In this system G will stand for glutamic acid; L, for lysine; T, for tyrosine; P, for phenylalanine; and A, for alanine. The subscripts will denote the mole percentage of each constituent amino acid. We have not used the more exact, but longer, notations developed by the workers in amino-acid sequence studies, because it would partially defeat the purpose of maximum information with maximum simplicity and clarity. The numbers and symbols for each polypeptide are given in Table 35.1.

Preparation.—The polypeptides were all made by random polymerization of the N-carboxy-anhydrides of the α-L-amino acids (*25*). The anhydrides were mixed at a 1 per cent concentration in benzene and the polymerization initiated with sodium methoxide at an anhydride:initiator ratio of 400. The polymerizations of $G_{44}L_{40}T_{16}$, $G_{59}L_{41}$, $G_{56}L_{43}T_{1}$, $G_{62}L_{33}P_{5}$, 11 and $G_{57}L_{34}P_{9}$ were stopped at 90 per cent of completion, because kinetic studies

Table 35.1—Composition, molecular weight, and helical content of polypeptides (*25*)

Polypeptide		Mole %				Molecular weight	$[\alpha]_D$	b_o	a_o	Per cent* helix
No.	Symbol	Glu	Lys	Phe	Tyr					
3	$G_{44}L_{40}T_{16}$	44.0	40.0	0	16.0	49,000	−61.0	0	−450	0
6	$G_{50}L_{34}T_{16}$	49.8	34.3	0	15.9	98,000	−60.3	−25	−511	0
7	$G_{56}L_{38}T_{6}$	56.4	37.8	0	5.8	110,000	−67.1	−92	−583	15
8	$G_{59}L_{41}$	59.0	41.0	0	0	101,000	−72.7	−131	−610	20
9	$G_{56}L_{43}T_{1}$	56.1	42.6	0	1.3	81,000	−66.4	−141	−545	20
10	$G_{62}L_{33}P_{5}$	61.6	33.4	5.0	0	63,000	−64.2	−96	−584	15
11†		58	41	1	0					
12	$G_{57}L_{34}P_{9}$	57.1	33.9	9.0	0	61,000	−68.8	−93	−576	15
16	L_{100}	0	100	0	0	324,000				
17	G_{100}	100	0	0	0	84,000				
		Glu	Lys	Ala	Tyr					
18†		96	0	0	4					
19†		0	70	30	0		−108	−58	−776	10
20†		0	96	0	4					
21†		70	0	30	0		−129	−9	−890	0
22†		42	28	30	0					

* From the b_o values in 0.15 M saline-phosphate buffer at pH 7.6 except for $G_{44}L_{40}T_{16}$ and $G_{50}L_{34}T_{16}$ which were measured in 0.15M carbonate at pH 10.0 (see text).
† From the composition of the starting mixture.

showed a slight deviation from first-order kinetics at this point. All of the other polymers, which were obtained commercially (Pilot Chemical Corp., Watertown, Mass.), had been made by allowing the reaction to go to completion. The reactions were stopped and the protecting groups removed by bubbling hydrogen bromide into the reaction mixture. The amino-acid compositions of all the polypeptides were approximately the same as those predicted from the composition of the starting mixture whether the reaction was stopped at 90 per cent or allowed to proceed all the way to completion. It was felt, however, that stopping the reaction at 90 per cent of completion prevented enrichment of the end of the polymer chain with one species of amino acid, as suggested by the deviation from first-order kinetics.

All of the polypeptides except $G_{44}L_{40}T_{16}$ and $G_{50}L_{34}T_{16}$ were soluble in saline-phosphate buffer at pH 7.6. This medium, which contained 0.11 M NaCl and 0.04 M phosphate, closely approximates physiological conditions. Because of the tendency to precipitate at lower pH's, $G_{40}L_{44}T_{16}$ and $G_{50}L_{34}T_{16}$ were dissolved in 0.15 carbonate buffer which made comparisons with other polypeptides difficult, because of the pH-dependence of the physical properties and immunochemical reactions. Polypeptide $G_{44}L_{40}T_{16}$ was the first one synthesized and its properties do not differ significantly, except for molecular weight, from those of $G_{50}L_{34}T_{16}$; hence it is given here only for completeness.

Following their preparation, the polymers were purified by acetone extraction and then dialysis against three changes of distilled water at pH 7.6, three changes of 0.05 M NaCl at pH 7.6, and finally three changes of distilled water at pH 7.6. They were then lyophilized and stored until use. The dialysis procedure is important for removing small peptides and for exchanging bromide counter-ions for chloride. This is important because there is a large amount of lysine in these polymers and the nature of its counter-ions will affect the hydrodynamic behavior markedly.

The removal of the carbobenzoxy and benzyl protecting groups was checked by infrared spectroscopy in D_2O at pD 7 to 8. The removal was established by the absence of an ester band at 1700–1730 cm^{-1}; this method should detect approximately one residue of protecting groups per 100 amino-acid residues. Because of the important biological information to be obtained from the polypeptides without any aromatic groups, the removal of the protecting groups was also checked by ultraviolet spectroscopy in saline-phosphate at pH 7.6; this should detect approximately one residue of aromatic groups per 10,000 amino-acid residues. Both methods showed the removal of all protecting groups within the limits of their sensitivity.

Physical properties.—All measurements were carried out in saline-phosphate buffer at *p*H 7.6 except for those of $G_{50}L_{34}T_{16}$, which had to be carried out in carbonate buffer at *p*H 10.0. The compositions, molecular weights, and optical rotatory dispersion data are presented in Table 35.1 (*25*).

The polypeptide compositions were determined in the Spinco Automatic Amino Acid Analyzer according to the method of Spackman, Stein, and Moore (*26*). The amount of aromatic amino acids in each polymer was determined spectroscopically, because of the loss of some tyrosine and phenylalanine by acid hydrolysis.

The molecular weights were determined from the Scheraga-Mandelkern equation (*27*) using a β value of 2.5×10^6, that for a random coil. The amount of helix in each polypeptide was too small to invalidate the assumption of a random coil. The partial specific volume of each residue was calculated to include the counter-ion, since the molecular weights are 30 per cent to 40 per cent too high if this is not done (*25*).

The optical rotatory dispersion measurements were done on a Rudolph Model 80S Spectropolarimeter using a General Electric AH6 high-pressure mercury arc over the wavelength range 320–620 mμ. The percent helix was calculated from the b_o values determined from the Moffitt equation with $\lambda_o = 212$ mμ (*28*) using a scale in which 0 was the value for a coil and -630, that for a complete helix. The b_o values give a better indication of the amount of helix than the $[\alpha]_D$, since the latter is subject to the effects of solvation, segmental interaction, and side-chain chromophore absorption.

The hydrodynamic data for polypeptides 7, 8, 9, 10, and 12 fitted the empirical equations:

$$\frac{S^{\circ}_{20,w}}{(1 - \bar{v}\rho)} = 0.046\ M^{0.47}$$

$$[\eta] = 0.0008\ M^{0.58}$$

The similarity of the hydrodynamic properties of these polymers, implied by their fitting these equations, gives further indication that the helical content and molecular weight distribution of each is quite similar. No measurement of the polydispersity of the polymers was made, but the ultracentrifuge pictures all showed single, sharp schlieren peaks.

The immune response

Immunization.—Groups of 8 to 10 albino rabbits weighing 2 to 2.5 kg each were bled for control sera and then immunized subcutaneously with 8 mg of polypeptide emulsified in complete Freund's adjuvant. The animals were bled and skin-tested 3 weeks later (first bleeding). Following this, they were given 8 mg of the polypeptide in saline-phosphate subcutaneously, rested two weeks, and then bled again (second bleeding). The groups receiving $G_{50}L_{34}T_{16}$, and L_{100} were not skin-tested, since the former was dissolved at *p*H 10, which caused skin reactions itself, and the latter was cytotoxic. Merthiolate (1:10,000) was added to the sera as a preservative and the sera were stored at 4° C. for two weeks to allow for the destruction of complement (*29*) and for nonspecific precipitation. Before use, the sera were centrifuged in a Model L Preparative Ultracentrifuge to remove lipid material as well as sediment. It is very important to get rid of all the fat to avoid nonspecific precipitation when studying very weak sera. None of the control sera or sera from rabbits immunized with Freund's adjuvant and saline alone showed any significant nonspecific precipitation.

Quantitation of the immune response.—The antibody content of each antiserum was determined by adding antigen to an aliquot of antiserum until no more precipitate was formed. The amount of antigen thus estimated and amounts somewhat greater and somewhat smaller were added to fresh aliquots of antiserum and the maximum antibody precipitated was taken as the antiserum titer. This second procedure was necessary in order to keep the total reaction volume constant and to avoid the Danysz effect (*30*) due to the successive addition of antigen. The antibody content of the serum thus determined agreed with the maximum determined from the precipitin curve to within 5 per cent and most frequently to within a smaller percentage. The amount of antibody in the precipitate can be determined directly from the absorbance at 277 mμ of the precipitate dissolved in 0.25 N acetic acid, since none of the polypeptides in antigenic amounts has any appreciable absorbance at this wave length. Using microcell attachments, this method can detect as little as 1 γ AbN/ml serum with a precision and accuracy of 10 per cent.

The skin-testing was carried out with 1 mg of the polypeptide used for immunization dissolved in 0.1 ml of buffer and the course of the reaction was followed by its gross and histological appearance. The reactions were classified as Type I Arthus reactions if they began within 2–3 hours, reached a maximum in 24–48 hours, and then faded within 5–7 days. This type was characterized by edema, hemorrhage, and necrosis. The reactions were classified as Type II Arthus reactions if they had a delayed onset, beginning in 6–9 hours, reached a maximum in 48–72 hours, and then faded within 5–7 days. This type was characterized by marked erythema, slight to moderate edema, and, rarely, slight hemorrhage and necrosis. The severity of both types of reactions was graded on a scale of 0 to ++++.

The data for the immune responses are given in Table 35.2 (*23*). It can be seen that the strongest antigen in terms of precipitating antibody production was $G_{56}L_{38}T_6$; $G_{59}L_{41}$, $G_{57}L_{34}P_9$, and $G_{56}L_{43}T_1$ were intermediate; and $G_{50}L_{34}T_{16}$ was the least antigenic. The true antigenicity of $G_{50}L_{34}T_{16}$ is difficult to ascertain, however, since the precipitin reactions were carried out at *p*H 10 where the precipitate is partially soluble. L_{100} and G_{100} did not elicit the production of any precipitating antibody. The antisera to L_{100} and G_{100} were tested, pooled and individually, with other polypeptides to determine whether antibody might be present that would not react with the homologous antigen but would cross-react. Both antisera failed to cross-react with $G_{56}L_{38}T_6$, $G_{59}L_{41}$, $G_{56}L_{43}T_1$, $G_{62}L_{33}P_5$, and $G_{57}L_{34}P_9$; the antisera to G_{100} also failed to cross-react with polypeptides 18 and 21.

The animals immunized with $G_{56}L_{38}T_6$ and $G_{57}L_{34}P_9$ gave strong Arthus skin reactions of the early onset type. Only the animal with the highest antibody titer to $G_{59}L_{41}$ gave a significant Arthus reaction, which was of the early onset type. All of the animals immunized with $G_{56}L_{43}T_1$ gave strong delayed Arthus reactions. This latter polypeptide gave quite unexpected results both with respect to its antibody titer and skin reactivity. It elicited approximately the same amount of antibody as $G_{59}L_{41}$, which had no aromatic amino acids, and gave an unusual skin reaction. Further studies are under way to repeat these findings with a new preparation of $G_{56}L_{43}T_1$.

Studies are in progress (*31, 32*) on a series of polypeptides (numbers 18–22), most of which contain only one charged species of amino acid and tyrosine or alanine. Their physical properties are listed in Table 35.1 and their immunological properties, in Table 35.2. Some polypeptides containing only one charged species are antigenic, but less so than those containing

Table 35.2—Immune response to synthetic polypeptides (*23*)

Polypeptide	Animals responding	Average γAbN/ml serum: First bleeding	Second bleeding	Arthus type	Reaction severity	Bands in gel: I	II
$G_{44}L_{40}T_{16}$	3/5	480*		Cytotoxic			
$G_{50}L_{34}T_{16}$	5/7	14	28	Cytotoxic		1	1
$G_{56}L_{38}T_6$	7/7	104	212	I	+++	1, 2	2
$G_{59}L_{41}$	5/6	63	70	I (one animal)	+++	2 (one serum)	2
$G_{56}L_{43}T_1$	8/8	40	41	II	++		
$G_{57}L_{34}P_9$	7/8	65	80	I	+++	1	1
L_{100}	0/8	0	0	Cytotoxic		0	0
G_{100}	0/8	0	0	0	0	0	0
18	5/8		5	0	0		
19	8/9		42	Cytotoxic			
20	4/10		7	Cytotoxic			
21	2/8		113, 7 (individual)	I (one animal)	++++		
22	8/8		56	I	++±		

* Hyperimmunized (see ref. *19*).

both positively and negatively charged amino acids; polypeptide 22 with glutamic acid, lysine, and alanine was as antigenic as $G_{59}L_{41}$. The polymers containing lysine elicited higher and more consistent antibody titers than those containing glutamic acid. When these animals showed skin reactions they were of the Arthus Type I variety. The only conclusions possible at the moment are that polypeptides containing only one charged species plus an uncharged one can be antigenic and lysine is somewhat more effective in this context than glutamic acid. The presence of an uncharged species in these polypeptides is important in view of the fact that polylysine (L_{100}) and polyglutamic acid (G_{100}) are not antigenic themselves (*6, 9, 23*).

Immunochemistry

Chemistry of the antibody-antigen reaction.—The amount of precipitate formed with antigen $G_{56}L_{38}T_6$ is quite dependent on the *p*H, ionic strength, dilution, and temperature (*23*). Less antibody is precipitated at *p*H 6.5 and 10.0 than at *p*H 7.6. As the NaCl concentration of the reaction mixture increases above 0.15 M, the amount of antibody precipitated decreases. When the precipitate is treated with 2 M NaCl, up to 50 per cent of the antibody can be dissociated without bringing any of the antigen into solution. Further treatment with 4 M NaCl brings up to 30 per cent more antibody into solution, but also brings up to 4 per cent of the antigen into solution. As the reaction volume increases above three times that of the antiserum, the amount of antibody precipitated in the antibody excess region remains constant, but the amount of antigen in the precipitate falls below 90 per cent of what it was at lower dilutions. When the precipitin reaction is incubated at 4° C., it is complete in approximately 15 hours, and the precipitate is only slightly soluble in saline-phosphate at this temperature. When it is incubated at 25° C., 20 per cent less antibody is precipitated; and when it is incubated at 37° C., 60 per cent less antibody is precipitated.

Thus, the optimal conditions for carrying out the precipitin reaction are: (*a*) *p*H 7.6 and NaCl concentration of 0.15 M, (*b*) total reaction volume of no more than three times that of the antiserum, (*c*) mixture and storage of the reactants at 4° C., (*d*) analysis of the precipitate after at least 15 hours at 4° C., and (*e*) washing the precipitate twice with 1.5 ml of saline-phosphate buffer at 4° C.

Precipitin reactions.—All of the antisera were tested by double diffusion in gel. Two antigenic systems were found in $G_{56}L_{38}T_6$ and $G_{59}L_{41}$ by gel diffusion and their precipitin curves, shown in Figures 35.1 and 35.2 (*23*), demonstrate that the precipitate does not completely dissolve in large antigen excess. Both of these findings suggest the presence of at least two antigenic systems. $G_{50}L_{34}T_{16}$ and $G_{57}L_{34}P_9$ showed only

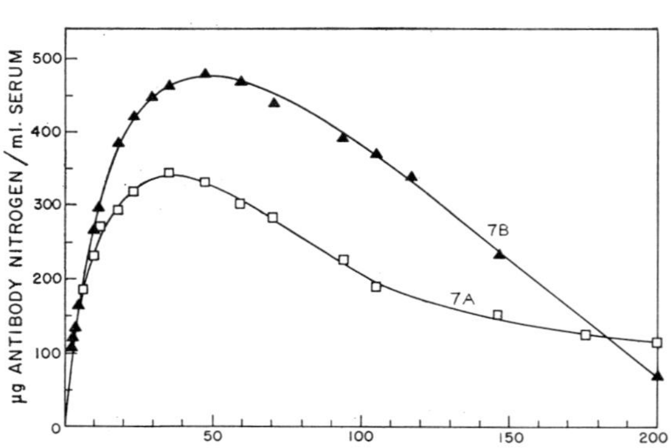

Fig. 35.1. Precipitin curves of $G_{56}L_{38}T_6$ and its antiserum in saline-phosphate buffer at *p*H 7.6. Curves 7A and 7B represent different antisera. The precipitate is not completely soluble in antigen excess, suggesting the presence of more than one antigenic system (*23*).

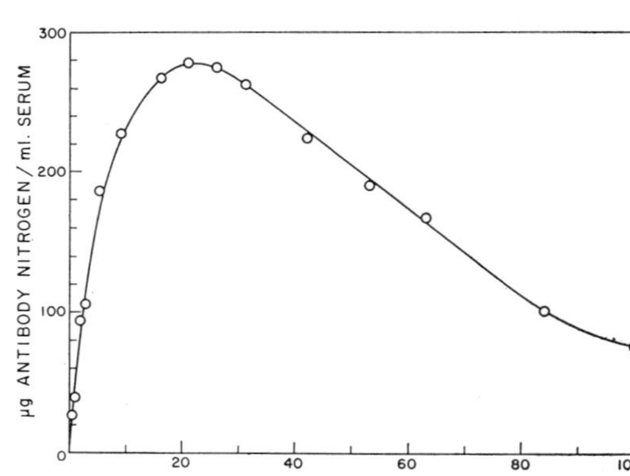

Fig. 35.2. Precipitin curve of $G_{59}L_{41}$ and its antiserum in saline-phosphate buffer at *p*H 7.6. The precipitate is not completely soluble in antigen excess, suggesting the presence of more than one antigenic system (*23*).

one band in gel diffusion and their precipitin curves, shown in Figures 35.3 and 35.4 (*23*), demonstrate almost complete dissolution of the precipitate in antigen excess. Thus, these polypeptides appear to have only one antigenic system. There is the possibility that a minor system in $G_{50}L_{34}T_{16}$ could be soluble at pH 10.0 where its reactions were carried out. In connection with the precipitin curves of the synthetic antigens, it should be pointed out that denatured proteins show some insolubility of the specific precipitate in antigen excess. Because of their lack of extensive conformational structure, these synthetic polypeptides resemble denatured proteins and this may be a factor in the incomplete solubility of their precipitates; it may completely account for the slight insolubility of $G_{50}L_{34}T_{16}$ and $G_{57}L_{34}P_9$.

Using radio-iodinated $G_{56}L_{38}T_6$, it was possible to determine the composition of the precipitate throughout most of the precipitin range. These data are summarized in Table 35.3. They show that the percentage of antigen in the precipitate decreases with addition of increasing amounts of antigen. The mole ratio of antibody:antigen, *Rm*, is quite high, indicating that there are a large number of reactive sites per polypeptide molecule; thus the valence sites must be located along the polypeptide chain and not just at the ends. Values of *Rm* of this magnitude have been found only with much larger native proteins (*33*, *34*).

Cross reactions.—All of the polypeptides con-

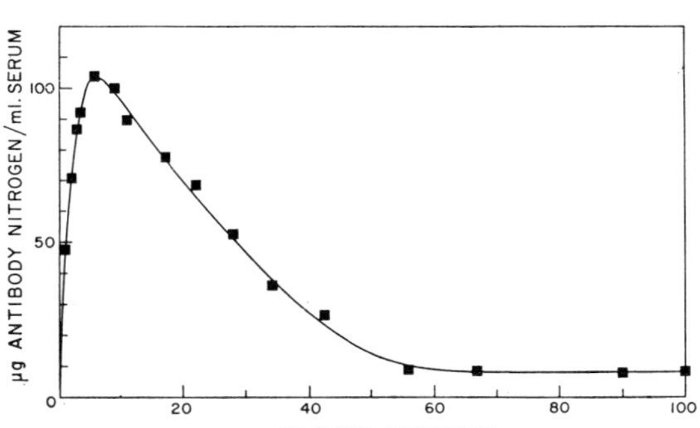

Fig. 35.3. Precipitin curve of $G_{50}L_{34}T_{16}$ and its antiserum in carbonate buffer at pH 10.0. The precipitate is almost completely soluble in antigen excess, suggesting the presence of only one antigenic system or the solubility of minor ones at pH 10.0 (*23*).

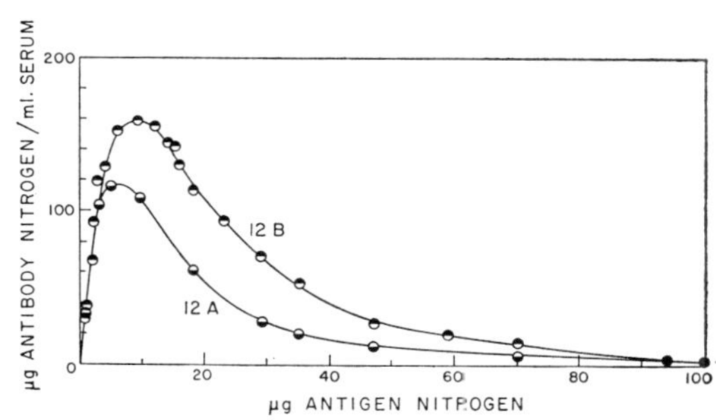

Fig. 35.4. Precipitin curves of $G_{57}L_{34}P_9$ and its antiserum in saline-phosphate buffer at pH 7.6. Curves 12A and 12B represent different antisera. The precipitate is almost completely soluble in antigen excess, suggesting the presence of only one antigenic system (*23*).

taining both glutamic acid and lysine among their constituents cross-reacted mutually to a large but variable extent. Gel diffusion studies showed that two precipitin lines were formed when $G_{56}L_{38}T_6$ and $G_{59}L_{41}$ reacted with each other's antisera. The cross-reaction precipitin curves for the antisera to $G_{50}L_{34}T_{16}$ and $G_{56}L_{38}T_6$ are shown in Figure 35.5 (*23*) and those for the antisera to $G_{59}L_{41}$ and $G_{57}L_{34}P_9$, in Figure 35.6 (*23*). The cross reactions with the antiserum to $G_{50}L_{34}T_{16}$ showed that the maximum antibody was precipitated by $G_{56}L_{43}T_1$, a lesser amount by $G_{56}L_{38}T_6$, and the least by $G_{59}L_{41}$ and $G_{62}L_{33}P_5$. The reaction with $G_{50}L_{34}T_{16}$ could not be studied, because the high pH required would invalidate the comparative interpretation of the results. Since the antiserum was produced against a polypeptide containing 16 per cent tyrosine, it would be expected that $G_{56}L_{38}T_6$ with 6 per cent tyrosine would have precipitated more antibody than $G_{56}L_{43}T_1$ with 1 per cent tyrosine. The fact that

Table 35.3—Precipitin curve using $G_{56}L_{38}T_6$—I^{131}

γAgN	γAbN/ml serum	% Ag in Ppt	*Rm*	Excess in supernatant	
				Ab	Ag
0.8	26	91	21.0	++	0
1.3	34	92	20.7	++	0
1.8	52	90	20.8	++	0
3.6	91	86	18.9	+	0
7.2	116	72	14.5	±	±
13.1	131	50	12.9	0	+
16.3	125	52	9.6	0	++
18.6	119	47	8.8	0	++

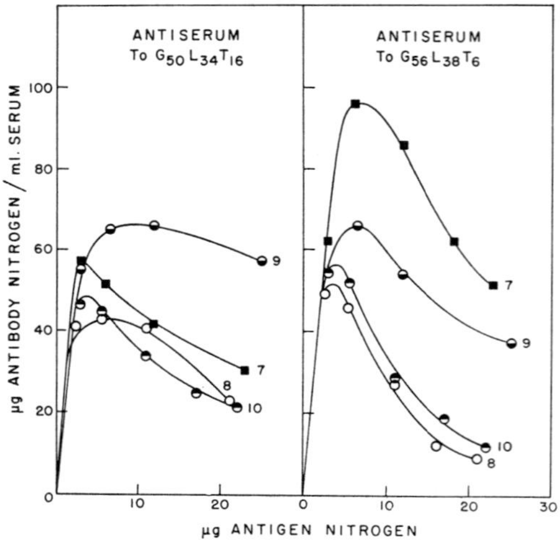

Fig. 35.5. Precipitin curves for the cross-reaction studies with the antisera to $G_{50}L_{34}T_{16}$ and $G_{56}L_{38}T_6$ in saline-phosphate buffer at pH 7.6. The numbers are used to designate the precipitin curves with the various antigens: $7 = G_{55}L_{38}T_6$, $8 = G_{59}L_{41}$, $9 = G_{56}L_{43}T_1$, and $10 = G_{62}L_{33}P_5$ (*23*).

this did not occur and the fact that $G_{62}L_{33}P_5$ with 5 per cent phenylalanine had no more effect than $G_{59}L_{41}$ with no aromatic amino acids indicates that a simple steric interpretation of the role of tyrosine with this antiserum is probably inadequate.

In the cross reactions with the antiserum to $G_{56}L_{38}T_6$, the homologous antigen caused the maximum precipitation, $G_{56}L_{43}T_1$ caused less, and $G_{59}L_{41}$ and $G_{62}L_{33}P_5$ again caused the least. Thus, in this case, the amount of antibody precipitated was less as the amount of tyrosine in the antigen decreased. Again the presence of phenylalanine did not enhance the ability of the antigen to precipitate antibody over that of the polypeptide containing no aromatic amino acids.

The cross reactions with the antiserum to $G_{59}L_{41}$ showed that the homologous antigen and $G_{56}L_{43}T_1$ were the least effective in precipitating antibody. Polypeptide 11, which contained 1 per cent phenylalanine, was the most effective and all of the other polypeptides were of comparable but somewhat lesser effect.

The situation with the antiserum to $G_{57}L_{34}P_9$ falls into a more regular pattern, in that the precipitating ability of the antigen decreases with decreasing phenylalanine content. $G_{56}L_{38}T_6$ was as potent as its phenylalanine analogue $G_{62}L_{33}P_5$, but $G_{56}L_{43}T_1$ had no more effect than $G_{59}L_{41}$. Thus, it is possible that simple steric considerations are more important in determining the role of phenylalanine in these synthetic antigens than they are in determining the role of tyrosine.

All of the antibody in the antisera is not precipitated in the cross reactions. Some of the antibody remains precipitable only by one of the other polypeptides, which is not always the homologous antigen, and some of the antibody remains permanently in solution after the initial cross reaction. The antibody that remains in solution may exist as a soluble complex with the first antigen added. Approximately 20 per cent to 40 per cent of the antibody remains nonprecipitable after the initial cross reactions with the antisera to $G_{50}L_{34}T_{16}$, $G_{56}L_{38}T_6$, and $G_{59}L_{41}$, whereas less than 10 per cent of the antibody in the antiserum to $G_{57}L_{34}P_9$ remains nonprecipitable after the initial cross reaction. $G_{59}L_{41}$ was the least able to form insoluble precipitates with any of the antisera and, together with $G_{56}L_{43}T_1$, it formed the greatest amount of nonprecipitable complexes in the cross reactions. In contrast to this behavior, the phenylalanine containing antigens $G_{62}L_{33}P_5$, polypeptide 11, and $G_{57}L_{34}P_9$ formed the smallest

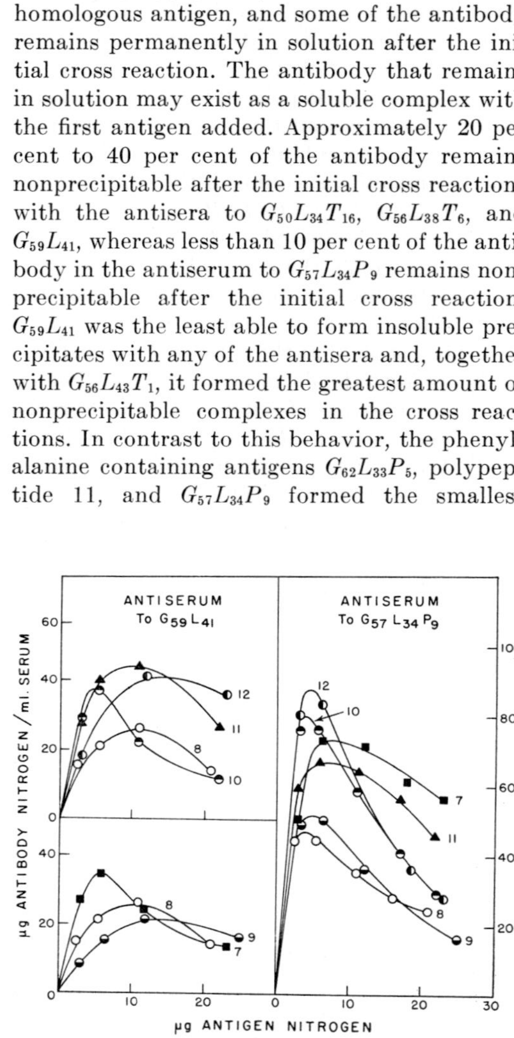

Fig. 35.6. Precipitin curves for the cross-reaction studies with the antisera to $G_{59}L_{41}$ and $G_{57}L_{34}P_9$ in saline-phosphate buffer at pH 7.6. The numbers are used to designate the precipitin curves with the various antigens as in Figure 35.5 and also: 11 contains 1 per cent phenylalanine and $12 = G_{57}L_{34}P_9$ (*23*).

amount of nonprecipitable complexes.

Thus the interplay of solubility factors and multiple antigenic systems makes the interpretation of these cross reactions difficult. However, one fact emerges rather consistently: there is a high degree of cross reactivity among all of these polypeptides.

Since glutamic acid and lysine are the major constituents of these polypeptides, further studies to elucidate their roles were undertaken. When the reaction between $G_{56}L_{38}T_6$ and its antiserum was inhibited by the dipeptide lysine-glutamate at a dipeptide:antigen ratio of 10^5:1, only 40 per cent of the total antibody was precipitated. Additional cross-reaction studies were done by reacting antisera to the polypeptides containing glutamic acid, lysine, and an aromatic amino acid with a polypeptide containing only glutamic acid and tyrosine. Conversely, antisera against polypeptides containing only glutamic acid or lysine and an uncharged amino acid were cross reacted with polypeptides containing both glutamic acid and lysine. The data for these cross reactions are given in Table 35.4. The antisera against $G_{56}L_{38}T_6$ and $G_{57}L_{34}P_9$, both of which contained glutamic acid and lysine, cross reacted fairly well with polypeptide 18 (glutamic acid and tyrosine). The antiserum to 19 cross reacted with the polypeptides containing both glutamic acid and lysine, but the antisera to 20 and 21 did not cross react with the glutamic acid–lysine copolymers. Although the combinations of cross reactions of the sort described above must be greatly extended and inhibition studies with peptides containing different numbers of amino acids must be performed, it appears from the data so far examined that the charged amino acids in the antigen played a prominent role in determining whether or not the precipitin reaction would take place between a given antibody and a group of broadly related antigens.

Table 35.4—Cross-reactions to study the effect of charged amino-acid residues

Polypeptide antigen	γAbN/ml Antiserum $G_{56}L_{38}T_6$*	$G_{57}L_{34}P_9$*	19†	20†	21†
$G_{56}L_{38}T_6$	78		42	0	0
$G_{59}L_{41}$			41	0	0
$G_{56}L_{43}T_1$			40	0	0
$G_{62}L_{33}P_5$			27	0	0
$G_{57}L_{34}P_9$		73			
18	24	25			
19			42		
20				7	
21					15

* Pooled antisera.
† Average of 8–10 individual antisera.

The complex data on the antigenicity and cross reactivity of each polypeptide are summarized qualitatively in Table 35.5. The intensity of the antigenicity and cross reactivity are graded on a scale of 0 to ++++.

Discussion

The present studies with synthetic polypeptide antigens have shown that the presence of aromatic amino acids is not necessary for the elicitation of precipitating antibody. Polypep-

Table 35.5—Qualitative summary of antigenicity and cross-reactivity

Antigenicity*	Immunizing antigen	Cross reactions with antisera to immunizing antigen: $G_{56}L_{38}T_6$	$G_{59}L_{41}$	$G_{56}L_{43}T_1$	$G_{62}L_{33}P_5$	11	$G_{57}L_{34}P_9$	19	20
+	$G_{50}L_{34}T_{16}$	+++	++	+++	++				
++++	$G_{56}L_{38}T_6$	++++	++	+++	++				
++	$G_{59}L_{41}$	++	+	+	++	++	++		
++	$G_{56}L_{43}T_1$								
+++	$G_{57}L_{34}P_9$	+++	++	++	++++	+++	++++		
0	L_{100}	0	0	0	0				
0	G_{100}	0	0	0	0				
±	18								
++	19	++	++	++	++			++	
±	20	0	0	0	0				±
±	21	0	0	0	0				
++	22								

* Estimated from the average γAbN/ml serum+skin reactivity.

tides $G_{59}L_{41}$ and 22, which have no aromatic amino acids, elicit antibody production consistently. Polylysine and polyglutamic acid are not antigenic, at least within the limits of our methods (1 γ AbN/ml serum), but the addition of alanine or tyrosine to the polypeptide chain makes them weakly antigenic. The copolymers that contain lysine appear to elicit antibody production more consistently and in somewhat larger amounts than their glutamic acid-containing counterparts. At this stage, then, it appears that the presence of charged groups is important in making a molecule antigenic. The effect of adding alanine or tyrosine to the polypeptide chain may be to decrease the charge density of the molecule as, indeed, may be the case in the copolymers containing both glutamic acid and lysine.

It is tempting to extend this hypothesis further. An antigen, or fragment thereof, has to reach the antibody-producing site in order to exert its effect; whether this is an instructive or selective process is of little moment to these speculations. If a transporting substance is involved, it would seem that the union between the antigen and the transporting substance would be of an electrostatic character. Thus, while the presence of charge would be important, too high a net charge might prevent such a union to a large extent, thereby accounting for the lack of antigenicity in polylysine and polyglutamic acid.

Since both glutamic acid and lysine seem important in the antigenic determinant sites of all the polypeptides that contain both of them, the same charge factors postulated in inducing antibody formation may also be involved in directing the specificity. Charge alone is too diffuse to account for the specificity of serological reactions, however, but it may be important in making steric factors in the antigen available to the antibody-synthesizing mechanism.

Some pertinent observations bearing on this hypothesis are the lack of extensive cross reaction between the antisera against polypeptides containing only glutamic acid and the polypeptides containing both glutamic acid and lysine, and between the antisera to polypeptides containing glutamic acid, lysine, and an aromatic amino acid and the polypeptide containing only glutamic acid and tyrosine. Also, $G_{62}L_{33}P_5$ does not cross react any better with antisera to $G_{56}L_{38}T_6$ than does $G_{59}L_{41}$; hence the presence of phenylalanine does not appear to exert any influence in this system. Both of these sets of observations suggest that the charged groups, as well as the steric factors, present when the antibody was formed played an important role in determining the nature of the antibody combining site.

The importance of electrostatic factors in the antibody-antigen reaction is indicated by *p*H and ionic-strength studies. The precipitates of both proteins (*3, 4, 35, 36*) and synthetic polypeptides (*23*), especially the latter, begin to dissolve below approximately *p*H 6.5 and above approximately *p*H 8.5. This suggests that both glutamic acid and lysine play prominent roles in the antibody-antigen union. Furthermore, Singer (*37, 38*) has dissociated the antibody-antigen complex by titrating through the *p*K of glutamic acid. The amount of precipitate formed in a reaction decreases as the ionic strength is increased and the formed precipitate can be dissociated to a large extent by 2–4 M NaCl; hence electrostatic factors must be important in holding the antibody and antigen together.

The aromatic amino acids, while not necessary for antigenicity, certainly play an important role in the direction of the immune process (*23*). They were found to alter the magnitude of the immune response in the antigens containing glutamic acid, lysine, and an aromatic amino acid. The presence of 6 per cent tyrosine ($G_{56}L_{38}T_6$) enhanced the amount of antibody produced and caused strong Arthus skin reactions of the early onset type. The presence of 1 per cent tyrosine ($G_{56}L_{43}T_1$), however, did not materially alter the amount of antibody formed from that for the antigens containing no aromatic amino acids ($G_{59}L_{41}$ and 22) and caused Arthus reactions of the delayed onset type. The presence of phenylalanine ($G_{57}L_{43}P_9$) did not greatly influence the amount of antibody produced, but caused the animals to show strong Arthus reactions with an early onset. Thus, it appears that the biological role of the aromatic amino acids may be involved in a regulatory process governing the magnitude of the immune response. The exact immunochemical role of the aromatic amino acids in the polypeptides studied remains to be elucidated, but it probably involves solubility as well as steric considerations.

From the studies presented here, it is indicated that the role of rigidity in antigenicity must be reconsidered (*39, 40*). Since aromatic amino acids are not necessary for the antigenicity of synthetic polypeptides, this source of rigidity is ruled out. The other type of rigidity present in these synthetic polypeptides is that due to the α-helix. While it is true that most of the antigenic polypeptides have 15 per cent to 20 per cent helix, lysine-alanine (number 19) has less than 10 per cent helix and glutamic acid–alanine (number 21) has no helix, within the limits of the method used, yet both are antigenic. Thus it appears that α-helical rigidity is not a factor in the antigenicity of the latter synthetic polypeptides. The presence of rigidity may be a factor in the antigenicity of native proteins, however, by stabilizing the juxtaposition of amino acids necessary for antigenicity, including probably polar ones which these studies show are important in eliciting the formation of antibodies and in directing their specificity.

Thus, the use of synthetic polypeptides in immunochemistry offers the unique advantage of control over the amino-acid composition of the antigen, thereby enabling one to study the ability of chosen amino acids to elicit an immune response and to determine the specificity of the antibody combining site.

References

1 K. Landsteiner, "Specificity of Serological Reactions," Harvard University Press, Cambridge, Mass., 1945.

2 M. Heidelberger and F. E. Kendall, *J. Exp. Med.* **61,** 563 (1935).

3 M. Heidelberger and F. E. Kendall, *ibid.* **62,** 467 (1935).

4 M. Heidelberger and F. E. Kendall, *ibid.*, **62,** 697 (1935).

5 M. A. Stahmann, H. Tsuyki, K. Weinke, C. Lapresle, and P. Grabar, *Compt. rend. soc. biol.*, **241,** 1928 (1955).

6 P. H. Maurer, *Proc. Soc. Exp. Biol. Med.* **96,** 394 (1957).

7 M. A. Stahmann, C. Lapresle, D. J. Buchanan-Davidson, and P. Grabar, *J. Immunol.* **83,** 534 (1959).

8 D. J. Buchanan-Davidson, E. E. Dellert, S. E. Kornguth, and M. A. Stahmann, *ibid.* **83,** 543 (1959).

9 D. J. Buchanan-Davidson, M. A. Stahmann, C. Lapresle, and P. Graber, *ibid.* **83,** 552 (1959).

10 D. J. Buchanan-Davidson, M. A. Stahmann, and E. E. Dellert, *ibid.* **83,** 561 (1959).

11 R. F. Clutton, C. R. Harington, and M. E. Yuill, *Biochem, J.* **32,** 1111 (1938).

12 J. H. Humphrey and M. E. Yuill, *ibid.* **33,** 1826 (1939).

13 F. Micheel and E. Schallenberg, *Hoppe-Seyler's Z. physiol. Chem.* **291,** 87 (1952).

14 M. Sela, E. Katchalski, and A. L. Olitzki, *Science* **123,** 1129 (1956).

15 M. Sela and F. Haurowitz, *Experientia* **14,** 91 (1958).

16 M. Sela and R. Arnon, *Biochem. J.* **75,** 91 (1960).

17 R. Arnon and M. Sela, *ibid.* **75,** 103 (1960).

18 P. H. Maurer, D. Subrahmanyam, E. Katchalski, and E. R. Blout, *J. Immunol.* **83,** 193 (1959).

19 T. J. Gill, III, and P. Doty, *J. Mol. Biol.* **2,** 65 (1960).

20 M. Sela and R. Arnon, *Bull. Research Council Israel* **9A,** 111 (1960).

21 M. Sela and R. Arnon, *Biochim. et Biophys. Acta* **40,** 382 (1960).

22 T. J. Gill, III, and P. Doty, *J. Clin. Invest.* **40,** 1042 (1961).

23 T. J. Gill, III, and P. Doty, *J. Biol. Chem.* **236,** 2677 (1961).

24 P. H. Maurer, *Federation Proc.* **20,** 21 (1961).

25 E. Friedman, T. J. Gill, III, and P. Doty, *J. Am. Chem. Soc.* **83,** 4050 (1961).

26 D. H. Spackman, W. A. Stein, and S. Moore, *Analytical Chem.* **30,** 1190 (1958).

27 H. A. Scheraga and L. Mandelkern, *J. Am. Chem. Soc.* **75,** 179 (1953).

28 W. E. Moffitt and J. T. Yang, *Proc. Nat. Acad. Sci.* **42,** 596 (1956).

29 W. E. Boyd, "Fundamentals of Immunology," Interscience Publishers, Inc., New York, 1956, p. 357.

30 J. Danysz, *Ann. inst. Pasteur* **16,** 331 (1902).

31 T. J. Gill, III, and P. Doty, studies in progress.

32 E. Friedman, T. J. Gill, III, and P. Doty, studies in progress.

33 S. B. Hooker and W. C. Boyd, *J. Immunol.* **30,** 33 (1936).

34 S. Malkiel and W. C. Boyd, *J. Exp. Med.* **66,** 383 (1937).

35 J. R. Marrack, "The Chemistry of Antigens and Antibodies," Med. Research Council (Brit.), Spec. Rep. Ser., No. 230 (1938).

36 V. R. Mason, *Bull. Johns Hopkins Hosp.* **33,** 116 (1922).

37 S. J. Singer and D. H. Campbell, *J. Am. Chem. Soc.* **77,** 3504 (1955).

38 S. J. Singer, L. Eggman, and D. H. Campbell, *ibid.* **77,** 4855 (1955).

39 F. Haurowitz, *Biol. Rev. Cambridge Phil. Soc.* **27,** 247 (1952).

40 M. Sela and R. Arnon, *Biochem. J.* **77,** 394 (1960).

Discussion of Paper 35: Immunological and Physico-Chemical Properties

PAUL MAURER

I should like to comment on some of the technical points in both Dr. Gill's and Dr. Sela's papers (35 and 33). When one is dealing with low levels of antibody the contribution of complement nitrogen has to be removed by another antigen-antibody specific precipitate. Storing rabbit serum for 2 weeks at 5° C. removes the hemolytic activity of the complement, but not the contribution of nitrogen to specific precipitate. Several years ago, we showed that rabbit serum which has been stored in the cold for several months to a year can still contribute 10–20 μg N per ml of serum.

When dealing with low levels of antibody of the order of 10 μg N per ml of serum, 15 hours at 0° C. is certainly not a long enough time to allow the precipitin reaction to go to completion. If there is no difference between the low values obtained after 15 hours and after 6 days, I would wonder about the specificity of the reaction. There are many reports in the literature showing that different amounts of antibody are precipitated after 1 day, 2 days, 6 days, 10 days when low levels of antibody are studied.

I think that in determining whether or not something is immunogenic the most sensitive test for antibody has to be employed. If one wants to do immunochemical investigations, then I believe that antisera that precipitate with the antigens are more advantageous. It all depends on what one is looking for.

In reply to Dr. Doty's comment that the guinea pig antibody which does not precipitate is univalent, I would like to comment that I do not believe this is so. There are differences in the properties of the antibodies produced in guinea pigs, horses, and rabbits. Oftentimes the antisera produced by these animals do not precipitate with the antigen, but it can be shown that they are not necessarily univalent antibody.

In answer to Dr. Gill's comment that he is interested in precipitating antibody and not complement fixing or agglutinating antibody, I'm sure he is aware of the work of Professor Heidelberger and coworkers establishing the unitarian concept of antibody activity, i.e., the same antibody that precipitates with antigens can also fix complement and agglutinate red blood cells which have the antigen on the surface. The antibody is the same, and one gives the antibody different names depending upon what property is being studied.

T. J. GILL, III

Some preliminary studies in our laboratory using starch gel electrophoresis and immunoelectrophoresis have shown the presence of at least two electrophoretic components in a copolymer of glutamic acid, lysine and tyrosine ($G_{56}L_{38}T_6$). This bears on the previous discussion on the nature of the products of random copolymerization and on the interpretation of the precipitin curves. In the latter case, the presence of antigen in the supernatant of the precipitates in the antibody-excess region may be due to the formation of soluble complexes of one antigenic system while still in the antibody-excess region of another antigenic system, the latter being the major system and thus determining the over-all shape of the precipitin curve.

PAPER 36

Biological Studies with Copolypeptides of L-Glutamic Acid and L-Lysine

E. R. BLOUT, S. FARBER, G. D. FASMAN, E. KLEIN, and M. NARROD

As a part of studies of the biological actions of synthetic high molecular weight polypeptides, their properties as plasma expanders have been explored. In our preliminary work and in that of others (*1*), it was found that polyglutamic acid and polylysine had profound effects on the formed blood elements and on the plasma proteins, which would render them undesirable as plasma expanders. Further exploration of these polymers in vivo (dogs) indicated that they were too toxic to warrant further animal studies. The activities of polyglutamic acid and of polylysine resembled, respectively, those of other large molecular weight polycations and polyanions.

In order to determine whether the biological effects were due to the charge of these materials, polypeptides containing various ratios of glutamic acid to lysine were synthesized (*2*). The effects in vitro of these materials on blood viscosity, electrophoretic mobilities of the plasma proteins, coagulation, sedimentation of red blood cells, and the phagocytic activity of polymorphonuclear white blood cells were studied.

On the basis of the findings in vitro, a copolymer of glutamic acid (60 mole per cent) and lysine (40 mole per cent) was selected for studies in vivo pertaining to plasma volume expansion, acute and chronic toxicity, and antigenicity. The toxic studies included hematological, coagulation, and clinical chemistry studies. The 6:4 copolymer (60 per cent glutamic acid, 40 per cent lysine) was found to be satisfactory as a plasma expander on dogs and to be well tolerated by dogs and mice (*3*). The material, however, had inhibitory effects on the coagulation mechanism in dogs and was shown to be capable of eliciting an antigenic response in rabbits (*4, 5*).

Current investigations are concerned with a number of more recently synthesized polypeptides in regard to their effects in vitro on the formed and nonformed blood elements, and their activities as potential antigens in rabbits. This work is preliminary to more extended studies in vivo.

Experimental

STUDIES IN VITRO

Polypeptides containing L-glutamic acid and L-lysine in mole ratios of 7:3, 6:4, 5:5, 4:6, and 54:46, respectively, were investigated (*2*). In addition, polypeptides of glutamic acid only and of lysine only (i.e., 100 per cent each) were studied. The estimated weight average molecular weights of these preparations ranged from 50,000 to 90,000.

A 3 per cent solution of each of the polypeptides in 0.85 per cent saline was prepared and the *p*H was adjusted to 7.4 unless otherwise stated. Saline of the same *p*H was used as the control. To three volumes of blood, one volume of the polypeptide solution was added in vitro. To plasm or serum, respectively, proportionate amounts of polypeptide solution were added, corresponding to the amounts which would be used in vivo for plasma expansion.

Blood was obtained from normal humans or healthy mongrel dogs. Human blood was used as samples obtained from single individuals, or as a pooled specimen (8–12 donors). Canine blood was used as individual samples. Unless stated otherwise, the data presented will refer to human blood.

RBC SEDIMENTATION RATE

The effects of the various polypeptides on the rate of sedimentation of red blood cells decreased as the ratio of glutamic acid to lysine approached unity. The 100 per cent lysine polymer produced almost instantaneous precipitation. It was thus not possible to determine the effects of this material on sedimentation of red blood cells in whole blood.

It was also noted that the degree of clumping of red blood cells paralleled the sedimentation rate. Clumping of red blood cells was minimized with copolymers containing glutamic acid and lysine in ratios of 5:5 and 6:4, respectively.

WHITE BLOOD CELL STUDIES

Morphology, ameboid movement, and phagocytosis were minimally affected by copolymers containing glutamic acid and lysine in ratios of 5:5 and 6:4, respectively. Higher proportions of lysine (60 per cent and 100 per cent) produced lysis of the polymorphonuclear leukocytes. Polyglutamic acid (100 per cent) did not produce lysis but inhibited ameboid movement and phagocytosis almost completely. The copolymer containing glutamic acid and lysine in a ratio of 7:3 increased the phagocytic activity of neutrophilic and eosinophilic polymorphonuclear leukocytes.

VISCOSITY STUDIES

The viscosity of plasma-polypeptide mixtures decreased as the ratio of glutamic acid to lysine approached unity. The copolymers containing glutamic acid and lysine in ratios of 5:5 and 6:4 produced slight alterations of intrinsic viscosity of plasma. As the proportion of glutamic acid was increased above 60 per cent, the intrinsic viscosity of plasma-polypeptide mixture increased. The addition of polyglutamic acid to plasma produced an increase in viscosity of more than 100 per cent.

ELECTROPHORESIS

Solution electrophoresis was carried out using plasma diluted with two parts of buffer (0.02 M phosphate and 0.15 M NaCl, *p*H 7.35), which contained 0.214 per cent of the polypeptides.

The patterns obtained indicated that the 6:4 glutamic acid:lysine copolymer disturbed the relative concentrations of the boundaries the least, in both the ascending and descending limbs. The calculated mobilities showed that the 6:4 copolymer increased the mobilities of all components. The descending patterns for the 5:5 copolymer and 7:3 copolymer indicated a considerable increase in the concentration of components traveling with the speed of the beta-globulins.

COAGULATION STUDIES

The usual methods for testing the coagulation mechanism were employed (*6*). Inhibition of coagulation was minimal or absent in the presence of the copolymer containing glutamic acid and lysine in a ratio of 6:4. As the proportions of glutamic acid to lysine were altered in either direction, inhibition of coagulation became progressively more marked. Inhibitory effects were of the same order in both canine and human blood, but were more marked in the former (Tables 36.1, 36.2, 36.3, 36.4).

Table 36.1—The effects of varying mole ratios of L-glutamic acid and L-lysine in copolypeptides on the recalcification time, the one-stage prothrombin time, and the thrombin time of canine plasma

Composition of polymer		Recalcification time		One-stage prothrombin		Thrombin time	
Glutamic acid (%)	Lysine (%)	Sec.	% P.E.*	Sec.	% P.E.*	Sec.	% P.E.*
100	0	570	8	25	7	33	6
70	30	380	6	20	8	27	12
60	40	98	11	14	7	18	8
50	50	172	7	14	5	19	5
40	60	ppt.	—	ppt.	—	ppt.	—
0	100	ppt.	—	ppt.	—	ppt.	—
Control	(Saline)	80	5	13	8	21	8

* % P.E. = percentage of probable error.

Table 36.2—The effects of varying mole ratios of L-glutamic acid and L-lysine in copolypeptides in the thromboplastin generation test (canine plasma and serum)

Composition of polymer		Minimum time for maximum activity (Min.)	Minimum clotting time of substrate (Sec.)
Glutamic acid (%)	Lysine (%)		
100	0	25	12
70	30	20	10
60	40	6	7
50	50	7	9
40	60	15	55
0	100	28	55
Control	(Saline)	5	7

Table 36.4—The effects of varying the mole ratios of L-glutamic acid and L-lysine in copolypeptides in the thromboplastin generation test (human plasma and serum)

Composition of polymer		Minimum time for maximum activity (Min.)	Minimum clotting time of substrate (Sec.)
Glutamic acid (%)	Lysine (%)		
100	0	25	34
70	30	16	26
60	40	5	14
50	50	5	19
40	60	20	5
0	100	no activity	—
Control	(Saline)	5	14

With polylysine (100 per cent) and the copolymer containing glutamic acid and lysine in a ratio of 4:6, precipitates were observed in all tests except in the thromboplastin generation test. Polyglutamic acid (100 per cent) was strongly inhibitory in the various coagulation tests.

Studies in vivo

The copolymer containing glutamic acid and lysine in the ratio of 6:4 was studied in dogs and mice. Polypeptide solutions were prepared by making a 3 per cent solution in isotonic saline. Commercial preparations of 6 per cent dextran in isotonic saline were used in control dogs. Sixty male mongrel dogs, weighing between 10 and 15 kg, were used. They were fasted overnight, then sedated with 2–4 mg morphine sulfate per kg, administered subcutaneously thirty minutes before the start of each experiment.

Plasma volume was measured by the dye dilution technique with Evans Blue (7). Measurements were made at the beginning of each experiment 1, 6, 12, and/or 24, 48, or 72 hours later. After the initial plasma volume determinations, normovolemic dogs were infused either through the jugular vein or the femoral vein with 40 ml of solution per kg of body weight. Dogs used for the acute hypovolemic studies, after the initial plasma volume determinations had been made, were bled to the extent of 40 ml/kg of their body weight via polyethylene cannula in the jugular vein, and were then immediately re-infused with polypeptide solution.

Normovolemic dogs.—The initial experiments were designed to ascertain whether the copolymer would expand the plasma volume of normovolemic dogs for a significant period of time, and whether or not the material was acutely toxic. The polypeptide solution was found to expand the plasma volume of the ten

Table 36.3—The effects of varying the mole ratios of L-glutamic acid and L-lysine in copolypeptides on the recalcification time, the one-stage prothrombin time, and the thrombin time of human plasma

Composition of polymer		Recalcification time		One-stage prothrombin		Thrombin time	
Glutamic acid (%)	Lysine (%)	Sec.	% P.E.*	Sec.	% P.E.*	Sec.	% P.E.*
100	0	1200	—	37	5	42	7
70	30	1080	7	30	7	34	5
60	40	320	6	12	5	21	3
50	50	750	8	17	8	21	3
40	60	ppt.	—	ppt.	—	ppt.	—
0	100	ppt.	—	ppt.	—	ppt.	—
Control	(Saline)	210	6	12	8	21	5

* % P.E. = percentage of probable error

dogs examined and this expansion was maintained in the normovolemic dogs for the 24-hour period of observation.

All of the animals survived without observable deleterious effects for periods of two to six weeks. Similar measurements were made in dogs when dextran was given for comparison.

Hypovolemic dogs.—Studies were then made on ten dogs to determine whether the polypeptide would re-establish and maintain an adequate plasma volume after a hypovolemic state has been induced. The plasma volume was maintained adequately for the 24-hour period in which the volume was measured. These animals showed no deleterious effects during an observation period of six weeks.

Cardiovascular studies in the intact dog.—Arterial blood pressure and right arterial pressure were measured in the pentobarbital-anesthetized intact dog. Twenty ml of polypeptide were injected at a time. There were no changes in arterial blood pressure. The right arterial pressure showed small transient rises following each injection.

Subacute toxicity.—Since acute toxicity was not apparent, subacute studies were undertaken in CAF-1 male mice. This was done by repeated intravenous injection of the copolymer in the same dose as that used on dogs (40 ml/kg of a 3 per cent solution in isotonic saline).

Ten groups of ten animals each were studied. Group 1 was given polypeptide three times a week for 2 weeks; group 2, polypeptide twice a week for 16 weeks; group 3, polypeptide once a week for 16 weeks; group 4, polypeptide once a week for 26 weeks; group 5, saline three times a week for 2 weeks; group 6, saline once a week for 16 weeks; group 7, dextran twice a week for 8 weeks; group 8, dextran twice a week for 16 weeks; group 9, dextran once a week for 16 weeks; group 10, dextran once a week for 26 weeks.

By the 9th to 10th day, five out of the ten animals in group 1 were in respiratory distress. At autopsy, marked congestion of the lungs was found. Groups 2 to 10 showed no deleterious effects.

Clinical-chemical studies

A group of 12 dogs was studied (4 hypovolemic and 8 normovolemic). The blood electrolytes (Na, K, Cl, Ca, and P), liver function (bilirubin, alkaline phosphatase, total protein, A/G ratio), nonprotein nitrogen and urea nitrogen, blood sugar and transaminase levels were determined prior to and at various periods following the infusion of the copolymer.

The changes observed in these determinations were minor or could be accounted for by the alterations induced in the circulating blood volume. In two out of the four hypovolemic animals, which had been kept at an arterial blood pressure of 30 mm Hg for one hour prior to infusion of the copolymer, marked elevations of the transaminase levels were observed.

Hematological studies

Marked changes in the number of circulating platelets and leukocytes occurred. The platelet count fell by 30 per cent to 50 per cent during the first 6 hours and then gradually returned to normal within 48 to 72 hours. The WBC rose by 80 per cent to 100 per cent within 6 hours, and returned to normal within 48 hours.

The coagulation mechanism was considerably altered in vivo. Whole blood clotting time and calcium clotting time were markedly increased, prothrombin consumption was reduced, and the generation of thromboplastin was inhibited. The prothrombin time remained unchanged and the thrombin time was markedly shortened.

Discussion

The data presented here indicate that the amino-acid composition of synthetic polypeptides is an important factor in determining their interaction with plasma proteins and with formed blood elements. The data suggest that the effects observed have a relationship to the net molecular charge. The degree to which the normal behavior of the plasma proteins and the formed elements was altered increased as the net charge of the polypeptide molecule was increased, either toward net negative charge by a relative excess of glutamic acid, or toward a net positive charge by an excess of lysine. It is thus apparent that the various parameters which were studied in these investigations of the effects on the formed and nonformed blood elements were each progressively more altered by increasing the net molecular charge.

The effects on all the systems examined were more pronounced with the polymer containing 100 per cent glutamic acid, than by that containing 70 per cent glutamic acid which, in turn, produced more alterations in the systems studied than did the copolymers with a glutamic acid content of 60 per cent or of 50 per cent, respectively. Analogous observations were made when the ratio of lysine was increased. A copolymer containing 56 mole per cent glutamic acid and 44 mole per cent lysine had minimal or no apparent effect in vitro on the various properties of the blood components studied.

Electrophoretic patterns indicated marked changes in the mobilities of the plasma proteins with polymers consisting of 100 per cent glutamic acid and 100 per cent lysine, respectively. These changes diminished as the molecular charge approached that of the copolymer containing six parts of glutamic acid to four parts of lysine. Free boundary electrophoresis indicated a progressive decrease in the alteration of electrophoretic mobility of the plasma proteins as charge equivalence was approached; all copolymers studied affected electrophoretic migration of the plasma proteins. The copolymer containing six parts of glutamic acid to four parts of lysine, however, did not distort the pattern, but led to an increase in the rate of migration of all the protein boundaries. These observations, however, remain tentative until the studies on the corresponding purified plasma protein fractions are completed.

The sedimentation rate and clumping of erythrocytes, while evidently related to molecular charge, did not correspond to observations made in regard to other macromolecules. Thus, dextran, polyvinyl pyrrolidone, fibrinogen, and gamma globulins with larger molecular weight, but having very small molecular charge, nevertheless, increase the sedimentation rate of erythrocytes markedly, even at concentrations considerably below those at which the synthetic polypeptides were studied. The copolymer consisting of six parts of glutamic acid to four parts of lysine, however, had no observable effect on the sedimentation rate of erythrocytes, and did not produce clumping of red blood cells.

It was of interest to note that polypeptides containing greater than 50 per cent of glutamic acid increased the rate at which the red blood cells took up and released oxygen. Although detailed measurements of O_2 content were not carried out, the intensity of this phenomenon appeared to decrease as the glutamic-acid content of the molecule decreased.

The effects on white-blood-cell morphology and ameboid movement paralleled the effects on other systems. Maximum disturbances occurred with the polymers consisting of lysine only and of glutamic acid only. These effects decreased as the net charge of the polymers was decreased. Maximum phagocytic activity and ameboid movement, however, were not observed at charge equivalence, but were observed with the 7:3 glutamic acid:lysine copolypeptide. Since ameboid movement and phagocytic activity represent responses to irritation, the net charge on the copolymer (glutamic acid to lysine, 7:3) may have provided enough stimulation to elicit an increased response, but not sufficient damage to prevent the response.

The effects in vitro on the coagulation system again paralleled, in degree, the alterations produced on other blood components; the more highly charged polymers produced more marked changes than the polymers with lower charges. Minimal or no changes were observed with the 6:4 glutamic acid: lysine copolypeptide. The minimal changes in regard to the coagulation systems were represented by a prolongation of the calcium clotting time of borderline significance in the human plasma.

In canine plasma, a significant difference in the calcium clotting time was not observed. Since pooled canine plasma, however, was not available, a definite conclusion must await corresponding studies. The prothrombin time, the thrombin time, and the thromboplastin generation tests revealed no changes in the presence of the copolymer containing glutamic acid and lysine in a ratio of 6:4. This would indicate that anti-hemophilic globulin (AHG), stable prothrombin conversion accelerator (SPCA), accelerator globulin (Acg), prothrombin plasma thromboplastin component (PTC), thrombin, and fibrinogen were not affected in vitro.

The effects on coagulation produced by the more highly net charged polymers, however, were marked and resembled those observed with other correspondingly charged macromolecules. Since, however, the synthesis of this type of polypeptide permits more precise regulation of the charge on the molecule than can be obtained with the natural or semi-synthetic products widely used in coagulation studies, this may represent a suitable means for further investigation of the interactions of macromolecules with coagulation factors.

It was apparent that the copolyglutamic acid:lysine 6:4 would expand plasma volume adequately and was not acutely toxic to dogs or mice. The expansion studies in dogs suggested that much of the material was still present twenty-four hours after it had been administered. If this occurred in mice also, as might be expected, then the daily repeated injections (group 1, subacute study) might lead to an increased blood volume with embarrassment of the cardiovascular system, thus resulting in congestion of the lungs. The fact that none of the other groups showed these symptoms, though a much larger total dose of the copolymer was given, suggested that the effects observed in the mice of group 1 were due to the frequency of administration and were not due to toxicity of the copolymer per se.

The lack of general toxicity of the 6:4 glutamic acid:lysine copolymer was further indicated by the clinical-chemical studies in normal and hypovolemic dogs. Thus the blood electrolytes, hepatic function, the nonprotein nitrogen, and the blood glucose levels of the blood did not appear to be affected by administration of the copolymer. Further studies are required to determine whether elevation of the plasma transaminase levels occurred as a result of the administration of the copolymer or was due to changes in blood volume. The level of the leukocytes was elevated and the peripheral platelet count was decreased; the first stage of the coagulation mechanism was markedly inhibited. Increased bleeding tendencies, however, were not noted. The reasons for the disturbances in hematological findings are not apparent. The reversal in vitro of the inhibition of the coagulation mechanism required the addition of large amounts of platelets. This suggests that the decrease in the number of platelets is not the only factor in the production of the defective coagulation status.

Summary and conclusions

1. Copolypeptides derived from L-glutamic acid and L-lysine in various proportions ranging from 100 per cent glutamic acid to 100 per cent lysine have been studied in vitro with respect to their effects on a number of properties of formed and nonformed blood elements: (1) viscosity of blood; (2) electrophoretic mobilities of the plasma proteins; (3) sedimentation of erythrocytes; (4) morphology, ameboid movement, and phagocytic activity of the leukocytes, and (5) the three stages of coagulation.

2. The degree to which the systems studied were altered in the presence of the various polypeptides increased with net molecular charge as a function of the amino-acid composition. Maximum deviations from normal were observed with polypeptides containing 100 per cent lysine or 100 per cent glutamic acid.

3. As a charge equivalence in the polypeptide was approached, the biological systems in vitro were less affected by them. The polypeptide containing six parts of glutamic acid to four parts of lysine had minimal effects on the calcium clotting time of human plasma. Other aspects of the coagulation mechanism (thromboplastin generation, prothrombin time, thrombin time) were within normal limits, and the behavior of the red blood cells, granulocytes, and nonformed blood elements did not appear to be qualitatively changed by this polypeptide.

4. The copolymer of glutamic acid and lysine in 6:4 ratio has been found to expand the plasma volume of normal and hypovolemic dogs, and to have a slight transient effect on right atrial pressure in the intact dog. There were no deleterious effects in normovolemic mice upon repeated administration of doses of 40 ml/kg of 3 per cent solution when administered once or twice a week, up to sixteen weeks.

5. Administration of the copolymer of lysine and glutamic acid in a ratio of 6:4 to intact and hypovolemic dogs did not have adverse effects on blood electrolytes, liver function tests, nonprotein nitrogen and glucose levels of the plasma. The white blood cell levels were doubled, and platelet levels were decreased by 30 per cent to 50 per cent following infusion. Considerable inhibition of the first stage of the coagulation mechanism was produced.

Acknowledgment

This work has been supported by the Office of the Surgeon General, Department of the Army.

References

1 M. Sela and E. Katchalski, see references cited in *Advances in Protein Chem.* **14,** 391 (1959).
2 E. R. Blout and M. Idelson, *J. Am. Chem. Soc.* **80,** 4909 (1958).
3 M. F. Narrod, *Federation Proc.* **19,** 106 (1960).
4 P. H. Maurer, D. Subrahmanyam, E. Katchalski and E. Blout, *J. Immunol.* **93,** 193 (1959).
5 P. H. Maurer, *Federation Proc.* **20,** 21 (1961).
6 "The Coagulation of Blood, Methods of Study," L. M. Tocantins (ed.), Grune and Stratton, N. Y., 1955.
7 M. I. Gregersen, *J. Lab. Clin. Med.* **23,** 423 (1938).

Discussion of Paper 36: Biological Studies with Copolypeptides

PAUL MAURER

Concerning the matter of antigenicity, it should be mentioned here that several 6:4 copolymers were studied for antigenicity in humans and that no antibody was detectable nor were there any positive skin reactions. I think we have to take into account the species differences in the immune responses against these materials.

Dr. Klein started his presentation by saying that one of the advantages of the glu-lys copolymer as a possible plasma expander is that it is metabolizable. I wonder whether Dr. Klein has any data on the breakdown of the copolymer in vivo. Is there any persistence of the polymer in tissues, etc.?

T. J. GILL, III

I would like to introduce a word of caution on the use of synthetic polypeptides in humans. Some studies in our laboratory indicate that rabbits respond to a primary stimulation by the synthetic polypeptides as if they had been previously exposed to a similar antigen. Professor Stahmann also observed a wide range of cross-reactivity between antisera to polyglutamic acid and native proteins. This evidence indicates that humans might form antibodies to synthetic polypeptides that might cross react with body proteins with possible dire results.

C. H. BAMFORD

I wonder if, in the biological studies reported so far, sufficient care has been exercised in the preparation of the polypeptides examined? The polymerization of NCA's is sensitive to the presence of impurities, and it is by no means easy to prepare polymers with reproducible physical properties, even in one laboratory. I am not referring merely to rather subtle features, such as molecular weight distributions, but also to average properties. It seems very desirable to carry out biological studies, in the first place, with very well characterized specimens, e.g., those which may be prepared by the chromatographic techniques described in this book, or failing that, with polymers made under the rigorous conditions necessary for kinetic work. Of course it may turn out that the biological results are not sensitive to the precise method by which the polymers are made, but, until this is established, it seems necessary to use reproducible materials.

Subject and Author Index